Handbook of Molecular Microbial Ecology

Handbook of Molecular Microbial Ecology

Editor:

Patricia Marques, Ph.D.

Delve Publishing

www.delvepublishing.com

Handbook of Molecular Microbial Ecology

Editor: Patricia Marques, Ph.D.

708 3rd Avenue, 6th Floor
New York, NY 10017
United States of America
www.delvepublishing.com

ISBN: 978-1-68095-827-0

Library of Congress Control Number: 2016951569

About the Editor

Patricia Marques, Ph.D.

Patricia obtained her PhD form University College Dublin in 2010. Her interests are on Microbiology and Parasitology. She is currently working as a Postdoc at University of Maryland Baltimore, USA on Chlamydia infections

Contents

List of Contributors

Surajit De Mandal
Department of Biotechnology, Mizoram University, Aizawl-796004, Mizoram, India

Amrita Kumari Panda
Department of Zoology, Kumaun University, Nainital-263002, Uttarakhand, India

Satpal Singh Bisht
Department of Zoology, Kumaun University, Nainital-263002, Uttarakhand, India

Nachimuthu Senthil Kumar
Department of Biotechnology, Mizoram University, Aizawl-796004, Mizoram, India

Anna Pei
Washington University, College of Arts and Sciences, St. Louis, Missouri, United States of America

Carlos W. Nossa
Department of Medicine, New York University School of Medicine, New York, New York, United States of America

Pooja Chokshi,
Tufts University College of Arts and Sciences, Medford, Massachusetts, United States of America

Martin J. Blaser
Department of Medicine, New York University School of Medicine, New York, New York, United States of America
Department of Microbiology, New York University School of Medicine, New York, New York, United States of America
Department of Veterans Affairs, New York Harbor Healthcare System, New York, New York, United States of America

Liying Yang
Department of Medicine, New York University School of Medicine, New York, New York, United States of America

David M. Rosmarin
Department of Medicine, New York University School of Medicine, New York, New York, United States of America

Zhiheng Pei
Department of Medicine, New York University School of Medicine, New York, New York, United States of America
Department of Veterans Affairs, New York Harbor Healthcare System, New York, New York, United States of America
Department of Pathology, New York University School of Medicine, New York, New York, United States of America
University of California, Berkeley, United States of America

Alexander W. Eastman
Southern Crop Protection and Food Research Centre, Agriculture and Agri-Food Canada, Government of Canada, London, ON, Canada
Department of Microbiology and Immunology, Schulich School of Medicine and Dentistry, University of Western Ontario, London, ON, Canada

Ze-Chun Yuan
Southern Crop Protection and Food Research Centre, Agriculture and Agri-Food Canada, Government of Canada, London, ON, Canada
Department of Microbiology and Immunology, Schulich School of Medicine and Dentistry, University of Western Ontario, London, ON, Canada

David M. Ward
Montana State University, Bozeman, Montana, USA

Melanie C. Melendrez
Montana State University, Bozeman, Montana, USA

Eric D. Becraft
Montana State University, Bozeman, Montana, USA

Christian G. Klatt
Montana State University, Bozeman, Montana, USA

Jason M. Wood
Montana State University, Bozeman, Montana, USA

Frederick M. Cohan
Wesleyan University, Middletown, Connecticut, USA

Linxiang Chen
Department of Preventive Medicine, Shantou University Medical College, Shantou, Guangdong, China,
Laboratory of Molecular Biology, Shantou Entry-Exit Inspection and Quarantine Bureau, Shantou, Guangdong, China

Ying Cai
Laboratory of Molecular Biology, Shantou Entry-Exit Inspection and Quarantine Bureau, Shantou, Guangdong, China

Guangbiao Zhou
Laboratory of Molecular Biology, Shantou Entry-Exit Inspection and Quarantine Bureau, Shantou, Guangdong, China

Xiaojun Shi
Department of Preventive Medicine, Shantou University Medical College, Shantou, Guangdong, China,

Jianhui Su
Laboratory of Molecular Biology, Shantou Entry-Exit Inspection and Quarantine Bureau, Shantou, Guangdong, China

Guanwu Chen
Laboratory of Molecular Biology, Shantou Entry-Exit Inspection and Quarantine Bureau, Shantou, Guangdong, China

Kun Lin
Department of Preventive Medicine, Shantou University Medical College, Shantou, Guangdong, China,

Susan M Huse
Josephine Bay Paul Center, Marine Biological Laboratory at Woods Hole, MBL Street, Woods Hole, MA 02543, USA.

Julie A Huber
Josephine Bay Paul Center, Marine Biological Laboratory at Woods Hole, MBL Street, Woods Hole, MA 02543, USA.

Hilary G Morrison,
Josephine Bay Paul Center, Marine Biological Laboratory at Woods Hole, MBL Street, Woods Hole, MA 02543, USA.

Mitchell L Sogin
Josephine Bay Paul Center, Marine Biological Laboratory at Woods Hole, MBL Street, Woods Hole, MA 02543, USA.

David Mark Welch
Josephine Bay Paul Center, Marine Biological Laboratory at Woods Hole, MBL Street, Woods Hole, MA 02543, USA.

John C. Wooley
Community Cyberinfrastructure for Marine Microbial Ecology Research and Analysis, California Institute for Telecommunications and Information Technology, University of California San Diego, La Jolla, California, United States of America,

Adam Godzik
Community Cyberinfrastructure for Marine Microbial Ecology Research and Analysis, California Institute for Telecommunications and Information Technology, University of California San Diego, La Jolla, California, United States of America,
Program in Bioinformatics and Systems Biology, Burnham Institute for Medical Research, La Jolla, California, United States of America,

Iddo Friedberg
Department of Microbiology, Miami University, Oxford, Ohio, United States of America,
Department of Computer Science and Software Engineering, Miami University, Oxford, Ohio, United States of America

Katharina J Hoff
Abteilung Bioinformatik, Georg-August-Universität Göttingen, Goldschmidtstr. 1, 37077 Göttingen, Germany and

Maike Tech
Abteilung Bioinformatik, Georg-August-Universität Göttingen, Goldschmidtstr. 1, 37077 Göttingen, Germany and

Thomas Lingner
Abteilung Bioinformatik, Georg-August-Universität Göttingen, Goldschmidtstr. 1, 37077 Göttingen, Germany and

Rolf Daniel
Abteilung Genomische und Angewandte Mikrobiologie, Georg-August-Universität Göttingen, Goldschmidtstr. 1, 37077 Göttingen, Germany

Burkhard Morgenstern
Abteilung Bioinformatik, Georg-August-Universität Göttingen, Goldschmidtstr. 1, 37077 Göttingen, Germany and

Peter Meinicke
Abteilung Bioinformatik, Georg-August-Universität Göttingen, Goldschmidtstr. 1, 37077 Göttingen, Germany and

Torsten Thomas
School of Biotechnology and Biomolecular Sciences & Centre for Marine Bio-Innovation, The University of New South Wales, Sydney, NSW 2052, Australia

Jack Gilbert
Argonne National Laboratory, 9700 South Cass Avenue, Argonne, IL 60439, USA
Department of Ecology and Evolution, University of Chicago, 5640 South Ellis Avenue, Chicago, IL 60637, USA

Folker Meyer
Argonne National Laboratory, 9700 South Cass Avenue, Argonne, IL 60439, USA
Computation Institute, University of Chicago, 5640 South Ellis Avenue, Chicago, IL 60637, USA

Lu Fan
School of Biotechnology and Biomolecular Sciences and Centre for Marine Bio-Innovation, University of New South Wales, Sydney, New South Wales, Australia

Kerensa McElroy
School of Biotechnology and Biomolecular Sciences and Centre for Marine Bio-Innovation, University of New South Wales, Sydney, New South Wales, Australia

Torsten Thomas
School of Biotechnology and Biomolecular Sciences and Centre for Marine Bio-Innovation, University of New South Wales, Sydney, New South Wales, Australia

F Meyer
Mathematics and Computer Science Division, Argonne National Laboratory, 9700 S Cass Avenue, Argonne, IL 60439, USA,
Computation Institute, University of Chicago, Chicago, IL 60637, USA and

D Paarmann
Computation Institute, University of Chicago, Chicago, IL 60637, USA and

M D'Souza
Computation Institute, University of Chicago, Chicago, IL 60637, USA and

R Olson
Mathematics and Computer Science Division, Argonne National Laboratory, 9700 S Cass Avenue, Argonne, IL 60439, USA

EM Glass
Mathematics and Computer Science Division, Argonne National Laboratory, 9700 S Cass Avenue, Argonne, IL 60439, USA

M Kubal
Computation Institute, University of Chicago, Chicago, IL 60637, USA and

T Paczian
Mathematics and Computer Science Division, Argonne National Laboratory, 9700 S Cass Avenue, Argonne, IL 60439, USA

A Rodriguez
Computation Institute, University of Chicago, Chicago, IL 60637, USA and

R Stevens
Mathematics and Computer Science Division, Argonne National Laboratory, 9700 S Cass Avenue, Argonne, IL 60439, USA,
Computation Institute, University of Chicago, Chicago, IL 60637, USA and

A Wilke
Computation Institute, University of Chicago, Chicago, IL 60637, USA and

J Wilkening
Mathematics and Computer Science Division, Argonne National Laboratory, 9700 S Cass Avenue, Argonne, IL 60439, USA,

RA Edwards
Mathematics and Computer Science Division, Argonne National Laboratory, 9700 S Cass Avenue, Argonne, IL 60439, USA,
Department of Computer Science, San Diego State University, 5500 Campanile Drive, San Diego, CA 92182, USA

Jens Roat Kultima
Structural and Computational Biology Unit, European Molecular Biology Laboratory, Heidelberg, Germany

Shinichi Sunagawa
Structural and Computational Biology Unit, European Molecular Biology Laboratory, Heidelberg, Germany

Junhua Li
Department of Science and Technology, BGI-Shenzhen, Shenzhen, Guangdong, China,
School of Bioscience and Biotechnology, South China University of Technology, Guangzhou, Guangdong, China

Weineng Chen
Department of Science and Technology, BGI-Shenzhen, Shenzhen, Guangdong, China

Hua Chen
Department of Science and Technology, BGI-Shenzhen, Shenzhen, Guangdong, China

Daniel R. Mende
Structural and Computational Biology Unit, European Molecular Biology Laboratory, Heidelberg, Germany

Manimozhiyan Arumugam
Structural and Computational Biology Unit, European Molecular Biology Laboratory, Heidelberg, Germany

Qi Pan
Department of Science and Technology, BGI-Shenzhen, Shenzhen, Guangdong, China

Binghang Liu
Department of Science and Technology, BGI-Shenzhen, Shenzhen, Guangdong, China

Junjie Qin
Department of Science and Technology, BGI-Shenzhen, Shenzhen, Guangdong, China

Jun Wang
Department of Science and Technology, BGI-Shenzhen, Shenzhen, Guangdong, China

Peer Bork
Structural and Computational Biology Unit, European Molecular Biology Laboratory, Heidelberg, Germany,
Max-Delbruck-Centre for Molecular Medicine, Berlin-Buch, Germany

Patrick D. Schloss
Department of Microbiology & Immunology, The University of Michigan, Ann Arbor, Michigan, United States of America

Daniel C. Richter
ZBIT- Center for Bioinformatics Tu¨bingen, University of Tu¨bingen, Tu¨bingen, Germany

Felix Ott
ZBIT- Center for Bioinformatics Tu¨bingen, University of Tu¨bingen, Tu¨bingen, Germany

Alexander F. Auch
ZBIT- Center for Bioinformatics Tu¨bingen, University of Tu¨bingen, Tu¨bingen, Germany

Ramona Schmid
Boehringer Ingelheim Pharma GmbH & Co. KG, Biberach, Germany

Daniel H. Huson
ZBIT- Center for Bioinformatics Tu¨bingen, University of Tu¨bingen, Tu¨bingen, Germany

Nadine Ziemert
Center for Marine Biotechnology and Biomedicine, Scripps Institution of Oceanography, University of California San Diego, La Jolla, California, United States of America,

Sheila Podell
Marine Biology Research Division, Scripps Institution of Oceanography, University of California San Diego, La Jolla, California, United States of America

Kevin Penn
Center for Marine Biotechnology and Biomedicine, Scripps Institution of Oceanography, University of California San Diego, La Jolla, California, United States of America,

Jonathan H. Badger
Microbial and Environmental Genomics, J. Craig Venter Institute, San Diego, California, United States of America,

Eric Allen
Marine Biology Research Division, Scripps Institution of Oceanography, University of California San Diego, La Jolla, California, United States of America

Paul R. Jensen
Center for Marine Biotechnology and Biomedicine, Scripps Institution of Oceanography, University of California San Diego, La Jolla, California, United States of America,

Weizhong Li
California Institute for Telecommunications and Information Technology, University of California, San Diego, La Jolla California 92093 USA

Francesc López-Giráldez
Department of Ecology and Evolutionary Biology, Yale University, New Haven, Connecticut, USA

Jeffrey P Townsend
Department of Ecology and Evolutionary Biology, Yale University, New Haven, Connecticut, USA

Damien M. O'Halloran
Department of Biological Sciences, The George Washington University, Science & Engineering Hall 5685, 800 22nd St N.W., Washington, DC 20052, USA
Institute for Neuroscience, the George Washington University, 636 Ross Hall, 2300 I St. N.W., Washington, DC 20052, USA

Yuzhen Ye
School of Informatics, Indiana University, Bloomington, Indiana, United States of America,

Thomas G. Doak
Biology Department, Indiana University, Bloomington, Indiana, United States of America

Yijun Sun
Interdisciplinary Center for Biotechnology Research,
Department of Electrical and Computer Engineering, University of Florida, Gainesville, FL 32610-3622

Yunpeng Cai
Department of Electrical and Computer Engineering, University of Florida, Gainesville, FL 32610-3622

Li Liu
Interdisciplinary Center for Biotechnology Research,

Fahong Yu
Interdisciplinary Center for Biotechnology Research,

Michael L. Farrell
Materials Technology Directorate, Air Force Technical Applications Center, 1030 S. Highway A1A, Patrick AFB, FL 32925-3002, USA

William McKendree
Materials Technology Directorate, Air Force Technical Applications Center, 1030 S. Highway A1A, Patrick AFB, FL 32925-3002, USA

William Farmerie
Interdisciplinary Center for Biotechnology Research

Tim Urich
Centre of Geobiology, Department of Biology, University of Bergen, Bergen, Norway
Department of Genetics in Ecology, Vienna Ecology Center, University of Vienna, Vienna, Austria

Anders Lanze´n
Computational Biology Unit, Bergen Center for Computational Science, University of Bergen, Bergen, Norway

Ji Qi
Center for Comparative Genomics and Bioinformatics, Pennsylvania State University, State College, Pennsylvania, United States of America

Daniel H. Huson
Center for Bioinformatics, Tu¨bingen University, Tu¨bingen, Germany

Christa Schleper
Centre of Geobiology, Department of Biology, University of Bergen, Bergen, Norway,
Department of Genetics in Ecology, Vienna Ecology Center, University of Vienna, Vienna, Austria

Stephan C. Schuster
Center for Comparative Genomics and Bioinformatics, Pennsylvania State University, State College, Pennsylvania, United States of America

Coralie Damon
Ecologie Microbienne, UMR CNRS 5557, USC INRA 1193, Universite´ de Lyon, Universite´ Lyon 1, Villeurbanne, France

Fre´de´ ric Lehembre
Ecologie Microbienne, UMR CNRS 5557, USC INRA 1193, Universite´ de Lyon, Universite´ Lyon 1, Villeurbanne, France

Christine Oger-Desfeux
Po^le Rho^ne-Alpes de Bioinformatique, Universite´ de Lyon, Universite´ Lyon 1, Villeurbanne, France

Patricia Luis
Ecologie Microbienne, UMR CNRS 5557, USC INRA 1193, Universite´ de Lyon, Universite´ Lyon 1, Villeurbanne, France

Jacques Ranger
Bioge´ochimie des Ecosyste`mes Forestiers, INRA centre de Nancy, Champenoux, France

Laurence Fraissinet-Tachet
Ecologie Microbienne, UMR CNRS 5557, USC INRA 1193, Universite´ de Lyon, Universite´ Lyon 1, Villeurbanne, France

Roland Marmeisse
Ecologie Microbienne, UMR CNRS 5557, USC INRA 1193, Universite´ de Lyon, Universite´ Lyon 1, Villeurbanne, France,

Florence Arsène-Ploetze
Génétique moléculaire, Génomique et Microbiologie, Université de Strasbourg, UMR7156 CNRS, Strasbourg

Philippe N. Bertin
Génétique moléculaire, Génomique et Microbiologie, Université de Strasbourg, UMR7156 CNRS, Strasbourg

Christine Carapito
Génétique moléculaire, Génomique et Microbiologie, Université de Strasbourg, UMR7156 CNRS, Strasbourg

A Ranjitha Dhanasekaran
Center for Integrated BioSystems, Computer Science Department, Utah State University, Logan, 84322-8700, USA.
Western Dairy Center, Department of Nutrition, Dietetics, and Food Sciences, Utah State University, Logan, 84322-8700, USA.

Jon L Pearson
Center for Integrated BioSystems, Computer Science Department, Utah State University, Logan, 84322-8700, USA.
University of California, Davis, School of Veterinary Medicine, 1089 Veterinary Medicine Dr., VM3B, Room 4023, Davis, CA, 95616, USA.

Balasubramanian Ganesan
Center for Integrated BioSystems, Computer Science Department, Utah State University, Logan, 84322-8700, USA.
Linda Crnic Institute for Down Syndrome, Department of Pediatrics, School of Medicine, University of Colorado Denver, 12700 E 19th Avenue, Aurora, CO, 80045,

Bart C Weimer
Spillman Technologies, 4625 West Lake Park Blvd, Salt Lake City, UT, 84120, USA.

Preface

Microbes are significant drivers of the world's ecosystems. Microbial ecology is the study of microbes in the environment and their interactions with each other. Microbes are the tiniest creatures on Earth, yet despite their small size, they have an enormous impact on us and on our environment. Microbial ecology can help answer some of our most practical questions such as "How can we improve our lives?" as well as basic questions such as "Why are we here?" Microbial ecology can show us our place in the cosmos -- how life originated and how it evolved, and how we are related to the great diversity of all other organisms. The study of microbial ecology can help us improve our lives through the use of microbes in environmental restoration, food production, and bioengineering of useful products such as antibiotics, food supplements, and chemicals. The study of these bizarre and diverse creatures that are everywhere yet nowhere to be seen is fascinating and a pursuit that appeals to the curiosity and playfulness in us. Most types of microbes remain unknown. It is estimated that we know less than 1% of the microbial species on Earth. Yet microbes surround us everywhere -- air, water, soil. The vast majority of the world's diversity and metabolic potential lies within micro-organisms, yet we are just beginning to understand and utilize this ultimate resource of biological diversity. Microbial populations are able to tolerate and to grow under varying environmental conditions, including habitats with extreme environmental conditions such as hot springs and salt lakes. Understanding the environmental factors controlling microbial growth and survival offers insight into the distribution of microorganisms in nature, and many studies in microbial ecology are concerned with examining the adaptive features that permit particular microbial species to function in particular habitats. The field of microbial ecology has been revolutionized in the past two decades by the introduction of molecular methods into the toolbox of the microbial ecologist. Handbook of Molecular Microbial Ecology starts by looking at general Microbial ecology in Chapter 1 and then introduces some strategies for genome analysis on Chapters 2 and 3. On Chapter 4, the new concept of species in the metagenomics era is discussed. It continues by looking at how metagenomics data is obtained (chapters 5 to 10) and goes on by exploring some of the available tools for analysis, like databases (chapters 11 and 12) and computer assisted (Chapters 13 to 21) . It ends by covering various –omics, metatranscriptomics (chapters 22 and 23), and metabolomics (chapter 24) analysis in a large variety of habitats, which could not previously have been analyzed without metagenomic methodology. *Handbook of Molecular Microbial Ecology*, gathering together some of the world's leading experts, provides an invaluable introduction to the modern theory and molecular methods used in studying microbial ecology. Molecular microbial ecology represents a recent development in research methods and consists of utilising techniques of molecular biology to investigate the ecology of microorganisms. The application of molecular techniques is contributing to our improved understanding of the ecology of microbial inoculants. Indeed, molecular microbial ecology offers new tools to facilitate the detection and identification of microorganisms in the environment. It opens new possibilities for the exploration of the biodiversity of microorganisms and thus represents a step forward concerning the exploitation of microbial diversity in air and water. In addition, molecular microbial ecology provides strategies and techniques to assess the impact of microbial inoculants on the crop (i.e. effectiveness of the inoculant) and the soil environment (i.e. biosafety of the inoculant).

Editor

Patricia Marques, Ph.D.

INTRODUCTION

Molecular microbial ecology represents a recent development in research methods and consists of utilizing techniques of molecular biology to investigate the ecology of microorganisms. The application of molecular techniques is contributing to our improved understanding of the ecology of microbial inoculants. Indeed, molecular microbial ecology offers new tools to facilitate the detection and identification of microorganisms in the soil environment. It opens new possibilities for the exploration of the biodiversity of microorganisms and thus represents a step forward concerning the exploitation of microbial diversity.

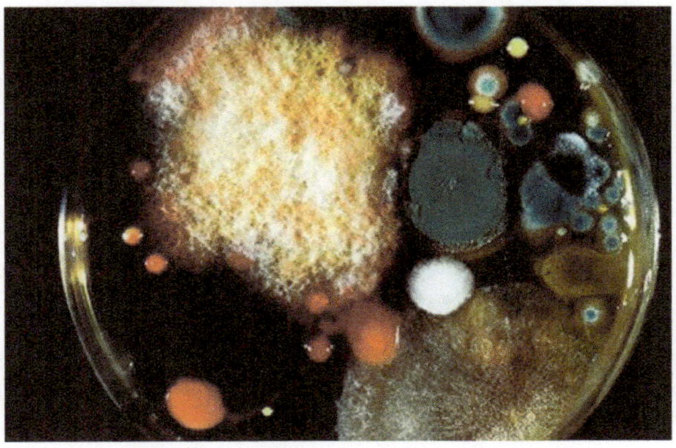

Microbial ecology (or environmental microbiology) is the ecology of microorganisms: their relationship with one another and with their environment. It concerns the three major domains of life—Eukaryota, Archaea, and Bacteria—as well as viruses. Microorganisms, by their omnipresence, impact the entire biosphere.

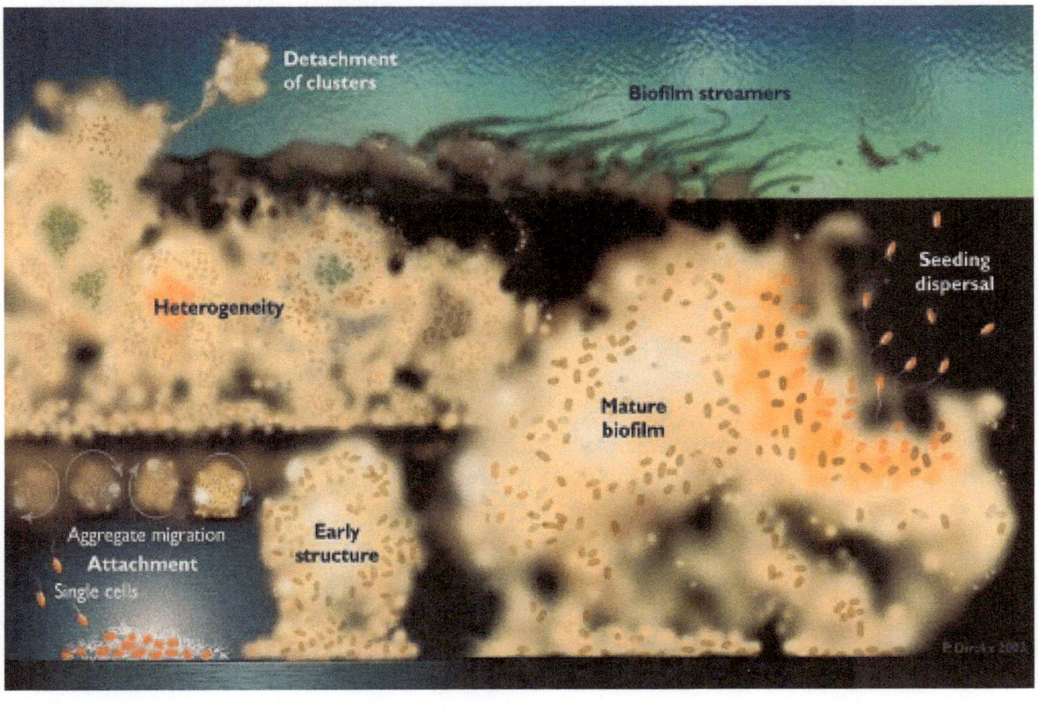

Microorganisms, by their omnipresence, impact the entire biosphere. Microbial life plays a primary role in regulating biogeochemical systems in virtually all of our planet's environments, including some of the most extreme, from frozen environments and acidic lakes, to hydrothermal vents at the bottom of deepest oceans, and some of the most familiar, such as the human small intestine. As a consequence of the quantitative magnitude of microbial life (Whitman and coworkers calculated 5.0×10^{30} cells, eight orders of magnitude greater than the number of stars in the observable universe) microbes, by virtue of their biomass alone, constitute a significant carbon sink. Aside from carbon fixation, microorganisms' key collective metabolic processes (including nitrogen fixation, methane metabolism, and sulfur metabolism) control global biogeochemical cycling. The immensity of microorganisms' production is such that, even in the total absence of eukaryotic life, these processes would likely continue unchanged.

History

While microbes have been studied since the seventeenth-century, this research was from a primarily physiological perspective rather than an ecological one. Martinus Beijerinck invented the enrichment culture, a fundamental method of studying microbes from the environment. He is often incorrectly credited with framing the microbial ecology idea that "everything is everywhere, but, the environment selects", which was started by Lourens Baas Becking. Sergei Winogradsky was one of the first researchers to attempt to understand microorganisms outside of the medical context—making him among the first students of microbial ecology and environmental microbiology—discovering chemosynthesis, and developing the Winogradsky column in the process.

Beijerinck and Windogradsky, however, were focused on the physiology of microorganisms, not the microbial habitat or their ecological interactions. Modern microbial ecology was launched by Robert Hungate and coworkers, who investigated the rumen ecosystem. The study of the rumen required Hungate to develop techniques for culturing anaerobic microbes, and he also pioneered a quantitative approach to the study of microbes and their ecological activities that differentiated the relative contributions of species and catabolic pathways.

Roles

Microorganisms are the backbone of all ecosystems, but even more so in the zones where photosynthesis is unable to take place because of the absence of light. In such zones, chemosynthetic microbes provide energy and carbon to the other organisms.

Other microbes are decomposers, with the ability to recycle nutrients from other organisms' waste products. These microbes play a vital role in biogeochemical cycles. The nitrogen cycle, the phosphorus cycle, the sulphur cycle and the carbon cycle all depend on microorganisms in one way or another. For example, the nitrogen gas which makes up 78% of the earth's atmosphere is unavailable to most organisms, until it is converted to a biologically available form by the microbial process of nitrogen fixation.

Due to the high level of horizontal gene transfer among microbial communities, microbial ecology is also of importance to studies of evolution

Microbial resource management

Biotechnology may be used alongside microbial ecology to address a number of environmental and economic challenges. For example, molecular techniques such as community fingerprinting can be used to track changes in microbial communities over time or assess their biodiversity. Managing the carbon cycle to sequester carbon dioxide and prevent excess methanogenesis is important in mitigating global warming, and the prospects of bioenergy are being expanded by the development of microbial fuel cells. Microbial resource management advocates a more progressive attitude towards disease, whereby biological control agents are favored over attempts at eradication. Fluxes in microbial communities has to be better characterized for this field's potential to be realized. In addition, there are also clinical implications, as marine microbial symbioses are a valuable source of existing and novel antimicrobial agents, and thus offer another line of inquiry in the evolutionary arms race of antibiotic resistance, a pressing concern for researchers.

Topics in microbial ecology

To find out more about how microbes evolve, about their tremendous diversity, their ecology, their unusual habitats, their role in bioremediation, recycling, food production and biotechnology, check out the page topics in microbial ecology.

Biodiversity

An estimated 1,000,000 bacterial species exist on this planet, according to the Global Biodiversity Assessment, yet fewer than 4500 have been described. The greatest genetic diversity of life comes from within the world of microorganisms, yet the least is known about them.

Diverse Habitats

Microbes inhabit the widest range of habitats from sub-freezing temperatures, to water hotter than boiling, from the rocks beneath our feet, to the atmosphere miles overhead, to the stuff between our toes, to the tops of mountains and to the deepest ocean trenches. For an introductory look at some diverse habitats that microbes call home, check out the Microbe Zoo web site at: http://commtechlab.msu.edu/sites/dlc-me/zoo/

Microbe Hunting

Hunting for new microbes is not as easy as taking a jeep trip in the outback with a pair of binoculars. By definition, microbes are invisible without the aid of a microscope, so the challenge to find new ones is great. The difficulty is compounded when one does look under the microscope to see two apparently similar bacteria which later prove to be not at all similar. For example: two bacteria may have the same rod shape, but one thrives in the presence of oxygen whereas the other one is killed by oxygen.

Microbial Biodiversity Working Group

The greatest diversity of life comes from within the world of microorganisms. To help sort out and understand this tremendous biodiversity, the Microbial Biodiversity Working Group was started. The goals of the working groups are generally to assess the advances, opportunities and needs of the field, consider society activities such as workshops, summary publications or new articles that highlight microbial biodiversity, and provide advice to national and international groups on biodiversity.

Topics in Microbial Ecology

Microbial ecology is a large field that includes many topics. Click on one of the pictures below to learn more about that particular aspect of microbial ecology.

- Evolution and Creation
- Biodiversity
- Exobiology
- Ecology
- Bioremediation
- Recycling
- Food Microbiology
- Biotechnology

Evolution and Creation

How did we get here? This is one of the first questions that many people ponder. The study of the origin and early evolution of microorganisms may provide information on the critical early stages of life. Microbes were the first living creatures on Earth, first appearing over 3.7 billion years ago, long before multicellular plants and animals, which evolved 600 million years ago. Because microbes are simple relative to humans, they provide a model for understanding how evolution works and insight into the origin of all life on Earth.

Biodiversity

Many years of evolution have created a stunning diversity of microbes. Understanding the richness of this awesome and mind-expanding microbial world puts our lives into perspective and gives us new respect for other beings. The huge numbers of microbial species are only beginning to be known. Of the more than one million species of bacteria suspected to exist, only about 4200 species of bacteria are described. The infant science of microbial systematics has only recently acquired the sophisticated tools required to assess this tremendous biodiversity. It is estimated that we know less than 1% of the microbial species on Earth.

Exobiology

Does life exist elsewhere in the universe? If there is life elsewhere in the universe, it is probably microbial. Fascination with extraterrestrial life has fueled enthusiasm for shows such as the "X-Files" and one of the most talked about news stories of 1996: evidence for past microbial life on Mars. While the evidence for microbes on Mars is intriguing, the debate continues about life on Mars, and the search goes on to detect life beyond our own planet. While we may never know whether life exists elsewhere in the universe, microbes will probably play a key role in helping us explore and colonize space.

Ecology

Ecology is the study of organisms in their environment. Ecology is one of the most complex scientific disciplines because of the myriad of organisms' interactions. Using microbes to study ecology may provide clues to how complex ecosystems operate because bacteria are small so large populations can be studied and quick to reproduce compared to larger organisms. A better understanding of ecology may help us better manage Earth and its dwindling renewable resources and to protect global ecological systems.

Global Ecology - Microbes play important global roles:

- Global Geochemical Cycles
- Carbon
- Nitrogen
- Iron,
- Sulfur
- El Niño - what role do microbes play in El Niño?
- Global Warming - The average temperature of the Earth has been steadily increasing over the past one hundred years. What role do microbes play in this global warming trend?
- Gaia - the Earth's biosystem is known to some as Gaia. Because microbes are responsible for most of the cycling of elements, they are the foundation of this system.

Human Ecology - The study of simple microbial communities may shed light on one of the most complex (and for most of us most fascinating) communities - the microbial community within a human being. We are hosts to billions of microbes which colonize our intestinal tracts, our skin, hair, teeth. Sometimes microbial colonization causes us problems when disease results. Some scientists have joked that humans are microbes' invention to move around.

Population Interactions - Microbes interact with plants and animals and each other in myriad ways with large plants and animals having hundreds of associations with different microbial species.

- plant-microbe interactions
- symbiosis - e. g. rhizobium which converts nitrogen to a form that can be used by plants
- disease
- animal-microbe interactions
- symbiosis - e.g. probiotics = beneficial gut microflora
- disease
- Microbe-microbe Interactions - e.g. lichens

Bioremediation

Environmental degradation is one of the worst threats facing humanity. Many environmental problems may be cured by microbes.

Bioremediation of toxic wastes: bioremediation, or the use of microbes to clean-up toxic wastes. Microbes are used to clean up the following:

- Oil spills
- Pesticides
- Industrial wastes

Recycling

Microorganism's recycle most organic wastes in nature into reusable resources. Humans have harnessed the power of microbes to recycle in the systems below:

- Waste treatment: Microbes are used in sewage treatment for nutrient recycling, methane recovery and disease control.
- Animal waste treatment: Microbes clean up waste produced by farm animals.
- Yard composting: Composting is now the fashionable practice (championed by Bette Midler among others) of converting kitchen and yard wastes into a rich soil amendment. Although worms have gotten good press for their contribution to composting, the real workhorses of composting are microbes.
- Municipal composting: Thousands of tons of leaves and grass clippings are composted in large municipal composting facilities.

Food Microbiology

All of us depend upon food crops for sustenance, and in turn these crops depend on healthy soils with its balanced microbial life. Much of the protein that we eat is the result of bacterial fixation of nitrogen from the air from microbes such as Rhizobium. Many of our foods are prepared with the aid of microbes, such as yeast in bread, and lactobacilli in yogurt.

Many of our food supplements, from vitamins, amino acids to flavor enhancers and preservatives, come from microbes.

Some of our foods are microbes, such as salted yeast paste found in Vegemite, and single cell protein.

Although not a food, microbes are used as probiotics, a digestion supplement that colonizes the intestines, preventing the colonization of bad, disease-causing microbes.

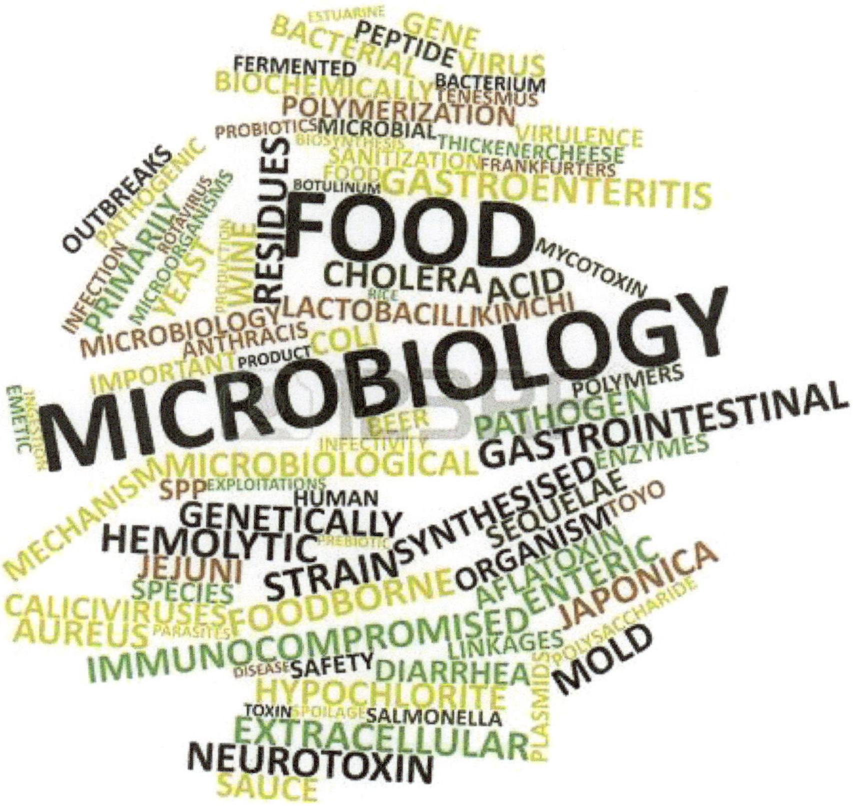

Biotechnology

Microbes are used as workhorses in the production of many compounds, from fuel, to pharmaceuticals, to chemicals. They are also used in mining, insect and disease control, genetic engineering and some are even used to make computer biochips.

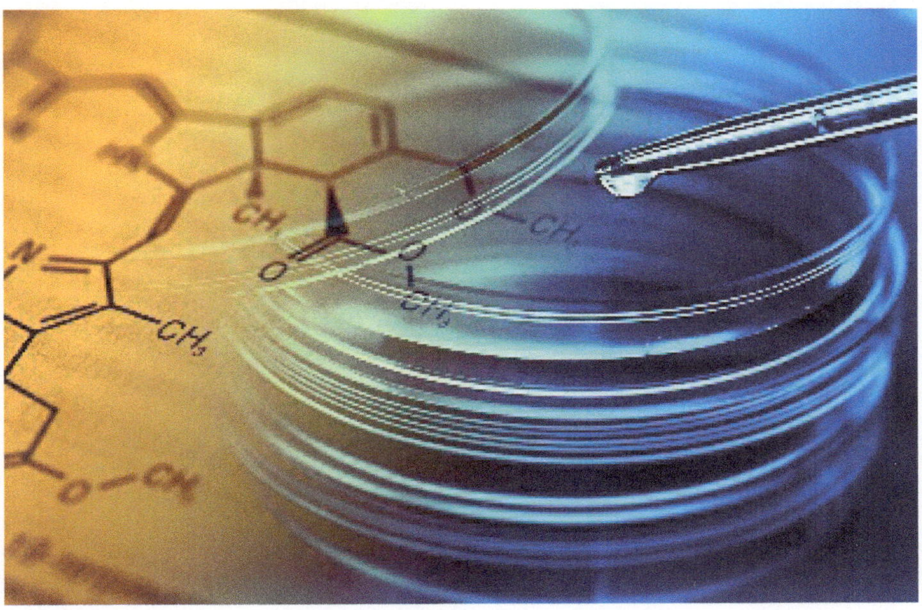

Fuel Production - ethanol production for automobile fuel-Mining - leaching of metals from ore-bearing rocks by microbes.

- 5% of world's copper ore is produced by bio-leaching
- Uranium is mined with help of bacteria

Biocontrol - Using microbes to combat pests is called biocontrol. One of the most popular forms of biocontrol is the use of Bacillus thuriengensis, a bacterium that produces a toxin that kills over 40 problem pests such as the gypsy moth. Computer Biochips - Microorganisms may some day be used to produce protein-based microprocessors with more switches than conventional microchips.Genetic Engineering - Some microorganisms are used to carry genes into other organisms.

SECTION-I: BACKGROUND CHAPTERS

CHAPTER 1

MICROBIAL ECOLOGY IN THE ERA OF NEXT GENERATION SEQUENCING

Surajit De Mandal, Amrita Kumari Panda, Satpal Singh Bisht and Nachimuthu Senthil Kumar

Keywords: Biological diversity, Metagenomics, Next generation sequencing

INTRODUCTION

Natural environments harbor a large number of microorganisms representing the major reservoir of undescribed biodiversity [1] and occupy diverse habitats from deep sea sediments to high up in the atmosphere [2]. But, most of the microbes are unculturable by standard techniques [3,4]. Thus, their ecology and functional roles are unknown and our understanding of the composition of the natural microbial world is therefore rudimentary. The reason for uncultivable nature may be the fact that they depend on other organisms for critical processes, fail to grow in vitro or have even become extinct in fossil records [5]. Some are metabolically active, even though they were unculturable in laboratory [6]. Culturing these microbes in artificial environment needs in-depth knowledge of the nutrient and growth requirements. Although some work has attempted to culture these uncultivable bacteria [7,8], the technique is time consuming and laborious and therefore needs some alternative methods to study them.

DNA sequence-based methods can overcome these problems since we can isolate the genetic material directly from live or dead cells from various environmental samples, allowing a new emerging field called metagenomics. Metagenomics involves the culture independent approach to study the unculturable microorganisms directly from an environmental samples such as soil [9,10], seawater [11], ground water [12], antarctic desert soil [13] etc. In the mid-1980s, Norman Pace [14] suggested a culture-independent approach which was later named as "metagenomics" by Handelsman and first appeared in publication in 1998. This field is also called as environmental genomics, ecogenomics or community genomics. In this technique, the isolation of environmental DNA directly resulted in the isolation of a number of novel genes and was designed with several practical gains such as the discovery of new genes and gene products that would lead to agricultural innovations, medicinal chemistry and industrial processes. It can also be used to reveal diversity patterns of microorganisms, horizontal gene transfer analysis, identification of novel metabolic pathways and examining genes/operons for desirable enzyme candidates (e.g., cellulases, chitinases, lipases, antibiotics) and other natural products [3,15,16].

Ribosomal RNA for the Identification of Uncultured Bacteria and its Community

Ribosomal RNA is largely used for the study of uncultured bacteria as it has several advantages. The 16S gene occurs in all living organisms with the notable exception of viruses and represents more than 80% of total bacterial RNA found in every living cell and generally stable and long enough (approx. 1500 nt), contains independently evolving domains, i.e. variable regions. They are shorter than 23s (approx. 2300 nt) and easier to sequence. The 16S rRNA gene structure consists of interspersed conserved and variable regions which is suitable for PCR amplification and sequencing. As a molecular chronometer, these molecules have preserved their evolutionary history. Most of the studies have been carried out with 16S rRNA and has been used as a phylogenetic marker as it is present in all prokaryotes and contains different hypervariable regions separated by highly conserved gene segments. They have highly conserved portion which carries the information on early evolutionary events and more recent changes are documented within less conserved position or stretches. The degree of divergence of present day rRNA sequence gives an estimate of their phylogenetic distance. They donot have any ortholog - paralog issues or horizontal gene transfer. It is easy to PCR amplify and they have a very slow rate of mutation. The new era of metagenomics was ushered in by studies using 16S rRNA as a phylogenetic marker of microbial taxa.

Citation: De Mandal S, Panda AK, Bisht SS, Kumar NS (2015) Microbial Ecology in the Era of Next Generation Sequencing. Next Generat Sequenc & Applic S1:001. doi:10.4172/2469-9853.S1-001.

Focusing on a small part of the microbial genome brings down the sequencing costs dramatically [17].

It is important to choose the appropriate variable region of the 16S rRNA for microbial ecological studies mainly for biogeography, metacommunity theory or microbiome analysis. Previous studies selected the hypervariable region which can exhibit sufficient phylogenetic signals to classify at the genus level [18]. Liu et al. [19] suggested for the 250 bp region between the V2 and V3 hypervariable regions for microbial ecological studies. The use of tags of the V3 and V6 regions in 454 pyrosequencing also provides sufficient taxonomic assignments [20]. Analysis by Wang et al. [21] proved that V2 and V4 regions can results best taxonomic assignment at the genus level, where as for the study of microbial community complex ecosystem it is recommended to use V7-V8 region [22,23].

Analysis of Metagenome using DGGE and TGGE

Sanger sequencing (first generation sequencing) was widely used for molecular biology studies with no or minimum errors. Based on the Sanger sequencing approach, different PCR based methods have been developed to study the microbial communities. Denaturing Gradient Gel Electrophoresis (DGGE) or Terminal Restriction Fragment Length Polymorphism (T-RFLP) has been developed to analyze the uncultured microbial communities in diverse environments [24,25]. These methods allow differentiating gene molecules (in polyacrylamide gels) based on the decreased electrophoresis mobility of a partially melted double stranded DNA molecule at the level of a single base containing a linear gradient of DNA denaturants (a mixture of urea and formamide) or a linear temperature gradient. However, these techniques present several drawbacks as the dominant populations are better revealed and bands from more than one species may be hidden behind a single band resulting in underestimation of bacterial diversity. The same isolate can have different identifiable bands because multiple copies of 16S rRNA gene are present in a single genome of species. The Sanger sequencing based methods present several drawbacks since it is expensive and cannot detect rare microorganisms [26,27].

Next Generation Sequencing Technology (NGS)

NGS allows massively parallel sequencing with thousands to millions of sequences in one experiment at considerably low cost compared to Sanger method. There are no requirements of bacterial cloning of DNA fragments and electrophoresis, since NGS libraries preparation process are carried out in a cell-free system and the sequence output is directly detected. However, it produces shorter reads with a higher error rate than Sanger sequencing [28-30]. Moreover, secondary structure and thermal stability of the genome affects the efficiency of PCR amplification of the genomic fragments. There are different NGS platforms (e.g. Illumina, Pacific Biosciences, Ion Torrent, SOLiD) with different principle and divergent features such as run times, yields and read lengths [31] (Table 1).

Technology (company)	*Amplification*	Chemistry	Sequencing method	Yield (Gb/run)	*Highest Average Read Length*	Error rate	Output file	Disadvantage	Advantage	Website
Illumina	Bridge amplification	Reversible dye terminator (seq-by-synthesis)	incorporation of fluorescent nucleotides	1- 60	300 bp (overlapping paired-end sequencing available	≥0.1	Fastq (Phred +64 & 33, Illumina +1.8)	Short reads and long runtime	High throughput and low cost	www.illumina.com
SOLiD (Life Technologies)	Emulsion PCR	Sequencing by ligation (SBL)	fluorescent short linkers	3	75 bp (paired-end sequencing available)	>0.06	Fastq (Phred +33)	Long runtime	Low error rate	www.appliedbiosystems.com
454 (Roche)	Emulsion PCR	Pyrosequencing (seq-by-synthesis)	incorporation of normal nucleotides	0.7	700 bp	1	SFF, fasta, fastq	High error rate in homopolymer	Long read	www.454.com
SMAT (Pacific Bio)	N/A (single molecule)	Single Molecule Real Time (SMRT™)	incorporation of fluorescent nucleotides	0.3-0.5	5,000 bp average; maximum read length ~22,000 bases	16	Fastq (Phred +33)	No PCR longest reads	High error	www.pacificbio.com
Ion Torrent (Life Technology)	N/A	Proton detection (seq-by-synthesis)	measuring pH change	1	400 bp (bidirectional sequencing available)	1	Fastq (Phred +33)	new	Short read	www.iontorrent. com

Table 1: Summary of the major next-generation sequencing platform.

Advances in NGS have revolutionized the field of microbial ecology. Any NGS study follow some common steps such as sample and metadata collection, DNA extraction, library construction, sequencing and read preprocessing followed by quantitative analysis and functional binning [32,33]. Collection of environment sample is the first step in any metagenomic study. But, the problem faced in microbial ecology studies are that no information is available about the amount of required sampling that will comprise entire population in an environment. Usually rarefaction curve, a plot of number of species versus number of individuals sampled, which typically defines the population richness and evenness is used to estimate the fraction of species sequenced.

Another important aspect for any successful metagenomic study is the extraction of high quality DNA from the environmental samples [34]. Environmental samples contain DNA in a variety of packages like free DNA, virus particle, and prokaryotic and eukaryotic cells. These can be suspended in water, bound to a solid matrix like soil, or encased in a biofilm or tissue. So extraction methods must be chosen carefully based on the medium and the DNA population of interest (Susannah Green Tringe) and DNA should come from all the microbial representative present in the sample without any shearing or contamination [35]. Although many commercial kits are available for metagenome isolation, it is required to develop our own method to optimize extraction and reducing bias caused by unequal lysis of different members of the soil microbial community [36,37]. Beat beating, sonication, detergents or enzymatic lysis have been used for the isolation purpose. Since some samples contain high humic acid, additional purification is required. Polyphenol compound often co-purified along with the DNA can enzymatically modify the isolated DNA [38]. One strategy may be the use of Skimmed milk which can aid in DNA extraction process from soil [39]. Multiple displacement amplification can be used to increase template DNA for samples with very less amount of DNA [40,41].

Two kind of NGS strategy is applied in microbial community analysis: 1) deep amplicon sequencing and 2) complete metagenome or metatranscriptome analysis. Shotgun sequencing is applied to know the function of the community. The analysis procedure involves assembly (merging overlapping short reads into contigs), binning (grouping reads or contigs into individual genomes and assigning the groups to specific species, subspecies, or genus), functional annotation (for the prediction of CDS, CRISPR, noncoding RNAs) and functional assignment to the predicted protein coding sequences. On the other hand, amplicon sequencing is used for community profiling using marker genes (e.g., 16S rRNA gene) in different ecosystems. Procedure involves denoising (filtered the 'noise' sequences from the raw data), Chimera detection (recombinants which are formed when prematurely terminated fragments reanneal to other template DNA during PCR), OTU clustering (alignment is done by either aligning query sequences against pre-aligned reference sequences (42) or using pairwise and multiple sequence alignments), Taxonomic classification (taxonomic assignment of OTUs) and Statistical analysis, (alpha and betadiversity analysis, multidimensional scaling and analysis of similarities (ANOSIM), hypothesis testing etc). A flow diagram for the analysis of metagenomic data is shown in Figure 1 and Table 2 represents metagenomics tools available for Microbial Ecology Studies.

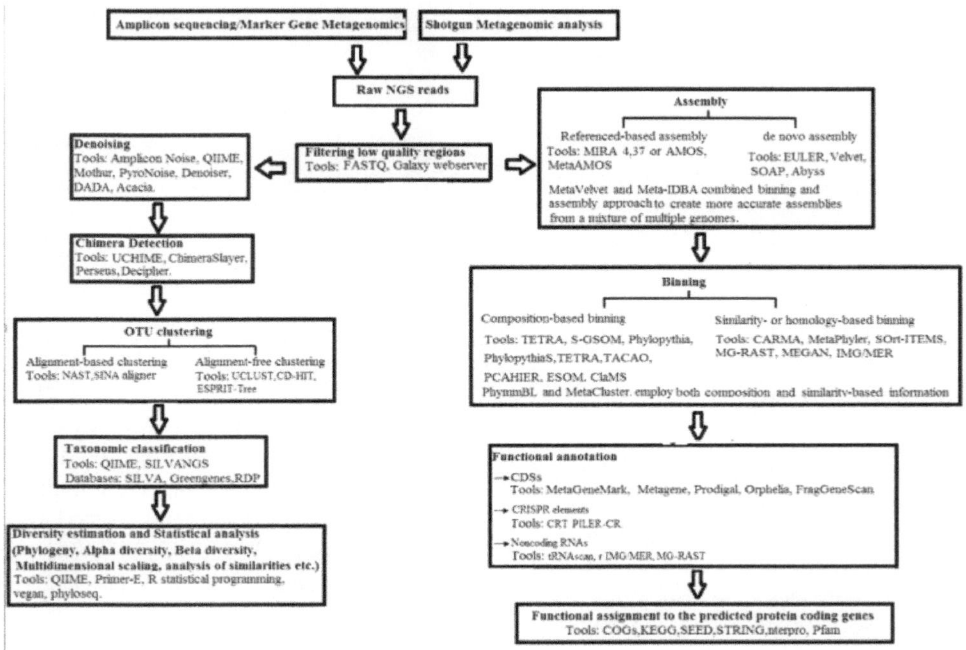

Figure 1: Flow diagram for the analysis of Metagenomics data.

Table 2: Metagenomics tools available for microbial ecology studies

Metagenomics tools	website
General Software package	
STAMP	http://kiwi.cs.dal.ca/Software/STAMP
CD-HIT-OUT	http://weizhong-lab.ucsd.edu/cd-hit-otu/
GAAS	http://sourceforge.net/projects/gaas/
Megan	http://www-ab.informatik.unituebingen.de/software/megan/ welcome.html
MetaPhlAn	http://huttenhower.sph.harvard.edu/metaphlan
MetaSim	http://www-ab.informatik.uni-tuebingen.de/software/metasim
Mocat	http://vm-lux.embl.de/~kultima/MOCAT/index.htm
Livermore Metagenomics Analysis Toolkit	https://computation-rnd.llnl.gov/lmat
Straine	http://www.hsls.pitt.edu/obrc/index.php?page=URL1221230034
Metagenome assembly	
Velvet	http://www.ebi.ac.uk/~zerbino/velvet/
Celera	http://www.cbcb.umd.edu/research/assembly.shtml#software
Metasim	http://ab.inf.uni-tuebingen.de/software/metasim/welcome.html #Download
Euler	http://nbcr.sdsc.edu/euler/ JAZZ
Mapping to reference genome	
Bowtie	http://bowtie-bio.sourceforge.net/index.shtml
BWA	http://bio-bwa.sourceforge.net/
MAPLE:	http://www.genome.jp/maple-bin/mapleSubmit.cgi?aType=sDirect
Quality analysis	
FastQC	http://www.bioinformatics.babraham.ac.uk/projects/download.html#fastqc
Prinseq	http://edwards.sdsu.edu/cgi-bin/prinseq/prinseq.cgi
Gene calling	
Genemark.	hmm http://exon.gatech.edu/GeneMark/metagenome/Prediction/

Microbial diversity Analysis	
MLST	http://www. mlst.net) http://www.mlst.net/
MOTHUR	http://www.mothur.org/
EstimateS	http://viceroy.eeb.uconn.edu/EstimateS/
QIIME	http://qiime.org/install/virtual_box.html
PHACCS	http://phaccs.sourceforge.net/
Binning	
TETRA	http://www.megx.net/tetra/index.html
Phylopathia	http://cbcsrv.watson.ibm.com/phylopythia.html
MEGAN	http://ab.inf.uni-tuebingen.de/software/megan/
CARMA	http://www.cebitec.uni-bielefeld.de/brf/carma/carma.html
Phymm	http://www.cbcb.umd.edu/software/phymm/
Functional Annotation	
MEX (Motif Extraction)	http://adios.tau.ac.il/SPMatch/
RAMMCAP	http://weizhong-lab.ucsd.edu/rammcap/cgi-bin/rammcap.cgi
Analysis of quantitative metagenomics data.	
FANTOM:	http://www.sysbio.se/Fantom/
Comparitive Metagenomics	
MEGAN	http://metagenomics.anl.gov/
MG-RAST	http://metagenomics.anl.gov
Camera	http://camera.calit2.net/
Shotgun FunctionalizeR	http://shotgun.math.chalmers.se/
MetaStats	http://metastats.cbcb.umd.edu/detection.html
UniFrac	http://bmf.colorado.edu/unifrac/
Galaxy	https://main.g2.bx.psu.edu/u/aun1/w/metagenomic-analysis
MetaMine	http://www.megx.net/metamine/
MetaLook	http://www.megx.net/metalook/index.php
IMG/M	http://img.jgi.doe.gov/cgi-bin/m/main.cgi
Online tools for NGS analysis	
Parallel Meta	http://www.computationalbioenergy.org/parallel-meta.html
SOrt-ITEMS	http://metagenomics.atc.tcs.com/binning/SOrt-ITEMS/
PANGEA	http://www.ohloh.net/p/pangea-plus
Genohub:	https://genohub.com/bioinformatics/17/metagenomic-analysis
Additional Metagenome analysis links	
EBI Metagenomics:	https://www.ebi.ac.uk/metagenomics
MetaPhlAn:	http://huttenhower.sph.harvard.edu/metaphlan
Amphora	Net http://pitgroup.org/amphoranet

Third-Generation Sequencing

Third-generation sequencing (SMRT sequencing) is based on the sequencing by synthesis approach and allows the detection of single molecules, but also enables real-time sequencing. The biggest advantage is that it allows 20,000 reads or more, with average read lengths of 5 kilobases. Moreover, it can also directly detect epigenetic modifications such as 4-methylctosine (mC), 5-mC and 6-methyladenine [42].

Application of NGS in Microbial Ecology

Analysis of microbial community composition and its function is a major objective in modern microbial ecology studies. NGS has already been applied in such studies in diverse ecosystem such as caves, forests, hot springs, deserts etc [43-47]. A pyrosequencing based approach revealed that the soil stratification and resource availability can impact the taxonomic and functional diversity of bacterial communities [48]. NGS based studies have proved that the agriculture management system can significantly affect the microbial diversity [49,50]. Chhabra et al. [51] identified genes and operons responsible for mineral phosphate solubilization in the barley rhizosphere using pyrosequencing. A NGS based study of soil bacterial communities showed that the conversion of primary Amazonian forest to pasture increases bacterial alphadiversity, but decreases beta diversity resulting in the homogenization of communities across space. This homogenization is driven by the loss of forest soil bacteria with restricted ranges (endemics) and results in a net loss of diversity [52]. Analysis using 227 million Illumina shotgun sequences from forest soil and 246 million from deforested soils revealed that bacterial taxonomic and functional adaptations at the bacterial community level is due to an increase in nutrient availability from slash and burn clearing of Amazon forest [53]. Analysis of 62081 pyrosequencing reads from 10 soil samples revealed that bacterial diversity and soil physicochemical properties did not show consistent changes along the elevation gradients in southwestern highlands of Saudi Arabia [54]. An NGS study carried out in temperate deciduous forest of Northeast Ohio, USA found that the vernal pool microbial communities may rely on their metabolic plasticity for growth and survival during limited resources [55]. Analysis of NGS reads shown that microbial functional activity increased throughout decomposition in spring, summer and winter, while it decreased in autumn in a Midwest temperate forest in Morris Bean Reserve of Greene County, Ohio, USA [56]. Pyrosequencing of V3-V4 regions described the bacterioplankton communities in a coastal Antarctic lake which is under long-term environmental change [57]. Fierer et al. [47] used Illumina shotgun sequencing in different ecosystem and found that desert microbial communities had high relative abundance of genes associated with osmoregulation and dormancy, whereas genes linked with nutrient cycling and catabolism of plant-derived organic compounds was less. This study also revealed that abiotic factors are more important in harboring the desert microbial communities. Another shotgun study detected identification of genes related to phytic acid utilization [58]. Analysis of total of 1,294,216 raw 16S rRNAV6 sequences revealed that Actinobacteria, Acidobacteria,

Nitrospirae, and Verrucomicrobia were abundant in nutrient-rich inner mangrove sediments, while Proteobacteria and Deferribacterias were high in outer mangrove sediments [59]. Based on pyrosequencing of the V2-V3 16S rRNA gene regions, Nacke et al. [60] found that soil bacterial community composition and diversity of the six analyzed management types showed significant differences between the land use patterns in grassland and forest ecosystems and more over bacterial community structure was largely driven by tree species and soil pH.

NGS technologies are improving day by day with more accuracy and longer sequence reads which are applied in many microbial ecology studies. Due to the continuous decrease in NGS cost, a huge number of environmental samples has been sequenced to solve research problems in microbial ecology. This paper will be helpful in understanding the basics of next generation sequencing as well as for designing new metagenomics projects. Future research with the help of NGS technologies will further unzip the unseen microbial diversity and its function useful in solving ecological problems.

ACKNOWLEDGEMENTS

The authors thank DBT, New Delhi for supporting Bioinformatics Infrastructure Facility, Mizoram University which has provided the essential facilities to carry out the work. The first author is thankful to DBT-State Biotech Hub, Mizoram University for the fellowship.

REFERENCES

1. Nichols DS, Sanderson K, Buia A, Kamp J, Halloway P et al. (2002) Bioprospecting and biotechnology in Antarctica. In: Jabour-Green J, Haward M (eds). The Antarctic: Past, Present and Future. Antarctic CRC. Res. Report, Hobart, 85-103.
2. Torsvik VL, Øvreås L (2011) DNA Reassociation Yields Broad-Scale Information on Metagenome Complexity and Microbial Diversity. In: Handbook of Molecular Microbial Ecology I. John Wiley & Sons, Inc., 3-16.
3. Handelsman J (2004) Metagenomics: application of genomics to uncultured microorganisms. MicrobiolMolBiol Rev 68: 669-685.
4. Kimura N (2006) Metagenomics: Access to unculturable microbes in the environment. Microb Environ 21: 201-215.
5. Tringe SG and Rubin EM (2005) Metagenomics: DNA sequencing of environmental samples. Nat Rev Genet 6: 805-814.
6. Rappe MS, Connon SA, Vergin KL, Giovannoni SJ (2002) Cultivation of the ubiquitous SAR11 marine bacterioplankton clade. Nature 418: 630-633.
7. Sait M, Hugenholtz P, Janssen PH (2002) Cultivation of globally distributed soil bacteria from phylogenetic lineages previously only detected in cultivation independent surveys. Environ Microbiol 4: 654-666.
8. Zwirglmaier K, Ludwig W, Schleifer KH (2004) Recognition of individual genes in a single bacterial cell by fluorescence in situ hybridization-RING-FISH. MolMicrobiol 51: 89-96.
9. Voget S, Steele HL, Streit WR (2006) Characterization of a metagenome-derived halotolerant cellulose. J Biotechnol 126: 26-36.
10. Waschkowitz T, Rockstroh S, Daniel R (2009) Isolation and characterization of metalloproteases with a novel domain structure by construction and screening of metagenomic libraries. Appl Environ Microbiol 75: 2506-2516.
11. Stein JL, Marsh TL, Wu KY, Shizuya H, DeLong EF (1996) Characterization of uncultivated prokaryotes: isolation and analysis of a 40-kilobase- pair genome fragment from a planktonic marine archeon J Bacteriol 178: 591-599.
12. Uchiyama T, Abe T, Ikemura T, Watanable K (2005) Substrate induced gene-expression screening of environmental metagenome libraries for isolation of catabolic genes. Nat Biotechnol 23: 88-93.
13. Heath C, Hu XP, Cary SC, Cowan D (2009) Identification of a novel alkaliphilic esterase active at low temperatures by screening a metagenomic library from Antarctic desert soil. Appl. Environ. Microbiol 75: 4657-4659.
14. Pace N, Stahl D, Lane D, Olsen G (1985) Analysing natural microbial populations by rRNA sequences. ASM News 51: 4-12.
15. Elend C, Schmeisser C, Leggewie C, Babiak P, Carballeira JD, et al. (2006) Isolation and biochemical characterization of two novel metagenome derived esterases. Appl Environ Microbiol 72: 3637-3645.
16. Vakhlu J, Kour A, Johri B (2008) Metagenomics: future of microbial gene mining. Indian J Microbiol 48: 202-215.
17. Pace NR (1997) A molecular view of microbial diversity and the biosphere. Science 276: 734-740.
18. Birtel J, Walser JC, Pichon S, Bürgmann H, Matthews B (2015) Estimating Bacterial Diversity for Ecological Studies: Methods, Metrics, and Assumptions. PLoS ONE 10- e0125356.
19. Liu Z, DeSantis TZ, Andersen GL, Knight R (2008) Accurate taxonomy assignments from 16S rRNA sequences produced by highly parallel pyrosequencers. Nucleic Acids Res 36: e120.
20. Huse SM, Dethlefsen L, Huber JA, Mark Welch D, Welch DM, et al. (2008) Exploring microbial diversity and taxonomy using SSU rRNAhypervariable tag sequencing. PLoS Genet 4: e1000255.
21. Wang Q, Garrity GM, Tiedje JM, Cole JR (2007) Naive Bayesian classifier for rapid assignment of rRNA sequences into the new bacterial taxonomy. Applied and Environment Microbiology 73: 5261-5267.
22. Kumar PS, Brooker MR, Dowd SE, Camerlengo T (2011) Target region selection is a critical determinant of community fingerprints generated by 16S Pyrosequencing.PLoS One6: e20956.
23. Nikolaki S, Tsiamis G (2013) Microbial Diversity in the Era of Omic Technologies. BioMed Research International 2013: 1-15.
24. Muyzer G, de Waal EC, Uitterlinden AG (1993) Profiling of complex microbial populations by denaturing gradient

gel electrophoresis analysis of polymerase chain reaction amplified genes encoding for 16S rRNA. Appl Environ Microbiol 59: 695-700.

25. Cancilla MR, Powell IB, Hillier AJ, Davidson BE (1992) Rapid genomic fingerprinting of Lactococcuslactis strains by arbitrarily primed polymerase chain-reaction with P-32 and fluorescent labels. Appl Environ Microbiol 58: 1772-1775.

26. Moyer CL, Dobbs FC, Karl DM (1995) Phylogenetic diversity of the bacterial community from a microbial mat at an active, hydrothermal vent system, Loihi Seamount, Hawaii. Appl Environ Microbiol 61: 1555-1562.

27. Sheffield VC, Beck JS, Stone EM, Myers RM (1992) A simple and efficient method for attachment of a 40-base pair, GC-rich sequence to PCR-amplified DNA. BioTechniques12: 386-387.

28. Rastogi G, Sani R (2011) Molecular Techniques to Assess Microbial Community Structure, Function, and Dynamics in the Environment. In: Springer New York 29-57.

29. Pop M, Salzberg SL (2008) Bioinformatics challenges of new sequencing technology. Trends Genet 24: 142-149.

30. Kircher M, Kelso J (2010) High-throughput DNA sequencing: Concepts and limitations. Bioessays 32: 524-536.

31. Quail MA, Smith M, Coupland P et al. (2012) A tale of three next generation sequencing platforms: comparison of Ion Torrent, Pacific Biosciences and IlluminaMiSeq sequencers. BMC Genomics 13: 341.

32. Metzker ML (2010) Sequencing technologies-the next generation. Nat Rev Genet 11:31-46.

33. Richter DC, Ott F, Auch AF, Schmid R, Huson DH (2008)MetaSim-A Sequencing Simulator for Genomics and Metagenomics. PLoS ONE 3: e3373.

34. KennedyJ, Marchesi, JR (2007) Metagenomic approach to exploit the biotechnological potential of the microbial consortia of marine sponges. ApplMicrobiolBiotechnol 75: 11-20.

35. Schmeisser C, Helen S, Wolfgang RS (2007) Metagenomics, biotechnology with non culturable microbes. ApplMicrobiolBiotechnol 75: 955-962.

36. Krsek M, Wellington EM (1999) Comparison of different methods for the isolation and purification of total community DNA from soil. J Microbiol Methods 39: 287-294.

37. Miller DN, Bryant JE, Madsen EL, Ghiorse WC (1999) Evaluation and optimization of DNA extraction and purification procedure for soil and sediment samples. Appl Environ Microbiol 65: 4715-4724.

38. Tsai YL, Olson BH (1992) Rapid method for separation of bacterial DNA from humic substances in sediments for polymerase chain reaction. Appl Environ Microbiol 58: 2292-2295.

39. Hoshino YT, Matsumoto N(2004)An improved DNA extraction method using skim milk from soils that strongly adsorb DNA. Microbes Environ 19: 13-19.

40. Eisen JA (2007) Environmental shotgun sequencing: its potential and challenges for studying the hidden world of microbes. PLoSBiol 5:e82.

41. Lasken RS (2009) Genomic DNA amplification by the multiple displacement amplification (MDA) method. BiochemSoc Trans 37: 450-453.

42. Murray IA, Clark TA, Morgan RD, Boitano M, Anton BP, et al. (2012) Themethylomes of six bacteria Nucleic Acids Res 40: 11450-11462.

43. De Mandal S, Sanga Z, Senthil Kumar N (2015) Metagenome sequencing reveals Rhodococcus dominance in Farpuk Cave, Mizoram, India, an Eastern Himalayan biodiversity hot spot region. Genome Announc 3:e00610-e00615.

44. De Mandal S, Panda AK, Lalnunmawii E, Bisht SS, Senthil Kumar N (2015) Illumina-based analysis of bacterial community in Khuangcherapuk cave of Mizoram, Northeast India. Genomics Data 5: 13-14.

45. De Mandal S, Sanga Z, Lalremsanga HT, Senthil Kumar N (2015) Bacterial diversity of Murlen National Park located in Indo-Burman Biodiversity hotspot region: A metagenomic approach. Genomics Data 5: 25-26.

46. Panda AK, Bisht SS, Senthil Kumar N, De Mandal S (2015) Investigations on microbial diversity of Jakrem hot spring, Meghalaya, India using cultivation independent approach. Genomics Data 4: 156-157.

47. Fierer N, LeffJW, Adams BJ, Nielsen UN, Scott TB, et al. (2012) Cross-biome metagenomic analyses of soil microbial communities and their functional attributes. PNAS 109: 21390-21395.

48. Uroz S, Ioannidis P, Lengelle J, Cébron A, Morin E, Buée M, et al. (2013) Functional Assays and Metagenomic Analyses Reveals Differences between the Microbial Communities Inhabiting the Soil Horizons of a Norway Spruce Plantation. PLoS ONE 8: e55929.

49. Roesch LF, Fulthorpe RR, Riva A, Casella G, Hadwin AK, et al. (2007) Pyrosequencing enumerates and contrasts soil microbial diversity. ISME Journal 1: 283-290.

50. Acosta-Martínez V, Dowd S, Sun Y, Allen V (2008) Tag-encoded pyrosequencing analysis of bacterial diversity in a single soil type as affected by management and land use. Soil Biology and Biochemistry 40: 2762-2770.

51. Chhabra S, Brazil D, Morrissey J, Burke JI, O'Gara F, et al. (2013) Characterization of mineral phosphate solubilization traits from a barley rhizosphere soil functional metagenome. Microbiologyopen 2: 717-724.

52. Rodrigues JLM, Pellizari VH, Mueller R, Baek K, Jesus EC, et al. (2013) Conversion of the Amazon rainforest to agriculture results in biotic homogenization of soil bacterial communities. ProcNatlAcadSci USA 110: 988-993.

53. Navarrete AA, Tsai SM, Mendes LW, Faust K, de Hollander M, et al. (2015) Soil microbiome responses to the short-term effects of Amazonian deforestation. MolEcol 24: 2433-2448.

54. Yasir M, Azhar E,Khan I,Bibi F, Baabdullah, et al. (2015) Composition of soil microbiome along elevation

gradients in southwestern highlands of Saudi Arabia. BMC Microbiol15:65

55. Pechal JL, Crippen TL, Tarone AM, Lewis AJ, Tomberlin JK, et al. (2013) Microbial Community Functional Change during Vertebrate Carrion Decomposition. PLoS ONE 8: e79035.

56. Carrino-Kyker S, Smemo K, Burke D (2013) Shotgun metagenomic analysis of metabolic diversity and microbial community structure in experimental vernal pools subjected to nitrate pulse. BMC Microbiol 13: 78.

57. Logares R, Lindström ES, Langenheder S, Logue JB, Paterson H et al., (2013) Biogeography of bacterial communities exposed to progressive long-term environmental change. The ISME Journal 7: 937-948.

58. Unno Y, Shinano T (2013) Metagenomic analysis of the rhizosphere soil microbiome with respect to phytic acid utilization. Microbes Environ 28:120-127.

59. Jiang XT, Peng X, Deng GH, Sheng HF, Wang Y, et al. (2013)Illumina sequencing of 16S rRNA tag revealed spatial variations of bacterial communities in a mangrove wetland. MicrobEcol 66: 96-104.

60. Nacke H, Thürmer A, Wollherr A, Will C, Hodac L,et al. (2011) Pyrosequencing-based assessment of bacterial community structure along different management types in german forest and grassland soils. PLoS ONE 6:e17.

CHAPTER 2

DIVERSITY OF 23S rRNA GENES WITHIN INDIVIDUAL PROKARYOTIC GENOMES

Anna Pei, Carlos W. Nossa, Pooja Chokshi, Martin J. Blaser, Liying Yang, David M. Rosmarin, and Zhiheng Pei

INTRODUCTION

Ribosomes are the large, ribonucleoprotein machinery in which proteins are synthesized. The structure of ribosomes is largely conserved amongst the three kingdoms of life. In prokaryotes, the ribosome is composed of two subunits: a large 50S subunit, and a small 30S subunit. The 50S subunit contains a 23S, a 5S rRNA, and more than 30 proteins. The 30S subunit contains a 16S rRNA plus 20 proteins. Precise spatial relationships may be essential for assembly of functional ribosomes, constraining ribosomal RNA genes from drastic change [1].

Because of functional constraints on the structure of rRNA gene sequences, certain regions in rRNA genes that are in contact with other components in the ribosome are conserved, while sequences between the conserved regions mutate at faster rates. The conserved regions are useful for determining distant relationships, while the "fast" regions are useful for distinguishing closely-related organisms. Horizontal gene transfer events that could falsify the evolutionary history of an organism are unlikely to occur in the highly constrained rRNA genes[2], [3], [4]. These features make rRNA genes the most suitable molecular chronometer for both phylogenetic analysis and taxonomic classification of cellular organisms [5].

In general, 16S and 23S rRNA based phylogenetic trees are in good agreement[6], [7], while 5S rRNA is considered to not contain sufficiently long sequences for statistically significant comparisons. The detailed and comprehensive phylogenies inferred from 16S as well as 23S rRNA sequence comparisons provide three primary domains [5], [8] the Bacteria, Archaea, and Eukarya. However, 23S rRNA has lost favor to 16S rRNA in phylogenetic analysis and is rarely used in taxonomic classification because of the lack of established broad-range sequencing primers and the difficulty of sequencing larger genes with early sequencing technology.

There has been renewed interest in the use of the 23S rRNA gene, driven by the decrease in sequencing costs with next generation DNA sequencing technology (454 sequencing) and demand from the new Roadmap Initiative in the Human Microbiome Project (http://nihroadmap.nih.gov/hmp/). Compared to 16S rRNA genes, 23S rRNA genes contain more characteristic sequence stretches due to a greater length, unique insertions and/or deletions, and possibly better phylogenetic resolution because of higher sequence variation[9]. A recent study indicated that 23S rRNA genes also contain conserved regions for designing broad-range primers with a similar degree of universality to the broad-range primers for 16S rRNA genes [10].

Gene redundancy is uncommon in prokaryotic genomes, yet the rRNA genes can vary from one to as many as 15 copies in a single genome [11]. Divergent evolution between rRNA genes in the same genome may corrupt the record of evolutionary history and obscure the true identity of an organism. When substantial variation occurs, use of rRNA gene sequences may lead to the artificial classification of an organism into more than one species. In this study, we performed a systematic survey of intragenomic variation of 23S rRNA genes in genomes representing 184 prokaryotic species.

MATERIALS AND METHODS

Annotation of 23S rRNA genes

Gene sequences were from the Complete Microbial Genomes database at the NCBI website (http://www.ncbi.nlm.nih.gov/genomes/MICROBES/Complete.html). For some species, more than one genome was available. Comparison of homologous 23S rRNA genes between these genomes did not

Citation: Pei A, Nossa CW, Chokshi P, Blaser MJ, Yang L, Rosmarin DM, et al. (2009) Diversity of 23S rRNA Genes within Individual Prokaryotic Genomes. PLoS ONE 4(5): e5437. doi:10.1371/journal.pone.0005437

show >3% sequence difference (data not shown). To avoid duplicating any species, only the most completely annotated genome was included for analysis. Lengths of 23S rRNA genes were identified by using experimentally defined 23S rRNA sequences from the closest relatives available and verified by 2° structure analysis based on minimizing free energy, using RNAstructure [12] and Rnaviz [13], with experimentally defined 23S rRNA or the consensus 23S rRNA models [14] used for reference. The number of copies of 23S rRNA genes present in a genome was determined by whole genome BLAST search based on the known 23S rRNA sequence.

Analysis of intragenomic diversity in 23S rRNA genes

Genomes that contained only a single 23S rRNA gene were not further analyzed. Copies of 23S rRNA genes from each remaining genome were aligned with Clustalw [15]. To calculate diversity, the number of revealed substitutions including point mutations and insertions/deletions (indels) was divided by the total number of positions, including gaps in the alignment.

Comparison of 2° structures

To compare two related 2° structures, a mismatch was defined as conserved if located in a loop, or located in a stem but causing GC:GU conversions or covariation resulting in no change in base-pairing. In contrast, a non-conserved mismatch altered base-pairing. Substitutions also were classified by the position-specific relative variability rate calculated from the consensus 23S rRNA model based on an alignment of 187 bacterial 23S rRNA genes [14]. Positions were classified as variable or non-variable positions, according to the substitution rate relative to the average substitution rate of all sites [14]. The relative substitution rate for a variable position $v>1$ indicates a substitution rate higher than that averaged for all sites in the 23S rRNA gene analyzed, while a conserved position had a relative substitution rate $v<1$; uncommon sites are positions occupied in <25% of organisms due to insertions. The expected variability for certain classes of positions was calculated from the consensus models. Differences between expected and observed variability was analyzed by Chi-Square analysis, considered significant if $p<0.05$.

Experimental verification of 23S rRNA gene sequences

Representative 23S rRNA gene sequences containing intervening sequences were experimentally verified by sequence-based PCR using allele-specific primers (Table S1). Bacterial genomic DNA was prepared from strains used in the specific bacterial genome projects. The borders of the 23S rRNA gene sequence and inserts were amplified and sequenced. The sequences obtained were compared with sequences of the corresponding 23S rRNA genes from the edited whole genomes to identify discrepancies.

RESULTS

rRNA gene database

In total, 342 complete prokaryotic genomes were available for analysis, 27 from Archaea and 316 from Bacteria. Of the 342 genomes, 73 redundant genomes were removed from the database, as were 85 genomes that contained only a single *rrn* operon. Remaining were 184 unique (10 Achaea, 174 Bacteria) species whose genomes contained multiple 23S rRNA genes (Table S2). The 184 species represented 11 phyla, of which Proteobacteria was the most abundant phylum (98 species) followed by Firmicutes (43 species), Euryarchaeota (10 species), Actinobacteria (9 species) (Table S2). The remaining 7 phyla were represented by only 24 species.

Diversity of 23S rRNA genes in the primary structure

The 184 genomes had a median of 4.57 23S rRNA genes/genome (range 1 to 15). Diversity was found in 113 genomes, with mean 0.40% and range from 0.01% to 4.04% (median 0.24%, interquartile range (IQR) 0.10%–0.52% sequence variation). The threshold for identification of outliers, calculated as IQR±1.5 IQR, was 0–1.15%. Using this threshold, 8 genomes were identified as having high intragenomic variation amongst their paralogous 23S rRNA genes (Table 1). The eight outliers were concentrated in the four most abundant phyla, four in Proteobateria, two in Firmicutes, one in Actinobacteria, and one in Euryarchaeota. The absence of outliers in the remaining seven phyla is likely because they were least represented in the dataset. In 4 genomes, *Carboxydothermus hydrogenoformans*, *Haloarcula marismortui*, *Shewanella oneidensis*, and *Streptococcus pyogenes*, there was on average 1.52% diversity due to 44.6 substitutions including 8.8 indels (Table 1). In 3 genomes, *Nocardia farcinica* (1.18% diversity, 37 point mutations, 1 indel), *Clostridium perfringens* (1.96% diversity, 16 point mutations, 41 indels), and *Salmonella typhimurium* (1.27% diversity, 16 point mutations, 41 indels), substitutions tended to concentrate within short segments of 23S rRNA genes causing complex rearrangement of the secondary structure (Figure 1), which will be described in detail separately. Highest diversity was observed in *Thermoanaerobacter tengcongensis*(4.04% diversity, 107 point mutations, 8 indels) that also contained intervening sequences (IVS) (Figure 2), which will be described below.

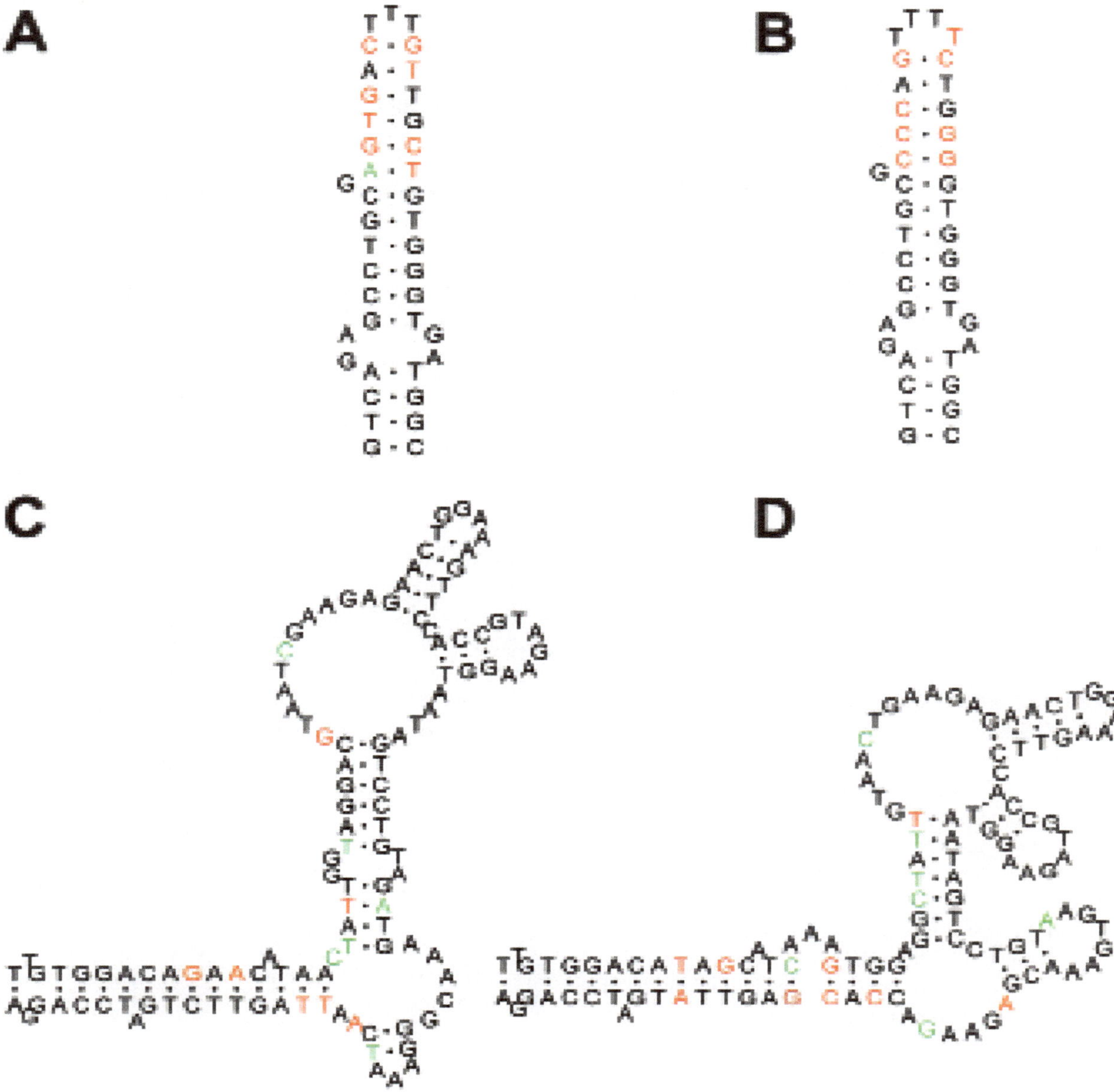

Figure 1: Conservation of 2° structure by complex rearrangement of base pairing and substitutions in *rrn23S* of *Nocardia farcinica* (A and B) and *Clostridium perfringens* (C and D).

Table 1: Genomes with significant intragenomic diversity among paralogous 23S rRNA genes.

Organism	Accession number	Copy	Alignment (nt)	Diversity (%)	Substitution*	Covariation	G:U→G:C	In-loop	Stem-loop transition	Indel
Carboxydothermus hydrogenoformans	CP000141	4	2921	1.38	38	4	3	25	7	2
Haloarcula marismortui	AY596297	3	2922	1.54	45	24	2	11	8	0
Shewanella oneidensis	AE014299	9	2894	1.17	34	22	4	5	3	1
Streptococcus pyogenes	AE004092	6	2912	1.66	48	22	8	11	7	9
Average		6.4	2912	1.52	44.8	14.4	4	14.4	8.2	8.8
Nocardia farcinica	AP006618	3	3132	1.18	37	Complex rearrangement**				1
Clostridium perfringens	BA000016	10	2905	1.98	57					41
Salmonella typhimurium	AE006468	7	3083	1.27	39					2
Thermoanaerobacter tengcongensis	AE008691	4	2985	4.04	115	24	19	62	10	8

*Substitution includes both point mutations and indels. **See Figure 1.
doi:10.1371/journal.pone.0080437.t001

Figure 2: Distribution of substituted positions and IVS in 23S rRNA genes of *T. tengcongensis*.

Effect on secondary structure

In *T. tengcongensis*, compared with the consensus 2° structure model of 23S rRNA [14] (Figure 2), 86 (74.8%) of the 115 mismatch positions occurred at highly variable positions - significantly higher than expected (40.4%, p<0.0001) (Table 2). Comparing the 2° structures of *rrnB23S* and *rrnC23S*, 105 of the 115 substitutions were conservative (62 in loops [including 4 insertions], 24 covariations, and 19 GU: GC conversions). Only 10 (9 point mutations and 1 single nucleotide insertion) of the 115 substitutions altered the 2° structure by causing stem-loop transition. These features indicate strong structural constraints, suggesting that *rrnC23S* may be functional.

Table 2: Observed and expected variability in 23S rRNA genes in *T. tengcongensis.*

Relative variability(i)	Observed substitution rate (%)	Expected substitution rate(%)	P value
i>1	78.4 (86/115)	40.4 (1175/2912)	<0.001
I<1	22.6 (26/115)	58.6 (1707/2912)	<0.001
Uncommon sites	2.6 (3/115)	1.0 (30/2912)	0.110

Besides *T tengcongensis*, conservation of the 2° structure also is apparent despite changes in the primary structure (Table 1) in other species. In *C. hydrogenoformans*, *H. marismortui*, *S. oneidensis*, and *S. pyogenes*, there are on average 44.6 substitutions per genome, of which only ~20% (8.2 substitutions) may potentially cause stem-loop transition. The effect of other substitutions are minimized by either covariations (14.4 substitutions), G:U to G:C transitions (4 substitutions) or in-loop substitutions (14.4). Complex rearrangements of the 2° structure likely occurred in *N. farcinica*, *C. perfringens*, and *S. typhimurium* as a result of changes in the primary structure (Figure 1). The complex rearrangement appears to be associated with indels. One example is between *rrnC23S* and *rrnB23S* of *Nocardia farcinica* in which a single deletion accompanied 8 point mutations in a segment of 40 nt. The topographic structure can be conserved by repositioning the bases *rrnB23S* although the resultant stem is one base pair shorter and the loop is one base larger (Figure 1A and 1B). Another example is the *rrnH23S* and *rrnD23S* of *C. perfringens* in which there are 13 indels and 8 point mutations in a segment of 115 nt. Although there was an apparent change in the 2° structure, rearrangement of the positions of the loops and the length of the stems appear to have minimized the effect on the topographic structure (Figure 1C and 1D). Variation of *rrn23S* in *S. typhimurium* has been previously reported [16].

Intervening sequences

Another type of alteration of 23S rRNA genes is insertion of an intervening sequence (Table 3). IVS was found in 7 genomes (Table 3). If the IVS are not included in calculations, the remaining sequences of paralogous 23S rRNA genes have <1% diversity in 6 of the 7 genomes except for *T. tengcongensis* whose diversity is due to both IVS-like insertions and a high level of substitutions.

Table 3: **Genomes with intervening sequences in 23S rRNA genes.**

Species	Accession #	No. of copies	Length Range (nt)	IVS (nt)	Diversity %	Type of IVS
Agrobacterium tumefaciens	AE007869.2	4	2233–2252	9,13	0.67	Short IVS
Deinococcus radiodurans	AE000513.1	3	2877–3944	1067	0.07	IVS encoding transposon
Nostoc sp.	BA000019.2	4	2828–4299	1471	0	IVS encoding transposon
Photobacterium profundum	CR354531.1	15	2894–2926	14,17	0.83	Short IVS
Photorhabdus luminescens	BX470251.1	7	2909–3021	112	0.96	IVS
Salmonella typhimurium	AE006468.1	7	2994–3093	83, 99	1.27	IVS
Thermoanaerobacter tengcongensis	AE008691.1	4	2881–2965	22, 62, 34	4.04	IVS and mutations

Transposon-mediated insertion of IVS

Nostoc sp PCC7120 contained four 23S rRNA genes; *rrnA23S* is 4299-nt, while the other three each are 2828-nt [17]. The variation was due to the presence of a 1471-nt insert in *rrnA23S* (Figure 3), including 15-nt imperfect inverted repeats at both ends of the insertion and a 963-nt ORF. The deduced protein is predicted to be a transposase in Clusters of Orthologous Group (COG) COG3547 (Expected value=7e^{-27}). The 303-amino acid of COG3547 consensus consists of both transposases 9 and 20. Blast of the 1471-nt insert against the genome of *Nostoc* sp PCC7122 revealed 11 other homologs, each with the terminal inverted repeats. Without the insertion, all four *rrn*23S copies were identical.

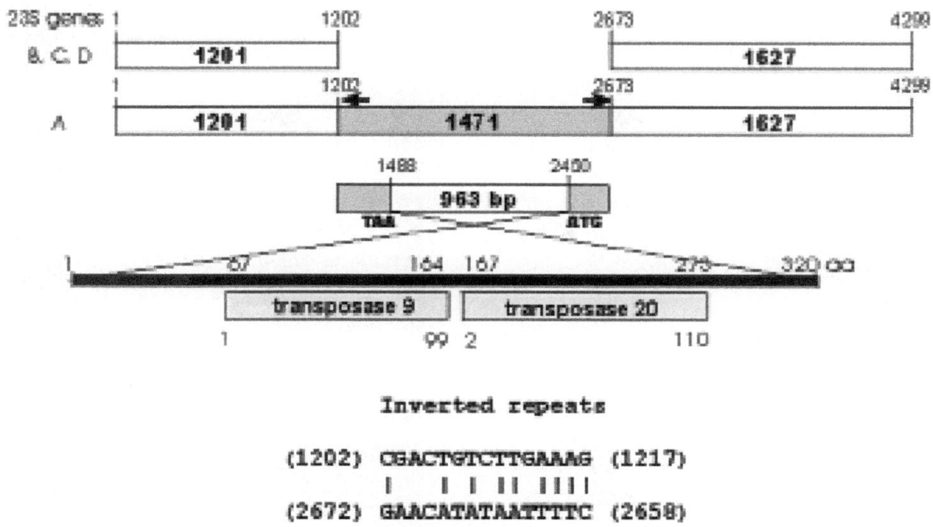

Figure 3: IVS in *Nostoc* sp (strain PCC 7120) 23S rRNA genes.

The 23S rRNA genes of *Deinococcus radiodurans* were only partially annotated (GenBank accession:AE000513) [18]. The full-length 23S rRNA genes were identified by search of the *D. radiodurans* genome using the unassigned RNA (GenBank accession:30749903) associated with the 50S ribosomal subunit [19]. Three putative 23S rRNA genes were identified. The 2° structures of the putative 23S rRNA genes were compared with the consensus prokaryotic 23S rRNA model to confirm identity and to define length [14].*rrnC23S* was 3944-nt while *rrnA23S* and *rrnB23S* were only 2877-nt (Figure 4). *rrnC23S* contains a 1067-nt insertion including a 519-nt ORF encoding a putative172-amino acid protein. The protein was predicted to be a transposase in COG3335 (Expected value=$8e^{-14}$) and 27-nt imperfect inverted repeats were present at both ends of the insert. BLAST of the 1067-nt insert *against the D. radiodurans* genome revealed two other nearly identical (>98% homology) copies, including the inserted repeats, located on plasmids MP1 and CP1.*rrnA23S* and *rrnB23S* were nearly identical. Excluding the insertion, only six single-nucleotide substitutions were observed between the three 23S rRNA genes.

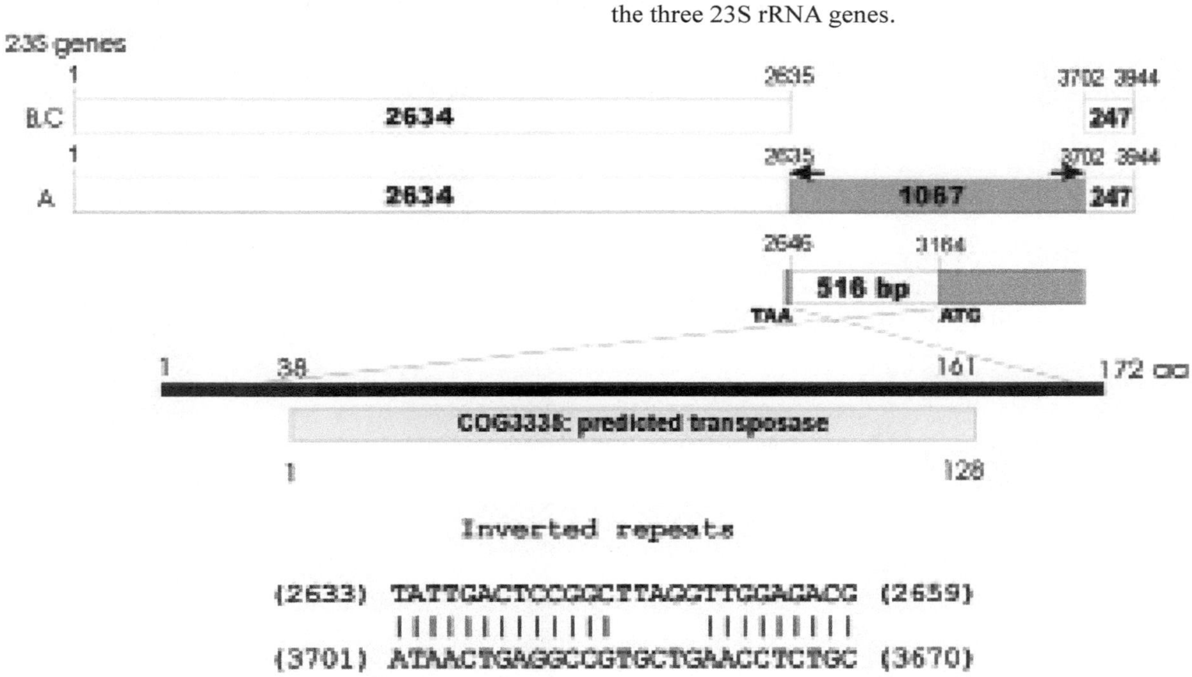

Figure 4: IVS in *Deinococcus radiodurans* 23S rRNA genes.

Small IVS

Salmonella typhimurium has seven 23S rRNA genes with sizes ranging from 2994 to 3093 bp[20]. Two insertions were found, 83-nt and 99-nt, consistent with previously described IVS in *S. typhimurium* [16] (Table 3). The largest 23S rRNA gene contains both inserts. The second largest 23S rRNA gene contains the 98-nt insert. The remaining five 23S rRNA genes only contain the 83-nt insert. The largest difference (1.27%, 37/2903) was found between *rrnB23S* and *rrnF23S* with 37 substitutions (Table 1).

Photorhabdus luminescens contains seven 23S rRNA genes [21]. *rrnC23S* and *rrnE23S* contain a shared 112-nt insert (Table 3). The other five 23S rRNA genes are 2909 nt in length. The insert is similar to those from *S. typhimurium* in size and in that it does not contain an ORF. No repetitive DNA was present in or near the insertion. Excluding the insert, the largest difference was observed between *rrnD23S* and *rrnE23S*, differing by 28 nt (0.96%). No sequences homologous to the insert were found in *P. luminescens* or GenBank.

Photobacterium profundum contains 15 paralogous 23S rRNA genes ranging from 2893–2926 nt in size (GeneBank accession: CR354531.1, unpublished). The size variation was caused by two IVS of 14 and 17 nt, respectively. Maximal diversity was 0.83% (24/2893) between *rrnC23S* and *rrnJ23S* (Table 3). No additional copy of the IVS was found outside of the 23S rRNA genes within the *P. profundum* genome.

Agrobacterium tumefaciens contains four copies of 23S rRNA genes, two on each of its linear and circular chromosomes, respectively [22]. The sizes of the 23S rRNA genes vary between 2233 and 2252 nt because the second 23S rRNA gene is shorter, caused by two small IVS of 9 and 13 nt, respectively (Table 3). Besides the IVS, 15 mutations were observed (0.67% diversity).

IVS and point mutations

T. tengcongensis had four 23S rRNA genes with sizes ranging from 2881 to 2965-nt, reflecting 3 major alleles [23]. The largest difference was between *rrnB23S* and *rrnC23S*, due to inserts of 22, 62, and 34-nt (Figure 2, Table 1) and 115 substitutions. The 22 and 62-nt inserts were present in *rrnA23S*and *rrnB23S*, respectively, the 34-nt insert in *rrnC23S*, and none in *rrnD23S*. No sequences homologous to the three inserts were found in GenBank. Excluding the 3 inserts, there remained differences of 4.0% (115/2884) between *rrnB23S* and *rrnC23S*.

Verification of 23S rRNA gene sequences by PCR-based sequencing

To examine the reliability of the whole genome sequences, we experimentally verified the IVS of *T. tengcongensis*, *D. radiodurans*, and *Nostoc sp*. The predicted junctions between the rRNA sequences and the IVS were present in all three strains tested. The sequences of cloned PCR products are identical to those published in the genome studies, providing validation of the presence of the diversity observed in the whole genome sequences. Because we did not experimentally verify the IVS of all species, our results are limited, although in total, the data suggest that the whole genome sequences are highly reliable.

DISCUSSION

This is the first large scale survey of intragenomic variation of paralogous 23S rRNA genes. We analyzed 23S rRNA genes from genomes representing 184 prokaryotic species to examine for evidence of ribosomal constraint of rRNA structures at the intragenomic level. Our findings support the hypothesis that individual 23S rRNA genes within a genome are conserved due to such structural constraints.

One level of ribosomal constraint on 23S rRNA genes was observed at the level of primary structure. In the large majority of the genomes (183/184) analyzed, variation of primary structure among paralogous 23S rRNA genes was small (<3%, which is the 16S *rrn*-based species boundary) [24]. In our dataset, 38% of the 184 genomes containing multiple paralogous 23S rRNA genes were invariable, similar to the finding in 16S rRNA genes [25]. Our findings of 0.4% average diversity for intragenomic variation of rRNA sequences among the 113 diversified genomes is higher than the reported 0.17% diversity among 16S rRNA genes[26], but will be more similar to the 16S rRNA genes diversity (0.25%) if all 184 genomes are included in the calculation (Table S2). Excluding *T. tengcongensis*, which has unusually high diversity among 23S rRNA genes, the maximal diversity among paralogous 23S rRNA genes is 1.96%, higher than the maximal variation of 1.26% among 16S rRNA genes [27]. In organisms containing multiple rRNA genes, the homogeneity of primary sequences is believed to be maintained through gene conversion by homologous recombination [2], [28]. The evolution of these paralogous rRNA genes is concerted in such a fashion that within a species the rRNA genes are nearly identical, whereas orthologous rRNA genes between species can vary considerably [2], [28]. However, our study suggests that homologous recombination is not always efficient in keeping the primary structure in check. In *T. tengcongensis*, the 4% difference (without considering the IVS) in primary structure between *rrnB23S* and *rrnC23S* is in agreement with the high % divergence reported between 16S rRNA paralogues [25]. It is tempting to attribute the unusual variation to the presence of IVS in the *T. tengcongensis* 23S rRNA genes, but variation amongst 23S rRNA genes in six other genomes containing IVS is much lower (Table 3). It appears that gene conversion or concerted evolution is one, but not the only, way to meet the requirement of ribosomal constraint.

Another level of ribosomal constraint was observed at the level of 2° structures. For example, the detailed analysis of the 23S rRNA genes of *T. tengcongensis* showed only in uncommon instance (10 of 115 point substitutions) a point substitution altered the 2° structure. This observation is in line with the current concept derived from comparing consensus rRNA sequences from closely related species [29], [30], [31], [32], [33]in that selection pressure from ribosomal constraints is to preserve 2° structures that are essential for the assembly of functional ribosomes [34]. In organisms with a single rRNA operon, ribosomal constraints could follow a simple birth-and-death model by linking the fate of a mutation to the viability of the organisms. However, in an organism with multiple rRNA operons, the nature of the selection pressure is unclear. In*Escherichia coli*, which has seven rRNA operons, after deletion of the other 6 operons only a single operon is sufficient to produce >50% of wild-type rRNA levels and support >50% of the wild-type growth rate [35]. This experiment suggests that in an organism with multiple rRNA operons, it may not be necessary for survival to keep all rRNA genes under ribosomal constraint. The observed tight constraints on individual 23S rRNA in *T. tengcongensis* needs a new explanation. *T. tengcongensis*, isolated from a fresh water hot spring in Tengchong, China, is a rod-shaped, gram-negative bacterium that grows anaerobically in extreme environments. It propagates at wide ranges of temperatures from 50° to 80°C and pH values between 5.5 and 9 [23]. It is possible that all four 23S rRNA genes are essential for survival of *T. tengcongensis* in its natural inhabit, though one might be sufficient under laboratory conditions. It also is possible that the divergence of 23S rRNA genes offers competitive advantage in a changing environment in nature and thus are selectively maintained. Ribosomal constraints could be met by maintaining conserved 2° structures through conserved point mutations under sufficient selective pressure to resist the tide of homologous recombination.

Our finding is consistent with the finding by Acinas *et al.* in *rrnC16S* of *T. tengcongensis* [25]. Both studies indicate that the 2° structures are conserved despite a high level of change in the primary structures and suggest that the *rrnC* operon is functional. Although Acinas *et al.* suggested that the rrnC operon could be brought into *T. tengcongensis* by horizontal gene transfer, the hypothetic donor has yet to be identified. The ribosomal constraint observed at the level of 2° structures is also evident in 23S rRNA genes in *C. hydrogenoformans*, *H. marismortui*, *S. oneidensis*, and *S. pyogenes*.

Complex rearrangement of the position of nucleotides, the position and size of loops, and the length of stems might be another mechanism to conserve the functional structure of an rRNA gene. The examples shown in*N. farcinica* and *C. perfringens* are beyond simple alteration or conservation of the 2° structures since there is no straightforward way to quantify the changes and their effects. Rather, this type of rearrangement seems to aim at conserving the topographic structure in the presence of drastic changes in the primary and secondary structures. However, an experimental approach will be required to address the relationship between the conservation of the topographic structure and function of an rRNA gene.

The concept of rRNA conservation has revolutionized exploration of microbial ecosystems and identifying the causes of diseases of unknown etiologies [36], [37], [38], [39]. The most common molecular sequences used for phylogenetic analysis are ribosomal RNAs. The 16S rRNA genes have become the most widely used sequences for phylogenetic analysis as seen by the vast and growing database of 16S rRNA genes [40],[41]. The choice of 16S rRNA as the most widely used sequences for phylogenetic analysis has been based on the simple assumption that 16S rRNA is optimal because 5S rRNA is too short while 23S rRNA sequencing requires extra effort without a commensurate increase in phylogenetic signal. However, misidentification has been reported. Strains from different species may have identical 16S rRNA gene sequences, and strains of one species may have 16S rRNA that differ by >4% while 3% is considered as the threshold for within-species dissimilarity [42], [43], [44]. On the other hand, the main obstacle limiting 23S rRNA from being used in classification of microorganisms has significantly diminished, as the cost for DNA sequencing is becoming less expensive. Evaluation of intragenomic variation of the rRNA genes has become possible since more and more sequences of whole microbial genomes have recently become available. It would be a logical expectation that the intragenomic diversity of a marker for taxonomic classification of microorganisms is below the level of interspecies diversity. We evaluated the suitability of 23S rRNA genes in this regard. Our study indicates that high-level variation among paralogous 23S rRNA genes might exceed the level of variation among 23S rRNA genes from different strains within a species but will not have a significant effect on the integrity of a 23S rRNA-based phylogenetic system, because its occurrence was rare. The insertions encountered when using 23S rRNA genes in phylogenetic analysis can be easily recognized and eliminated in comparisons of homologous sequences.

Intervening sequences (IVS) were frequently observed in 23S rRNA genes. In the seven species in which IVS were identified, five species contain small IVS that did not contain any ORFs but in *D. radiodurans* and*Nostoc* sp, IVS were large and contained ORFs encoding putative transposases. Experimental studies have demonstrated that IVS in 23S rRNA does not interfere with its functions[45]. The 23S rRNA of *Salmonella typhimurium* LT2 are known to carry IVS at two sites, helix-25 and helix-45, which are excised by RNase III during rRNA maturation, resulting in rRNA which is fragmented, but nevertheless functional [45]. Fragmentation of the 23S rRNA also has been observed in diverse Eubacteria [46]. 23S rRNAs with insertions have been shown to rescue an *E. coli* strain in which all chromosomal rRNA operons were deleted[47].

ACKNOWLEDGMENTS

We thank Mr. Meisheng Zhou for assistance in verification of sequence diversity.

http://nihroadmap.nih.gov/hmp/; Roadmap Initiative in Human Microbiome Project.

http://www.ncbi.nlm.nih.gov/genomes/MICROBES/Complete.html; Completed Microbial Genomes database.

http://rtfm.arb-home.de; The ARB project (latin, "arbor"=tree) home page.

REFERENCES

1. Doolittle WF. Phylogenetic classification and the universal tree. Science. 1999;284:2124–2129.
2. Eickbush TH, Eickbush DG. Finely orchestrated movements: evolution of the ribosomal RNA genes. Genetics. 2007;175:477–485.
3. Santoyo G, Romero D. Gene conversion and concerted evolution in bacterial genomes. FEMS Microbiol Rev. 2005;29:169–183.
4. Gurtler V. The role of recombination and mutation in 16S–23S rDNA spacer rearrangements. Gene.1999;238:241–252.
5. Woese CR. Bacterial evolution. Microbiol Rev. 1987;51:221–271.
6. De Rijk P, Van de Peer Y, Van den Broeck I, De Wachter R. Evolution according to large ribosomal subunit RNA. J Mol Evol. 1995;41:366–375.
7. Cedergren R, Gray MW, Abel Y, Sankoff D. The evolutionary relationships among known life forms. J Mol Evol. 1988;28:98–112.
8. Woese CR, Kandler O, Wheelis ML. Towards a natural system of organisms: proposal for the domains Archaea, Bacteria, and Eucarya. Proc Natl Acad Sci U S A. 1990;87:4576–4579.
9. Ludwig W, Schleifer KH. Bacterial phylogeny based on 16S and 23S rRNA sequence analysis. FEMS Microbiol Rev. 1994;15:155–173.

10. Hunt DE, Klepac-Ceraj V, Acinas SG, Gautier C, Bertilsson S, et al. Evaluation of 23S rRNA PCR primers for use in phylogenetic studies of bacterial diversity. Appl Environ Microbiol. 2006;72:2221–2225.

11. Klappenbach JA, Saxman PR, Cole JR, Schmidt TM. rrndb: the Ribosomal RNA Operon Copy Number Database. Nucleic Acids Res. 2001;29:181–184.

12. Mathews DH, Disney MD, Childs JL, Schroeder SJ, Zuker M, et al. Incorporating chemical modification constraints into a dynamic programming algorithm for prediction of RNA secondary structure. Proc Natl Acad Sci U S A. 2004;101:7287–7292.

13. De Rijk P, De Wachter R. RnaViz, a program for the visualisation of RNA secondary structure. Nucleic Acids Res. 1997;25:4679–4684.

14. Wuyts J, Van de Peer Y, De Wachter R. Distribution of substitution rates and location of insertion sites in the tertiary structure of ribosomal RNA. Nucleic Acids Res. 2001;29:5017–5028.

15. Thompson JD, Higgins DG, Gibson TJ. CLUSTAL W: improving the sensitivity of progressive multiple sequence alignment through sequence weighting, position-specific gap penalties and weight matrix choice. Nucleic Acids Res. 1994;22:4673–4680.

16. Mattatall NR, Sanderson KE. Salmonella typhimurium LT2 possesses three distinct 23S rRNA intervening sequences. J Bacteriol. 1996;178:2272–2278.

17. Kaneko T, Nakamura Y, Wolk CP, Kuritz T, Sasamoto S, et al. Complete genomic sequence of the filamentous nitrogen-fixing cyanobacterium Anabaena sp. strain PCC 7120. DNA Res. 2001;8:227–253.

18. White O, Eisen JA, Heidelberg JF, Hickey EK, Peterson JD, et al. Genome sequence of the radioresistant bacterium Deinococcus radiodurans R1. Science. 1999;286:1571–1577.

19. Schlunzen F, Zarivach R, Harms J, Bashan A, Tocilj A, et al. Structural basis for the interaction of antibiotics with the peptidyl transferase centre in eubacteria. Nature. 2001;413:814–821.

20. McClelland M, Sanderson KE, Spieth J, Clifton SW, Latreille P, et al. Complete genome sequence of Salmonella enterica serovar Typhimurium LT2. Nature. 2001;413:852–856.

21. Duchaud E, Rusniok C, Frangeul L, Buchrieser C, Givaudan A, et al. The genome sequence of the entomopathogenic bacterium Photorhabdus luminescens. Nat Biotechnol. 2003;21:1307–1313.

22. Wood DW, Setubal JC, Kaul R, Monks DE, Kitajima JP, et al. The genome of the natural genetic engineer Agrobacterium tumefaciens C58. Science. 2001;294:2317–2323.

23. Bao Q, Tian Y, Li W, Xu Z, Xuan Z, et al. A complete sequence of the T. tengcongensis genome.Genome Res. 2002;12:689–700.

24. Stackebrandt E, Goebel BM. Taxonomic Note: A Place for DNA-DNA Reassociation and 16S rRNA Sequence Analysis in the Present Species Definition in Bacteriology. 1994. pp. 846–849.

25. Acinas SG, Marcelino LA, Klepac-Ceraj V, Polz MF. Divergence and redundancy of 16S rRNA sequences in genomes with multiple rrn operons. J Bacteriol. 2004;186:2629–2635.

26. Stewart FJ, Cavanaugh CM. Intragenomic variation and evolution of the internal transcribed spacer of the rRNA operon in bacteria. J Mol Evol. 2007;65:44–67.

27. Coenye T, Vandamme P. Intragenomic heterogeneity between multiple 16S ribosomal RNA operons in sequenced bacterial genomes. FEMS Microbiol Lett. 2003;228:45–49.

28. Liao D. Gene conversion drives within genic sequences: concerted evolution of ribosomal RNA genes in bacteria and archaea. J Mol Evol. 2000;51:305–317.

29. Zwieb C, Glotz C, Brimacombe R. Secondary structure comparisons between small subunit ribosomal RNA molecules from six different species. Nucleic Acids Res. 1981;9:3621–3640.

30. Stiegler P, Carbon P, Ebel JP, Ehresmann C. A general secondary-structure model for procaryotic and eucaryotic RNAs from the small ribosomal subunits. Eur J Biochem. 1981;120:487–495.

31. Curtiss WC, Vournakis JN. Quantitation of base substitutions in eukaryotic 5S rRNA: selection for the maintenance of RNA secondary structure. J Mol Evol. 1984;20:351–361.

32. Hori H, Osawa S. Evolutionary change in 5S rRNA secondary structure and a phylogenic tree of 352 5S rRNA species. Biosystems. 1986;19:163–172.

33. Gutell RR, Larsen N, Woese CR. Lessons from an evolving rRNA: 16S and 23S rRNA structures from a comparative perspective. Microbiol Rev. 1994;58:10–26.

34. Higgs PG. Compensatory neutral mutations and the evolution of RNA. Genetica. 1998;102–103:91–101.

35. Asai T, Zaporojets D, Squires C, Squires CL. An Escherichia coli strain with all chromosomal rRNA operons inactivated: complete exchange of rRNA genes between bacteria. Proc Natl Acad Sci U S A.1999;96:1971–1976.

36. Pei Z, Bini EJ, Yang L, Zhou M, Francois F, et al. Bacterial biota in the human distal esophagus. Proc Natl Acad Sci U S A. 2004;101:4250–4255.

37. Eckburg PB, Bik EM, Bernstein CN, Purdom E, Dethlefsen L, et al. Diversity of the human intestinal microbial flora. Science. 2005;308:1635–1638.

38. Paster BJ, Boches SK, Galvin JL, Ericson RE, Lau CN, et al. Bacterial diversity in human subgingival plaque. J Bacteriol. 2001;183:3770–3783.

39. Relman DA. The search for unrecognized pathogens. Science. 1999;284:1308–1310.

40. DeSantis TZ, Hugenholtz P, Larsen N, Rojas M, Brodie EL, et al. Greengenes, a chimera-checked 16S rRNA gene database and workbench compatible with ARB. Appl Environ Microbiol. 2006;72:5069–5072.

41. Cole JR, Wang Q, Cardenas E, Fish J, Chai B, et al. The Ribosomal Database Project: improved alignments

and new tools for rRNA analysis. Nucleic Acids Res. 2009;37:D141–145.

42. Vandamme P, Harrington CS, Jalava K, On SL. Misidentifying helicobacters: the Helicobacter cinaedi example. J Clin Microbiol. 2000;38:2261–2266.

43. Oberreuter H, Charzinski J, Scherer S. Intraspecific diversity of Brevibacterium linens, Corynebacterium glutamicum and Rhodococcus erythropolis based on partial 16S rDNA sequence analysis and Fourier-transform infrared (FT-IR) spectroscopy. Microbiology. 2002;148:1523–1532.

44. Harrington CS, On SL. Extensive 16S rRNA gene sequence diversity in Campylobacter hyointestinalis strains: taxonomic and applied implications. Int J Syst Bacteriol. 1999;49 Pt 3:1171–1175.

45. Burgin AB, Parodos K, Lane DJ, Pace NR. The excision of intervening sequences from Salmonella 23S ribosomal RNA. Cell. 1990;60:405–414.

46. Evguenieva-Hackenberg E. Bacterial ribosomal RNA in pieces. Mol Microbiol. 2005;57:318–325.

47. Yokoyama T, Suzuki T. Ribosomal RNAs are tolerant toward genetic insertions: evolutionary origin of the expansion segments. Nucleic Acids Res. 2008;36:3539–3551

CHAPTER 3

DEVELOPMENT AND VALIDATION OF AN rDNA OPERON BASED PRIMER WALKING STRATEGY APPLICABLE TO DE NOVO BACTERIAL GENOME FINISHING

Alexander W. Eastman and Ze-Chun Yuan

Keywords: Genome finishing, ribosomal DNA, contigs assembly, bacterial genomics, second-generation sequencing (SGS)

INTRODUCTION

The advent of high-throughput second-generation sequencing platforms has allowed for a monumental leap in the affordability and feasibility of genomic studies (Bentley, 2010; Delseny et al., 2010). Applications of these second-generation sequencing technologies have been especially helpful and disruptive during the characterization of microorganisms, where whole genome sequencing has become a routine step in bacterial strain characterizations (Medini et al., 2008; Maclean et al., 2009).

Four second-generation sequencing platforms are commonly employed for *de novo* sequencing of bacterial genomes; Illumina, 454 Roche, SOLiD and Ion Torrent. With the exception of SOLiD, second-generation sequencing technologies determine the sequence of short, fragmented DNA molecules through detection of nucleotides incorporated during synthesis of a complementary strand. Each technology varies in the specifics of the sequencing chemistry, read-length, library preparation and output accuracy, and the advantages and disadvantages of each specific technology have been reviewed previously (Miller et al., 2008; Glenn, 2011; Quail et al., 2012; McGinn and Gut, 2013; Morey et al., 2013).

In second-generation sequencing platforms, sequencing reads are assembled by computational assignment of nucleotide identities. Where applicable, software trims the known adaptor sequences from the DNA sequences, and removes bases and reads that do not meet quality thresholds and *in silico* assembly algorithms are used to assemble the remaining reads into contiguous sequences (contigs) (Pop et al., 2004; Horner et al.,2009; Gritsenko et al., 2012). Detailed methods using the power of second-generation sequencing platforms for various experimental aims are available; however there are few examples in the literature describing the finishing procedures for bacterial genomes. *De novo* assembly of a bacterial genome results in multiple contigs ranging in length from a few hundred to thousands of base pairs, commonly referred to as a draft genome (Tsai et al., 2010; Wetzel et al., 2011). Taken alone, contigs contained within a draft genome offer no spatial orientation relative to each other and are numbered in descending order by size. In addition, draft genomes are error-prone, containing frame shifts, missing genes/sequence, rearrangements, ambiguous bases, and sequencing artifacts and do not always accurately reflect genome structure (Poptsova and Gogarten,2010; Ricker et al., 2012). The increased accuracy and fidelity of completely sequenced and finished genomes are preferential to draft genomes for later use in systems biology and the various "omics" fields.

Paenibacillus polymyxa, the type species of the *Paenibacillus* genus, have been isolated from diverse environments including marine sediments, hot-springs and soils world-wide and are commonly found in association with plant hosts. *P. polymyxa* strains are best known for promoting plant fitness through pathogen antagonism, nutrient cycling,

Citation: Eastman AW and Yuan Z-C (2015) Development and validation of an rDNA operon based primer walking strategy applicable to de novo bacterial genome finishing. Front. Microbiol. 5:769. doi: 10.3389/fmicb.2014.00769.

and priming of plant defenses (Lal and Tabacchioni, 2009). Recently, interest for potential applications in agriculture and renewable bio-fuel production has driven the resurgence in studies of *P. polymyxa* biology (Yu et al., 2011; Eastman et al., 2014b). Besides our sequencing of *P. polymyxa* strain CR1 (here forth, CR1), complete genomes of four other *P. polymyxa* strains (E681, SQR-21, M1, and SC2) are publically available among numerous draft sequences (Kim et al., 2010; Ma et al.,2011; Niu et al., 2011; Eastman et al., 2014a; Li et al., 2014).

Here we detail the sequencing, assembly, and finishing of the CR1 genome using a modified primer walking strategy developed to minimize efforts for gap closure. Our approach simplifies the sequencing procedure for repetitive rDNA operons, thereby speeding up genome sequencing, assembly, and finishing. Furthermore, we illustrate the most common approach to bacterial genome sequencing and offer an overview of the challenges of bacterial genome finishing.

METHODS AND RESULTS

Second-Generation Sequencing

Whole genome sequencing using second-generation sequencing technologies is a powerful tool for microbiologists looking to identify the potential genetic basis of observed phenotypes. Prior to proceeding with completing a draft bacterial genome, the quality of the sequencing run and draft assembly must be taken into careful consideration. Adequate genome coverage of the sequencing read libraries are critical for assembling a genome with high fidelity. Typically, 40~50× coverage is needed for genome sequencing and higher coverage (>50×) does not necessarily increase assembly strength. If the sequencing depth is low it is advisable to re-sequence the genome since the power of second-generation sequencing highly relies on high coverage to overpower the limitations of short read lengths and comparatively low individual base accuracy. Currently the Q_{30} score is the *de facto* standard for measuring the accuracy of second-generation reads, a Q_{30} score represents a 1/1000 chance of an incorrect base identification (Ewing and Green, 1998; Ewing et al., 1998). It is important to note that Q_{30} is presented as a percentage of total bases in the sequencing run that reach the threshold and typically, Q_{30} scores for a high-quality run are >80%. In the case of the CR1 genome, sequencing was performed using the Illumina MiSeq platform. A short-insert read library was generated with a target insert size of 400 bp using the NexteraXT DNA sample preparation kit and a mate-pair library was generated using the Nextera mate-pair sample preparation kit. MiSeq 2× 150 bp sequencing yielded a 140× genome coverage short-insert read library (2.9 million reads) and a 107× genome coverage mate pair read library (2.2 million reads). Adaptor sequences were removed, mate-pair reads with insert sizes >400 base pairs were filtered and the two libraries were merged generating a 40× genome coverage merged library with a mean mate-pair insert size of 1.25 kb and a Q_{30} score of 86% (Figure 1).

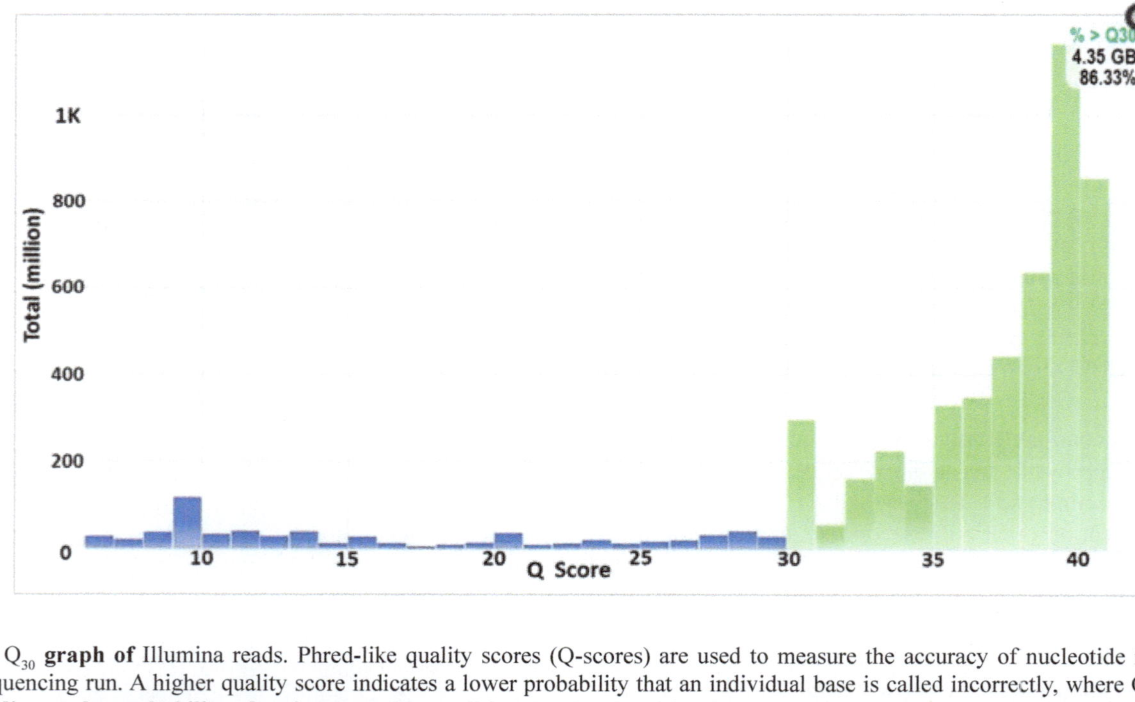

Figure 1: Q_{30} **graph of** Illumina reads. Phred-like quality scores (Q-scores) are used to measure the accuracy of nucleotide identity data from a sequencing run. A higher quality score indicates a lower probability that an individual base is called incorrectly, where $Q = -10_{log}10$ (e). Q_{20} **indicates the probability of an incorrect base call is 1 in 100;** Q_{30}, 1 in 1000; Q_{40}, 1 in 10,000. The green proportion of the graph represents bases with a high quality score (defined as Q_{30}) identified during the run. The whole genome *de novo* **sequencing of** *Paenibacillus polymyxa* **strain CR1 had 86.33% =** Q_{30} **bases.**

Genome Draft Assembly of *P. polymyxa* Strain CR1

Concomitant with the advances of second-generation sequencing, developments of *in silico* assembly programs has yielded numerous bioinformatics tools designed to minimize and compensate for sequencing errors inherent to each respective assembly platform. Each algorithm varies the length of matching sequence required before two reads are considered overlapping (called the k-mer length), which has a dramatic effect on the generated contig assembly. A short k-mer length results in a more contiguous assembly at the cost of accuracy and fidelity, while a long k-mer length results in a highly fragmented but more accurate draft assembly (Gibbons et al., 2009). Recent tools have been developed to integrate the outputs from multiple assembly programs each with different k-mer lengths into a single set of contigs with higher strength and accuracy. The advantage of integrating the contigs generated by multiple programmes stems from the ability to combine the highly accurate, long k-mer assemblies with the less accurate but more contiguous short k-mer assemblies. In the case of CR1 genome sequencing, three separate contig assembly programs, ABySS with a k-mer length of 67 bp (Birol et al., 2009), Velvet with a k-mer length of 31 bp (Zerbino and Birney, 2008), and SOAPdenovo with a k-mer length of 55 bp (Luo et al., 2012), were run independently, generating three separate contig assemblies that are then integrated into a final draft assembly by CISA (Lin and Liao, 2013). The draft assembly of the CR1 genome contained 38 contigs with an N_{50} value of 1.5 Mb (N_{50} = the size of contig where 50% of the total bases are accounted for when combined with all longer contigs), representing an excellent draft genome for finishing. It may be wise to re-sequence the genome prior to finishing if the sequencing read library has poor genome coverage, a low Q_{30}-score or the draft genome is highly fragmented (>100 contigs), especially in cases where no closely related strains or species are available as reference genomes for scaffold assembly.

Regardless of which technology is used to generate the draft genomic data, the procedure for genome assembly and finishing are more or less analogous (Nagarajan et al., 2010). Unfortunately, no automated/high-throughput and universally applicable method has been developed for finishing of *de novo*bacterial genomes and Sanger sequencing remains the most commonly used approach (Schuster, 2008; Hurt et al., 2012). Finishing of bacterial genomes typically involves five major steps after the assembly of a draft genome (Figure;2); locating and sequencing of ambiguous bases, scaffold assembly of contigs, investigation of contig rearrangements, sequencing of contig gaps, and genome annotation.

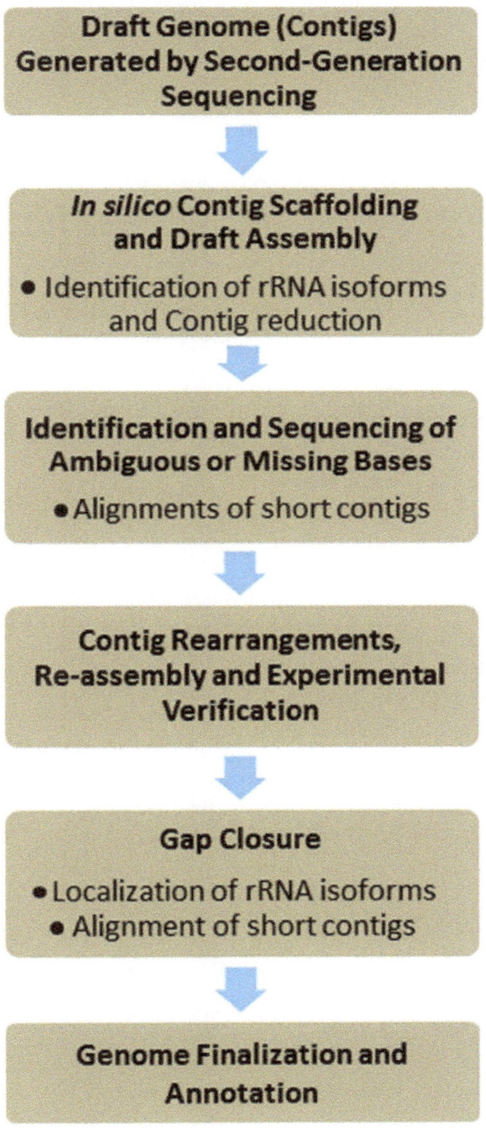

Figure 2: Flow chart for *de novo* **bacterial genome finishing. The assembly process is initially simplified by removal of** contigs comprised entirely of rRNA genes. The draft genome is then scaffolded against other closely related strains and species to identify potential gaps, rearrangements, and missing sequences. Ambiguities in structure and nucleotide identity are targeted for Sanger re-sequencing. Furthermore, large scale rearrangements are carefully investigated through a combination of PCR and sequencing to ensure mistakes are not incorporated during draft genome assembly. Finally gaps between contigs are sequenced and the genome is assembled and annotated.

Reduction of rRNA contigs by *in silico* Alignment

Repetitive genetic elements are known to be causative of misassembled contigs during *de novo* genome assembly (Ricker et al., 2012). Bacterial genomes encode multiple, highly-similar operons encoding 16S, 23S, and 5S ribosomal subunits, commonly referred to as rDNA operons (Rainey et al., 1996). Manual inspection of contig ends determined that various short contigs (<2 kb) corresponded to isoforms of rRNA genes. Contigs comprised of multi-copy repetitive elements such as rDNA have the potential to be assembled from sequencing reads corresponding to distant genomic loci and are thus prone to being chimeric. We employed Sanger sequencing with universal 16S and 23S rDNA primers (Table1) to generate consensus sequences for both the 16S and 23S genes (Anzai et al., 2000; Hunt et al., 2006). Alignment of the consensus sequences against the 38 contigs from the CR1 draft genome allowed for masking of 10 short contigs (<2 kb) that corresponded to various isoforms of rRNA genes. Thus, assembly and masking of the short contigs containing highly similar rDNA into consensus 16S and 23S genes reduced the set of contigs from 38 to 28. Within this set of 28 contigs, 24 of the remaining ends (of a possible 56 ends) contained some fragment of 5S, 16S, or 23S rRNA genes.

Table 1: Universal 16S and 23S primers for rapid rRNA operon gap sequencing

Primer name	Sequence	Gene	Direction	References
8F	AGAGTTTGATCCTGGCTCAG	16S	Forward	Anzai et al., 2000
1492R	CGTTACCTTGTTACGACTT	16S	Reverse	Anzai et al., 2000
127F	CYGAATGGGRVAACC	23S	Forward	Hunt et al., 2006
2241R	ACCGCCCCAGTHAAACT	23S	Reverse	Hunt et al., 2006
U1	TGGGATACCACCCTGATCGT	16S	Reverse	This study
U2	GTTTGGGCTAATCCGCGTTC	16S	Forward	This study
U3	CCGTCACACCACGAGAGTTT	23S	Reverse	This study
U4	GTCCGCCGCTAGGTTGATTA	23S	Forward	This study

Scaffolding of Draft Genome

The explosion of second-generation sequencing projects has dramatically increased the number of published genomes from all walks of life, especially prokaryotes. Accordingly the availability of a closely related sequence facilitates scaffolding of newly obtained draft genomes against a previously completed, closely related genome, identifying putative gaps and variability in the sequence for experimental investigation. The multiple genome alignment software Mauve iteratively compares segments of DNA to user-provided genomes by characterizing regions of high similarity as local collinear blocks, which can then be used to reorder the contigs to generate the best assembly compared to any user-provided reference genome (commonly referred to as scaffolding) (Darling et al., 2010, 2011). In the case of the CR1 genome assembly, the draft genome with 28 remaining contigs was aligned using progressiveMauve against the completely sequenced genome of *P. polymyxa* strain E681 which showed the highest nucleotide-level similarity to CR1 (Kim et al., 2010). With the reference genome as scaffold, we were able to draft the final CR1 genome assembly and identify putative contig gaps for further sequencing and closure. For *de novo* sequencing projects where the bacterium is the first member of its species to be sequenced, the next closest related species may be used for scaffold assembly. For example, alignments of the CR1 draft genome against the complete *Paenibacillus terrae* strain HPL-003 genome (Shin et al., 2012) also resulted in a contig assembly that approximated the alignment against *P. polymyxa* strain E681 (Figure 3).

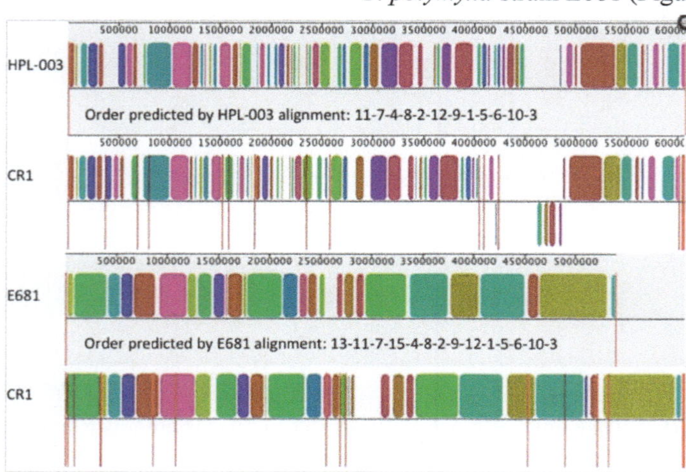

Figure 3: CR1 draft genome scaffold assemblies with related reference genomes. The local collinear block (LCB) plot was generated with the contig reordering tool within Mauve using default parameters. The name of the strain represented is listed

to the left of each LCB plot. HPL-003 represents*Paenibacillus terrae* **strain HPL-003, a closely related species to *P. polymyxa*.** Conversely, E681 represents*P. polymyxa* **strain E681, the closest related *P. polymyxa* strain to CR1. The** contigs contained within the *P. polymyxa* **strain CR1 draft genome are reordered to approximate the LCB plot of the above completely finished genome. Global alignments are visualized as LCBs, which represent regions with high levels of nucleotide similarity between genomes. LCBs are colored according to homology to LCBs of the compared genome. LCBs drawn below the horizontal correspond to inversions relative to the reference genome.**

Targeted Sequencing of Ambiguous Bases

Reduction of a highly fragmented draft genome comprised of numerous contigs to a single contiguous sequence is the most noticeable feature of finished genomes (Mardis et al., 2002). However, genomes sequenced using paired-end runs and assembled using Velvet or ABySS occasionally include series of ambiguous bases within the assembled contigs, represented by N nucleotides in FASTA formatted data (Mardis et al., 2002; Zerbino and Birney, 2008; Birol et al., 2009). The lengths of ambiguous base stretches can range from three to hundreds of bases in length and are determined computationally during read assembly. These ambiguous base stretches can result from mate-pair reads without sufficient internal coverage, nucleotide polymorphisms and individual bases within reads that do not meet quality thresholds. Nevertheless, ambiguous bases should be targeted for Sanger sequencing to fill in the missing bases and can be treated as short, pre-defined contig gaps during sequencing. The draft genome (28 contigs) of CR1 contained a total of 198 ambiguous base stretches, between 9 and 769 bp in length, distributed irregularly throughout the genome (Figure4). The generated draft assembly and locations of ambiguous bases were annotated in Artemis (Rutherford et al., 2000) and visualized using DNAplotter (Carver et al., 2009). To identify ambiguous bases, 198 sets of primers flanking the 5′ and 3′ ends of each ambiguous base stretch were designed. Each primer set was used to amplify the ambiguous base stretch by PCR using the following cycle; 95°C for 1 min followed by 35 cycles of 95°C for 30 s, varied annealing temperature dependent on T_m of the primer pair for 45 s, 72°C for 60 s followed by a final extension at 72°C for 5 min. To ensure each PCR product corresponded to the intended ambiguous bases location, 198 PCR products were individually sequenced bi-directionally and aligned against the CR1 draft genome (28 remaining contigs) to verify the location of each ambiguous base stretch and fill in the missing sequence. Interestingly, sequences of 10 ambiguous base stretches matched with the sequence of 10 small contigs (<1 kb in size), indicating these 10 small contigs actually fit within these 10 ambiguous base stretches. Thus, sequencing of ambiguous base stretches reduced the set of contigs from 28 to 18, where 10 small contigs merged into 10 ambiguous base stretches.

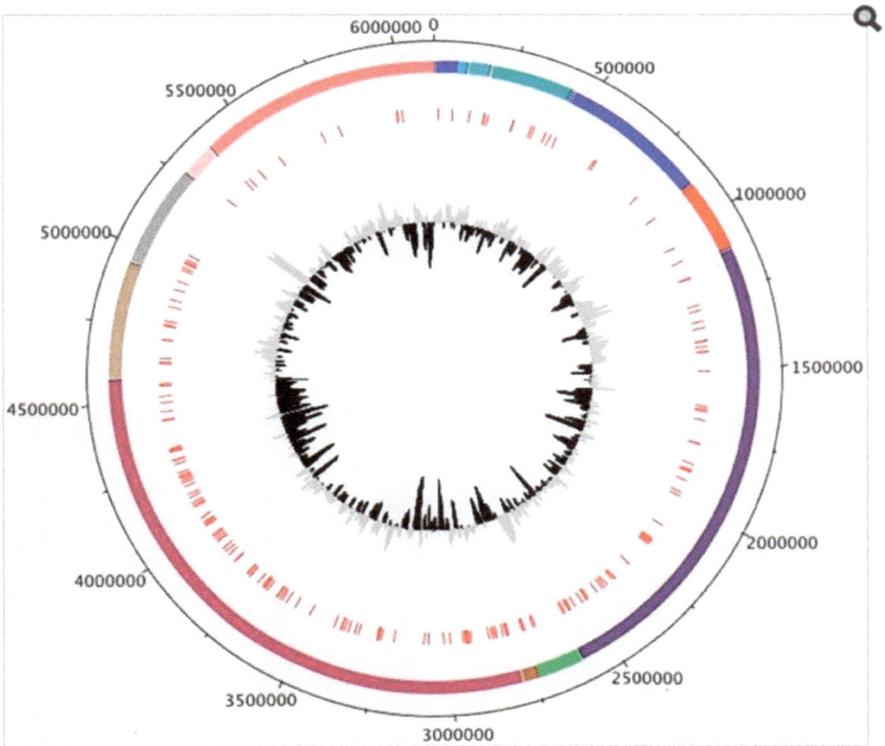

Figure 4: Circular representation of CR1 scaffold assembly showing locations of ambiguous bases. From inside to outside; 1st ring – G+C mol% content, 2nd ring – locations of ambiguous bases (red), 3rd ring – final assembly of draft contigs where separate contig from the draft assembly are labeled with different colors, 4th ring – genome size in base pairs. Annotations were performed in Artemis and visualized using DNAplotter.

Novel Gap Closure Approach

For gap closure of the CR1 draft genome, gaps (average length of 7.5 kb) between the 18 remaining contigs were identified through Mauve alignments against the complete genomes of *P. polymyxa* strain E681 and *P. terrae* strain HPL-003 (Figures3,.4). Contig gaps were amplified from CR1 genomic DNA using oligonucleotides designed to hybridize between 250 and 350 bp from the 5′ and 3′ ends of each contig. PCR conditions for amplification of contig gaps were as follows; 95°C for 1 min followed by 40 cycles of 95°C for 30 s, varied annealing temperature dependent on T_m of the primer pair for 45 s, 72°C for 300 s followed with a final extension at 72°C for 20 min. All PCR amplifications of contig gaps were performed using Phusion® Taq polymerase from NEB and products were excised and gel-purified. Absence of contaminating genomic DNA in gel-purified contig gap product was confirmed using primer sets corresponding to distant genomic DNA locations.

During our masking of contigs comprised entirely of rDNA we noticed the prevalence of fragments of rDNA at contig boundaries. Since rRNA operons are multi-copy in any given bacterial genome, universal primers targeting 16S and 23S rRNA genes (Table 1) were used for PCR to determine if a candidate amplified gap (gel-purified PCR products as template) contained rDNA (Figures 5,6). Through this method, we identified 12 out of the 18 remaining amplified contig gaps contained an rRNA operon. Their prevalence and localization to contig boundaries is likely a consequence of short read sequencing technology that cannot sequence across the entirety of an rRNA operon. For instance, the gaps between contigs 10 and 3 (10/3), contigs 5 and 6 (5/6), and contigs 8 and 4 (8/4) contain both 16S and 23S rDNA, while gap between contigs 12 and 9 (12/9) does not contain any rDNA (Figure 6). These results further confirmed our earlier assumption that the repetitive nature of rRNA operons was preventing a more complete *in silico* assembly.

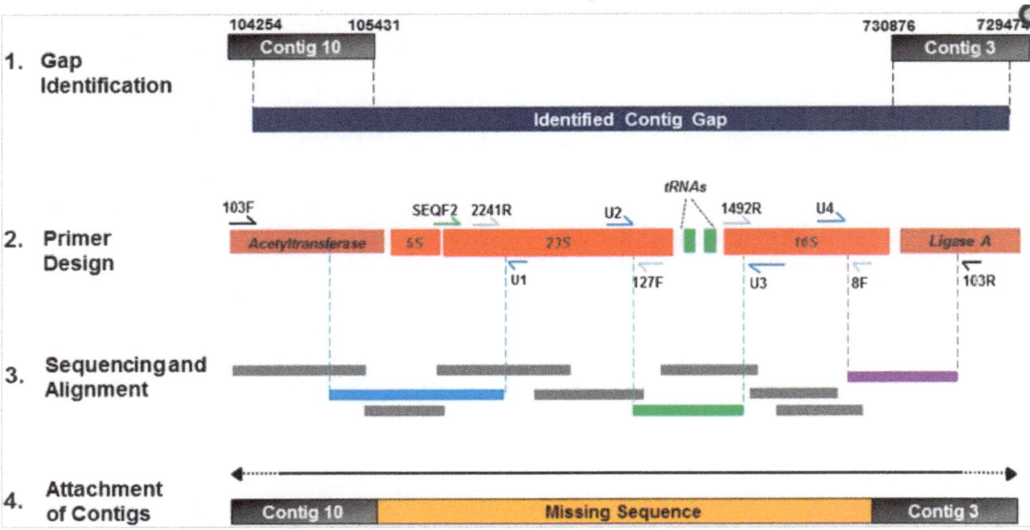

Figure 5: Innovative primer walking strategy for sequencing of repetitive rRNA operons. Putative gaps between contigs were identified by the contig-reordering tool within Mauve against other completely sequenced *P. polymyxa* **strains. Primers U1, U2, U3, and U4 were designed to amplify the** rRNA operon as depicted, allowing for rapid resolution of the rRNA gene flanking sequences and intragenic regions. Primers depicted were used as sequencing primers for Sanger sequencing of unknown regions and the resulting sequences were overlapped to identify missing sequence. Sequences of the depicted primers are presented in Table 1.

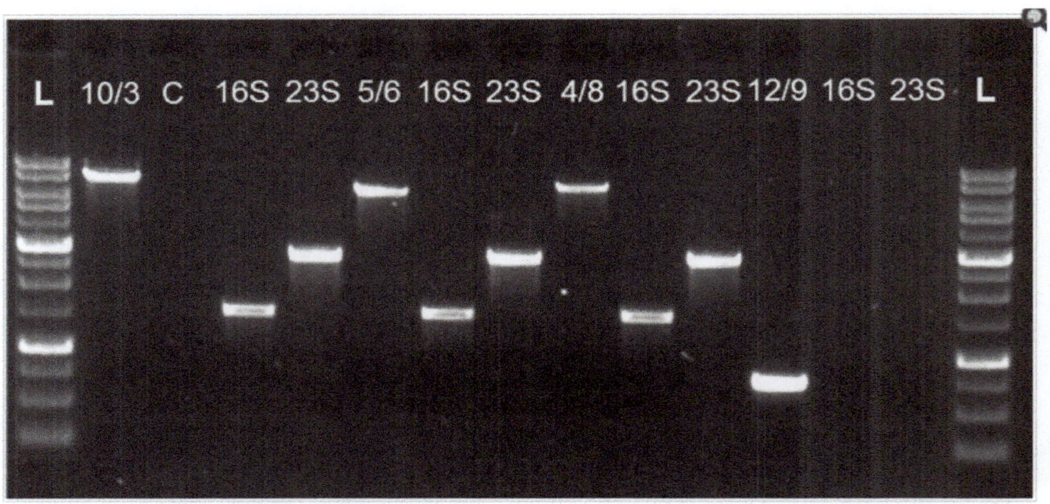

Figure 6: Confirmation of the presence of rDNA operons located within contig gaps. PCR products joining the ends of two contigs were used as templates for PCR to determine presence of rRNA genes within the gaps. Gaps that tested positive for rRNA operons were closed using our innovative rRNA gap sequencing procedure as illustrated in Figure 4. Gaps that did not contain rRNA operons required traditional primer

walking to determine the missing sequence. PCR products labeled as 16S and 23S rRNA genes were amplified from the respective preceding PCR product corresponding to the following gaps; 10/3 – gap between contig 10 and contig 3, 5/6 – gap between contig 5 and contig 6, 8/4 – gap between contig 8 and contig 4, 12/9 – gap between contig 12 and contig 9. L represents a 1 kb molecular weight ladder and C is a representative template control utilizing distant primers not expected to be contained within the 10/3 gap to confirm template purity.

Traditional primer walking involves Sanger sequencing using a primer complementary to the end of known sequence, resulting in approximately 750 bp to 1 kb of readable sequence per sequencing run. New primers are then designed to hybridize to the determined sequence and the process repeats itself until the gap is filled. To speed up gap closure, we designed several sets of rRNA gene specific primers (Primers U1, U2, U3, U4) that hybridize to the 5′ and 3′ end of both the 16S and 23S genes (Table1) so that the entire rDNA operon contained in the contig gaps can be sequenced simultaneously including; 16S and 23S genes, the 16S/23S intragenic sequence, and gene flanking regions (Figure5). Therefore, for each rDNA operon, six total sequencing reactions were performed in parallel (equivalent to three stages of primer walking). By taking advantage of the redundant nature of rDNA operons, 10 PCR amplified contig gaps containing rDNA operons were sequenced in parallel by this rDNA based primer walking approach using the designed rRNA gene specific primers (Table 1). In the majority of cases this strategy was sufficient to fill the contig gap in a single step, dramatically increasing the speed of the primer walking procedure. In cases where sequencing results did not reach 16S/23S genes or where the contig gap did not contain 16S/23S genes, traditional primer walking was necessary to fill the remaining gaps.

Investigation of Rearrangements

One of the most common errors present in draft genomes are rearrangements, a result of incorrect assignment of overlapping reads (Salzberg and Yorke, 2005). Typically, the PCR amplification of ambiguous base stretches during genome finishing is straightforward. For the CR1 genome, difficulties generating PCR product for an ambiguous base stretch located in contig 4 of the draft assembly suggested either an error in the scaffold assembly or a rearrangement in the draft genome. Analyzing low similarity regions in the Mauve alignments of the draft genome also indicated potential rearrangements in the CR1 draft genome assembly (compare local collinear block composition between CR1 and *P. polymyxa* strain E681, Figure.3). To pinpoint the location of the rearrangements, 10 kb windows of nucleotide sequence from the CR1 draft genome flanking the putative rearrangements were aligned using BLASTn against completely sequenced genome of *P. polymyxa* strain E681. Taken together, the results from Mauve alignments and BLASTn alignments suggested a large insertion of the assembled contigs of 11, 13, 7, and 15 immediately adjacent to an ambiguous base stretch in contig 4. Figure7 offers a schematic representation of the final *P. polymyxa*strain CR1 genome assembly compared to the originally identified *in silico* assembly and scaffold. To verify the contig rearrangement, we designed the primers 13B and 445N5 to specifically amplify the joint region of contig 13 and the ambiguous base stretch in contig 4, as well as specific primers 15B and 445N3 that flanking the joint region of contig 15 and the ambiguous base stretch in contig 4 (Table2). As expected, DNA sequences from the generated PCR products confirmed that the rearrangement present in contig 4 of the draft genome was an assembly error. The PCR product from primers 13B and 445N5, (gap between contig 4 and contig 13) was under 1 kb and was immediately filled by direct sequencing. Interestingly, the PCR product from primers 15B and 445N3 was quite large and contained both 16S and 23S rDNA and was verified and sequenced using the same rDNA based primer-walking strategy described above (Figure,7, lane 4/15). Resolution of the rearrangement uncovers an approximately 200 kb assembly error in the draft genome, and these types of large-scale rearrangements are commonly present in published draft genomes (Latreille et al., 2007).

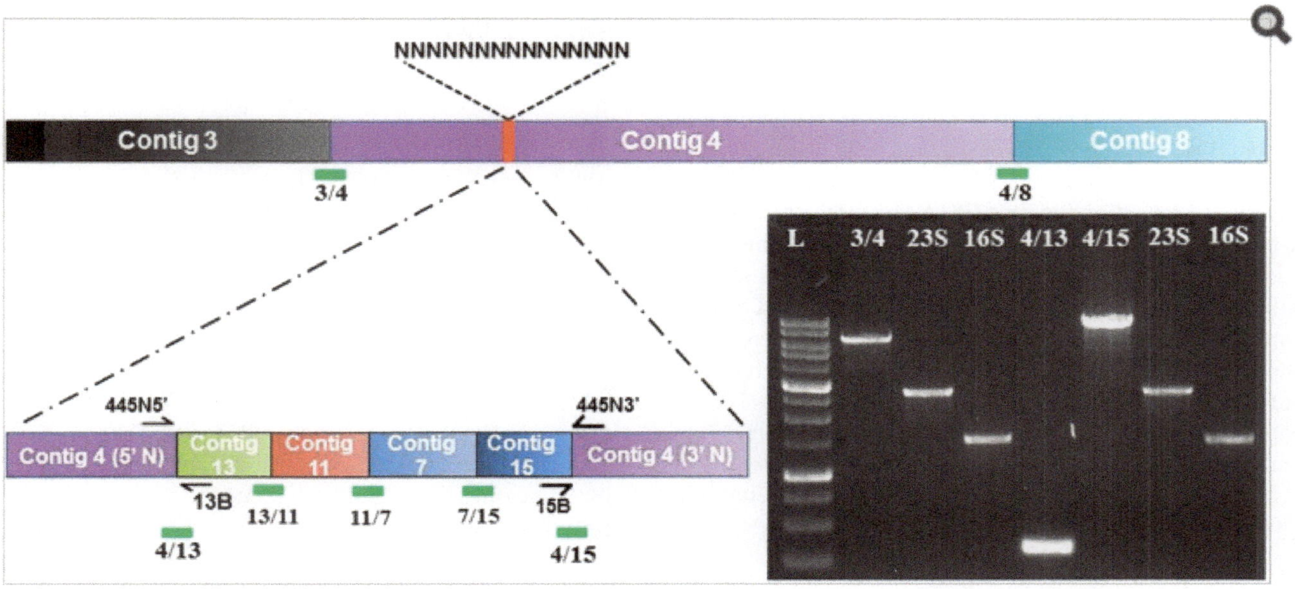

Figure 7: Identification and resolution of rearrangement in draft contig assembly. The combined contigs of 13, 11, 7, and 15 was predicted to be located within an ambiguous base stretch based on Mauve alignments, which was confirmed by PCR amplification of the adjoining ends of contigs 4(5′N) to contig 13 and contig 15 to contig 4 (3′N) as shown by lanes 4/13 and 4/15, respectively. L represents a 1 kb molecular weight

ladder, 16S and 23S genes were amplified from the preceding PCR product correspond to the following contig gaps; 3/4 – gap between contig 3 and contig 4, 4/15 – PCR product corresponding to insertion of contig 15 within an ambiguous base stretch of contig 4, 4/13 – PCR product corresponding to insertion of contig 13 within an ambiguous base stretch of contig 4.

Table 2: Specific Primers targeting identified contig gaps for amplification

Contig gap	Forward primer	Reverse primer
5–1	AATAGACGGGTCTTCCTGCTTATAC	CGATACAACAGCCGACATTTGATTA
5–6	ACCCTAATCGACTGCTTAATCTTGT	CATCTATTGAACTCATTGAACGGGG
6–10	GATGAACCAAAACCTCACCAAGAAT	CGATTGGATCAAGATATTCGCTACG
10–3	GCATCCACAATACGACCCATAAAAT	GAATGTAGGACGAACGGGTAAAATC
3–4	CGGATTACCTCAAAAGGATTGGATG	AAAGAACCATCATGCACAGACAAAT
12–1	GTGCCGTTCTAATGTGTAGCTTATC	AGATGGATAGTAGTCAGGCAAATCC
9–2	GTATATCGGGGAAAGAGTAGGCAAT	GGTGTTTGTGTTGTAAGCTGTATGA
8–4	TTGTTTGATACGTTTGGTACCTTGG	TCTCCAAGTCAGCGTTCGTC
13–11	GACATGGTTGATTGAAAGTGACTGG	AGTGCTCAGCATGGAAGCAA
11–7	TGGTCAATGTAAAACGCAATCTTCA	CCATCATATCCGGGCACCAA
7–15	GCACTGAATAATCCCATTCTCAACC	TGAGGCAACAAGAATCCGCT
4–15	ACATGTTGCTTTCTTTTGCTGA	GGACAACCAGGATACCGCAA
4–13	AACCTGCTGATTATGCGGCT	AGTGCTTGCAAAGTTGGCTC

As a final validation of genome assembly, the sequence of the 38 contigs from the original draft assembly, including the contigs containing 16S and 23S rRNA gene fragments that were merged earlier, were aligned against the final contiguous assembly of the CR1 genome, which confirmed coverage of the entire draft in the final sequence.

Genome Annotation

The data contained in gene and protein annotations is ultimately the primary reasoning for genome sequencing. Automated *in silico* annotation suites rely on open reading frame predictions and homology searches against databases of characterized proteins. A multitude of automated annotation pipelines exist, each with their own respective algorithms, strengths, and limitations, and careful consideration must be taken when deciding which annotation suite to be used (Richardson and Watson, 2013). To provide consistency with other publically available genomes we utilized the NCBI Prokaryotic Genome Automatic Annotation Pipeline (PGAAP) for the annotation of the complete CR1 genome. The PGAAP uses a combination of Hidden Markov Model based gene prediction and a sequence similarity search against Entrez Protein Clusters, the Conserved Domain Database, and the Clusters of Orthologous Groups database. Annotations are automatically cross-referenced by homology to proteins contained in the NCBI database and are assigned standard locus tags, protein/gene names, and protein domain annotations. Annotation of the *P. polymyxa*strain CR1 genome using the PGAAP revealed 5283 protein-coding sequences, 36 rRNAs, 87 tRNAs and 103 pseudogenes. Despite the wealth of knowledge on structure-function relationships in prokaryotes, over 20% of the genes in the CR1 genome correspond to a COG category of general function prediction or unknown function (Eastman et al., 2014b).

DISCUSSION

Regardless of the breakthroughs in sequencing technologies in the preceding decade and the power of second-generation sequencing platforms, generation of a complete, finished bacterial genome remains a difficult task. Draft genome contiguity is limited by the presence of repetitive genetic elements longer than short-insert and mate-pair read lengths, such as multiple copy and highly similar rDNA operons and tRNAs. Many sequencing and assembly errors inherent to second-generation sequencing technologies are present in draft assemblies as evidenced by a large number of ambiguous base stretches (Figure4) and rearrangements (Figure7) in the CR1 draft genome. The identification of rRNA gene fragments in a large proportion of contig ends in the draft CR1 genome reinforce previous findings that these types of repetitive genetic elements are a major limiting factor for *in silico* assembly (Rainey et al., 1996; Ricker et al., 2012). In addition, ribosomal subunit intragenic sequences are variable within an individual genome and it cannot be assumed that all rDNA operons have an identical composition (Figures8, 9). Thus, the variety of features that need to be addressed during genome finishing requires a large amount of labor and resources to intensely scrutinize the draft genome sequence and experimentally confirm the correct assembly, arrangement, and structure.

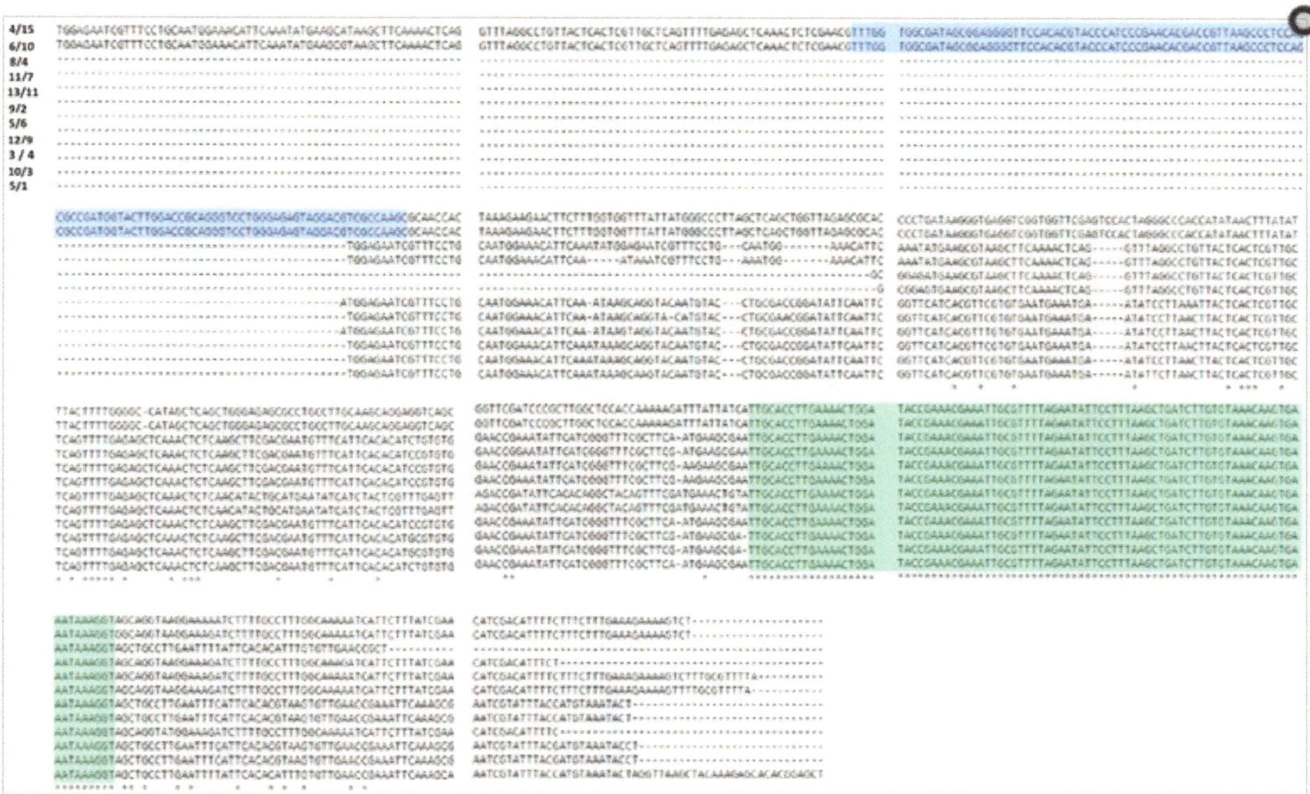

Figure 8: ClustalΩ alignment of the 16S and 23S ribosomal subunit intragenic region of the CR1 genome. Numbers to the left of sequence data represent the specific contig gap where the ribosomal subunit DNA was located. CR1 encoded ribosomal subunit DNA was unanimously localized to gaps from the draft genome. Intragenic DNA sequences were obtained using primers U2 and U3 and sequenced by Sanger sequencing and aligned by ClustalΩ using default parameters. Sequence highlighted in blue corresponds to 5S ribosomal subunit DNA, highlighted in green represents a highly conserved *P. polymyxa* **species-specific 89** nt sequence.

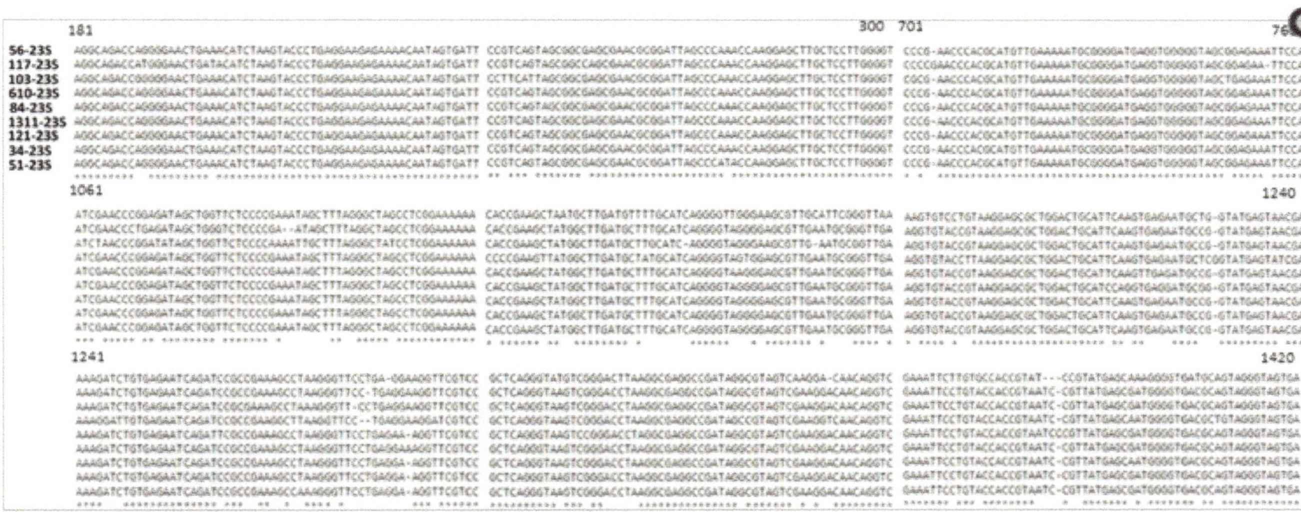

Figure 9: Clustal **Ω alignment of 23S ribosomal subunit regions from CR1 genome. Numbers to the left of sequence data represent the specific** contig gap where the ribosomal subunit DNA was located. Numbers above the nucleotide sequence data represent the position of each 60 nt frame. Regions not shown in the figure correspond to stretches of identical nucleotides between the depicted genes.

The continued development and availability of simple and effective genome finishing strategies is paramount to accommodate the persistent onslaught of advances in sequencing technologies. Without concomitant advances in genome finishing strategies we cannot fully explore the nuances of evolution, phylogenetics, and prokaryotic genomic structure that can only be obtained from completed genomes.

As the costs of *de novo* genome sequencing continue to decrease, *de novo* genome sequencing will become increasingly common in the characterizations of bacterial strains. Based on the repetitive and conserved nature of rDNA operons we designed a procedure that allows for precise identification and sequencing of rDNA sequences within contig gaps. Using our procedure multiple primer walking sequencing reactions can be performed in parallel to target repetitive rDNA operons, thereby effectively reducing the time and resources needed for gap closure and assembly of scaffolded bacterial genomes. Despite recent developments in long read and single molecule sequencing technologies, due to the low base calling accuracy and high costs it is unlikely these technologies will completely replace Illumina sequencing for bacterial genome sequencing in the near future (Quail et al., 2012). Our approach increases the feasibility and speed of genome sequencing and offers an economical and technically simple strategy for microbial genome finishing, especially when available manpower and resources is a limiting factor in the decision to finish a genome.

Nucleotide Sequence Accession Numbers

The complete genomic sequences of strains referenced in the text are publically available in the NCBI GenBank database with the following accession numbers; *P. polymyxa* strain CR1 - NC_023037.1, *P. polymyxa* strain E681 - NC_014483.1, *P. terrae* strain HPL-003 - NC_016641.1.

AUTHOR CONTRIBUTIONS

Alexander W. Eastman and Ze-Chun Yuan conceived and designed the study and drafted the manuscript. Alexander W. Eastman performed the data collection and analysis. Both authors approved the final manuscript.

Conflict of interest statement

The authors declare that the research was conducted in the absence of any commercial or financial relationships that could be construed as a potential conflict of interest.

ACKNOWLEDGMENTS

We would like to thank Naeem Nathoo and Brian Weselowski for editing of the manuscript and technical support. This research was funded by the Agriculture and Agri-Food Canada, Growing Forward-AgriFlex funding (RBPI number 2555), and conducted by the authors as part of their duties. We really appreciate Dr. Frank T. Robb, University of Maryland, USA, for editing of our manuscript. We also appreciate Dr. Jonathan H. Badger from J. Craig Venter Institute, USA, and Dr. Viktoria Shcherbakova from Russian Academy of Sciences, Russia for reviewing our manuscript and their valuable comments to improve the manuscript.

REFERENCES

1. Anzai Y., Kim H., Park J.-Y., Wakabayashi H., Oyaizu H. (2000). Phylogenetic affiliation of the Pseudomonads based on 16S rRNA sequence. Int. J. Syst. Evol. Microbiol. 50, 1563–1589. 10.1099/00207713-50-4-1563
2. Bentley S. (2010). Taming the next-gen beast. Nat. Rev. Microbiol. 8, 161. 10.1038/nrmicro2322
3. Birol I., Jackman S. D., Nielsen C. B., Qian J. Q., Varhol R., Stazyk G., et al. . (2009). *De novo*transcriptome assembly with ABySS. Bioinformatics 25, 2872–2877. 10.1093/bioinformatics/btp367
4. Carver T., Thomson N., Bleasby A., Berriman M., Parkhill J. (2009). DNAPlotter: circular and linear interactive genome visualization. Bioinformatics 25, 119–120. 10.1093/bioinformatics/btn578
5. Darling A. E., Mau B., Perna N. T. (2010). progressiveMauve: multiple genome alignment with gene gain, loss and rearrangement. PloS ONE 5:e11147. 10.1371/journal.pone.0011147
6. Darling A. E., Tritt A., Eisen J. A., Facciotti M. T. (2011). Mauve assembly metrics. Bioinformatics27, 2756–2757. 10.1093/bioinformatics/btr451
7. Delseny M., Han B., Ie Y. (2010). Plant science high throughput DNA sequencing: the new sequencing revolution. Plant Sci. 179, 407–422. 10.1016/j.plantsci.2010.07.019
8. Eastman A. W., Heinrichs D. E., Yuan Z.-C. (2014b). Comparative and genetic analysis of the four sequenced *Paenibacillus polymyxa* genomes reveals a diverse metabolism and conservation of genes relevant to plant-growth promotion and competitiveness. BMC Genomics 15:851. 10.1186/1471-2164-15-851
9. Eastman A. W., Weselowski B., Nathoo N., Yuan Z.-C. (2014a). Complete genome sequence of*Paenibacillus polymyxa* CR1, a plant growth-promoting bacterium isolated from the corn rhizosphere exhibiting potential for biocontrol, biomass degradation, and biofuel. Genome Announ. 2, e01218-13. 10.1128/genomeA.01218-13
10. Ewing B., Green P. (1998). Base-calling of automated sequencer traces using Phred. II. Error probabilities. Genome Res. 8, 186–194. 10.1101/gr.8.3.175
11. Ewing B., Hillier L., Wendl M. C., Green P. (1998). Base-calling of automated sequencer traces using Phred. I. Accuracy assessment. Genome Res. 8, 175–185. 10.1101/gr.8.3.175
12. Gibbons J. G., Janson E. M., Hittinger C. T., Johnston M., Abbot P., Rokas A. (2009). Benchmarking next-generation transcriptome sequencing for functional and evolutionary genomics. Mol. Biol. Evol.26, 2731–2744. 10.1093/molbev/msp188
13. Glenn T. C. (2011). Field guide to next-generation DNA sequencers. Mol. Ecol. Resour. 11, 759–769. 10.1111/j.1755-0998.2011.03024.x
14. Gritsenko A. A., Nijkamp F. J., Reinders M. J. T., de Ridder D. (2012). GRASS: a generic algorithm for scaffolding next-generation sequencing assemblies. Bioinformatics 28, 1429–1437. 10.1093/bioinformatics/bts175
15. Horner D. S., Pavesi G., Castrignano T., Onorio De Meo P. D., Liuni S., Sammeth M., et al. . (2009).Bioinformatics approaches for genomics and post genomics applications of next-generation sequencing. Brief. Bioinformatics 11, 181–197. 10.1093/bib/bbp046
16. Hunt D. E., Klepac-Ceraj V., Acinas S. G., Gautier C., Bertilsson S., Polz M. F. (2006). Evaluation of 23S rRNA PCR primers for use in phylogenetic studies of bacterial diversity. Appl. Environ. Microbiol. 72, 2221–

2225. 10.1128/AEM.72.3.2221-2225.2006

17. Hurt R. A., Brown S. D., Podar M., Palumbo A. V., Elias D. A. (2012). Sequencing intractable DNA to close microbial genomes. PloS ONE 7:e41295. 10.1371/journal.pone.0041295

18. Kim J. F., Jeong H., Park S.-Y., Kim S.-B., Park Y. K., Choi S.-K., et al. . (2010). Genome sequence of the polymyxin-producing plant-probiotic rhizobacterium *Paenibacillus polymyxa* E681. J. Bacteriol.192, 6103–6104. 10.1128/JB.00983-10

19. Lal S., Tabacchioni S. (2009). Ecology and biotechnological potential of *Paenibacillus polymyxa*: a minireview. Indian J. Microbiol. 42, 2–10. 10.1007/s12088-009-0008-y

20. Latreille P., Norton S., Goldman B. S., Henkhaus J., Miller N., Barbazuk B., et al. . (2007). Optical mapping as a routine tool for bacterial genome sequence finishing. BMC Genomics 8:321. 10.1186/1471-2164-8-321

21. Li S., Yang D., Qiu M., Shao J., Guo R., Shen B., et al. . (2014). Complete genome sequence of*Paenibacillus polymyxa* SQR-21, a plant growth-promoting rhizobacterium with antifungal activity and rhizosphere colonization ability. Genome Announ. 2, e00281-14. 10.1128/genomeA.00281-14

22. Lin S.-H., Liao Y.-C. (2013). CISA: contig integrator for sequence assembly of bacterial genomes.PloS ONE 8:e60843. 10.1371/journal.pone.0060843

23. Luo R., Liu B., Xie Y., Li Z., Huang W., Yuan J., et al. . (2012). SOAPdenovo2: an empirically improved memory-efficient short-read *de novo* assembler. GigaScience 1, 1–6. 10.1186/2047-217X-1-18

24. Ma M., Wang C., Ding Y., Li L., Shen D., Jiang X., et al. . (2011). Complete genome sequence of*Paenibacillus polymyxa* SC2, a strain of plant growth-promoting rhizobacterium with broad-spectrum antimicrobial activity. J. Bacteriol. 193, 311–312. 10.1128/JB.01234-10

25. Maclean D., Jones J. D. G., Studholme D. J. (2009). Application of 'next-generation' sequencing technologies to microbial genetics. Nat. Rev. Microbiol. 7, 287–296. 10.1038/nrmicro2088

26. Mardis E., McPherson J., Martienssen R., Wilson R. K., McCombie R. W. (2002). What is finished, and why does it matter? Genome Biol. 12, 669–671. 10.1101/gr.032102

27. McGinn S., Gut I. G. (2013). DNA sequencing – spanning the generations. New Biotechnol. 30, 366–372. 10.1016/j.nbt.2012.11.012

28. Medini D., Serruto D., Parkhill J., Relman D. A., Donati C., Moxon R., et al. . (2008). Microbiology in the post-genomic era. Nat. Rev. Microbiol. 6, 419–430. 10.1038/nrmicro1901

29. Miller J. R., Delcher A. L., Koren S., Venter E., Walenz B. P., Brownley A., et al. . (2008).Aggressive assembly of pyrosequencing reads with mates. Bioinformatics 24, 2818–2824. 10.1093/bioinformatics/btn548

30. Morey M., Fernández-Marmiesse A., Castiñeiras D., Fraga J. M., Couce M. L., Cocho J. A. (2013). A glimpse into past, present, and future DNA sequencing. Mol. Genet. Metab. 110, 3–24. 10.1016/j.ymgme.2013.04.024

31. Nagarajan N., Cook C., Di Bonaventura M., Ge H., Richards A., Bishop-Lilly K. A., et al. . (2010). Finishing genomes with limited resources: lessons from an ensemble of microbial genomes. BMC Genomics 11:242. 10.1186/1471-2164-11-242

32. Niu B., Rueckert C., Blom J., Wang Q., Borriss R. (2011). The genome of the plant growth-promoting rhizobacterium *Paenibacillus polymyxa* M-1 contains nine sites dedicated to nonribosomal synthesis of lipopeptides and polyketides. J. Bacteriol. 193, 5862–5863. 10.1128/JB.05806-11

33. Pop M., Kosack D. S., Salzberg S. L. (2004). Hierarchical scaffolding with bambus. Genome Res. 14, 149–159. 10.1101/gr.1536204

34. Poptsova M. S., Gogarten J. P. (2010). Using comparative genome analysis to identify problems in annotated microbial genomes. Microbiology 156, 1909–1917. 10.1099/mic.0.033811-0

35. Quail M. A., Smith M., Coupland P., Otto T. D., Harris S. R., Connor S. R., et al. . (2012). A tale of three next generation sequencing platforms: comparison of Ion Torrent, Pacific Biosciences and Illumina MiSeq sequencers. BMC Genomics 13:341. 10.1186/1471-2164-13-341

36. Rainey F. A., Ward-Rainey N. L., Janssen P. H., Hippe H., Stackebrandt E. (1996). *Clostridium paradoxum* DSM730AT contains multiple 16s rRNA genes with heterogeneous intervening sequences. Microbiology 142, 2087–2095. 10.1099/13500872-142-8-2087

37. Richardson E. J., Watson M. (2013). The automatic annotation of bacterial genomes. Brief. Bioinformatics 14, 1–12. 10.1093/bib/bbs007

38. Ricker N., Qian H., Fulthorpe R. R. (2012). The limitations of draft assemblies for understanding prokaryotic adaptation and evolution. Genomics 100, 167–175. 10.1016/j.ygeno.2012.06.009

39. Rutherford K., Parkhill J., Crook J., Horsnell T., Rice P. (2000). Artemis: sequence visualization and annotation. Bioinformatics 16, 944–945. 10.1093/bioinformatics/16.10.944

40. Salzberg S. L., Yorke J. A. (2005). Beware of mis-assembled genomes. Bioinformatics 21, 4320–4321. 10.1093/bioinformatics/bti769

41. Schuster S. C. (2008). Next-generation sequencing transforms today's biology. Nat. Methods 5, 16–18. 10.1038/nmeth1156

42. Shin S. H., Kim S., Kim S. Y., Young H., Cho S. J., Kim D. R., et al. . (2012). Genome sequence of *Paenibacillus terrae* HPL-003, a xylanase-producing bacterium isolated from soil found in forest residue. J. Bacteriol. 194, 1266. 10.1128/JB.06668-11

43. Tsai I. J., Otto T. D., Berriman M. (2010). Improving draft assemblies by iterative mapping and assembly of short reads to eliminate gaps. Genome Biol. 11, R41. 10.1186/gb-2010-11-4-r41

44. Wetzel J., Kingsford C., Pop M. (2011). Assessing the benefits of using mate-pairs to resolve repeats in *de novo*

short-read prokaryotic assemblies. BMC Bioinformatics 12:95. 10.1186/1471-2105-12-95

45. Yu B., Sun J., Bommareddy R. R., Song L., Zeng A.-P. (2011). Novel (2R, 3R)-2, 3-butanediol dehydrogenase from potential industrial strain *Paenibacillus polymyxa* ATCC 12321. Appl. Environ. Microbiol. 77, 4230–4233. 10.1128/AEM.02998-10

46. Zerbino D. R., Birney E. (2008). Velvet: algorithms for *de novo* short read assembly using de Bruijn graphs. Genome Res. 18, 821–829. 10.1101/gr.074492.107

SECTION II: THE SPECIES CONCEPT

CHAPTER 4

METAGENOMIC APPROACHES FOR THE IDENTIFICATION OF MICROBIAL SPECIES

David M. Ward, Melanie C. Melendrez, Eric D. Becraft, Christian G. Klatt, Jason M. Wood and Frederick M. Cohan

Keywords: Metagenomic Approaches, For microbial species identification, Species identification, and molecular methods, Individual sequence variants of species populations, Metagenomic databases, Investigating existence of natural genetically coherent populations

INTRODUCTION

In the most recent iteration of our long-term molecular studies of hot spring cyanobacterial mat communities [Ward et al., 1998, 2002, 2006], we have developed and employed several genomics and metagenomics approaches to investigate the existence and character of species populations [Ward et al., 2008, 2011]. The aim of this chapter is to provide a concise overview of these approaches and the inferences to which they have led us, relying on several original papers to report the results in detail [Klatt et al., 2011; Melendrez et al., 2011a,b; Becraft et al., 2011].

We do not believe that there is a single, universal way to conceptualize microbial species. Rather, we believe that, as with plant and animal species, different taxa form species differently depending on how diversity is generated within the taxon and on how nature acts upon this diversity [Cohan and Perry, 2007; Ward et al., 2008]. Our investigations have been of unicellular cyanobacteria (Synechococcus spp.), which are predominant members of a mat community living in 50 to ~72° C regions of the effluent channels of alkaline siliceous hot springs in Yellowstone National Park. The evidence in hand suggests that these organisms have undergone speciation through adaptation to parameters that vary in the environments they inhabit [Ward, 1998, 2006]. The Stable Ecotype Model is one of many models created to explain these observations in an evolutionary context [Cohan and Perry, 2007; Ward et al., 2008], and it is currently the best explanation for how these organisms have speciated [Ward and Cohan, 2005; Ward et al., 2006]. The lessons learned in our studies should be broadly applicable, since there is evidence of adaptive speciation in many taxa [Field et al., 1997; Bielawski et al., 2004; Koeppel et al., 2008; Connor et al., 2010; and see Ward, 2006].

To observe the native cyanobacterial inhabitants in this mat community, it was necessary to shift from cultivation-dependent approaches, which led to the false impression that Synechococcus lividus was the dominant cyanobacterium in the mat [Ferris et al., 1996], to cultivation-independent molecular approaches, which revealed the genetic signatures of the truly dominant and dramatically unrelated collection of Synechococcus spp. (the A/B lineage; Fig. 13.1A) [Ward et al., 1990]. The inference that these Synechococcus SSU rRNA sequences should be considered to be associated with different ecological species, defined as populations of individuals that occupy the same unique niche [van Valen, 1976; Ward, 1998], was based on observations of the patterning of this molecular diversity relative to natural spatiotemporal gradients of environmental difference. For instance, SSU rRNA-based diversity was found to be distributed along environmental gradients that occur along the flow path away from the hot spring source (e.g., parameters such as changing temperature and nutrients [Ferris and Ward, 1997]). More specifically, Synechococcus. SSU rRNA genotypes A", A', A, B', and B

Citation: David M. Ward, Melanie C. Melendrez, Eric D. Becraft, Christian G. Klatt, Jason M. Wood, and Frederick M. Cohan, Metagenomic Approaches for the Identification of Microbial Species, DOI: 10.1002/9781118010518.ch13.

were found to be differently distributed along the effluent flow path between the upper temperature limit of mat formation at ~72° C downstream to ~50° C (Fig. 13.1A). Demonstration of adaptations to different temperatures in genetically relevant Synechococcus isolates [Allewalt et al., 2006] helped us understand, at least in part, the basis for this distribution along the flow path.

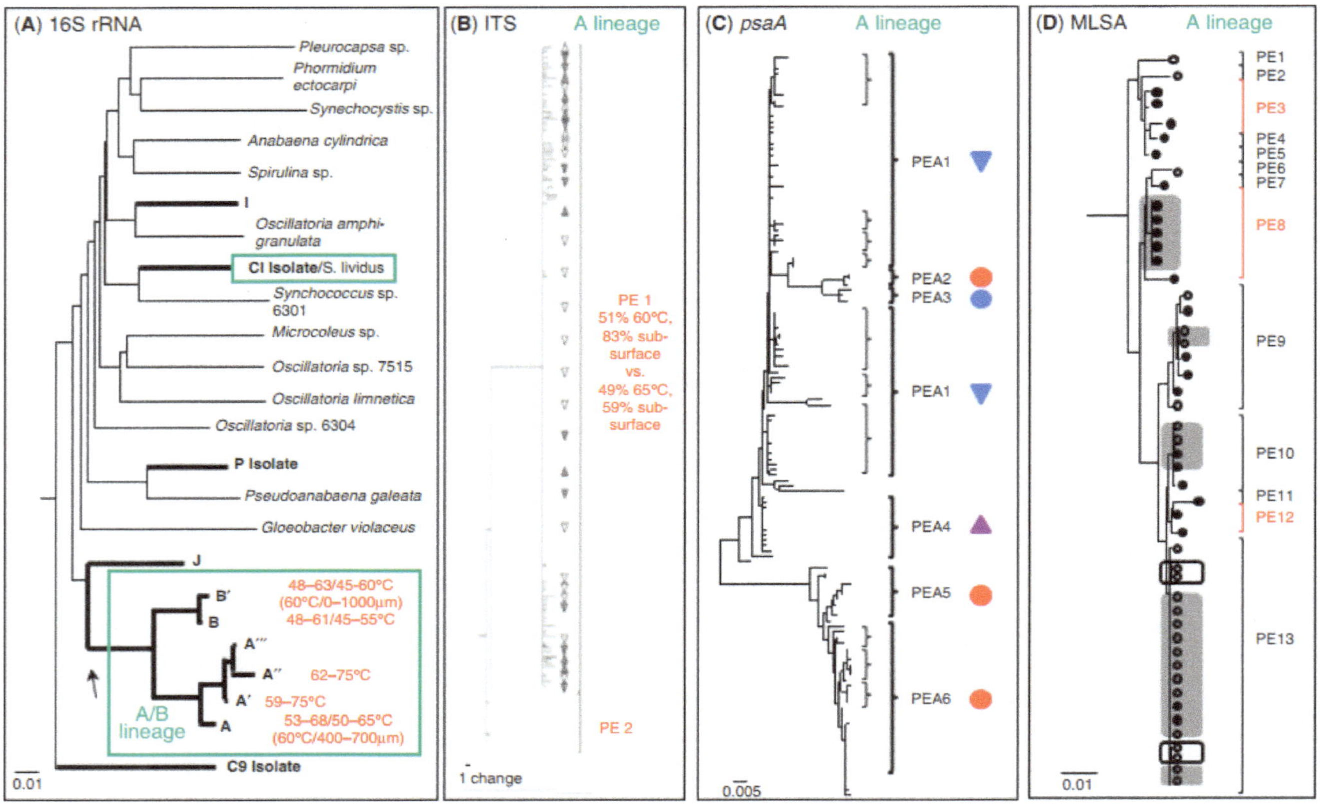

Figure 13.1: Evolutionary ecology of mat Synechococcus spp. based on sequence variation in (A) SSU rRNA (modified from Ward et al. [1998]), (B) 16S–23S rRNA internal transcribed spacer (modified from Ward et al. [2006]), (C) psaA (modified from Becraft et al. [2011]) and (D) multilocus sequence analysis (from [Melendrez et al., 2011a], showing how the number of putative ecotypes (PE) predicted by Evolutionary Simulation analysis rises with the molecular variation offered by these loci or sets of loci. PEs are designated by vertical brackets; red brackets in panel D indicate PEs comprised only of clones collected at 65° C. In panel C, triangles indicate surface and subsurface association as on B; colors indicate predominant collection temperature: blue, 60° C; purple, 63° C; red, 65° C; Brackets within PE brackets represent subclades. In panel A, bold lines indicate sequences retrieved from Octopus and Mushroom Spring mats or isolates therefrom. In panel B, triangles indicate where sequences were collected (upward-pointing, top green layer surface; downward-pointing, subsurface; clear, 60° C; shaded, 65° C). In panel D, the neighbor joining tree shown is based on five concatenated loci (rbsK, PK, lepB, hisF, CHP and aroA) for equal numbers of Synechococcus sp. A-like BACs from samples collected at 60° C (open circles) or 65° C (closed circles). Information about temperature and vertical distribution of genotypes and putative ecotypes is summarized in red (in panel A from Ferris and Ward [1997] and Ramsing et al. [2000]; in panel B from Ward et al. [2006]; in panel C from Becraft et al. [2011]). Scale bars indicate the number of fixed point mutations per sequence position (unlabeled) or the number of base differences (labeled changes).

For instance, isolates with SSU rRNA genotypes A, B; and B were found to have progressively lower growth temperature ranges, consistent with the temperatures at which they are found in situ (Fig. 13.1A). Similarly, differential SSU rRNA genotype distributions within the upper few-millimeter-thick photic zone in the vertical dimension suggested possible adaptations to parameters such as light intensity and quality and/or chemical parameters that vary during the diel light cycle [Ramsing et al., 2000]. More specifically, at ~60° C, Synechococcus sp. genotype A occurred as a subsurface population ~400–700 μm beneath the mat surface, whereas B' was the only Synechococcus genotype detected near the mat surface (Fig. 13.1A). We found, however, that it was necessary to use faster-evolving molecular markers, such as the 16S–23S rRNA internal transcribed spacer region, to identify very closely related, but differently distributed, Synechococcus spp. [Ferris et al., 2003]. For instance, at 68° C, Synechococcus spp. populations with the same SSU rRNA genotype (A'), but with distinct internal transcribed spacer sequences, were found at different mat depths. Because these populations were both genetically and ecologically distinct, we hypothesized that they represent two differently adapted species populations. In general, internal transcribed spacer region analyses led to an increase in species predicted (Fig. 13.1B and see below), making apparent the need for using even faster-evolving, protein-encoding genes to view how variation among individual members of a Synechococcus population might be grouped into true ecologically distinct species populations.

THEORY-BASED PREDICTION OF SYNECHOCOCCUS SPECIES POPULATIONS

The grand challenge of species identification using molecular methods is to determine which individual sequence variants belong to which species populations. Various molecular cutoffs have been suggested for species demarcation [Wayne et al., 1987; Goodfellow et al., 1997; Stackebrandt and Ebers, 2006; Konstantinidis and Tiedje, 2005; see also Chapter 11, Vol. I], but these do not account for the reality that species populations may be younger or older, and thus they are cutoffs of convenience, often calibrated to match the variation found among strains of traditional named species [Cohan, 2002], and not natural biological groupings of variants. For the prediction of hot spring Synechococcus spp., we used an evolutionary simulation based on the Stable Ecotype Model, called Ecotype Simulation [ES; Koeppel et al., 2008]. This simulation models the evolution of a population into a number of putative ecotypes (i.e., ecological species) by determining the most likely history of (i.e., order and timing of) diversity-purging periodic selection events, adaptive ecotype formation (i.e., ecological speciation) events, and genetic drift that explains the actual phylogeny under investigation. Thus, ES demarcates species based on the evolutionary history of the gene (and taxon) being analyzed.

When used to analyze variation in protein-encoding loci in samples collected at 60° C and 65° C sites in Mushroom Spring (Yellowstone National Park) mat, ES predicted up to 14 times more ecological species populations than were predicted from 16S rRNA variation (i.e., A-like and B' -like species, see Fig. 13.1C], the number depending on the molecular resolution of the gene [Melendrez et al., 2011b,; Becraft et al., 2011]. The most well-sampled populations are dominated by a single sequence variant, which may be of central importance, because this is likely to be the variant that was the driver [and hence survivor] of the most recent periodic selection event in the ecotype population and, thus, the founder of the current species clade. Where a predicted species population has more than one dominant variant, we hypothesize that ES has demarcated species too conservatively

The ecotypes (i.e., ecological species) predicted by ES are taken as putative until evidence is obtained to confirm their ecological uniqueness as well as the ecological interchangeability of its members. Importantly, the species populations predicted by ES exhibit unique distributions relative to flow and vertical gradients [Melendrez et al., 2011b; Becraft et al., 2011] (Fig. 13.1C). Some have also been found to exhibit unique gene expression patterns and to respond, as populations, to environmental change [Becraft et al., unpublished]. Our expectation is that the genomes of individuals of a true species population, while variable, should contain niche-specifying adaptations, coded by unique alleles and/or genes, which determine the unique ecological character of all members of the population [Ward et al., 2008; Cohan and Koeppel, 2008].

CULTIVATION-INDEPENDENT MULTILOCUS SEQUENCE ANALYSIS

To better account for the possible influence of recombination, we developed a cultivation-independent, multilocus sequence analysis approach, based on the use of bacterial artificial chromosome cloning to sample multiple loci in the genomic neighborhood of SSU rRNA genes [Melendrez et al., 2011a]. ES also predicted from multiple concatenated loci obtained from the same two mat samples a number of species comparable to that observed with the highest-resolving single loci, even though the recombination/mutation ratio is estimated to be about 2.6–14.4 in these Synechococcus populations [Melendrez et al., 2011a]. Many of these predicted species populations contained dominant variants with identical sequences at up to 7 loci and were comprised of variants that came exclusively or predominantly from either a 60° C or a 65° C sample (Fig. 13.1D), suggesting their ecological distinction. Apparently, recombination in the genes analyzed has been unable to erode the clusters of phylogenetically similar variants that represent the ecological species, although it may have displaced some variants in the loci under investigation to positions in the phylogeny outside of this clade in cases where homologous recombination has occurred with a phylogenetically distant individual. An organism containing such a molecular variant may nevertheless belong to the same species population, because it is only excluded from clades that are based on analysis of genes that have been exchanged with distantly related organisms.

METAGENOMIC ASSEMBLY AND SPECIES POPULATIONS

We have simultaneously examined metagenomic databases to investigate the existence of natural genetically coherent populations, using an assembly approach that is based on molecular sequence similarity, rather than any theoretical basis [Klatt et al., 2011]. Sanger sequences from the ends of 2- to 12-kbp inserts of clones obtained from samples collected at average temperatures of ~60° C and ~65° C in the Mushroom Spring and Octopus Spring mats (same samples as above) were assembled using the Celera assembler. Oligonucleotide frequency analysis of these assemblies, followed by cluster analysis using the K-means algorithm, resulted in a single cluster containing both Synechococcus spp. A-like and B' -like populations [Klatt et al., 2011] (Fig. 13.2). Thus the current level of species resolution is poor compared to that provided by population genetics studies described above lumping all members of the Synechococcus A/B lineage. We are currently investigating databases with much deeper coverage, using alternative clustering algorithms, to test whether it is possible to assemble metagenomic segments of Synechococcus spp. populations predicted by single and multi-locus analyses combined with ES.

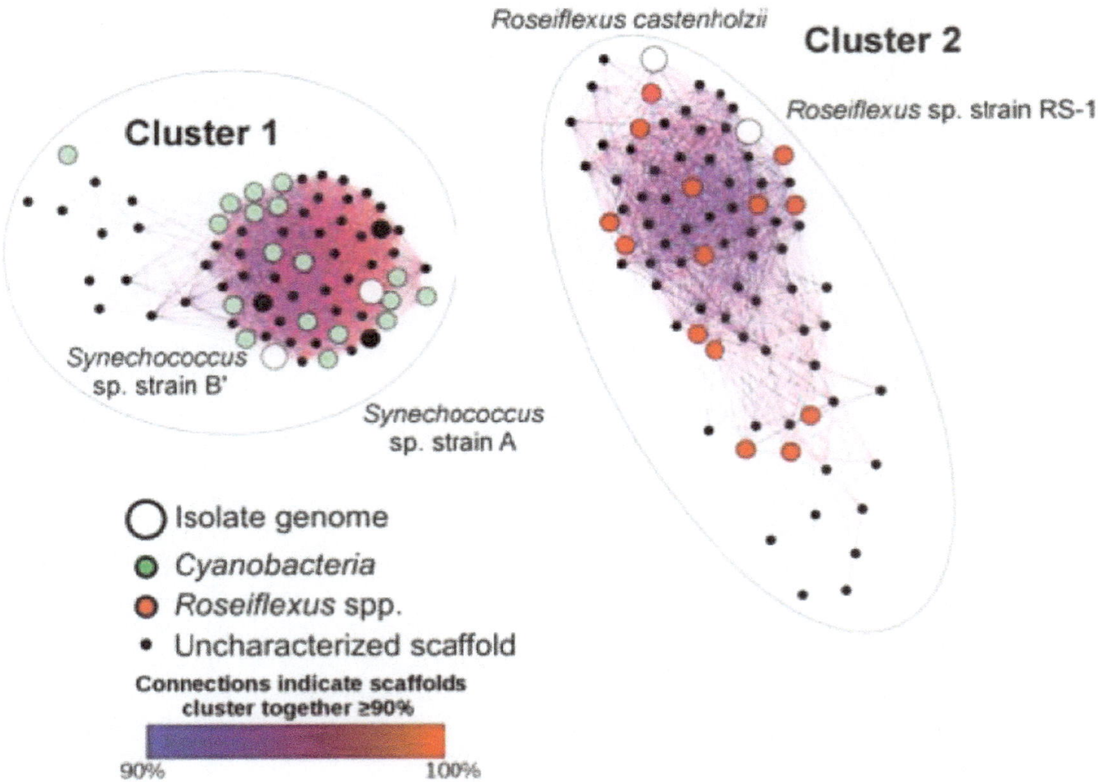

Figure 13.2: Network map showing core cyanobacterial scaffold cluster observed in Celera assemblies of Mushroom Spring and Octopus Spring mat samples collected at 60◦ and 65◦ C. The Roseiflexus spp. cluster is included to exemplify the discrete separation of cyanobacterial cluster from seven noncyanobacterial clusters observed. Scaffolds with similar oligonucleotide frequency profiles that group together in the same cluster in ≥90% of 100 trials are indicated by connecting lines, whose color reports the percentage that scaffolds group together as defined by the scale bar. Isolate genomes included in this analysis are indicated by large white circles, whereas metagenomic scaffolds that contain characterized phylogenetic marker genes of cyanobacteria are shown as green medium-sized circles. Large color-shaded ovals were drawn by hand to demarcate the different clusters. Genomes and scaffolds that are not linked by lines to other scaffolds formed linkages in

CONCLUSIONS

It is abundantly clear that we are approaching a solid understanding of the nature of species populations for hot spring Synechococcus. However, it is also clear that, using present approaches, we are unable to resolve these populations through metagenomic assembly. This may never be possible given the many factors that make assembly difficult (e.g., repetitive sequence elements, conserved genes). Nevertheless, comparison of metagenomic clones with genomes of cultivated Synechococcus spp. strains A and B' genomes did reveal genes not found in the isolate, but present in native populations, that might have adaptive character [Bhaya et al., 2007; Klatt et al., 2011], but such analyses depend on the availability of genomes from isolates that are genetically relevant to the community [Ward et al., 2011].

ACKNOWLEDGMENTS

This research was supported by long-term support from NSF (currently from the Frontiers in Integrative Biology Research Program, EF-0328698 and the IGERT Program in Geobiological Systems, DGE 0654336), NASA (most recently from the Exobiology Program, NAG5-8824,- 8807 and NX09AM87G) and the U.S. Department of Energy (DOE), Office of Biological and Environmental Research (BER), as part of BER's Genomic Science Program (GSP). This contribution originates from the GSP Foundational Scientific Focus Area (FSFA) at the Pacific Northwest National Laboratory (PNNL), contract no. 112443. We appreciate the support of the National Park Service personnel at Yellowstone National Park.

REFERENCES

1. Allewalt JP, Bateson MM, Revsbech NP, Slack K, Ward DM. 2006. Effect of temperature and light on growth and photosynthesis of Synechococcus isolates typical of those predominating in the Octopus Spring microbial mat community. Appl, Environ. Microbiol. 72:544– 550.
2. Becraft, ED, Cohan FM, Kuhl M Jensen S ¨ , Ward DM. 2011. Distribution patterns of ecological species of Synechococcus inhabiting the microbial mat of Mushroom Spring. Yellowstone National Park. In preparation.
3. Bhaya D., Grossman AR, Steunou A-S, Khuri N, Cohan FM, et al. 2007. Population level functional diversity in a microbial community revealed by comparative genomic and metagenomic analyses. ISME J . 1:703– 713
4. Bielawski JP, Dunn KA, Sabehi G, Bej´ a O` . 2004. Darwinian adaptation of proteorhodopsin to different light intensities in the marine environment. Proc. Natl. Acad. Sci. USA 101:14824– 14829.
5. Cohan FM. 2002. What are bacterial species? Ann. Rev. Microbiol. 56:457– 487.

6. Cohan FM, Koeppel AF. 2008. The origins of ecological diversity in prokaryotes. Curr. Biol. 18:R1024–R1034.

7. Cohan FM, Perry EB. 2007. A systematics for identifying the fundamental units of bacterial diversity. Curr. Biol. 17:R373–R386.

8. Connor N, Sikorski J, Rooney AP, Kopac S, Koeppel, et al. 2010. The ecology of speciation in Bacillus. Appl. Environ. Microbiol. 76:1349– 1358.

9. Ferris MJ, Ward DM. 1997. Seasonal distributions of dominant 16S rRNA-defined populations in a hot spring microbial mat examined by denaturing gradient gel electrophoresis. Appl. Environ. Microbiol. 63:1375–1381.

10. Ferris MJ, Ruff-Roberts AL, Kopczynski ED, Bateson MM, Ward DM. 1996. Enrichment culture and microscopy conceal diverse thermophilic Synechococcus populations in a single hot spring microbial mat habitat Appl. Environ. Microbiol. 62:1045– 1050.

11. Ferris, MJ, Kuhl M ¨ , Wieland A, Ward DM. 2003. Different lightadapted ecotypes in a 68◦ C Synechococcus mat community revealed by analysis of 16S–23S intervening transcribed spacer variation. Appl. Environ. Microbiol. 69:2893– 2898

12. Field KG, Gordon D, Wright T, Rappe, Urbach E, Vergin K, Giovannoni SJ. 1997. Diversity and depth-specific distribution of SAR11 cluster rRNA genes from marine planktonic bacteria. Appl. Environ.Microbiol. 63:63–70.

13. Goodfellow M, Manfio GP, Chun J. 1997. Towards a practical species concept for cultivable bacteria. In Claridge MF, Dawah HW, Wilson MR, eds. Species: Units of Biodiversity. London: Chapman and Hall, pp. 25–59.

14. Klatt CG, Wood JM, Rusch DB, Bateson MM, Hamamura N, Heidelberg JF, Grossman AR, Bhaya D, Cohan FM, Kuhl M ¨ , Bryant DA, Ward DM. 2011. Community Ecology of Hot Spring Cyanobacterial Mats: Predominant Populations and their Functional Potential. ISME J . (submitted)

15. Koeppel A, Perry EB, Sikorski J, Krizanc D, et al. 2008. Identifying the fundamental units of bacterial diversity: A paradigm shift to incorporate ecology into bacterial systematics. Proc. Natl. Acad. Sci. USA 7:2504– 2509.

16. Konstantinidis KT, Tiedje JM. 2005. Genomic insights that advance the species definition for prokaryotes. Proc. Natl. Acad. Sci. USA 102:2567– 2572.

17. Melendrez MC, Cohan FM, Ward DM. 2011a. Cultivationindependent multi-locus sequence analysis of Synechococcus populations inhabiting a hot spring cyanobacterial mat. In preparation.

18. Melendrez MC, Lange RK, Cohan FM, Ward DM. 2011b. Influence of molecular resolution on sequence-based discovery of ecological diversity among Synechococcus populations in an alkaline siliceous hot spring microbial mat. Appl. Environ. Microbiol. 77: 1359– 1367

19. Ramsing NB, Ferris MJ, Ward DM. 2000. Highly ordered vertical structure of Synechococcus populations within the one-millimeter thick photic zone of a hot spring cyanobacterial mat. Appl. Environ. Microbiol. 66:1038– 1049.

20. Stackebrandt E, Ebers J. 2006. Taxonomic parameters revisited: Tarnished gold standards. Microbiol Today 33:152– 155.

21. van Valen L. 1976. Ecological species, multispecies, and oaks. Taxon 25:233– 239

22. Ward DM 1998. A natural species concept for prokaryotes Curr. Opin. Microbiol. 1998; 1:271– 277.

23. Ward DM 2006. A macrobiological perspective on microbial species. Microbe 1:269– 278.

24. Ward DM, Cohan FM. 2005. Microbial diversity in hot spring cyanobacterial mats: Pattern and prediction. In Inskeep B, McDermott T, eds. Geothermal Biology and Geochemistry in Yellowstone National Park . Bozeman, MT: Montana State University Thermal Biology Institute, pp. 185–201.

25. Ward DM, Weller R, Bateson MM. 1990. 16S rRNA sequences reveal numerous uncultured microorganisms in a natural community. Nature 345:63– 65.

26. Ward DM, Bateson MM, Ferris MJ, Nold SC. 1998. A natural view of microbial biodiversity within hot spring cyanobacterial mat communities. Microbiol. Mol. Biol. Rev. 62:1353– 1370.

27. Ward DM, Nubel U ¨ , Papke RT, McKitrick MC. 2002. Natural history of microorganisms inhabiting hot spring microbial mats: Clues to the origin of microbial diversity and implications for microbiology and macrobiology. In Staley JT, Reysenbach A-L, eds. Biodiversity of Microbial Life: Foundation of Earth's Biosphere. New York: John Wiley & Sons, pp. 25– 48.

28. Ward DM, Bateson MM, Ferris MJ, Kuhl M ¨ , et al. 2006. Cyanobacterial ecotypes in the microbial mat community of Mushroom Spring [Yellowstone National Park, Wyoming] as species-like units linking microbial community composition, structure and function. Philos. Trans. R. Soc., London [Ser. B] 361:1997– 2008.

29. Ward DM, Cohan FM, Bhaya D, Heidelberg JF, et al. 2008. Genomics, environmental genomics and the issue of microbial species. Nat. Heredity 100:207– 219.

30. Ward DM, Klatt CG, Wood J, Cohan FM, Bryant DA. 2011. Functional genomics in an ecological and evolutionary context: Maximizing the value of genomes in systems biology. In Burnap RL, Vermaas W, eds. Functional Genomics and Evolution of Photosynthetic Systems. Dordrecht: Springer; in press.

31. Wayne LG, Brenner DJ, Colwell RR, Grimont DD, et al. 1987. Report of the ad hoc committee on reconciliation of approaches to bacterial systematics. Int. J. Syst. Bacteriol. 37:463– 464.

SECTION III: METAGENOMICS

CHAPTER 5

RAPID SANGER SEQUENCING OF THE 16S rRNA GENE FOR IDENTIFICATION OF SOME COMMON PATHOGENS

Linxiang Chen, Ying Cai, Guangbiao Zhou, Xiaojun Shi, Jianhui Su, Guanwu Chen, Kun Lin

INTRODUCTION

To date, conventional Sanger sequencing technology is sometimes used in a few diagnostic laboratories, however, it remains time-consuming and laborious. In this article, we have improved the conventional Sanger sequencing and validated it for detecting and genotyping the most common pathogens, including *Pseudomonas aeruginosa*, *Staphyloccocus aureus* and*Escherichia coli*. We presented this protocol and it described a new combination of SYBR Green I real-time polymerase chain reaction (PCR) and Sanger sequencing of DNA collected and extracted through *Whatman* FTA® cards. The bacterial 16S ribosomal RNA gene was used for PCR amplification and subsequent sequencing. Sample collection and DNA preparation for PCR in this assay involve directly use of FTA® cards instead of commercial kits, boiling, phenol-chloroform extraction and ethanol precipitation, or also using FTA® cards but should be prior cleaned with purification reagent or sterile water in previous studies [1]–[4]. *Whatman* FTA® paper is a commercial product that provides a remarkably easy way to collect, preserve and purify genomic DNA from bacteria, consisting of filter paper impregnated with a proprietary mix of chemicals that serve to lyse cells, prevent the growth of bacteria, protect the DNA in the sample, and can be stored at room temperature for even as long as 50 years [5]. Though it has been widely used for PCR, few researches reported its utility of pathogens sequencing typing, we would give a confirmation here. The common 16S rRNA sequencing technique in diagnostic laboratories is still currently based on the conventional Sanger sequencing method, called "first generation sequencing", involving PCR amplification, product qualitative detection and separation by gel electrophoresis, purification of the amplicon through ethanol precipitation, sequencing by an amplification reaction and final capillary electrophoresis. Due to time-consuming, laborious, high operation skills requirement and potential hazard of *ethidium bromide* in agarose gel electrophoresis, the first generation sequencing technique has not been commonly used in most diagnostic laboratories [6]. To save time and reduce workload, we make improvement and propose a new combined protocol involving direct sequencing of the product generated by diagnostic SYBR Green?real-time PCR. The PCR product is diagnosed via the amplifying curve, and specificity of the product is determined by analysis of the melting curve, avoiding the step of agarose gel electrophoresis. In addition, we optimized all hands-on instrument steps by using modern reagents, by means of sequencing 16S rRNA gene of reference and clinical pathogenic strains, we validated the applicability and also found the shortcomings of 16S rRNA gene sequencing method for identification.

Citation: Chen L, Cai Y, Zhou G, Shi X, Su J, Chen G, et al. (2014) Rapid Sanger Sequencing of the 16S rRNA Gene for Identification of Some Common Pathogens. PLoS ONE 9(2): e88886. doi:10.1371/journal.pone.0088886.

MATERIALS AND METHODS

Ethics Statement

The study protocol was approved by the Human Ethical Committee of Shantou University Medical College and Shantou Central Hospital, China. The patient records/information was anonymized and de-identified prior to analysis.

The Evaluation of Improved Sanger Sequencing and Compared to Conventional Sanger Sequencing with Small Samples

Tested strains

To save time and cost for comparing these two methods, we only target 12 pathogenic strains, including 3 reference strains, ATCC.27853 *Pseudomonas aeruginosa* (American Type Culture Collection), AS.26003 *Staphylococcus aureus* and AS.44113 *Escherichia coli* (China General Microbiological Culture Collection Center), and 9 clinical isolates (had been identified by using conventional biochemical tests by microbiologists), each of 3 *Pseudomonas aeruginosa*, 3*Staphylococcus aureu* and 3 *Escherichia coli.*

Preparation of bacterial suspension and DNA processed in each method.

The clinical bacterial strains were isolated and the reference strains were rejuvenated. Both of them used conventional cultural methods, then the suspensions of pathogen strains were made at proper concentrations. DNA prepared for improved method was performed referred to Menassa et al. [7]. In brief, after vortexing thoroughly, 50 microliters of suspension were dropped onto a FTA® card and were allowed to permeate evenly through the paper. All cards were then allowed to air-dry at room temperature so as to inactivate pathogens by the reagents within the cards. For conventional method, DNA was processed as Corless et al. [8] described but needs some modification, briefly, pipetting all the bacterial suspensions each of 100 μl to 900 μl sterile distilled water, centrifugation at 12,000×g for 3 min prior to remove the 900 μl supernatant, repeating this step one more time and the residual 100 μl mixture which contains bacteria were boiled at 100°C for 10 min to release DNA, after slightly centrifugation, the supernatant can be stored at 4°C and prepared for PCR using.

SYBR Green ? Real-time 16S rDNA PCR by improved method.

Punch one disk with appropriate diameter from the sample spot on the FTA® card and place the disk into a real-time PCR reaction tube for direct SYBR Green?PCR, which contained 10 μl of SYBR Green?PCR Master Mix reagent (Takara), 1 μl each of 2 μM stocks of universal bacteria 16S rRNA gene forward and reverse primers (forward: 5′-TGGA GAGTTTGATCCTGGCTCAG-3′; reverse: 5′-TACCGCGGCTGCTGGCAC-3′) [9], and 9 μl of water. PCR was performed in an Roche LightCycler 480 system thermocycler (Roche) with an initial step of 2 min at 95°C, followed by 35 cycles of 10 s at 95°C, 20 s at 60°C, and 40 s at 72°C. Fluorescent signal intensities were recorded during the end of the elongation phase in each cycle. To interpret the data, the Cp (cross point) values for each sample and negative control were calculated by using analysis mode of "Abs Quant/2nd Derivative Max". Besides, we regarded the results as potentially positive (amplified products existed) if the Cp value cycle of amplifying curves was <30, while the melting curve of the amplicon presented a single melting peak. If there were two or more melting peaks in a melting curve, the products would be considered impure and unreliable. Specially, to seek for the optimum size of FTA® card disk for PCR in this assay, we used three sizes, which were 0.5-mm, 1.2-mm and 2.0-mm of FTA® of AS.26003 *Staphylococcus aureus* strains for direct PCR. Furthermore, to quest the best bacteria concentration dropping onto the FTA® card, a standard curve, including a linear range of known quantification from 6×10^4 to 6×10^9 CFU ml^{-1} of AS.26003 *Staphylococcus aureus*strains, was constructed. Thereafter formal experiments started when the above-mentioned optimal condition had been affirmed.

Traditional PCR and products qualitative detection through agarose gel electrophoresis by conventional method.

The processed DNA extraction was placed in PCR tubes, with the following components added and the final volume adjusted to 20 μL with sterile double distilled water: 100 nM each primer, 800 μM dNTPs, 1.5 mM $MgCl_2$&2.5 U Taq polymerase. Using Roche LightCycler 480 the specimens were heated to 96°C for 10 min followed by 35 cycles of 96°C for 10 s, 59°C for 20 s, 72°C for 30 s, with a final extension step of 72°C for 10 min. The reaction products were held at 4°C until use within 24 h. The PCR products were visualised using a 1.5% agarose gel with ethidium bromide staining. A DNA marker of known DNA fragment sizes (2000 bp ladder) was run along side the specimens to aid in identification of the products. Electrophoresis in Tris-borate-EDTA buffer was performed at 100 V for 40 minutes, and photographed under UV light illumination (UNIVERSAL HOOD.365 nm), when visual band were observed at about 500 bp fragment, PCR succeeded.

Processing of PCR products and subsequent sequencing PCR by two methods.

In order to eliminate potentially adverse effects of PCR reagents on subsequent Sanger sequencing, we diluted the PCR reaction both of two methods 5-fold by adding 2 μl PCR product to 8 μl water. Then sequencing PCR reaction was performed in a 20 μl final volume containing 4 μl of BigDye Terminator v3.1 Sequencing Buffer (5×) (Applied Biosystems), 2 μl of 1 μM amplify PCR backward primer, 4 μl of BigDye Mix, 9 μl water and 1 μl of diluted PCR product,and run in a Veriti 96-Well Fast Thermal Cycler (Applied Biosystems) using the following parameters: denaturation at 98°C for 2 min, followed by 25 cycles of 96°C for 10 s, annealing at 50°C for 5 s and extension at 60°C for 4 min.

Purification of sequencing PCR products and capillary electrophoresis by two methods.

In the improved method, Sanger sequencing products was purified using the BigDye XTerminator purification kit (Applied Biosystems). SAM solution (45 µl) and BigDye XTerminator solution (10 µl) were respectively added and premixed in each 0.2 ml tube. Then 5 µl Sanger sequencing product was added in each tube, vortexed for 5 min, centrifuged at 2000×g for 2 min. Then 10 µl supernatant in each tube was transferred into a plate and covered with septa. After a pulse spin, the plate was mounted in a 3130 Genetic Analyzer (Applied Biosystems) using default module BDx_StdSeq50_POP7_1 optimized to a 3 s injection. Then the sequences were automatically compiled using Sequencing Analysis 5.3.1 software (Applied Biosystems). While in conventional method, Sanger sequencing products were purified by using traditional ethanol precipitations referring to Peattie [10]. Briefly, added 35 µl 100% ethanol solution and 2 µl to each 0.2 ml tube which contains 5 µl products, centrifuged at 12000×g for 20 min then carefully discarded all the supernatants. Added 50 µl 75% ethanol solution to each tube to further eliminate impurities, discarded all the supernatants after 2 min. Air-dried products in the tubes for 20 minutes. Added 10 µl Hi-Di™ Formamide to each tube and vortex as needed, then centrifuged at 2000×g for 1 min and following a final step of denaturation at 96°C for 3 min. All the 10 µl mix in each tube was transferred into the plate and processed as improved method described.

Nucleotide blast analysis in the Genbank database for species or genus identification in two methods.

Sequences obtained were blasted with the GenBank database (http://www.ncbi.nlm.nih.gov/. 16S Ribosomal RNA Sequences (Bacteria and Archaea)) for species or genus assignment. The highest identity was selected as the identified species or genus.

Improved Sequencing Protocol for Practical Application with Clinical Samples

90 pathogen strains, comprised of 30 samples each of *Pseudomonas aeruginosa,Staphyloccocus aureus* and *Escherichia coli*, were isolated from in-patients admitted to the Shantou Central Hospital (Shantou, China) between May 2012 and July 2012, and identified at the species level also using conventional culture and phenotypic methods by microbiologists before PCR and sequencing. And all the following procedures were performed as the section of improved method in 2.1.2?2.1.7 described.

RESULTS

Optimized Tests of Improved Sanger Sequencing Protocol

In our optimized text, we found that no matter whether 1.2-mm and 2.0-mm disks were dropped with either a higher concentration or a lower concentration of suspension, neither of them could produce an interpretable Cp value in amplification curves, it suggesting 0.5-mm was the most suitable option (Figure 1A). When a series of known quantification from 6×10^4 to 6×10^9 CFU ml^{-1} in 0.5-mm card of AS.26003 *Staphylococcus aureus* strains were built to SYBR Green I PCR, a linear relationship between the Cp and the logarithm of concentration was observed (Figure 1B). The amplification efficiency calculated from these data was 1.98, very close to the theoretical maximal yield 2.

A

Samples				Results		
Include	Color	Pos	Name	CP	Concentration	Stand...
✔		A3	-6_0.5mm	>30.00		
✔		B3	-6_1mm			
✔		C3	-6_2mm			
✔		D3	-1_0.5mm	25.33		
✔		E3	-1_1mm			
✔		F3	-1_2mm			

B

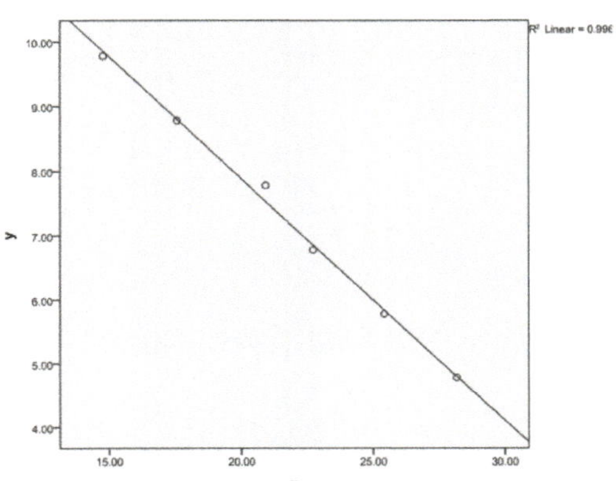

Figure 1: A: Three sizes of diameter punch (0.5-mm, 1-mm, 2-mm) involving higher and lower bacteria concentrations for PCR. −6 stands for 60 CFU ml^{-1}, −1 stands for 6×10^6 CFU ml^{-1}; B: Standard curve of real-time quantitative PCR of 6 decimal dilutions of representative Staphylococcus aureus ranging from 6×10^4 to 6×10^9 CFU ml^{-1}. http://dx.doi.org/10.1371/journal.pone.0088886.g001

The slope of the standard curve is −0.37, and the correlation coefficient is 0.97, generating a regression equation Y(Log concentration)=-0.37X(Cp) +15.442. According to the Cp values and regression equation, we suggest that the best concentration on 0.5-mm FTA® disk should range from 6×10^4 to 6×10^7 CFU ml^{-1}, for the corresponding Cp values were from 20.92 to 28.17. However, either higher or lower concentrations are not recommended, since higher concentration is difficult to prepare and easy to generate cross-contamination, while lower concentration is not sufficient to be amplified.

Comparison Results from 12 Specimens by Using the Two Methods

In this improved method, after the first PCR step, the amplification curves and melting curves of all 12 strains were showed in Figure 2. The curves showed that all strains were performed perfectly, as amplification curves presented Cp value <30, and the melting curves presented a single sharp peak. Instead, the negative control (PCR mix excluding DNA) appeared as a line in which no Cp value was observed. Sometimes, equivocal peaks were observed in curves, but non adverse impact would affect the identified results. In conventional method, PCR products formation was visually confirmed through agarose gel electrophoresis (Figure 3). From lane 1 to lane 12, the specific about 500-bp products were detected and presented an unambiguous visible band in each lane as we expected. In lane 13, it was noteworthy that a faint band was observed and located at about 20 bp fragment, and we confirmed that this minute quantity of amplicon was probably universal primers dimmer, however, it could not be sequenced by itself, and also did not interfere with the16S rRNA fragment and subsequent sequencing. Finally, all the 12 specimens in two methods were successfully sequenced and presented explicit chromatograms for unambiguously distinguishing (the representative diagram of sequence chromatogram and quality referred to Figure 4). In order to evaluate the sequencing quality, we analysed their detailed qulity parameters through DNA Sequencing Analysis software 5.1 (Applied Biosystems). According to the manufacturer's instructions, KB basecaller of sequence generates QV from 1 to 99, with typical high quality bases scoring from 20 and above, typical medium quality bases ranging from 15 to 19, while low quality base scoring less than 15. The LOR (length of read), which were defined as the usable range of high-quality or high-accuracy base sequence, determined by quality values, were 477.4 and 477.8. The average base numbers with low QV were respectively 66.9 and 46.3, high QV were 414 and 444.3. Herein we defined two derived parameters: PLQ (percentage of low QV bases) as percentage of base numbers with low QV to LOR, and PHQ (percentage of high QV bases) as percentage of base numbers with high QV to LOR. The average PLQs of the two methods were 14.0% and 9.69%, average PHQs were 86.7% and 93.0%, sample score (the average quality value of the bases in the clear range sequence for that sample) were 34.1 and 38.1. Statistical texts (using Wilcoxon Matched-Pairs Signed-Ranks Test or Matched-Pairs t-text) showed that all the differences had statistical significance (P<0.05) in PLQ, PHQ and sample score, and we considered that the sequences quality from conventional method was superior to the improved method (Table 1). However, even so, they had no impact on identification results when submitted to Genbank for blasting, in other word, although statistical significance was found in comparison of sequences quality, the blasting results from two methods were still correct and consistent, which respectively 99% or 100% matched the three kinds of strains recorded as NR_026078.1/, NR_037007.1/and NR_074891.1/from NCBI. Notably, the 6th sample *Escherichia coli* had another more similar matching item NR_074894.1 (Table 2), and we would give explanations below.

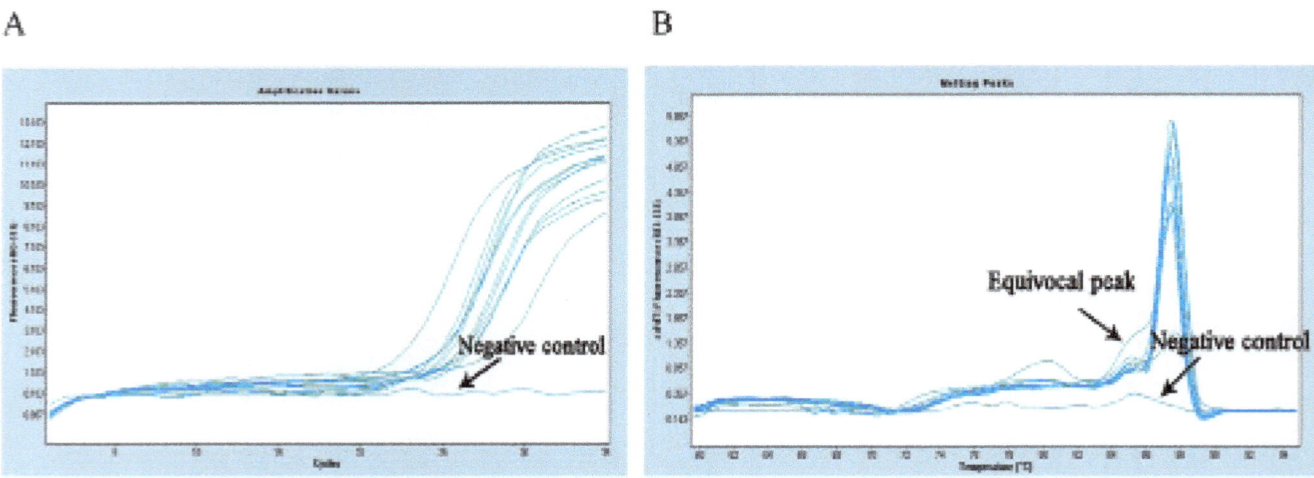

Figure 2: Amplification curves (A) and melting curves (B) of 12 specimens and negative control from improved method. http://dx.doi.org/10.1371/journal.pone.0088886.g002

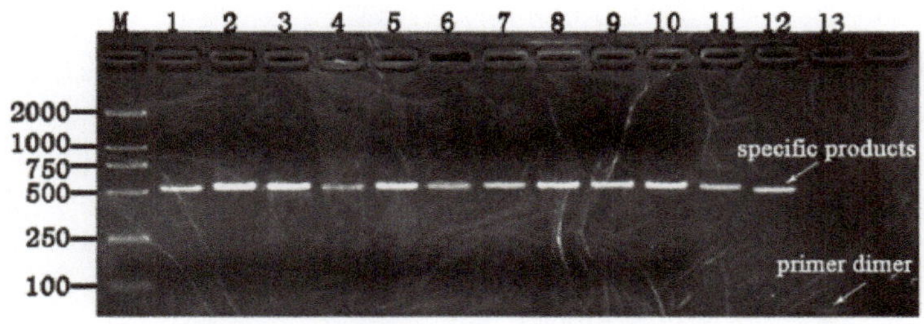

Figure 3: Ethidium-bromide–stained agarose gel of PCR reaction products from 12specimens through conventional method. Lane M, 100-base-pair (bp) DNA ladder; lane 1, reference strain of *Escherichia coli*; lane2, reference strain of *Pseudomonas aeruginosa*; lane 3, reference strain of*Staphylococcus aureus*; lane 4 to lane 12, representative clinical strain of *Escherichia coli 1-3*, *Staphylococcus aureus 1-3* and *Pseudomonas aeruginosa 1-3*; lane 13, negative control without DNA template. http://dx.doi.org/10.1371/journal.pone.0088886.g003

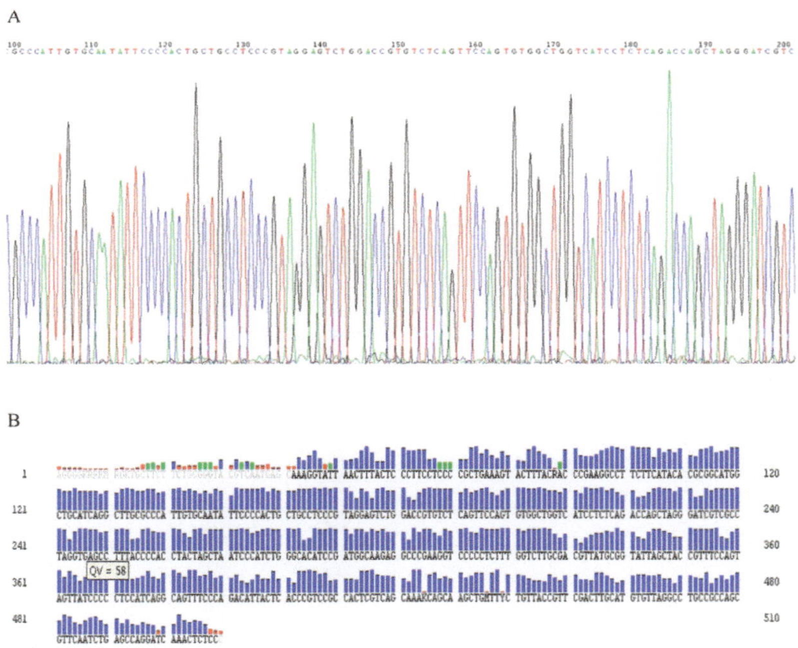

Figure 4: Sequence chromatogram (A) and sequence quality evaluation (B) from clinical Staphylococcus aureus strain 1. Red stands for low quality, green for medium quality, blue for high quality. http://dx.doi.org/10.1371/journal.pone.0088886.g004

Table 1: Sequence quality from the 12 samples in two methods. http://dx.doi.org/10.1371/journal.pone.0088886.t001

Samples description	Improved Sanger sequencing/Conventional Sanger sequencing					
	LOR	Low QV (base numbers)	PLQ (percentage of low QVs.%)	High QV (base numbers)	PHQ (percentage of high QVs.%)	Sample score
1. AS.44113 *Escherichia coli*	463/492	84/37	18.1/7.5	360/454	77.8/92.3	27/37
2. ATCC.27853 *Pseudomonas aeruginosa*	466/466	78/39	16.9/8.4	401/436	86.1/93.6	34/42
3. AS.26003 *Staphylococcus aureus*	501/469	51/39	10.2/8.3	433/437	86.4/93.2	38/34
4. *Escherichia coli 1*	490/479	56/47	11.4/9.8	432/450	88.2/93.9	35/37
5. *Escherichia coli 2*	469/491	85/53	18.1/10.8	401/449	85.5/91.4	36/38
6. *Escherichia coli 3*	476/476	96/36	20.2/7.56	394/460	82.8/96.6	36/42
7. *Staphylococcus aureus 1*	487/478	56/75	11.5/15.7	427/401	87.8/83.9	31/30
8. *Staphylococcus aureus 2*	488/486	68/47	13.9/9.67	418/441	85.7/90.7	29/32
9. *Staphylococcus aureus 3*	500/464	37/57	7.4/12.3	470/437	94.0/94.2	37/34
10. *Pseudomonas aeruginosa 1*	482/469	60/48	12.5/10.2	432/449	89.6/95.7	37/43
11. *Pseudomonas aeruginosa 2*	485/478	51/44	10.5/9.2	434/456	89.5/95.4	34/43
12. *Pseudomonas aeruginosa 3*	422/485	81/33	19.2/6.8	366/462	86.3/95.3	35/45
Population mean	477.4/477.8	66.9/46.3	14.0/9.7	414/444.3	86.7/93.0	34.1/38.1
Statistical text. P value	-	-	<0.05*	-	<0.05*	<0.05#

*using Wilcoxon Matched-Pairs Signed-Ranks Test.
#using Matched-Pairs t-text.
doi:10.1371/journal.pone.0088886.t001

Table 2: Sequencing results of the 12 specimens separately performed by two methods. http://dx.doi.org/10.1371/journal.pone.0088886.t002

Reference strains			Sequence with highest blastn scores	
Description (cases)	Source	(average) Valid Sequence Length (after manually called) improved method/conventional method	%identity improved method/ conventi-onal method	Accessions/Description
Reference strains				
ATCC.27853 *Pseudomonas aeruginosa*	American Type Culture Collection	411/422	99%/99%	NR_026078.1/*Pseudomonas aeruginosa*
AS.26003 *Staphylococcus aureus*	China General Microbiological Culture Collection Center	408/420	100%/100%	NR_037007.1/*Staphylococcus aureus*
AS.44113 *Escherichia coli*		394/410	99%/99%	NR_074891.1/*Escherichia coli*
Clinical strains				
Pseudomonas aeruginosa (3)	urine, pus, sputum or faeces from in-patient	400/412	99%/99%	NR_026078.1/*Pseudomonas aeruginosa*
Staphylococcus aureus (3)		412/422	100%/100%	NR_037007.1/*Staphylococcus aureus*
Escherichia coli (2)		395/405	99%/99%	NR_074891.1/*Escherichia coli*
Escherichia coli (1)		401/414	99%/99%	NR_074894.1 *Shigella sonnei*

doi:10.1371/journal.pone.0088886.t002

Other Miscellaneous Comparison of Two Sequencing Protocols

For the 12 specimens (Table 3), the successful rate for my first experiments was 100%, of course, it should be caution enough for operator performing, especially in conventional method. The workload, time consumption, and the cost per batch with 12 samples were respectively light versus heavy, 8 h versus 11 h and $420 versus $400. Obviously, it was more labor-saving and time-saving if using improved Sanger sequencing, while an advantage in conventional Sanger sequencing was that it cost less. However, we would rather recommend the former method than the latter, which was an inconvenient job indeed.

Table 3: Comparison of several items between two methods with 12 specimens. http://dx.doi.org/10.1371/journal.pone.0088886.t003

	Comparison items			
Methods	successful rate (once experiment)	workload	Time consumption per batch	Cost per batch
Improved Sanger sequencing	100%	Light	8 h	$420
Conventional Sanger sequencing	100%	Heavy	11 h	$400

doi:10.1371/journal.pone.0088886.t003

Results of 90 Clinical Isolates by Using the Improved Sequencing Protocol

Among the 90 real-time PCR amplifications performed on the experimental isolates, all amplification curves were considered as positive with Cp values ranged from 20.15 to 29.55 (mean: 25.12). From the 90 melting curves, 70 showed a single peak with a Tm value of 88°C as reference strains', so the corresponding products were regarded as the purest products and were the most suitable for subsequent sequencing. The other 20 showed dual peaks including a high main peak with a Tm of 88°C, and a minor equivocal peak with Tm of approximate 85°C (Figure 5). Since the equivocal peaks were much smaller than the main peaks, we considered that the amount of "unknown products" was far less than the 16S rRNA fragment products, and inferred they would not interfere with the following sequencing process [11]. Furthermore, after agarose gel electrophoresis, the signals produced by "unknown products" were relatively weak or were only seen in the absence of DNA as input.

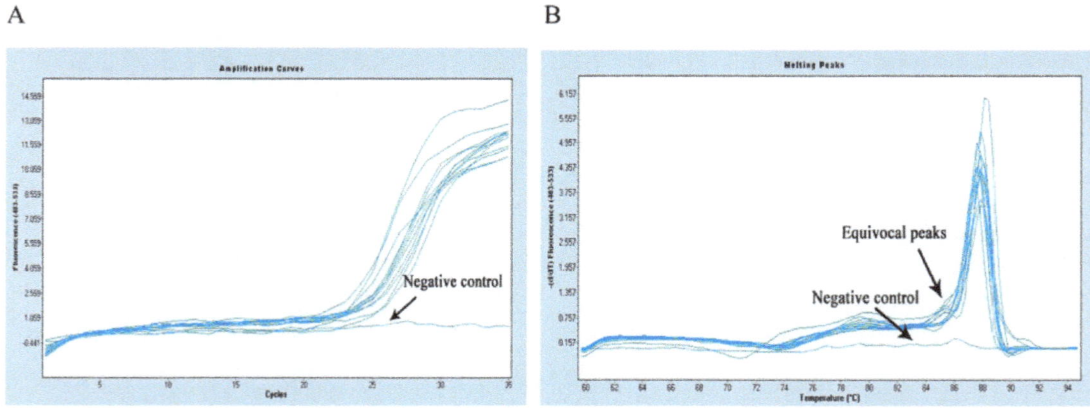

Figure 5: Amplification curves (A) and melting curves (B) of partial experimental strains. http://dx.doi.org/10.1371/journal.pone.0088886.g005

After screening based on their amplification curves, all of the clinical samples entered into a rapid and simplified sequencing pipeline. 90 PCR products were ultimately sequenced and gave valid read length between a minimum of 367 nucleotides and a maximum of 481 nucleotides, with a mean ± standard deviation length of 404.5±20.5 bases, and some ambiguous bases that could be manually called. Besides, the sequence quality value showed that average base numbers with low QV were 68.3, high QV were 420, LOR were 475.6, PLQ were 14.4%, PHQ were 88.4% and sample score was 35.2≥20, ≤1% probability of a miscalled base. According to Tewari D et al. [12], isolates were categorized into either a species with ≥99% match, a genus with ≥95% match, or a higher taxon with <95% having criteria described previously. Therefore, in 90 experimental samples, identification of *Pseudomonas aeruginosa*and *Staphylococcus aureus* to the species level of the best match in the Genbank database corresponded to 100% of the organisms identified by conventional microbiological methods. But from 30 *Escherichia coli* samples, the best matching strains with significant alignment, respectively were 8 *Escherichia coli* (26.7%), 10 *Shigella sonnei* (33.3%) and 12 *Shigella dysenteriae* (40%) (Table 4).

Table 4: Identification results of clinical pathogen strains. http://dx.doi.org/10.1371/journal.pone.0088886.t004

Species identified by conventional techniques (strain numbers)	Valid Sequence Length (after manually called)	Sequence with highest blastn scores	
		%identity	Accessions/Description
Pseudomonas aeruginosa (30)	383–412	99%–100%	NR_074828.1 *Pseudomonas aeruginosa*
Staphylococcus aureus (30)	386–481	99%–100%	NR_037007.1 *Staphylococcus aureus*
Escherichia coli (8)	396–415	99%	NR_074891.1 *Escherichia coli*
Escherichia coli (10)	367–412	99%	NR_074894.1 *Shigella sonnei*
Escherichia coli (12)	371–477	99%	NR_074892.1 *Shigella dysenteriae*

doi:10.1371/journal.pone.0088886.t004

Specially, for the 30 specimens of *Escherichia coli* strains with discordant blasting, sequence similarity would be assessed further by constructing a relatedness diagram (phylogenetic tree) using a minimum of 1,000 bootstrap trees. 31 *Escherichia coli* sequences (30 clinical specimens and 1 reference specimens) and 3 corresponding best matching sequences (*Escherichia coli*, NR_074891.1; *Shigella sonnei*, NR_074894.1; *Shigella dysenteriae*, NR_074892.1) from Genbank would join the construction (Figure 6). It is noticeable that in this phylogenetic tree, the 3 sequences from Genbank were too similar to be separated, and were unable to get close to those specimens that were respectively matched, and compared to the previous study using 16S rRNA gene sequence to construct Phylogenetic tree between *Shigella* and *Escherichia coli* [13], the giving results verified the results of our experiment.

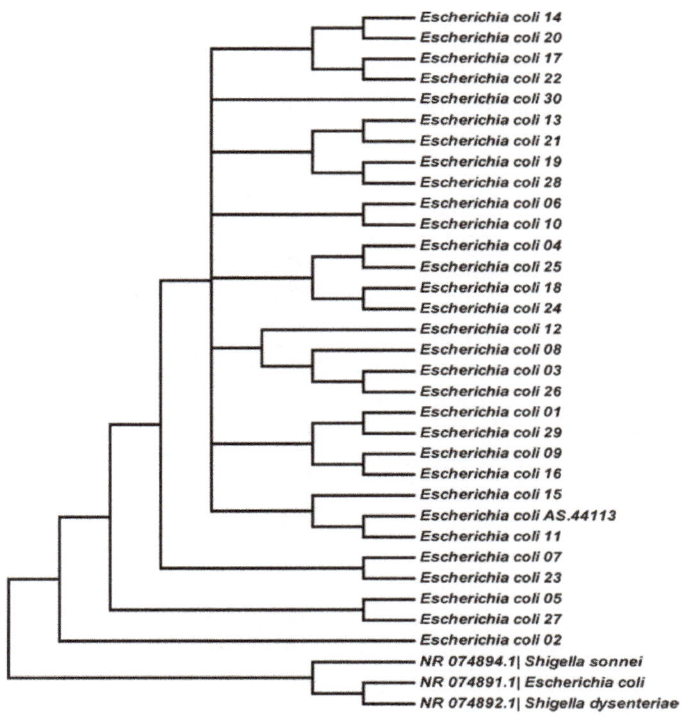

Figure 6: Dendrogram showing the phylogenetic relationships among *Escherichia coli* sequences based on 16S rDNA partial genes. The tree was constructed by using the neighbor-joining method with Mega 5.02. http://dx.doi.org/10.1371/journal.pone.0088886.g006

DISCUSSION

In recent years, instead of traditional biochemical methods, Sanger sequencing is becoming increasingly popular [14], however, the traditional processes are still relatively troublesome, but these drawbacks will be overcame after our improvement. In the comparative test, when evaluating the sequences quality of both methods, with parameters of PLQ, PHQ and sample score, statistical differences were found, we submitted all the sequences to NCBI for blasting, however, the highest blastn scores in two methods were equal and the identification results were both correct and consistent. These convincing data have illustrated the utility of the improved Sanger sequencing we proposed. In addition, the turnaround time required for real-time PCR, Sanger sequencing, capillary electrophoresis reaction and data analysis plus labor consumption needed to complete 500 bp sequencing in improved method which was less than 8 h per batch, including 12 specimens, while additional 3 h per batch should be provided in the latter method. The cost per batch was $36 for DNA captured by FTA® card, and $384 for amplification reagents, sequencing, purification of products and capillary electrophoresis (excluding instrument and labor costs), while it would cost less in the latter method due to avoiding some reagent. However, much more time and labor should be needed during the procedure of DNA extraction, agarose gel electrophoresis, and products purification. Even worse, in the processing of agarose gel electrophoresis, we were unable to prevent the toxicity of ethidium bromide which is a kind of strong carcinogen. While SYBR Green I added in PCR reaction was able to check the effectiveness of PCR safely, save time and reduce workload as well. The DNA templates for PCR, FTA® card with bacterial suspension was directly PCR amplified in improved method, while DNA extracting from bacterial suspension in the latter method, no doubt, the former choice was a more convenient method, and it also reduced the risks of microbial contamination. In the step of products purification, improved method just needed reagent for simple mix and slight centrifugation, while the extraordinary laborious operation, such as oft-repeated high speed centrifugation and oft-repeated removing supernatant carefully, were essential in the conventional method.

Additionally, we had applied expanded specimens to assess the utility of our new improved method. SYBR Green ? releases intense fluorescence only when combined with double-stranded DNA, but does not emit detectable fluorescence, which were detected upon DNA denaturation, it is a non-specific indicator dye [15]. For this reason, the combination of primers and SYBR Green ?leads to some equivocal melting curves, but their Cp values still remain at an acceptable level, and agarose gel electrophoresis results of the corresponding products also emerged as a well-defined pattern of bands, so we still have sequenced them. Though, compared to others, the final 20 chromatograms appeared to be devoid of some additional discernible bases, with a QV larger than 20, high-quality sequences were still acquired, and matches were still obtained when submitted to the Genbank blast system, supporting the report that some interference within products was not completely eliminated or impacted by primer formation [16].

From the identification results of pathogenic strains, we learn that partial 16S rRNA gene sequencing is a suitable tool for *Staphylococcus aureus* and *Pseudomonas aeruginosa* identification, which have produced consistent results with conventional culture methods as others have done [17], [18]. However, 30 *Escherichia coli* specimens generated 3 blast results of *Shigella sonnei*, *Shigella dysenteriae* and *Escherichia coli*, and the 16S rDNA-based phylogenetic tree suggested that it was difficult to distinguish each of them. It has been demonstrated by other researchers that there are many similarities in many respects between some *Shigella* (e.g. *Shigella dysenteriae*) and *Escherichia coli* (e.g. Enteroinvasive *Escherichia coli*), such as clinical symptoms, biochemical characteristics and antigens [19]. In fact, previous study showed that a few *Escherichia coli* have been assigned to a different genus (e.g., *Shigella flexneri*), based primarily on their distinct clinical presentation and their importance as human pathogens [20]. A research by Pupo et al. [21], analyzing sequence variation in housekeeping genes, also showed that most *Shigella* serotypes fall into three clusters within *Escherichia coli*, proving that, it is indeed difficult to distinguish *Shigella* from*Escherichia coli*. So the false identification results in some *Escherichia coli* of our specimens may attribute to the false classification of *Escherichia coli* sequences, which were virtually *Shigella* sequences submitted to GenBank by other researchers.

Compared with conventional Sanger sequencing, our improved protocol has emerged as a faster and more convenient method to identify those common bacteria. However, it also should be applied cautiously. Firstly, although sequencing is particularly helpful in situations where organisms are difficult to characterize by using conventional culture methods, but 1 to 14% of the isolates remain unidentified after testing [22]. Secondly, the variable regions, as a foundation for discriminating bacteria, only distributing V1–V3 in the first 500 bp area, is one third of full-length of 16S gene (V1–V10) [23]. This system uses universal primers to amplify and sequence a 500 bp fragment from the 5′-terminus of the 16S rRNA gene [24], but only a mean of 404 bp is read, because the first approximately 100 bp had to be manually discarded owing to residual SYBR Green?left over from PCR products, and was difficult to be removed by purification kit. Consequently the V1, distributed in the first 104 bp, have to be discarded and hence slightly impaired the discrimination ability of the sequencing chromatogram. Lastly, though SYBR Green?does not require specific probes to be developed, as is the case for some other detection chemistries. However, the detection specificity of SYBR green I assays depends entirely on the PCR primers [25], suggesting that it is important to ensure the high specificity of primers, so negative control in PCR reaction should be needed.

In the future, 16S rRNA gene sequencing will continue to be the gold standard for identification of most bacteria [26], and better automation of such an improved technology may put it into routine use in large microbiology laboratories. The assay described here is a suitable tool for sequencing identification of *Pseudomonas aeruginosa* and *Staphyloccocus aureus* faster and more conveniently, but it is not completely accurate to discriminate *Escherichia coli* and*Shigella* strains. Under optimal conditions, the protocol can be applied for any PCR and sequence-based analysis after proper modification. The time-consumption and the cost remain acceptable for most laboratories, and will become further reduced as the technology becomes more widely adopted and refined. However, despite the fact that it is difficult to accurately assign some particular isolates to a specific species, assigning to a certain genus can successfully assist the further research [27].

ACKNOWLEDGMENTS

We are very grateful to Shantou University Medical College, Shantou Entry-Exit Inspection and Quarantine Bureau and Shantou Central Hospital as well as all the people participating in this work. Many thanks to Dr. Jianjun Zhang for correcting language.

REFERENCES

1. Gustavsson I, Lindell M, Wilander E, Strand A, Gyllensten U (2009) Use of FTA card for dry collection, transportation and storage of cervical cell specimen to detect high-risk HPV. Journal of Clinical Virology 46: 112–116. doi: 10.1016/j.jcv.2009.06.021
2. Mullen M, Howard D, Powell R, Hanrahan J (2009) A note on the use of FTA™ technology for storage of blood samples for DNA analysis and removal of PCR inhibitors. Irish Journal of Agricultural and Food Research: 109–113.
3. Schuurman T, De Boer RF, Kooistra-Smid A, van Zwet AA (2004) Prospective study of use of PCR amplification and sequencing of 16S ribosomal DNA from cerebrospinal fluid for diagnosis of bacterial meningitis in a clinical setting. J Clin Microbiol 42: 734–740. doi: 10.1128/jcm.42.2.734-740.2004
4. Smith L, Burgoyne LA (2004) Collecting, archiving and processing DNA from wildlife samples using FTA® databasing paper. BMC ecology 4: 4.
5. Aye KS, Matsuoka M, Kai M, Kyaw K, Win AA, et al. (2011) FTA card utility for PCR detection of Mycobacterium leprae. Jpn J Infect Dis 64: 246–248.
6. Fukushima H, Tsunomori Y, Seki R (2003) Duplex real-time SYBR green PCR assays for detection of 17 species of food- or waterborne pathogens in stools. J Clin Microbiol 41: 5134–5146. doi: 10.1128/jcm.41.11.5134-5146.2003
7. Menassa N, Bosshard PP, Kaufmann C, Grimm C, Auffarth GU, et al. (2010) Rapid detection of fungal keratitis with DNA-stabilizing FTA filter paper. Invest Ophthalmol Vis Sci 51: 1905–1910. doi: 10.1167/iovs.09-3737
8. Corless CE, Guiver M, Borrow R, Edwards-Jones V, Kaczmarski EB, et al. (2000) Contamination and sensitivity issues with a real-time universal 16S rRNA PCR. J Clin Microbiol 38: 1747–1752.
9. Hall L, Doerr KA, Wohlfiel SL, Roberts GD (2003) Evaluation of the MicroSeq system for identification of mycobacteria by 16S ribosomal DNA sequencing and its integration into a routine clinical mycobacteriology laboratory. J Clin Microbiol 41: 1447–1453. doi: 10.1128/jcm.41.4.1447-1453.2003
10. Peattie DA (1979) Direct chemical method for sequencing RNA. Proc Natl Acad Sci U S A 76: 1760–1764. doi: 10.1073/pnas.76.4.1760
11. Han XY, Pham AS, Tarrand JJ, Sood PK, Luthra R (2002) Rapid and accurate identification of mycobacteria by sequencing hypervariable regions of the 16S ribosomal RNA gene. Am J Clin Pathol 118: 796–801. doi: 10.1309/hn44-xqym-jmaq-2edl
12. Tewari D, Cieply S, Livengood J (2011) Identification of bacteria recovered from animals using the 16S ribosomal RNA gene with pyrosequencing and Sanger sequencing. Journal of Veterinary Diagnostic Investigation 23: 1104–1108. doi: 10.1177/1040638711425583
13. Fukushima M, Kakinuma K, Kawaguchi R (2002) Phylogenetic analysis of Salmonella, Shigella, and Escherichia coli strains on the basis of the gyrB gene sequence. J Clin Microbiol 40: 2779–2785. doi: 10.1128/jcm.40.8.2779-2785.2002
14. Killgore G, Thompson A, Johnson S, Brazier J, Kuijper E, et al. (2008) Comparison of seven techniques for typing international epidemic strains of Clostridium difficile: restriction endonuclease analysis, pulsed-field gel electrophoresis, PCR-ribotyping, multilocus sequence typing, multilocus variable-number tandem-repeat analysis, amplified fragment length polymorphism, and surface layer protein A gene sequence typing. J Clin Microbiol 46: 431–437. doi: 10.1128/jcm.01484-07
15. Willi B, Meli ML, Luthy R, Honegger H, Wengi N, et al. (2009) Development and application of a universal Hemoplasma screening assay based on the SYBR green PCR principle. J Clin Microbiol 47: 4049–4054. doi: 10.1128/jcm.01478-09
16. Vliegen I, Jacobs JA, Beuken E, Bruggeman CA, Vink C (2006) Rapid identification of bacteria by real-time amplification and sequencing of the 16S rRNA gene. J Microbiol Methods 66: 156–164. doi: 10.1016/j.mimet.2005.11.005
17. Becker K, Harmsen D, Mellmann A, Meier C, Schumann P, et al. (2004) Development and evaluation of a quality-controlled ribosomal sequence database for 16S ribosomal DNA-based identification of Staphylococcus species. J Clin Microbiol 42: 4988–4995. doi: 10.1128/jcm.42.11.4988-4995.2004
18. Bhattacharya D, Sarma PM, Krishnan S, Mishra S, Lal B (2003) Evaluation of genetic diversity among Pseudomonas citronellolis strains isolated from oily sludge-contaminated sites. Appl Environ Microbiol 69: 1435–1441. doi: 10.1128/aem.69.3.1435-1441.2003
19. Li Y, Liu D, Cao B, Han W, Liu Y, et al. (2006) Development of a serotype-specific DNA microarray for identification of some Shigella and pathogenic Escherichia coli strains. J Clin Microbiol 44: 4376–4383. doi: 10.1128/jcm.01389-06
20. Lan R, Reeves PR (2001) When does a clone deserve a name? A perspective on bacterial species based on population genetics. Trends Microbiol 9: 419–424. doi: 10.1016/s0966-842x(01)02133-3
21. Pupo GM, Lan R, Reeves PR (2000) Multiple independent origins of Shigella clones of Escherichia coli and convergent evolution of many of their characteristics. Proceedings of the National Academy of Sciences 97: 10567–10572. doi: 10.1073/pnas.180094797

22. Janda JM, Abbott SL (2007) 16S rRNA gene sequencing for bacterial identification in the diagnostic laboratory: pluses, perils, and pitfalls. J Clin Microbiol 45: 2761–2764. doi: 10.1128/jcm.01228-07

23. Petti CA (2007) Detection and identification of microorganisms by gene amplification and sequencing. Clin Infect Dis 44: 1108–1114. doi: 10.1086/512818

24. Cloud JL, Conville PS, Croft A, Harmsen D, Witebsky FG, et al. (2004) Evaluation of partial 16S ribosomal DNA sequencing for identification of Nocardia species by using the MicroSeq 500 system with an expanded database. J Clin Microbiol 42: 578–584. doi: 10.1128/jcm.42.2.578-584.2004

25. Monnet C, Correia K, Sarthou AS, Irlinger F (2006) Quantitative detection of Corynebacterium casei in cheese by real-time PCR. Appl Environ Microbiol 72: 6972–6979. doi: 10.1128/aem.01303-06

26. Woo PCY, Ng KHL, Lau SKP, Yip K, Fung AMY, et al. (2003) Usefulness of the MicroSeq 500 16S ribosomal DNA-based bacterial identification system for identification of clinically significant bacterial isolates with ambiguous biochemical profiles. J Clin Microbiol 41: 1996–2001. doi: 10.1128/jcm.41.5.1996-2001.2003

27. Woo PCY, Cheung EYL, Leung K, Yuen K (2001) Identification by 16S ribosomal RNA gene sequencing of an Enterobacteriaceae species with ambiguous biochemical profile from a renal transplant recipient. Diagn Microbiol Infect Dis 39: 85–93. doi: 10.1016/s0732-8893(01)00206-1

CHAPTER 6

ACCURACY AND QUALITY OF MASSIVELY PARALLEL DNA PYROSEQUENCING

Susan M Huse, Julie A Huber, Hilary G Morrison, Mitchell L Sogin and David Mark Welch

BACKGROUND

Direct interrogation of microbial genomes based upon comparisons of orthologous gene sequences or metagenomic surveys provides a means to assess the diversity of microbial communities without requiring the cultivation of microbes in the laboratory. Since the cost of cloning DNA templates and capillary-based DNA sequencing constrains the number of sequences included in most of these investigations, the detection of low abundance taxa demands surveys that are many orders of magnitude larger than those reported in the literature. Massively parallel pyrosequencing on the Roche GS20 system developed by 454 Life Sciences offers a means to more extensively sample molecular diversity in microbial populations. It is now possible to generate hundreds of thousands of short (100-200 nucleotide) DNA sequence reads in a few hours without requiring the preparation of sequence templates by conventional cloning. In the near future, technical advances will likely increase the number and length of sequence reads.

Pyrosequencing technology relies upon enzyme cascades and CCD luminescence detection capabilities to measure the release of inorganic pyrophosphate with every nucleotide incorporation [1]. The GS20 system takes advantage of DNA capture beads that contain, on average, one single-stranded template, which is amplified to millions of copies in an oil emulsion PCR (emPCR). The beads are then distributed on a solid-phase sequencing substrate (a PicoTiterPlate™) with 1.6 million wells that can each hold a bead and additional reagents, including polymerase, luciferase, and ATP sulfurylase. Microfluidics cycle each of the four nucleotide triphosphates over the PicoTiterPlate™, and incorporation of a nucleotide releases pyrophosphate, the substrate for a luminescence reaction, which is recorded with a cooled CCD camera. The record of intensity of each flow of a nucleotide is a flowgram, analogous to a chromatogram that reports the order of A, C, G and T residues from a DNA sequencing template. Flowgram values correspond to the homopolymer length for that base. The average number of wells with detectable sequencing templates is about 450,000, which produces about 200,000 usable reads. This new methodology brings with it different sources of error to traditional dideoxy capillary sequencing. Since the nucleotide triphosphates are flowed one at a time, substitutions are less likely than with traditional methods. However, it is sometimes difficult to resolve the intensity of luminescence produced when a homopolymer is encountered. The result can be ambiguity of homopolymer length, particularly for longer homopolymers. In addition, insufficient flushing between flows can cause single base insertions (carry forward events) usually near but not adjacent to homopolymers. Insufficient nucleotides within a flow can cause incomplete extension within homopolymers. Generally, an excess of intermediate flowgram values indicates a poor quality read [2]. The GS20 software makes corrections for carry forward and incomplete

Citation: Susan M Huse, Julie A Huber, Hilary G Morrison, Mitchell L Sogin and David Mark Welch, Accuracy and quality of massively parallel DNA pyrosequencing, doi:10.1186/gb-2007-8-7-r143.

extensions (CAFIE); it shortens reads from the 3' end until fewer than 3% of the remaining flowgram values are of intermediate value, and it removes reads if the trimming falls below a threshold length. The software identifies as ambiguous flow cycles in which no flowgram value was greater than 0.5. If 5% or more of the flow cycles for a read are ambiguous, the read is removed.

The assembly of many overlapping pyrosequencing reads can produce highly accurate consensus sequences [3, 4]. Wicker *et al.* [5] compared assemblies of the barley genome produced by reads from GS20 pyrosequencing and from ABI dideoxy sequencing. Both methods produced consensus sequences with error rates of approximately 0.07% at each consensus position. Gharizadeh *et al.* [6] compared pyrosequences with Sanger dideoxy methods for 4,747 templates. Comparisons of the traditional capillary sequences with the 25-30 nucleotide pyrosequence reads demonstrated similar levels of read accuracy. Assemblies of massively parallel pyrosequencing reads of plastid genomes from *Nandina* and *Platanus* exhibited overall error rates of 0.043% and 0.031%, respectively, in the consensus sequence [4]. The generation of consensus sequences to improve accuracy, however, is generally not appropriate for studies that seek information about natural variation from every read. For example, in metagenomic [7] or PCR amplicon [8] libraries from environmental DNA samples, each sequence read can theoretically represent DNA from a distinct gene from a complex mixture of microbial genes.

A viable but imperfect alternative to building consensus sequences for metagenomic and diversity investigations is to identify and remove pyrosequencing reads that are likely to be incorrect. For example, Gilbert *et al.* [9], in a study of ancient woolly mammoth mitochondrial DNA, removed pyrosequencing reads that were not 98% identical to previously sequenced mammoth mitochondrial DNA sequences, assuming that they must be poor quality. Dostie *et al.* [10] sequenced an amplicon library and discarded reads in which the PCR primer was not recognized by BLAST. These studies removed 15% and 7% of their reads, respectively, but it is not clear that these statistics improved the quality of the remaining data.

To explore error modalities, we used the GS20 system to generate more than 340,000 reads from a PCR amplicon library that was prepared from a collection of 43 reference templates of known sequence. Each reference template contains a distinct ribosomal RNA gene (rDNA), including the V6 hypervariable region from a collection of 43 divergent bacteria [11]. Differences between pyrosequences and their cognate reference sequences identified signatures of low quality data.

RESULTS

Read Accuracy

We obtained 340,150 reads that passed the GS20 quality filters, that is, flowgrams for each read: contained the correct base key at the start (a portion of the 454 primer used to differentiate reads from internal quality control sequences); included at least 84 flows; had fewer than 5% of flow cycles resulting in an ambiguous base call (N); and had fewer than 3% of flowgram values between 0.5 and 0.7 [12]. We aligned each read to its reference sequence using an optimized Needleman-Wunsch algorithm. Our data included 159,981 total errors over 32,801,429 bases. The error rate, defined as the number of errors (miscalled bases plus inserted and deleted bases) divided by the total number of expected bases, was 0.49%. As shown in Table 1, 39% of these errors correspond to homopolymer effects, including extension (insertions), incomplete extensions (deletions) and carry forward errors (insertions and substitutions). Carry forward occurs when an incomplete flush of base flow results in a premature incorporation of a base. The presence of homopolymers tends to increase the likelihood of both carry forward and incomplete extension with the GS20 sequencer [12]. Insertions were the most common type of error (36% of errors) followed by deletions (27%), ambiguous bases, Ns (21%), and substitutions (16%). It should be noted that the V6 region does not contain long or frequent homopolymers. The errors did not correlate significantly with distance along the sequence ($R^2 < 0.03$), indicating that, for the retained reads, there was not significant degradation of sequencing quality as the run progressed. Similar to the results of Margulies *et al.* [2], the error rate for test fragments (sequences used for GS20 diagnostics and included in all runs as part of the reagent stream), was much lower than for the reads, 0.1% bases in error in our experiment.

Table 1: Types of error

Error type	Number of occurrences	Percent of errors	Error rate
Insertions	58,337	36%	0.18%
Homopolymer extension and CAFIE	32,858	20%	0.10%
Not associated with homopolymers	25,479	16%	0.08%
Deletions	43,107	27%	0.13%
Incomplete homopolymer extension	13,868	9%	0.04%
Not associated with homopolymers	29,239	18%	0.09%
Mismatches	25,281	16%	0.08%
Homopolymer extension and CAFIE	16,725	10%	0.05%
Not associated with homopolymers	8,556	5%	0.03%
Ambiguous base calls (N)	34,184	21%	0.10%
Read errors	Number of occurrences	Cumulative percent of reads	Percent of reads

Reads with no errors (perfect match)	279,468	82%	82%
Reads with no more than one error	35,813	93%	11%
Reads with no more than two errors	11,651	96%	3%
Reads with more than two errors	13,218	100%	4%

Errors were classified as insertions, deletions, mismatches (substitutions) and ambiguous base calls (Ns). We further classified insertions, deletions and mismatches by their association with homopolymer effects. Deletions corresponding to an adjacent base are considered incomplete extension. Insertions and mismatches of the same base as an adjacent base are extensions. Insertions and mismatches that are the same as an upcoming homopolymer with no more than two intervening bases are considered carry forward errors.

Two transition mismatches, A to G and T to C, were more frequent than other mismatches, but the reverse transitions, G to A and C to T, were not. Nearly 70% of the homopolymer extensions were A/T. After correcting for the relative number of homopolymers of each type (A/T versus C/G) in the reference sequences (45% and 55%, respectively), the prevalence of A/T extensions was 24% higher than expected, and C/G was concomitantly lower. Although substantially less common, the carry-forward errors, where a base is inserted or substituted ahead of a homopolymer run of the same base (for example, GACTGGG could become GACGTGGG with a carry forward insertion of a G) reflected the same relative increase of A/T over C/G.

Read quality

While the errors were evenly distributed along the length of the reference sequences, they were not evenly distributed among reads. Of the reads, 82% had no errors, 93% had no more than a single error, and 96% had no more than 2 errors. Conversely, a small number of reads, fewer than 2%, contained a disproportionate number of errors that account for nearly 50% of the miscalls for the entire dataset (Figure 1). We identified several diagnostic features that correlate well with the presence of a large number of errors in a read. These include average quality, length, and the presence of ambiguous base calls.

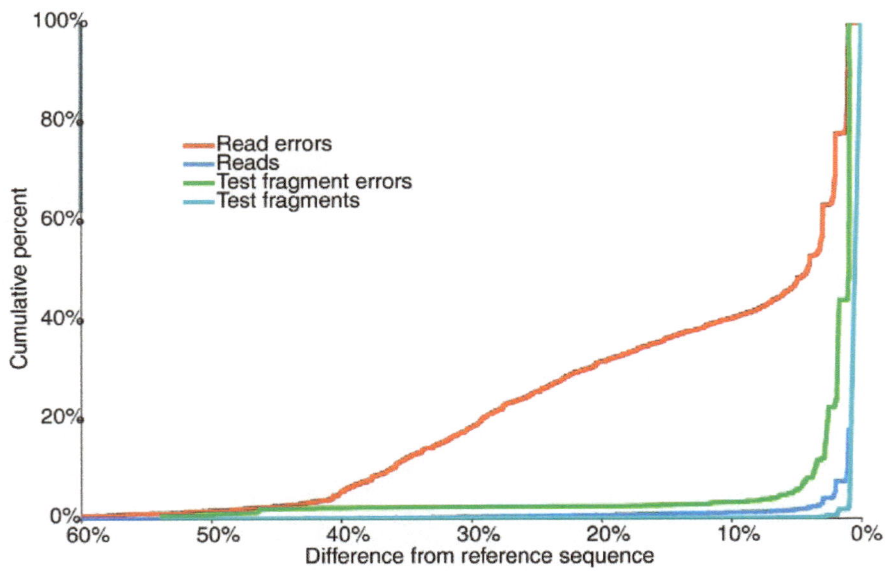

Figure 1: Low quality reads contribute disproportionately to the overall error rate. The graph shows the proportion of reads and test fragments at each percent difference from their reference sequence (individual error rate) and the proportion of errors contributed by reads and test fragments at each given difference (cumulative error rate). The vast majority of both experimental and test fragment reads contain few or no errors; only 5% of all reads and 0.6% of test fragments differ from their reference sequence by 2% or more. The experimental reads that have errors, however, are likely to have a large number of errors and thus be quite different from their reference sequence. For instance, 40% of all errors are from the 1% reads differing by at least 10% from their reference sequence. The GS20 test fragments, by contrast, show far fewer very low-quality sequences: only 3% of the test fragment errors are from sequences at least 10% different from their reference.

The output from the GS20 includes a quality score for every position in a read, and we found that the average quality score of a read is inversely proportional to the number of errors in that read (Figure 2a). In our data, reads with an average quality score above 25 had very few errors, but the number of errors per sequence was noticeably higher for any read where the average of quality scores fell below 25 (Figure 2b). However, the quality score of a position is not a measure of the confidence that the correct base is called at that position, as with a traditional PHRED score [13]. Instead, the GS20 quality score is a measure of confidence that the homopolymer length at that position is correct [2]. The quality scores drop as the homopolymer length increases, even when those homopolymers are correctly called (Figure 2c). Thus, as the number and length of homopolymers in a sequence increase, the average quality score decreases regardless of error (Figure 2d). This is evident even for V6 regions of rDNAs that contain relatively few and short homopolymers. The range of average quality scores is very narrow, from 24.7 to 26.4, compared to the range of quality scores for longer homopolymers, which range from 5 to 26. Predicting an average quality cutoff may not be feasible with data of mixed or unknown homopolymer lengths and frequencies.

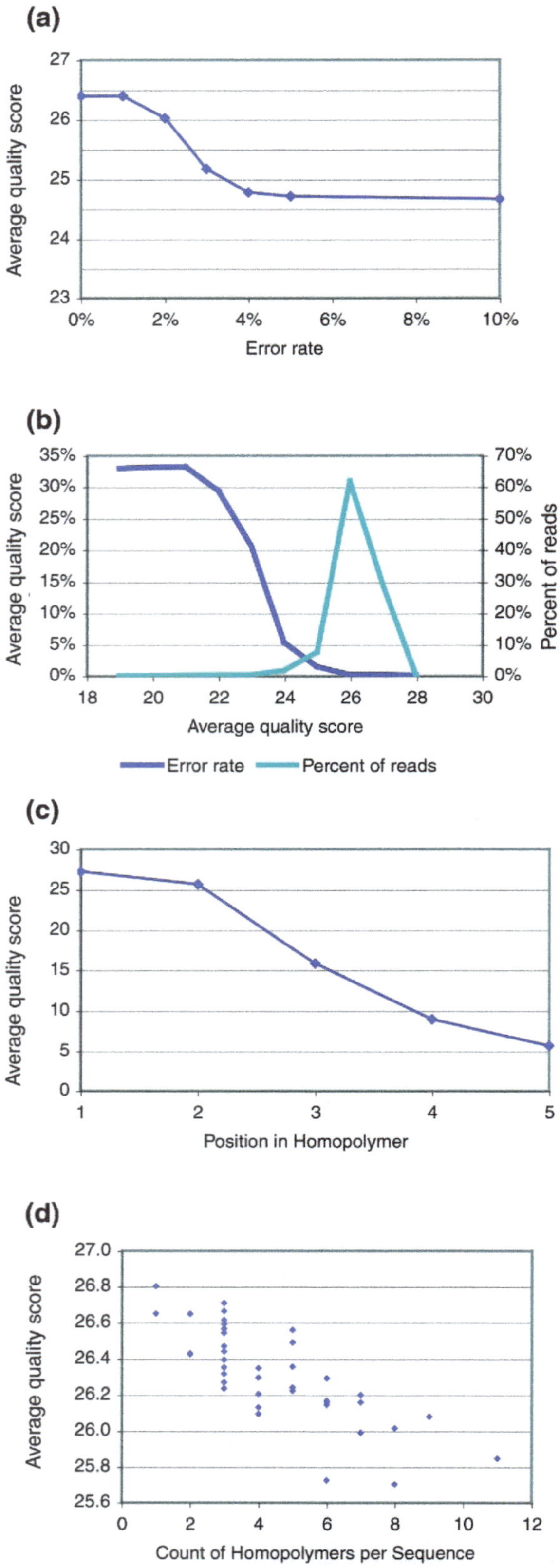

Figure 2: Quality scores are of limited use in predicting accuracy of unknown sequences. The quality scores reported by the GS20 software correlate with decreased confidence in calling the correct homopolymer length rather than the accuracy of the called bases. **(a, b)** The average quality score of reads decreases as the number of errors in the read increases. **(c)** The average quality score as a function of position in the homopolymer: as the length of the homopolymer increases, the quality scores decrease, for both correctly and incorrectly called bases. **(d)** The average quality scores of perfect reads containing differing numbers of homopolymers. The average quality scores decrease with the number of homopolymers. Our sequences contain only short homopolymers, primarily 3-mers. As the length and frequency of homopolymers increases, the expected quality scores will decrease. Without *a priori*knowledge of the number and length of homopolymers in a particular read, it will be difficult to assess an appropriate quality threshold - a low threshold may not cull data adequately and a high threshold may remove homopolymeric regions.

Reads that are much shorter or much longer than their predicted length based on the GS20 flow order also had a very large proportion of errors, while those within a few bases of the expected length had very few errors (Figure 3). We compared read length and average error rate for each of the reference sequences and found that the distribution of high-quality reads peaked at the optimal length as well as a few common shorter lengths, while reads not associated with these clear peaks have much larger average error rates.

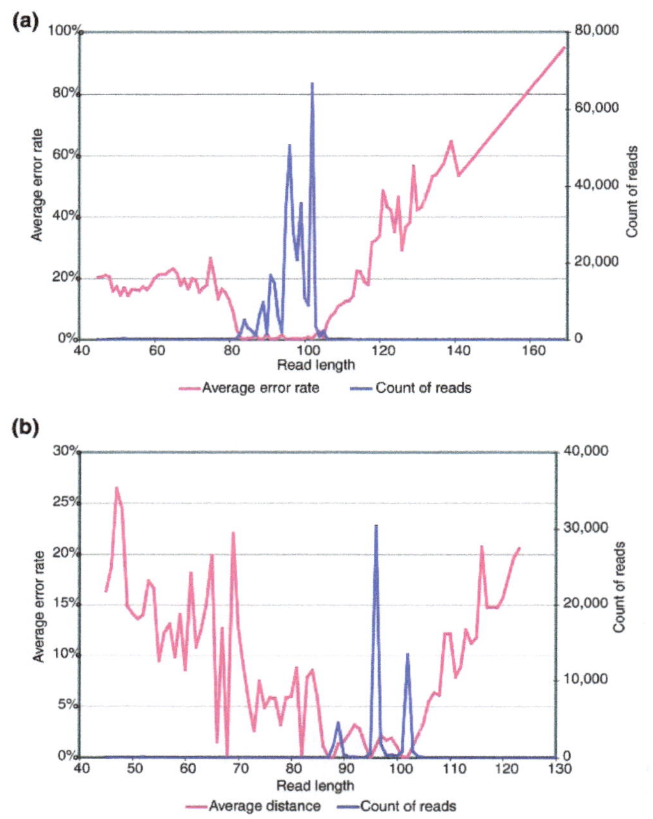

Figure 3: Error rates increase as read length diverges from predicted. The graphs show the average difference from the reference sequence for all reads of a given length, and the distribution of read lengths for all reads. The majority of reads peak at a few specific lengths. The number of reads beyond the peaks shown are too few to appear on the graph; however, they contain many more errors than the reads of the majority length(s). Perfect reads peak at only a few specific lengths. Sequences that fall outside of these lengths are unlikely to be truncated sequences or to have sequenced beyond the end of the primer. Instead they tend to be low-quality reads of spurious sequences. **(a)** The average error rate of sequences at each length for 56,700 reads of reference sequence 517. **(b)** The average error rate of sequences at each length for all reads combined. Even with a mixture of sequence lengths, the reads outside of the peak lengths are highly error prone.

WE observed a strong correlation between the presence of ambiguous base calls and other errors in a read (Figure 4). The presence of even a single ambiguous base in a read correlates strongly with the presence of other errors. Ambiguous base calls result from a failure of the GS20 system to identify any base at an individual position throughout an entire flow cycle. The GS20 bioinformatics pipeline removes reads having more than 5% of the flows resulting in an N; however, this can still result in reads with as many as 9 ambiguous bases calls, and many thousands with 1-2 ambiguous calls. We found that a more stringent threshold than that incorporated into the quality filtering software can substantially improve the quality. After excluding all reads containing ambiguous bases, the error rate is 0.24%, whereas the error rate of reads containing one or more Ns is 4.7% (n = 19,223, 5.6% of all reads) accounting for 54% of the errors. We reexamined the data from each of the two runs separately and found that although their initial error rates did differ slightly (0.58% versus 0.32%), they were essentially the same after removal of reads containing Ns (0.23% and 0.27%, respectively).

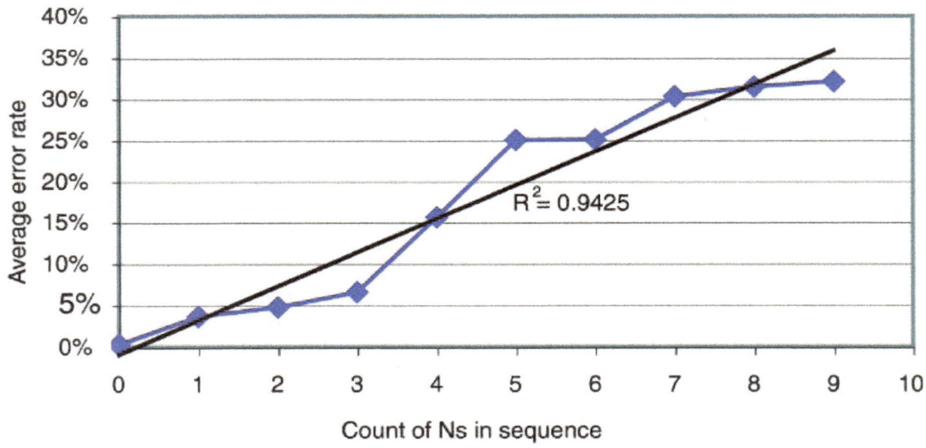

Figure 4: The presence of Ns correlates well with sequencing errors. The presence of an N in the sequence indicates the GS20's inability to accurately call a base at that position within the sequence. The number of other sequencing errors (substitutions, insertions and deletions) within a sequence read correlates with number of uncalled bases. By removing all reads that contain one or more Ns, the overall sequencing error rate drops substantially.

We compared error rates in the proximal primer region with error rates in the sequence following the primer. We found a positive, non-linear correlation between the two error rates (not shown). Primers with three or more errors are indicative of low-quality sequences downstream of the primer, but primers with fewer errors are not a good predictor of the overall error of the read. In our experiment, 3% of the reads contained inexact matches to the proximal primer, but removing them only reduced the error rate from 0.49% to 0.45%: most of this improvement reflects elimination of incorrect base calls in the primer sequence. When all reads with Ns, primer errors, and long or short read lengths were removed the total error rate was reduced to 0.16%. The remaining errors correspond to insertions (51%), deletions (33%) and substitutions (16%).

We looked for spatial patterns of read errors using the xy location of each read on the plate (available in the GS20 output files). There can be some minor edge effects but these are not consistent across runs and do not represent a large enough portion of the reads under normal conditions to be a useful tool. However, we note that the total number of errors due to edge effects will be greater in experiments that use 4 or 16 lane gaskets, as the total number of edges will be proportionally larger.

DISCUSSION

The massively parallel pyrosequencing of emulsification PCR-based templates holds great promise for revolutionizing high-throughput sequencing. However, there is concern over the potentially high degree of error, particularly for applications that cannot rely on consensus of large assemblies. Consensus-based projects would also benefit from a lower error rate, as fewer sequences would be required to build a reliable consensus. The original description of the 454 Life Sciences system reported an error rate for shotgun library reads of four bases per hundred nucleotide positions. The test fragment data in that publication had much lower error rates despite the fact that the test fragments have extensive homopolymers and are designed to be difficult to read correctly. This discrepancy indicates that the basic method of pyrosequencing, luminescence detection and flow intensity resolution, is sound, and suggests that the higher error with experimental data may come from the experimental manipulation of the sequences prior to pyrosequencing. Margulies *et al.* [2] suggested that this may be due to multiple sequences binding to an individual bead prior to the emPCR amplification, resulting in a heterogeneous amplification pool. The GS20 quality filters will eliminate sequences from beads that contain two highly divergent DNA templates but the software will attempt to interpret flowgrams from a single bead that contains two similar but non-identical sequences. Unlike shotgun genomic data, V6-tag data may have large numbers of highly similar sequences. It is, therefore, even more important in V6-tag and metagenomic sequencing to remove reads that may result from multi-templated beads.

We conducted an in-depth analysis of experimentally generated GS20 reads by sequencing an amplicon library made from a set of clones of known sequence. We found an error rate (incorrect bases/total number of expected nucleotides) of 0.49%, considerably lower than that reported by Marguiles *et al.* [2] but still higher than they or we found for test fragment data. Significantly, we found that the errors in our experimental reads were not randomly distributed across all reads: 86% of the reads contain no errors, while reads that differ from the reference sequence by more than 4% contained nearly 50% of the errors (Figure 1). In contrast, errors were much more randomly distributed in our test fragments, where 50% of the errors were from those fragments that differ by less than 1% from the reference. A multi-templated bead would frequently have multiple bases at a position, which could cause indeterminate flows - neither base having ample luminescence to clearly register. The convergence of the error rates of the two separate sequencing runs when the reads containing Ns were removed is consistent with the multi-templated bases as the primary source of error. The error distribution across reads and the similarity of error rates for reads with no Ns are consistent with a high general accuracy of the pyrosequencing method and poor resolution of a small number of beads with a heterogeneous amplification population.

If heterogeneous templates on a single bead represented a major contribution to observed errors in a low quality read, we anticipate a disproportionate number of errors would occur in sequences that correspond to low abundance templates in the original emPCR reactions; if a low-frequency strand shares a bead with another sequence, the other sequence is likely to be different. In contrast, high-frequency sequences are more likely to be contaminated by an identical sequence. Our data match this pattern. The removal of the bulk of the errors via the removal of reads with ambiguous bases is also consistent with multi-templated beads. All of our reads shared the same proximal primer, and would, therefore, sequence with few errors in the primer, even on multi-templated beads. Other experiments with heterogeneous primers may find primer fidelity to be more useful at identifying low-quality reads.

A significant decrease in the heterogeneous amplification population (HAP) between the original reported experiment and ours is likely given the improvements in the protocol developed by 454 Life Sciences. Unfortunately, it suggests that the highest single source of error in an emPCR-based pyrosequencing experiment may vary from experiment to experiment. The evidence from our two separate sequencing runs, however, is that the removal of reads with Ns is a good surrogate for the removal of reads from multi-templated beads. Nonsynchronized extension of fragments will also produce Ns. These will also be culled when reads with Ns are removed. Advances in pyrosequencing that would reduce the occurrence of multi-templated beads, reduce nonsynchronized extension, or better identify these errors in the base-calling software could significantly improve the overall accuracy of the technology.

Our analysis of the distribution of the types of error in pyrosequencing of emPCR libraries suggests ways of identifying and removing these HAP-hazards: reads with a disproportionately large number of errors are disproportionately likely to contain ambiguous bases (Ns) and to be aberrantly long or short. Short reads arise from short fragments on emPCR beads, but also, and perhaps more likely, from sequential deletion of a read by the software in the GS20 machine, which successively trims bases presumed to be in error from the end of reads. The more bases trimmed, the more likely the entire read is of dubious quality. These errors may be multi-templated beads of similar sequence, or nonsynchronized extension of the templates. Once fragments lose synchrony, they will have successively more errors as the read extends. A reduced threshold for removing reads that have bases trimmed from the end might remove many of these poor-quality short reads. Many sequencing projects cannot judge an appropriate read length, although it is always possible to detect and remove short reads. Reads of aberrant length represented only a small fraction, approximately 1%, of our data, and most of these reads, >60%, also contain Ns.

CONCLUSION

Our analysis of the GS20 sequencing error rate of the V6 region of bacterial rRNA genes shows a marked improvement over the original error rates published by Margulies *et al* [2]. The largest source of errors may be due to multi-templated beads, and enhancements to both the chemistry protocol for the GS20 and the built-in bioinformatics software may account for the change in error rates. Our results highlight that a small proportion of low quality reads, presumably from multi-templated beads, are responsible for the majority of sequencing errors. The ability to identify and remove these reads is the best way to improve the accuracy of the entire dataset. It is not a replacement for assigning quality scores to detect the position of miscalled bases. The interpretation of chromatograms by programs such as PHRED [13] employs quality scores that reflect the probability of any type of base call error. Although it uses the same scale, the GS20 software generates quality values based on the probability of homopolymer extension rather than probability of a correct base call.

Regardless of the ultimate cause of poor reads, the presence of even a single ambiguous base (N) was an effective indicator of low-quality sequence. The removal of all reads containing one or more Ns can drastically improve the overall quality of the remaining dataset, reducing the error rate from about 0.5% to about 0.25% (Table 2). For our data, this strategy eliminated only 6% of the total reads. By excluding approximately 1% of all reads whose lengths lie outside of the main distribution, as well as those with inexact matches to the primer, the error rate for the V6-tag data dropped to less than 0.2%. The pyrosequencing technology provides such a large number of reads that the elimination of even 10% or more of the reads in a data set should be more than offset by the increase in quality of the remaining reads.

Table 2: Identifying low-quality reads and their contribution to the error rate

Data selection	Percent of reads	Error rate
All reads	100.0%	0.49%
Reads with no Ns	94.4%	0.24%

Reads with one or more Ns	5.6%	4.7%
Reads with length ≥81 and ≤108	98.8%	0.33%
Reads with length <81 or >108	1.2%	18.9%
Reads with no Ns and length ≥81 and ≤108	93.3%	0.20%
Reads with no proximal errors	97.0%	0.45%
Reads with fewer than three proximal errors	>99.99%	0.48%
Reads with more than three proximal errors	<0.01%	12.2%
Reads with no Ns and length ≥81 and ≤108 and no proximal errors	90.6%	0.16%

Removing reads with Ns is the most effective means we found of removing low-quality data and improving the error rates. Read lengths that are either longer or shorter than expected, and are outside the peak of common reads, also correlate strongly with incorrect reads.

Our strategy for detecting low quality reads circumvents the need to generate consensus sequences for improving data quality in massively parallel pyrosequencing experiments of environmental DNA. Our criteria for detecting reads with errors allows for the acquisition of pyrosequencing data in the context of molecular ecology that can surpass the accuracy of traditional capillary methods.

MATERIALS AND METHODS

Generation of 1 kb clone library and selection of clones for pyrosequencing

DNA was extracted according to Huber *et al.* [14] from diffuse flow hydrothermal vent samples as described in Sogin *et al.* [8]. PCR primers were designed using ARB software [15] to target the bacterial 16S rDNA. The primers used were 337 F (5' CAN CCT ACG GGN GGC NGC) and 1391R (5' GAC GGG CGG TGW GTN CA). The amplification mix contained 5 units *Pfu* Turbo polymerase (Stratagene, La Jolla, CA, USA), 1× *Pfu* reaction buffer, 200 μM dNTPs (Pierce Nucleic Acid Technologies, Milwaukee, WI, USA), and 0.2 μM each primer in a volume of 100 μl. Environmental DNA (3-10 ng) was added to 3 separate 30 μl amplification mixes. A positive control (*Marinobacter aquaeolei* genomic DNA) and two negative controls (no DNA and genomic DNA from the archaeon *Methanococcus jannaschii*) were also run. An initial denaturation step of 3 min at 94°C was followed by 30 cycles of 94°C for 30 s, 55°C for 45 s, and 72°C for 2 minutes. The final extension step was 72°C for 2 minutes. Following PCR, three reactions for each sample were combined, purified, and concentrated using the MinElute PCR Purification Kit (Qiagen, Valencia, CA, USA) according to the manufacturer's instructions. PCR product quality was assessed on a 0.8% agarose gel stained with ethidium bromide and ligated with pCR4-TOPO vector for 20 minutes at room temperature and transformed with TOP10 electrocompetent cells according to the manufacturer's instructions (Invitrogen, Carlsbad, CA, USA). Colonies for each library were randomly selected and grown in SuperBroth with 50 mg/ml kanamycin in 96 deep-well blocks overnight. Alkaline lysis template preparation was carried out on cell pellets using the RevPrep Orbit II (Genomic Solutions, Ann Arbor, MI, USA) or the Biotech RoboPrep2500 (MWG Biotech, Ebersberg, Germany). The 1,000 base-pair amplicons were sequenced bidirectionly using primers T3 (5'-ATT AAC CCT CAC TAA AGG GA) and T7 (5'-TAA TAC GAC TCA CTA TAG GG), and on an ABI 3730 × l genetic analyzer. Sequences were aligned with MUSCLE (with parameters -diags and -maxiters 10) [16] and manually manipulated in the BioEdit 7.0.1 program [17]. Distance matrixes were calculated using quickdist [8], and taxonomic identities determined using RDP-II Classifier [18]. Sequences were trimmed to include only the V6 region of the gene, the distance matrix re-calculated, and from this analysis, 43 divergent sequences were chosen for further experimentation. The average length of the V6 region for these clones was 101 bases, ranging from 95 to 109, with one longer reference of 183 bases. The 16S rDNA sequences are deposited at GenBank under accession numbers DQ909092, DQ909128DQ909132, DQ909133, DQ909142, DQ909144, DQ909158, DQ909184, DQ909202, DQ909204, DQ909218, DQ909223, DQ909224, DQ909248, DQ909251, DQ909253, DQ909266, DQ909274, DQ909337, DQ909368. DQ909392, DQ909396, DQ909400, DQ909414, DQ909423, DQ909438, DQ909440, DQ909465, DQ909474, DQ909498, DQ909513, DQ909519, DQ909538, DQ909603, DQ909618, DQ909631, DQ909662, DQ909688, DQ909702, DQ909706, DQ909719. DQ909727, DQ909753.

Generation of known V6 amplicon library

We treated each plasmid with plasmid-safe DNAase (Epicentre, Madison, WI, USA) to remove *Escherichia coli*genomic DNA and confirmed that each plasmid produced an amplification product of the expected size with primers targeting the V6 region of the bacterial rDNA according to Sogin *et al.* [8]. We then pooled the individual plasmids and amplified with the primers that flank the V6 region of rRNA genes according to Sogin*et al.* [8]. We assessed the product quality using a BioAnalyzer Agilent DNA 1000 LabChip following the manufacturer's instructions. Three reactions were combined, purified, and concentrated using the MinElute PCR Purification Kit (Qiagen). The final amplicon library was sequenced independently by 454 Life Sciences and our own lab. Both labs used the Roche Genome Sequencer 20 (GS20) according to the manufacturer's specifications [2]. The original GS20 output files as text are available in Additional data files 3-5.

Error rate calculations

We combined the data from both sequencing runs for a total of 340,150 reads (226,150 and 114,000), with an average read length of 94.5 nucleotides and a total of 32,816,656 bases. These sequences are available in fasta format in Additional data file 2. To determine the reference sequence source of each pyrosequencing read, we ran a separate multiple sequence alignment of each individual read against the 43 reference sequences using MUSCLE [16] (default options plus maxiters 2, diags). We calculated the number of sequence differences between each read and the reference sequences to determine the reference sequence to which each read mapped most closely. All subsequent error calculations are based on comparing reads to their assigned reference sequence.

The overall error rate is the number of errors in a read divided by the length of sequence. Specifically, we calculated errors in several ways. In all methods, each base mismatch or N in the test sequence counts as an error, and a terminal gap caused by a GS20 read terminating before the end of the reference does not count as an error. In the first and second methods each base of an insertion or deletion counts as one error. In the third method, insertions or deletions are counted by homopolymer runs. If a TAAA is inserted, it is counted as two insertions, one single-T and one multi-A insertion. The denominator for the first method was the read length. The denominator for the second and third methods was the length of reference sequence minus any discounted terminal gaps. All error rate calculations produced essentially the same results. We report error rates using all base errors (not by homopolymer run) divided by the expected length (reference sequence length minus terminal gaps).

The error rates were calculated for each sequence in a pairwise comparison of the pyrosequencing read and the reference sequence to which it was assigned. We used the Needleman-Wunsch algorithm [19] for these pair-wise alignments because it selects for the best possible alignment given its run parameters. Using a set of 100 sequences and a matrix of Needleman-Wunsch run parameter combinations, we found that a gap opening penalty of 5.75 and a gap extension penalty of 2.75 minimized the calculated error rate. Error rates, reference sequences and read sequences were imported into a MySQL database for storage and analysis.

ADDITIONAL DATA FILES

The following additional data are available with the online version of this paper. Additional data file 1 is a fasta file of the 43 known sequences used. Additional data file 2 is a gzip-compressed fasta file of the sequences output by the GS20. These sequences correspond to those included in Additional data files 3, 4, 5 but include only the final sequence information. Additional data files 3, 4, 5 are three compressed text files representing the text translations of the original GS20 binary output (sff) files for all of the sequencing used in the analysis, including sequence, flowgram and other run information. GS20 data are reported by region of the PicoTiterPlate™; we sequenced three plate regions.

ACKNOWLEDGEMENTS

We thank Richard Fox and Jonathan Neff for computing facility support. This work was supported by National Aeronautics and Space Administration Astrobiology Institute Cooperative Agreement NNA04CC04A (to MLS), subcontracts from the Woods Hole Center for Oceans and Human Health from the National Institutes of Health and National Science Foundation (NIH/NIEHS 1 P50 ES012742-01 and NSF/OCE 0430724-J Stegeman PI to HGM and MLS), grants from the WM Keck Foundation and the G Unger Vetlesen Foundation (to MLS), and a National Research Council Research Associateship Award (to JAH).

REFERENCES

1. Ronaghi M, Uhlén M, Nyrén P: DNA sequencing: a sequencing method based on real-time pyrophosphate. Science. 1998, 281: 363-365. 10.1126/science.281.5375.363.
2. Margulies M, Egholm M, Altman WE, Attiya S, Bader JS, Bemben LA, Berka J, Braverman MS, Chen YJ, Chen Z, et al: Genome sequencing in microfabricated high-density picolitre reactors. Nature. 2005, 437: 376-380.
3. Goldberg SMD, Johnson J, Busam D, Feldblyum T, Ferriera S, Friedman R, Halpern A, Khouri H, Kravitz SA, Lauro FM, et al: A Sanger/pyrosequencing hybrid approach for the generation of high-quality draft assemblies of marine microbial genomes. Proc Natl Acad Sci USA. 2006, 103: 11240-11245. 10.1073/pnas.0604351103.
4. Moore MJ, Dhingra A, Soltis PS, Shaw R, Farmerie WG, Folta KM, Soltis DE: Rapid and accurate pyrosequencing of angiosperm plastid genomes. BMC Plant Biol. 2006, 6: 17-10.1186/1471-2229-6-17.
5. Wicker T, Schlagenhauf E, Graner A, Close T, Keller B, Stein N: 454 sequencing put to the test using the complex genome of barley. BMC Genomics. 2006, 7: 275-10.1186/1471-2164-7-275.
6. Gharizadeh B, Herman ZS, Eason RG, Jejelowo O, Pourmand N: Large-scale pyrosequencing of synthetic DNA: A comparison with results from Sanger dideoxy sequencing. Electorphoresis. 2006, 27: 3042-3047. 10.1002/elps.200500834.
7. Edwards RA, Rodriguez-Brito B, Wegley L, Haynes M, Breitbart M, Peterson DM: Using pyrosequencing to shed light on deep mine microbial ecology. BMC Genomics. 2006, 7: 57-10.1186/1471-2164-7-57.
8. Sogin ML, Morrison HG, Huber JA, Welch DM, Huse SM, Neal PR, Arrieta JM, Herndl GJ: Microbial diversity in the deep sea and the underexplored "rare biosphere". Proc Natl Acad Sci USA. 2006, 103: 12115-12120. 10.1073/pnas.0605127103.
9. Gilbert MTP, Binladen J, Miller W, Wiuf C, Willerslev E, Poinar H, Carlson JE, Leebens-Mack JH, Schuster SC: Recharacterization of ancient DNA miscoding lesions: insights in the era of sequencing-by-synthesis. Nucleic Acids Res. 2007, 35: 1-10. 10.1093/nar/gkl483.
10. Dostie J, Richmond TA, Arnaout RA, Selzer RR, Lee WL, Honan TA, Rubio ED, Krumm A, Lamb J, Nusbaum C, et al: Chromosome Conformation Capture Carbon Copy (5C): A massively parallel solution for mapping interactions between genomic elements. Genome Res. 2006, 16: 1299-1309. 10.1101/gr.5571506.
11. Kysela DT, Palacios C, Sogin ML: Serial Analysis of V6-ribosomal sequence tags (SARST-V6): A method for efficient, high-throughput analysis of microbial community composition. Environmental Microbiol. 2005, 7: 356-364. 10.1111/j.1462-2920.2004.00712.x.

12. Roche Applied Sciences: GS20 Data Processing Software Manual. 2006, Penzberg: Roche Diagnostics GmbH

13. Ewing B, Green P: Base-calling of automated sequencer traces using Phred. II. Error probabilities. Genome Res. 1998, 8: 186-194.

14. Huber JA, Butterfield DA, Baross JA: Temporal changes in archaeal diversity and chemistry in a mid-ocean ridge subseafloor habitat. Applied Environmental Microbiol. 2002, 68: 1585-1594. 10.1128/AEM.68.4.1585-1594.2002.

15. Ludwig W, Strunk O, Westram R, Richter L, Meier H, Yadhukumar , Buchner A, Lai T, Steppi S, Jobb G, et al: ARB: a software environment for sequence data. Nucleic Acids Res. 2004, 32: 1363-1371. 10.1093/nar/gkh293.

16. Edgar RC: MUSCLE: multiple sequence alignment with high accuracy and high throughput. Nucleic Acids Res. 2004, 32: 1792-1797. 10.1093/nar/gkh340.

17. Hall TA: BioEdit: a user-friendly biological sequence alignment editor and analysis program for Windows 95/98/NT. Nucleic Acids Symp Series. 1999, 41: 95-98.

18. Cole JR, Chai B, Farris RJ, Wang Q, Kulam SA, McGarrell DM, Garrity GM, Tiedje JM: The Ribosomal Database Project (RDP-II): sequences and tools for high-throughput rRNA analysis. Nucleic Acids Res. 2005, 33: D294-296. 10.1093/nar/gki038.

19. Needleman SB, Wunsch CD: A general method applicable to the search for similarities in the amino acid sequence of two proteins. J Mol Biol. 1970, 48: 443-453. 10.1016/0022-2836(70)90057-4.

CHAPTER 7 A PRIMER ON METAGENOMICS

John C. Wooley, Adam Godzik, Iddo Friedberg

INTRODUCTION

For most of its history, life on Earth consisted solely of microscopic life forms, and microbial life still dominates Earth in many aspects. The estimated 5×10^{30} prokaryotic cells inhabiting our planet sequester some 350–550 Petagrams (1 Pg=10^{15} g) of carbon, 85–130 Pg of nitrogen, and 9–14 Pg of phosphorous making them the largest reservoir of those nutrients on Earth [1]. Bacteria and archaea live in all environments capable of sustaining other life and in many cases are the sole inhabitants of extreme environments: from deep sea vents with temperatures of 340°C to rocks found in boreholes 6 km beneath the Earth's surface. Bacteria, archea, and microeukaryotes dominate Earth's habitats, compound recycling, nutrient sequestration, and, according to some estimates, biomass. Microbes are not only ubiquitous, they are essential to all life, as they are the primary source for nutrients, and the primary recyclers of dead matter back to available organic form. Along with all other animals and plants, the human condition is profoundly affected by microbes, from the scourges of human, farm animal, and crop pandemics, to the benefits in agriculture, food industry, and medicine to name a few. We humans have more bacterial cells (10^{14}) inhabiting our body than our own cells (10^{13}) [2],[3]. It has been stated that the key to understanding the human condition lies in understanding the human genome [4],[5]. But given our intimate relationship with microbes [6], researching the human genome is now understood to be a necessary though insufficient condition: sequencing the genomes of our own microbes would be necessary too. Also, to better understand the role of microbes in the biosphere, it would be necessary to undertake a genomic study of them as well.

The study of microbial genomes started in the late 1970s, with the sequencing of the genomes of bacteriophages MS2 [7] and φ-X174 [8]. In 1995 microbiology took a major step with the sequencing of the first bacterial genome *Haemophilus influenza* [9]. The genomes of 916 bacterial, 1,987 viral, and 67 archaeal species are deposited in GenBank release 2.2.6. Having on hand such a large number of microbial genomes has changed the nature of microbiology and of microbial evolution studies. By providing the ability to examine the relationship of genome structure and function across many different species, these data have also opened up the fields of comparative genomics and of systems biology. Nevertheless, single organism genome studies have limits. First, technology limitations mean that an organism must first be clonally cultured to sequence its entire genome. However, only a small percentage of the microbes in nature can be cultured, which means that extant genomic data are highly biased and do not represent a true picture of the genomes of microbial species [10]–[12]. Second, very rarely do microbes live in single species communities: species interact both with each other and with their habitats, which may also include host organisms. Therefore, a clonal culture also fails to represent the true state of affairs in nature with respect to organism interaction, and the resulting population genomic variance and biological functions.

New sequencing technologies and the drastic reduction in the cost of sequencing are helping us overcome these limits. We now have the ability to obtain genomic information directly from microbial communities in their natural habitats. Suddenly, instead of looking at a few species individually, we are able to study tens of thousands all together. Sequence data taken directly from the environment were dubbed the metagenome [13], and the study of sequence data directly from the environment—metagenomics [14].

However, environmental sequencing comes with its own information-restricting price tag. In single organism genomics practically all of the microbe's genome is sequenced, providing a complete picture of the genome. We know from which species the DNA or RNA originated. After assembly, the location of genes, operons, and transcriptional units can be computationally inferred. Control elements and other cues

Citation: Wooley JC, Godzik A, Friedberg I (2010) A Primer on Metagenomics. PLoS Comput Biol 6(2): e1000667. doi:10.1371/journal.pcbi.1000667.

can be identified to infer transcriptional and translational units. Consequently, we achieve a nearly complete and well-ordered picture of all the genomic elements in the sequenced organism. We may not recognize all the elements for what they are, and some errors may creep in, but we can gauge the breadth of our knowledge and properly annotate those areas of the genome we manage to decipher.

In contrast, the sequences obtained from environmental genomic studies are fragmented. Each fragment was obviously sequenced from a specific species, but there can be many different species in a single sample, for most of which a full genome is not available. In many cases it is impossible to determine the true species of origin. The length of each fragment can be anywhere between 20 base pairs (bp) and 700 bp, depending on the sequencing method used. Short sequence reads that are dissociated from their original species can be assembled to lengths usually not exceeding 5,000 bp; consequently, the reconstruction of a whole genome is generally not possible. Even the reconstruction of an entire transcriptional unit can be problematic. In addition to being fragmented and incomplete, the volume of sequence data acquired by environmental sequencing is several orders of magnitude larger than that acquired in single organism genomics.

For these reasons, computational biologists have been developing new algorithms to analyze metagenomic data. These computational challenges are new and very exciting. We are entering an era akin to that of the first genomic revolution almost two decades ago. Whole organism genomics allows us to examine the evolution not only of single genes, but of whole transcriptional units, chromosomes, and cellular networks. But more recently, metagenomics gave us the ability to study, on the most fundamental genomic level, the relationship between microbes and the communities and habitats in which they live. How does the adaptation of microbes to different environments, including host animals and other microbes, manifest itself in their genomes?

For us humans, this question can strike very close to home, when those habitats are our own bodies and the microbes are associated with our own well-being and illnesses: almost every aspect of human life, as well as the life of every other living being on the planet, is affected by microbes. We now have the experimental technology to understand microbial communities and how they affect us, but the sheer volume and fragmentary nature of the data challenge computational biologists to distill all these data into useful information.

In this article we shall briefly outline some experimental, technological, and computational achievements and challenges associated with metagenomic data, from sequence generation and assembly through the various levels of metagenomic annotation. We will also discuss computational issues that are unique to environmental genomics, such as estimating the metagenome size and the handling of associated metadata. Finally, we will review some studies highlighting the advantages of metagenomic-based research, and some of the insights it has enabled.

SAMPLING

Sample Size and Number of Samples

The first step in a metagenomic study is to obtain the environmental sample. Samples should represent the population from which they are taken. The problem in microbial ecology is that we are unable to see the organisms we are trying to capture. How many samples are enough?

To estimate the fraction of species sequenced, rarefaction curves are typically used. A rarefaction curve plots the number of species as a function of the number of individuals sampled. The curve usually begins with a steep slope, which at some point begins to flatten as fewer species are being discovered per sample: the gentler the slope, the less contribution of the sampling to the total number of operational taxonomic units or OTUs. For microbial samples, different OTUs are typically characterized by 16S (prokaryotic) or 18S (eukaryotic) rDNA, and are also referred to as ribotypes. Classification is rarely done in the field, so some initial estimate of species diversity by a pilot study or previous studies is desirable to gauge the number of samples needed to get a comprehensive picture of the OTUs in the sampled habitat. More of this will be discussed in the "Species Diversity" section below.

Filtering

When filtering an environmental sample, as with any kind of filtering, the goals are: (1) get as much as you can of what you want and (2) leave out as much as you can from what you do not want. So if we are interested in bacteria only, our goal would be to filter out the smaller viroid particles, and the usually larger protists. Of course, this process will leave in the lysogenic phages and prophages, which are integrated in bacterial genetic material, as well as mimivirus particles, which are as large as some bacteria. On the other side of the size scale, small protists and large bacteria may overlap in size, making a full size-based separation impossible. Also, filamentous forms of bacteria that grow in multicellular colonies may also be filtered out owing to colony size exceeding that of the filter's pores.

Computational filtering can be used after sequencing. Genomic material that is obviously within the clades of interest can be filtered in using similarity searches against annotated sequence databases. Care must be taken, though, with false negatives: relevant genomic material may be filtered out in this fashion simply because homologs have never been deposited in existing databases. Another option would be to search for obviously false-positive sequence motifs, e.g., eukaryote material when only prokaryote material is to be analyzed. This technique can also be used to detect sample contamination.

Recording Metadata

Keeping strict and comprehensive records of metadata is as important as the sequence data. Metadata are the "data about the data": where the samples were taken from, when, and under which conditions. In microbial ecology, this commonly refers to physical, chemical, and other environmental characteristics of the sample's location. For example, an ocean sample metadata will typically include sampling date and time, depth, salinity, light intensity, geographical coordinates, pH, soluble gases, etc. In clinical microbiology, metadata would refer to the pathology, medical history, and vital statistics of the patient as well as the exact location and tissue from which the sample was taken, the sampling conditions, and so on.

Many metagenomic studies are driven by discovery and data mining, rather than by hypothesis. These studies seek statistically significant correlations between the metagenomic data and the habitat-associated metadata, which may lead to biologically significant discoveries. There is therefore a need to provide metadata in a form that is standard, comprehensive, and amenable to computation. For example, semantic information should be provided, wherever possible, in ontological form. A description of the environmental context and the experimental methods used is vital to enable comparative studies. As we shall see, genes or even "gene-less" sequence signatures are linked to habitats rather than to species. Finally, sequencing technology is rapidly improving, and the adoption of new sequencing methods will require the adoption of descriptors of those methods such as sequence coverage, quality, assembly programs that were used, and so on.

The Genomic Standards Consortium (http://gensc.org/) is an international group working to standardize the description of genomes and metagenomes and the exchange of genomic data and metadata. In a recent publication, a standard for the Minimum Information of Genomic and Metagenomic (MIGS/MIMS) metadata was suggested for adoption [15], and an associated markup language, the Genomics Contextual Data Markup Language or GCDML is under active development [16]. It is the consortium's aim that the MIGS/MIMS shall be adopted by journals as a publication requirement when genomic or metagenomic data are being deposited, akin to standards such as MIAME for microarray data [17] or PDB/mmCIF for structural biology data[18].

SEQUENCING

First, Second, and Third Generation Sequencing

Until recently, prokaryotic genomes have been typically sequenced using Sanger shotgun [19],[20] sequencing. The first step is shearing the DNA content of a genomic clone into random fragments, hence the "shotgun." The fragments are then cloned into plasmid vectors that are grown in monoclonal libraries to produce enough genomic material for sequencing. The DNA is then sequenced using dye-termination methods. Repetition of this process ensures that all parts of the studied genome are sequenced, several times over. Assembly software is then used to assemble the sequence fragments into the whole genome. Theoretically any genome shorter than 5 Mbp can be assembled this way, although regions with large repeats tend to frustrate assembly algorithms. Therefore, regions with large repeats are often not incorporated into the whole genomic picture, leaving some gaps. Another disadvantage of shotgun sequencing is the "cloning bias." Some genes cannot be incorporated into the library vector, usually because of toxicity to the vector expressing them [21]. This inability to be incorporated is typically mitigated by using more than one organism for cloning, or by using sequencing techniques that do not require cloning (see below) in second generation sequencing.

In metagenomics, shotgun sequencing is done in the same manner as in clonal culture genomics. However, the raw genomic material does not come from a single organism: it comes from a community of microbes, hence the name environmental shotgun sequencing or ESS. Depending on our ability to sample, this DNA may provide only a partial genomic picture of the organisms in the environment, since the genomic material from the more abundant species dominates the sample. To obtain a better picture of the species composing the community, 16S rDNA or 18S rDNA for prokaryotic and eukaryotic samples, respectively, are sequenced separately using universal primers, see Figure 1. It should be noted that when using primers for rDNA to classify OTUs in an environmental sample, there are choices to be made regarding the primer sequence, especially when the studied OTU composition is expected to differ significantly from most known species, the so-called rare biosphere [22],[23]. In this case, there is the possibility that the primers used will be too different from the rDNA in the sample, which would result in many OTUs not being identified [24],[25].

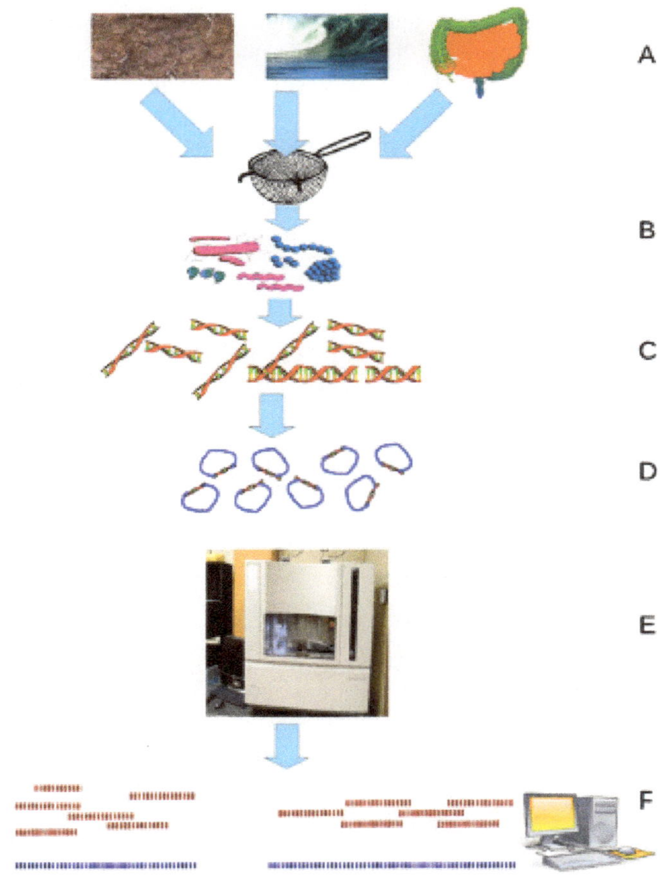

Figure 1: Environmental Shotgun Sequencing (ESS). (A) Sampling from habitat; (B) filtering particles, typically by size; (C) DNA extraction and lysis; (D) cloning and library; (E) sequence the clones; (F) sequence assembly. http://dx.doi.org/10.1371/journal.pcbi.1000667.g001

Second generation sequencing methods have been rapidly gaining ground and are replacing Sanger sequencing for small sized genomes and environmental genomics. A common denominator among second generation methods is the generation of "polymerase colonies" or polonies [26],[27]. Polonies are PCR amplicons derived from a single molecule of nucleic acid. Thousands to millions of polonies, each with an effective reaction size of 10^{-9} l to 10^{-12} l can be amplified simultaneously, generating templates for sequencing. Following that, enzymatic reactions can be performed in parallel to sequence the nucleic acid material in the polonies. Polony-based methods produce considerably more sequences than Sanger sequencing, but those sequences are much shorter. Furthermore, each polony-based method has its own anomalies that should be accounted for when processing the data. See Table 1 for a comparison between the yield, fragment length, and run times of the different sequencers.

Table 1: Comparison of different sequencing technologies, taken from [34]. http://dx.doi.org/10.1371/journal.pcbi.1000667.t001

Sequencer	ABI 3730	Roche 454	Solexa[a]	SOLiD (mp, frag)[b]	HeliScope[c]
Read length	600–900	400–500	75–100	50	25–35
Run time	6–10 h	10 h	2–10 d	(4–7 d,8–14 d)	h
Yield (Mbp)	0.01	1	3,000–4,500/d	(400, 1,800)	105–140/h
Cloning bias	Yes	No	No	No	No
Mate pair information	Yes	No	Yes	Yes	No

[a]Based on the GA IIx. See full specifications at: http://www.illumina.com/systems/genome_analyzer.ilmn.
[b]mp, mate pair (frag, fragment). See http://products.appliedbiosystems.com/ SOLiD 3 Plus System.
[c]See: http://www.helicosbio.com/Products/HelicosregGeneticAnalysisSystem/HeliScopetradeSinglemoleculeSequencer/tabid/87/Default.aspx.
doi:10.1371/journal.pcbi.1000667.t001

In pyrosequencing (Figure 2) [28],[29], methods such as Roche 454 [30] sequencing is performed by polymerase extension of a primed template. Single nucleotide species are added at each cycle. If the particular nucleotide species added to the polymerase reaction pairs with the one on the template, the incorporation causes luciferase-based light reaction. The reaction chamber is then washed, and the cycle repeated. Several hundreds of thousands of wells containing material for sequencing are typically used in a single reaction. Second is the inability to read long mononucleotide repeats correctly.

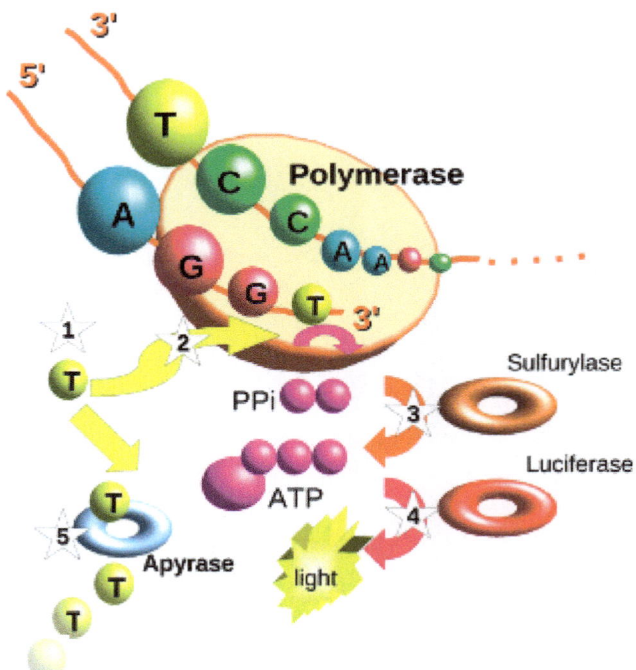

Figure 2: Pyrosequencing. Single stranded DNA template is first hybridized with the sequencing primer and mixed with the enzymes along with the two substrates adenosine 5′-phosphosulfate (APS) and luciferin. In each cycle, (1) one of the four nucleotides (dTTPi, in this case) is then added to the reaction. (2) If the nucleotide is complementary to the base in the template strand then the DNA polymerase incorporates it into the growing strand. (3) Pyrophosphate (PPi)—in an amount equal in molarity to that of the incorporated nucleotide—is released and converted to ATP by sulfurylase in the presence of APS. (4) ATP then serves as a substrate to luciferase, causing a light reaction. Photon emission is in equimolar quanta to the amount of nucleotide incorporated in a given cycle. (5) The excess nucleotides are degraded by apyrase. http://dx.doi.org/10.1371/journal.pcbi.1000667.g002

ABI SOLiD and Illumina GAII sequencers produce even shorter reads: 25–100 bp, but very large volumes of DNA per sequencing run. As we shall see in the "Assembly" section below, despite the individual short read lengths, these technologies provide a viable alternative for sequencing whole genomes, by sheer volume of DNA sequenced. For further reading on second generation sequencing see [31]–[34].

Third generation sequencing, loosely defined as technology that is capable of sequencing long sequences without amplification, is in advanced development. There are encouraging signs that this technology might be available as early as 2011 [35]–[37].

ASSEMBLY

When sequencing a whole genome, the reads are assembled into progressively longer contiguous sequences or contigs, and finally to the whole genome. Dealing with genomic data, we are used to analyzing long stretches of contiguous sequence data. This analysis lets us find not only open reading frames, but also operons, operational transcriptional units, their associated promoter elements, and transcription factor binding sites. Longer elements such as pathogenicity islands, and other mobile genetic elements are evident only when large fractions of the genome are assembled. The gain of information correlates with the length of the genomic elements. Table 2 shows the length of a genomic sequence, and the information that may be gleaned from it.

Table 2: The information contained in different lengths of genomic DNA. http://dx.doi.org/10.1371/journal.pcbi.1000667.t002

Sequence Length (bp)	Genome Element
25–75	SNPs, short frameshift mutations
100–400	Short functional signatures
500–1,000	Whole domains, single domain genes
1,000–5,000	Short operons, multidomain genes
5,000–10,000	Longer operons, some *cis*-control elements
>100,000	Prophages, pathogenicity islands, various mobile insertion elements
>1,000,000	Whole prokaryotic chromosome organization

doi:10.1371/journal.pcbi.1000667.t002

In contrast, in all but the most species-poor metagenome, a full assembly is not possible—first, because the sampling is incomplete, and many if not all species' genomes are partially sampled, if at all; second, because the species information itself is incomplete, and it is difficult to map individual reads to their species of origin. Therefore, the analysis of genomic elements using metagenomic data is generally limited to the first three or four rows in Table 2.

In this section, we will discuss assembly of metagenomic data, how information is extracted from partial assemblies, and how the extent of information gained can be estimated.

Metagenomic Sample Coverage

Coverage.

Coverage of a genome is defined as the mean number of times a nucleotide is being sequenced. Thus, 5× coverage means that each nucleotide in the genome is sequenced a mean number of five times. If we could sequence a genome in a single read, then 1× coverage would suffice for sequencing. Shorter read lengths (25–700, depending on sequencing technologies, see Table 1), necessitate more coverage, to ensure all reads overlap, and that those overlaps are unique enough to reconstruct the genome by assembling the fragments. If we treat DNA shearing and sequencing as random events, and our ability to detect and overlap between two truly overlapping reads does not vary between clones (when those are used), then we can use a Poisson distribution model to estimate the number of reads required to sequence an entire genome. This model is given by the Lander-Waterman equation [38]:

$$C = \frac{L \times N}{G}$$

Where L is the read length, N is the number of reads, G is the genome length, and C is coverage as described above. The fraction of sequence covered would be given as:

$$P_0 = 1 - e^{-C} = 1 - e^{-(LN/G)}$$

To get the number of reads sequencing fraction P_0 of the genome

$$N = -\frac{log(1 - P_0)}{L} \times G$$

In an environmental sample containing l species, the metagenome size G_m is:

$$G_m = \sum_{i=1}^{l} n_i G_i$$

Where G_i is the size of any given genome in a sample containing l genomes, and n_i the number of copies of genome g_i.

However the species that constitute the sample appear in different frequencies in the metagenome. Therefore a metagenome of size G_m composed of genomes of sizes G_1 throughG_k can be viewed as a sum of fractions. Each component genome of size G_i constitutes a fraction of G_m:

$$\hat{G}_m = p_1 G_m + p_2 G_m + \ldots + p_l G_m$$

and:

$$\sum_{i=1}^{l} p_i = 1$$

Where p_i is the fraction of copies of the genome of species i in the sample and G_i is the size of the genome of species i.

Using species-specific gene markers, usually small ribosomal subunit rDNA, it is possible to estimate the species diversity in the sample, and provide an estimate of the different p_i values. Nevertheless, full or sometimes even adequate coverage (as judged by the rarefaction curve) of a species-rich environmental sample may be unattainable, especially for the genomes of the less represented species [39]–[42]. We expand upon this subject in the "Species Diversity" section.

Jeroen Raes and his colleagues have suggested an effective genome size or EGS measure that includes multiple plasmid copies, inserted sequences, and associated phages and viruses[43]. EGS uses the density (counts per megabase) of single copy marker genes to extrapolate the EGS.

$$EGS = \frac{a + b \times L^{-c}}{x}$$

Where L is the read length, x is the marker gene density, a, b, and c are empirical parameters empirically derived from 154 simulated metagenomes and found to be 21.2, 4,230, and 0.733, respectively. Raes and colleagues derived this formula from several different metagenomes, providing a useful measure of central tendency for genome size using a metagenomic sample. Note that a, b, and c were derived from simulated metagenomes, Therefore, care must be taken in using the EGS formula above, since the parameters given only provide a snapshot of a particular simulation. It is probably better to use EGS as a framework, in conjunction with a metagenomic simulator such as MetaSim [44] to generate parameters more compatible with population estimates in one's own research. MetaSim enables the creation of a simulated genome from regular genomic files; this makes it useful for testing and assessing the performance of other programs that manipulate and analyze metagenomic data, such as assembly or annotation programs.

Metagenome Assembly

In a genome project of a single organism or clone we can be certain that all extracted DNA fragments belong to the same genome, barring contaminants and extrachromosomal DNA. That is not the case when a metagenome is concerned. As we have just seen, coverage is usually incomplete, since environmental sequence sampling rarely produces all the sequences required for assembly. Furthermore, there is also the danger of assembling sequences from different OTUs, creating interspecies chimeras. Phrap, Forge, Arachne [45], JAZZ [46], and the Celera Assembler [47] are all assembly programs that were developed for single genome assembly from Sanger sequencing. They seem to provide good results even when assembling metagenomic sequence data from Sanger sequencing [48]. Most of these algorithms use mate-pair information for the assemblies. This information is used in assembly to check the scaffolds or the assembled intermediaries between raw reads and whole chromosomes. These assembly algorithms represent each read as a vertex and each detected overlap as an edge between the overlapping vertices. Finding the correct assembly is cast as a Hamiltonian path finding problem, for finding a path in a graph where each vertex is visited once (see Figure 3A–3C).

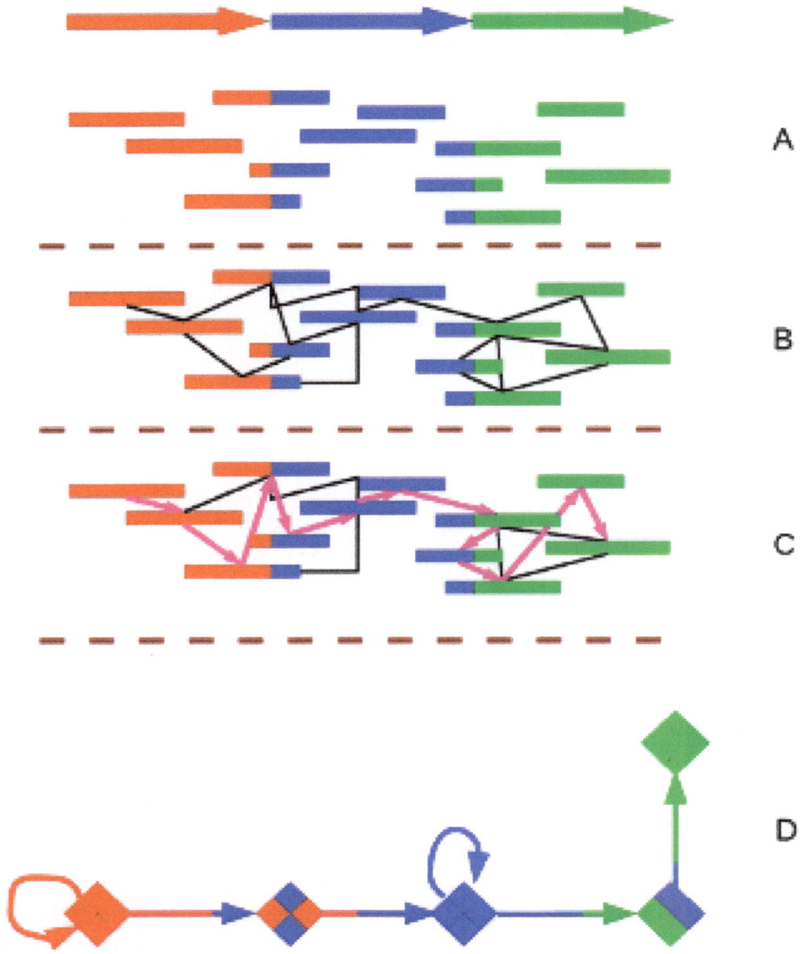

Figure 3: Fragment assembly. (A–C) Hamiltonian. (A) A sequence with overlapping reads; (B) Each read is represented as a vertex, with edges connecting the overlapping vertices; (C) the assembly solution is a Hamiltonian path (all vertices are visited, no vertex is visited more than once) through the resulting graph; (D) For short reads assembly, each vertex is a *k*-mer (or a hashed collection of *k*-mers), and the reads are threaded between vertices as edges. The solution is a Eulerian path, where each edge is visited once. Repeats are merged into a single edge. For detailed algorithms see [49], [50], [53]–[55]. http://dx.doi.org/10.1371/journal.pcbi.1000667.g003

For short reads, however, this technique is not suitable. To establish adequate coverage, short reads need to be produced in large quantities, and their short lengths means that there are many identical, or nearly identical, reads. The plethora of reads makes representing the vertices as single reads impossible. Another problem is that the sheer volume of reads makes the graph large and unmanageable. The solution to a Hamiltonian path is an NP-complete problem, meaning that the time necessary for a solution grows exponentially with the number of nodes. So while it is possible to solve for a relatively low number of reads as are produced using Sanger sequencing, the problem becomes intractable with the large amounts of sequence data from second generation sequencers.

One solution is for the vertices to represent *k*-mer words with the reads themselves being the edges connecting the vertices. Since the vertices represent *k*-mers rather than reads, the high number of reads and their redundancy does not affect the number of nodes. Repeats exist in the graph only once, with links to the different start and end points. Searches for overlaps are simplified, as overlapping reads are mapped onto the same edge and can easily be followed simultaneously. Finally, since the reads are represented as edges rather than vertices, the solution is a Eulerian path, where each edge is visited once. Unlike a Hamiltonian path, a linear-time algorithm to solve a Eulerian path does exist, making the assembly problem tractable for large number of reads.

The EULER assembler [49],[50] was the first to present this technique using de Bruijn graphs. De Bruijn graphs are *n*-dimensional graphs of *m* symbols. For metagenomic assembly, m=4 (A,T,G,C) and $n \simeq k$-mer length. Theoretically, there are m^n vertices, but the dimensionality can be greatly reduced by hashing the reads in the dataset to be assembled (see Figure 3). Other variations have since been published, adapting to short (100–200) [51],[52] and very short read lengths [53]–[55]. EULER and VELVET are available for download. Recently, Ye and Tang developed an assembly method that finds putative open reading frame (ORF) regions first, and then assembles those regions. This method, dubbed ORFome assembly, increases assembly accuracy for ORF regions at the expense of losing noncoding regions. Nevertheless, for many practical purposes this method is very useful, because it appears to have a better recovery rate, for coding regions only, than regular, whole genome assemblers [56]. For recent reviews on computational assembly methods see [57],[58].

GENE CALLING

Genes are the basic functional unit in the genome, which may constitute larger functional units such as operons, transcriptional units, and functional networks. Again, the incomplete and fragmentary nature of metagenomic data presents challenges to identifying genes. With Sanger random shotgun sequencing, whole genomes are rarely assembled, and in species-rich environments, many reads remain as singletons rather than being joined in contigs. In the Global Ocean Sampling (GOS) data, which were Sanger-sequenced, the mean number of whole reading frames per assembly is 4.7 [59].

Gene finding algorithms are trained to find whole ORFs and take into account information gleaned from large genomic stretches. For metagenomic data, however, this information is unavailable. Despite such drawbacks, Mavromatis and colleagues have shown that for a high complexity metagenomic dataset, gene prediction on assemblies can be as accurate as 85% of the originally predicted genes in the constituting genomes. For a low complexity set this goes up to 90% [48].

For genes with known homologs, BLASTing (using the Basic Local Alignment Search Tool)[60],[61] against known databases is a common approach. This approach informs of the existence of gene family members within a metagenome. BLAST cannot be used to find new families and new genes that have no homologs in known databases. For that, ab initio gene prediction tools are used. Those tools are mostly based on supervised learning and statistical pattern recognition methods. Most models use Markov models or Hidden markov models. Genemark.hmm is a program that uses inhomogeneous Markov models based on monocodon frequency analysis for gene calling [62]. When applied to metagenomic data, however, those methods lose sensitivity, because they often fail to identify partial ORFs that may be part of true genes. This is especially true when conventional gene calling methods are applied to raw Sanger fragments rather than to assemblies. Unsupervised methods are therefore required.

Yooseph and colleagues [59],[63] have used a different approach to gene finding when analyzing the global ocean survey data. They began with simple ORF identification of consecutive translatable regions that translate to at least 60 amino acids (aa). They then clustered those sequences using an all-against-all BLAST search, identifying clusters containing nonredundant sequences. In the next step, shadow ORFs were eliminated. Shadow ORFs are false ORFs in a different reading frame than the true ORF, but they overlap the true ORF and hence may be mistaken for a coding region. Yooseph and colleagues handled this by clustering all ORF candidates in the same reading frame and selecting the larger cluster as the one containing true ORFs, discarding the other ones as shadow ORFs. Finally, they removed ORF families with a KaKs K_a/K_s ratio that is close to 1. The rationale for this step is that putative proteins that are seemingly under no selective pressure (positive or negative) are probably falsely identified. Gene families coding for proteins under selective pressure are expected to have a

$$K_a/K_s \gg 1$$

Or

$$K_a/K_s \ll 1.$$

It has been argued that one drawback of the incremental clustering method is that it increases specificity at the expense of sensitivity; that is, it may have an excess of false negatives due to the removal of putative ORFs that do not cluster well or do not cluster at all in the database [64]. As of today, however, there has not been a thorough comparative evaluation of gene calling methods on first or second generation sequence data.

SPECIES DIVERSITY

Measuring Diversity

In the "Sample Size" we discussed using 16S/18S rDNA for phylotyping and assessing species coverage using a rarefaction curve. Microbial ecology has many tools for assessing species diversity. Rarefaction curves are used to estimate the coverage obtained from sampling, see Figure 4. α-diversity, β-diversity, and γ-diversity are all well-established diversity indices used in ecology, including microbial ecology. α-diversity is the biodiversity in a defined habitat or ecosystem; β-diversity compares species diversity between habitats; γ-diversity is the total biodiversity over a large region containing several ecosystems. Here we will discuss the application of these indices to metagenomic data.

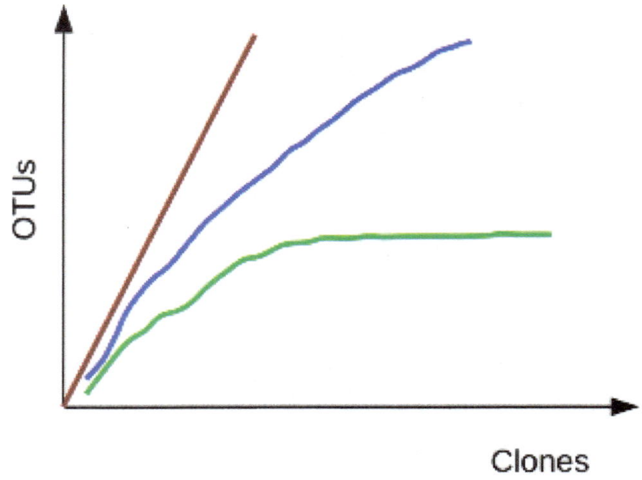

Figure 4: Rarefaction curves. Green, most or all species have been sampled; blue, this habitat has not been exhaustively sampled; red, species rich habitat, only a small fraction has been sampled. http://dx.doi.org/10.1371/journal.pcbi.1000667.g004

One way to calculate α-diversity is by using Shannon›s index:

$$H'_{\alpha} = -\sum_{i=1}^{S} p_i \ln p_i$$

where:

$$p_i = \frac{n_i}{N}$$

Where S is the total number of OTUs, n_i is the number of clones in each OUT, and N is the total number of individuals. p_i is the relative abundance of each OTU.

$$H_{\alpha max} = \ln S.$$

Using different sequence markers for OTU identification.

It should be noted that using 16S/18S rDNA as a proxy for OTU identification and counting is not without problems. First, rDNA has been criticized as an OTU marker, and evidence of horizontal gene transfer involving rDNA may confound its reliability even more [65]. Second, 16S rDNA may exist in multiple different sequence copies in a single bacterium: this would cause a variance in both the estimated individual bacterial count, and OTU numbers. It is commonly accepted that the mean number of bacterial ribosomal operons per genome is 4.1[66], but in a recent publication it has been shown that 16S rDNA gene copy numbers may vary between 1 and 15 [67],[68]. Alternative markers, such as single copy housekeeping genes have been suggested as alternative or complementary species and population tally markers for bacterial genomes. The *rpoB* gene is a strong candidate [69], but *amoA*, *pmoA*,*nirS*, *nirK*, *nosZ*, and *pufM* have also been suggested in different contexts [67],[70]. The housekeeping functionality of these genes makes them less susceptible to horizontal gene transfer. However, these studies have shown that on a finer level the use of housekeeping genes does improve upon 16S rDNA alone, the use of 16S rDNA as a marker for OTU identification and count is still sufficiently accurate for many purposes. The use of housekeeping genes for OTU classification is primarily for those cases when 16S rDNA provides a lower resolution than when a high diversity of species is expected. Another case where a housekeeping gene is preferable to 16S rDNA is when the variation in the housekeeping gene matches the acceptable taxonomy better than the variation in the rDNA sequences. The use of non-rDNA phylogenetic markers has been applied to metagenomic data, showing that certain microbial communities evolve faster than others [71].

Epidemiologists classify bacterial serovars for pathogen verification using Multilocus Sequence Typing (MLST) [72],[73]. MLST is a technique by which several standardized housekeeping genes are selected for OTU typification. There is an online resource for MLST, including a database for OTU identification (http://www.mlst.net/). MLST has been used successfully in some metagenomic studies [74]. However, MLST appears to be more useful for a finer level substrain typification, rather than OTUs.

In the same vein, 18S rDNA can have different count numbers in microeukaryotes, with an even larger copy number variation between species than 16S rDNA counts in prokaryotes. Care must be taken to account for this copy number variation when assessing the cell count in eukaryotic samples [75],[76].

There are several software packages we found very useful for biodiversity analysis. The first is a general purpose population analysis software, EstimateS (8.0) [77]. EstimateS contains a rich set of biodiversity analysis modules, but for microbial analysis it requires preprocessing of sequence data to transform it into generic population data. MOTHUR [78] is tailored towards microbial diversity analysis and provides tools for transforming sequence data to population data. It is not as rich in functional modules as EstimateS, but for most diversity analyses (rarefaction curves, standard estimate indices) it is more than adequate. QIIME, an extension of PyCogent [79], is in beta, but testing by one of us (IF) has shown it to be a very powerful and versatile package for analysis of genomic and metagenomic microbial ecology data (http://qiime.sourceforge.net). A more specialized software geared to the analysis of viral metagenomic data is PHACCS [80].

Binning

We wish to know not only who populates the sample, but also what the different OTUs are doing. We must therefore associate sequence data with the OTU of its origin. This analysis is called binning (placing the sequence in its correct "bin" or OTU). In many cases, suitable phylogenetic marker genes are missing either because rDNA sequences may be unsuitable (as in virus analyses), or may have been undersampled.

Here we will examine two binning strategies: composition-based binning and phylogenetic binning.

Composition-based binning.

The GC content of bacterial genomes is being used routinely for higher-level systematics [81]. With the advent of ESS data, a finer resolution for classifying or binning sequences is called for. Markov models based on *k*-mer frequencies have shown to be quite powerful for statistical analyses of DNA sequences [82]. For example, tetranucleotides are being used by the TETRA[83] program in the following fashion. There are 4^4=256 possible DNA tetranucleotides. For each tetranucleotide $t_i = [n_1, n_2, n_3, n_4]$, an expected frequency

$E(t_i)$ can be calculated by means of a maximal-order Markov model:

$$E(t_i)=\frac{O([n_1,n_2,n_3])\times O([n_2,n_3,n_4])}{O([n_2,n_3])}$$

Where O is the observed count of the sub-trimers and dimer of the tetramer.

The level of over- and underrepresentation of each tetranucleotide is evaluated using z-scores:

$$\sigma(O(t_i))$$

$$=E(t_i)\frac{(O([n_1,n_2])-O([n_1,n_2,n_3]))\times(O([n_2,n_3])-O([n_2,n_3,n_4]))}{O([n_2,n_3])^2}$$

$$Z(t_i)=\sqrt{\frac{(O(t_i)-E(t_i))^2}{\sigma(O(t_i))}}$$

Where $\sigma(O(t_i))$ is the variance in the tetranucleotide t_i.

Composition-based binning is not error-free. The closer the OTUs in the studied metagenome and the more numerous they are, the higher is the frequency of misclassification errors. The strength of k-mer–based binning is that there are no reference sequences required for the actual binning: all the information is intrinsic. This makes k-mer a powerful tool for binning ORFan sequences: sequences that have few or no homologs and therefore no known function. Therefore, TETRA is independent of existing genomic data, since it does not require any training. PhyloPythia [84] is a supervised method that trains a set of support vector machines (SVMs) to bin sequences of a length greater than 1 kb, and thus not suitable for binning second generation sequences. It performs best when a training set is similar in phylotypic composition to the training set. Growing Self Organizing Maps or GSOM [85] and Seeded GSOM or S-GSOM [86] use a variant of the machine learning algorithm self-organizing maps. S-GSOM improves upon GSOM by extracting the flanking sequences of highly conserved 16S rDNA from the metagenome and using them as seeds to assign other reads on the basis of their compositional similarity. Both use frequencies of di- to penta-nucleotides for binning assignment.

Another composition-based method is codon-usage. An old technique in genomics, codon usage, can also be used for binning metagenomic data. Different species use different codon frequencies to encode the same amino acids, and this observation can be exploited to classify ORF sequences. Shani Tzahor and colleagues have developed a composite supervised method that uses both TETRA and codon usage statistics to classify fragments in the 100–300-bp range [87].

TETRA is available for download, and PhyloPyhtia is available as a Web site, with a downloadable version available by request. GSOM/S-GSOM does not seem to be available at this time.

Similarity-based binning.

Another way to bin sequences is to find similarities to reference sequences that can be used to build a tree. This technique is useful when most sequences in the sample have significant similarities to reference sequences from known OTUs. Given an unannotated sequence A, and two annotated reference sequences B and C, and using the similarity function sim, let us consider the case where we have $sim(A,B)>sim(A,C)$; then, the sequence A will be placed on a node in the tree between B and C, and, in the case considered, closer to B. MEGAN [88] implements this method by reading a BLAST file output. Typically, the output is from the metagenomic reads or assemblies against nr, or any other sequence database that has a phylogenetic tree associated with it. MEGAN then assigns each read to the lowest common ancestor on the phylogenetic tree. This allows all sequences that have a homolog in nr to be assigned. Predicted gene sequences, having no homologs, are aggregated into their own single node on the tree. CARMA [89] is somewhat similar to MEGAN, but uses Pfam [90]as its source for taxonomic classification. It should be noted that a precise assignment to an OTU may not be possible in many cases. Nevertheless, unless it is an ORFan, the sequence can be placed in the species tree. The resulting picture of sequences on the species tree can provide an overview of the dominant species in the sample. Phymm [91] uses interpolated Markov models to characterize variable length DNA sequences by their phylogenetic grouping, unlike other methods. Phymm is trained on existing OTUs and learns which nucleotide length is best for classification. Also, Phymm does not leave reads unclassified, although that may impact its overall accuracy if there are many reads that cannot be accurately binned to any phylogenetic group.

As far as the usability of these software, CARMA will run on Unix-like environments, and its installation requires some third party software, and a rudimentary knowledge of Perl and MySQL. MEGAN runs in a Java virtual machine, and thus runs on almost out of the box Java-enabled platforms; it does require an installation of National Center for Biotechnology Information (NCBI)-formatted taxonomic reference database for lowest common ancestor mapping. Also, CARMA can run its own BLAST, whereas MEGAN requires a previously generated BLAST output as its input.

Functional Annotation

Having assembled the metagenome and identified putative ORFs we would now like to understand the functional potential of the microbial community from where we derived the metagenome: what are these microbes capable of doing as a community? The first level of functional annotation is assigning biological functions to the ORFs. This task is highly challenging when applied to regular genomic data [92], and the challenge is compounded in metagenomic data where many ORFs are partial, and a large fraction have no annotated homologs. The second level would be discovering genes that constitute biological networks, such as metabolic pathways, in the data. The latter task is hampered by our inability to accurately associate each annotated ORF with a single species, which means it is sometimes hard to determine which component of a network comes from which organism. Nevertheless, binning can help to some extent. As we shall see in the "Case Studies" section below, several studies have been carried out and led to the successful discovery of complementary metabolic pathways from microbes that constitute a community.

In metagenomic samples the probability of not calling all genes is higher than in a fully assembled genome, since many ORFs may be partial, and thus invisible to regular gene calling software that require a full ORF. Therefore, one strategy for functional annotation would be to skip the gene calling step altogether. Instead, simply use six-frame translations on the reads provided. If the translations are reasonably long they may be ORFs. Even if they are short, but they are cut short because of being at the edge of a contig, they may still be partial ORFs. Now these putative partial ORFs can be searched for motifs, HMM profiles, and other sequence signatures that may indicate functionality. The rationale is that the probability of calling a false ORF that also includes a known sequence signature is negligible. Some metagenomic annotation programs use this rationale. For example, Motif EXtraction (MEX)[93] is an unsupervised motif creation method that is successful in identifying enzymes in genomic and metagenomic data [94],[95]. Short, enzyme-specific peptides are identified in an unsupervised learning stage. They are subsequently associated with certain functions, in the supervised learning stage. The reason an unsupervised stage takes place is because, in many cases, new motifs can be identified within ORFans, even though their functional association may be unknown.

Even unassembled single reads (singletons) may be used to infer functional information, being long enough to find short motifs or significant BLAST hits. BLASTing singletons and annotating the results without assembly or postassembly has its use. Two versatile and useful annotation pipelines for metagenomics that implement the annotation principles outlined above are MG-RAST [96] and RAMMCAP [97]. MG-RAST accepts a 454 dataset as input, normalizes it (removes artefactual duplicate sequences, a known problem with 454 sequencing), and then performs gene calling and annotation by a variety of sequence similarity searches (mainly BLAST) against various sequence databases, including 16S rDNA. It then produces statistics on species associations and on metabolic pathway associations using the SEED subsystems database as its guideline. RAMMCAP uses the fast clustering algorithm CD-HIT [98] to cluster translated ORFs by high sequence similarity. The rationale is that many similar putative ORFs strengthen the hypothesis that they are indeed real ORFs. Optionally, CD-HIT also serves to reduce the volume of data to be annotated by picking representatives from identical or nearly identical sequences and annotating only the representative sequences. The annotation is then transfered to the highly similar sequence in each similarity-based cluster. The sequences are then compared to the profile HMM databases Pfam [90] and TIGRfam [99] using HMMer (http://hmmer.janelia.org/) for functional annotation.

Comparative Metagenomics

Comparing two or more metagenomes is necessary to understand how genomic differences affect, and are affected by, the abiotic environment. There are several sequence-based traits that can be compared: GC content was compared between marine and soil samples [59], microbial genome size [43], taxonomic [71], and functional content (e.g., [100]). Many comparative analyses, pairwise or multiple, make use of ordination statistics as when several metagenomic datasets are involved, or when several types of metadata are hypothesized to affect the observed compositions of the metagenomic populations. Principal component analysis (PCA) and nonmetric multidimensional scaling (NM-MDS) are typically used to visualize the data and to reveal which factors affect the observed data most (e.g., [101],[102]).

We mentioned MEGAN before as a binning software. MEGAN can also be used to compare the OTU composition of two or more frequency-normalized samples [103],[104]. MG-RAST provides a comparative functional and sequence-based analysis for uploaded samples, whereas IMG/M provides similar analysis for metagenomes that exist in the IMG/M site [105]. RAMMCAP also provides the ability to compare metagenomes. Other software used for the comparison of microbial populations based on phylogenetic data are UniFrac [106] and MetaStats [107], the latter being suitable for preprocessed clinical metagenomic data. Galaxy, an online workbench for the analysis of genomic data, can also perform some comparative metagenomic analysis, as well as taxonomic mapping [108]. Shotgun FunctionalizeR [109] is a stand-alone analysis tool for metagenomics samples written in R [110]. The megx.net resource includes include Meta Mine [111] for annotating genes using neighboring ORF information, and MetaLook [112] for organization of sequences using customized habitat criteria. CAMERA (http://camera.calit2.net) offers to BLAST the user's sequences against 40 existing genomic and metagenomic datasets. CAMERA also serves as an archive for select metagenomic datasets generated by marine microbial research funded by the Gordon and Betty Moore Foundation. All of these sites appear to be in a state of flux, with promised new functionalities to be added soon and with datasets constantly being updated.

We mentioned the importance of standardized recording of metadata in the Recording Metadata section above. Comparative analysis is where the importance of metadata comes into play: in order to properly compare between different environments, we need a common vocabulary describing the abiotic components. To date we do not know of software that provides a comparison between metadata or a comparative correlation between metadata and sequence data, although several such comparisons have been performed (see "Case Studies" section below).

APPLICATIONS

In this section we will discuss a few studies involving metagenomics. We chose these studies because each one illustrates a different insight that is derived from using metagenomics.

Correlations between Environmental Data and Metadata

The study of the effects of the environment on microbes is as old as microbiology itself. Antoni van Leeuwenhoek noted that the "animalcules" scraped from his mouth and that he viewed under his microscope were gone or were immobile after he drank hot coffee. Leeuwenhoek was the first to describe a correlation between temperature change and organism viability [113]. Ever since then, microbe species distribution, genetics, pathogenicity, virulence, colonization—indeed every aspect of microbial life—has been correlated with habitat traits such as temperature, salinity, pH, nutrient content, etc. Traits of host-borne microbes have been correlated with the host species, age, habitat, behavior, feeding habits, host organs chosen for settlement/pathogenicity, and, of course, clinical symptoms and many other traits.

With the advent of metagenomics, we are now able to study the genomic potential of a bacterial community and how it is affected by and affects its habitat. Many metagenomic studies have looked to some extent at correlations between sequence data, environment, and environmental attributes in an attempt to gain biological insight. One notable study by Turnbaugh and colleagues looked at the connection between the gut microbiome and obesity. The authors discovered that the metagenome in obese mice was enriched in carbohydrate active enzymes over that of lean mice. A separate biochemical experiment confirmed that the microbiome in obese mice has a larger energy harvesting capacity than in lean mice. They concluded that the gut microbiome contributes to obesity through this feed-forward cycle [100].

Studies such as those presented above looked at bivariate correlations: obesity and carbohydrate active enzyme enrichment. One recent study by Gianoulis and colleagues suggests how to locate multivariate correlations between metagenomic data and environmental attributes [114]. At the same time, environmental factors may combine in unexpected ways revealing new insights. Gianoulis and colleagues have identified covariation in amino acid transport and cofactor synthesis in nutrient-poor ocean areas, suggesting that limiting amounts of cofactor can (partially) explain increased import of amino acids in nutrient-limited conditions.

Understanding Symbiosis

In many cases, symbiotic bacteria living in an animal host consist of a small number of species, which are often phylogenetically distant. Because they are few species and the phylogenetic distance makes their sequences relatively easy to bin, metagenomics is useful for studying symbionts. Eisen and his colleagues sequenced ESS data from bacterial symbionts living in the glassy-winged sharpshooter, which is an insect that lives solely on tree sap, a nutrient poor diet. By binning the ESS data they inferred that one symbiont synthesizes amino acids for the host insect, while another synthesizes cofactors and vitamins [115]. Not only that, but the symbiont providing the vitamins lacks some amino-acid synthetic pathways, and the symbiont providing the amino-acid synthetic lacks the ability to synthesize the vitamins. Thus, both symbionts complement each other's metabolic deficiencies, as well as feeding their host. Another study of the marine gutless worm *Olavius algarvensis* has revealed the different roles of its four symbionts in generating nutrients and processing the worm's waste [116]. None of the symbionts in the insect or in the worm study could be cultured under the reported conditions. Metagenomics thus became the chosen avenue for these studies.

Enriching Gene Families

Another type of study enabled by metagenomics is the search for new members of a gene family. Metagenomics has opened up the floodgates of genomic material. Consequently the laborious hen-pecking for exemplars to enrich a studied gene family from known cultured species, has been replaced by the laborious computational filtering of appropriate exemplars from millions of environmental sequences. The previously small bacterial Eukaryotic Protein Kinase Like (ELK) family has been enriched several folds by the Global Ocean Sampling (GOS) project. Many new members of known families were identified, as well as new families. Within the protein sequences, four new residues of unknown function were found to be conserved, setting the stage for future functional studies of this family [117].

Metagenomics and Environmental Virology

Outnumbering living microbes, viruses are the most abundant biological entity on Earth: there are an estimated 10^{30} tailed bacteriophages in the biosphere [118]. In marine environments, viruses constitute 94% of all nucleic-acid containing particles, although owing to their small size they are estimated to constitute only 5% of the biomass. Metagenomic studies have enriched our knowledge of viral diversity and the role viruses play as facilitators of microbial genetic diversity. Sequence similarity analyses of viral metagenomic data have shown that approximately 90% of the sequences have no similarity to GenBank sequences, telling of an underrepresentation of viral sequence data in sequence databases [119].

Transduction—the transfer of genetic material via a viral vector—is known to be a strong contributer to genetic diversity in prokaryotes. Metagenomic studies help us assess the magnitude of virally contributed genetic diversity. For example, the existence of photosynthetic genes in cyanophages—viruses infecting cyanobacteria—has been known for some time[120],[121]. However, metagenomic studies have revealed the extent of this phenomenon: it is estimated that 60% of the psbA genes, a component of Photosystem I, in surface water are of phage origin. Another metagenomic study revealed the existence of whole photosynthetic cassettes in cyanophages, which may increase host fitness by supplementing and enhancing existing cyanobacterial photosystems. The latter findings were enabled by the metagenomic data from Global Ocean Sampling (GOS). Surveying these data using simple sequence similarity analyses and chromosomal gene location have revealed the existence of Photosystem I genes in cyanophages, and the extent of their distribution [94],[122].

Clinical virology also stands to benefit from metagenomic analysis [123]. Indeed, recent molecular-based discoveries of highly prevalent viral infections caused by anellovirus [124] and GBV-C [125] highlight the need for a better understanding of the human viral flora.

The computational analysis of viral metagenomic data is particularly challenging. First, viruses may exist as a chromosomal insert, such as prophages, which are incorporated in the host genome. This incorporation confuses the ability to distinguish viral genomic elements from the host. Furthermore, when filtering exclusively for viral particles, prophage elements are lost. Second, viruses have no distinct phylogenetic marker gene, equivalent to the small ribosomal subunit rRNA in prokaryotes or eukaryotes. The lack of a consensual marker gene hampers phylogenetic and diversity analysis. Third, as stated above, most viral genes have no annotated homolog in sequence databases, which impedes functional analysis and indeed the identification of viral genes for what they are. Indeed, by some estimates the majority of ORFans in the biosphere is due to lateral gene transfer of viral origin [126] and the fact that phage-induced lateral gene transfer contributes in a major way to microbial diversity [127].

THE FUTURE

We are in the midst of the fastest growing revolution in molecular biology, perhaps in all of life science, and it only seems to be accelerating. Sanger sequencing has been with us for over three decades. High-throughput 3730 sequencing has been around for 8 years, Roche 454 instrumentation has been available for 6 years, and Illumina GA for 3 years. The latter two methods have enabled us to generate more sequence data than Sanger sequencing has. We are still coming to grips with the large volume of data, and how to analyze it. Assembly, quality control, binning, and annotation all require ingenious algorithms combined with the latest computational power. It appears that sequencing technology is changing almost faster than the associated computational techniques can keep up. There are many indications that within a few years, short-read second generation sequencing may be outdated. Third generation sequencing that will enable the sequencing of a single chromosome in a single pass with few or no fragments should be established very soon [35],[36]. Does this plausible obsolescence of second generation sequencing change current metagenomic computational challenges? For some applications, assembly algorithms may be less warranted, but for species-rich samples, we may not be able to rely solely on third-generation sequencing for good sampling. Coverage assessment, gene finding, binning, and annotation will still be necessary.

The BASE technology from Oxford Nanopore is able to differentiate between cytosine and methyl-cytosine during sequencing [37]. Methylation acts as a primitive immune system in bacteria [128], and as an expression control mechanism in eukarya [129]. This additional epigenetic information has been mostly unavailable in sequencing projects due to an inability to obtain it in a high-throughput fashion. Pyrosequencing already offers a capability for quantitative methylation [130] and in all likelihood methylation data will be soon made available routinely along with the four base data, and the associated bioinformatics would need to address that.

Another growing problem is that of data management. Sequencing centers are working to equip themselves with computational infrastructure to meet the flow of sequence data. However, many research institutes who request the sequencing do not have the computational infrastructure needed to deal with analysis and long-term storage of these data. The sheer volume of data raises new constraints on its transfer and analysis. These challenges would have to be met by concerted efforts of life scientists, computer scientists, engineers, and funding agencies [131],[132]. Genomic data tell us what an organism is capable of doing, i.e., its genomic potential. What it is actually doing at a given time-frame is discovered by examining transcription (mRNA) and translation (protein) data. In the world of microbial communities, those studies have been dubbed metatranscriptomics and metaproteomics, respectively. These two fields are outside the scope of this review, but note that they too are very much in a development boom, technologically and computationally [133]–[135].

We hope this primer has been useful and informative. Because computational metagenomics is changing rapidly, we call upon the readers of this article who are knowledgeable in the subject to use the comment section of *PLoS Computational Biology* to provide updated information.

Box 1. Glossary of terms

Binning Clustering sequences based on their nucleotide composition or similarity to a reference database

Contig A set of overlapping DNA segments

Coverage (in sequencing) The mean number of times a nucleotide is sequenced in a genome

ESS Environmental Shotgun Sequencing

K_a/K_s The ratio of the rate of nonsynonymous substitutions (K_a) to the rate of synonymous substitutions (K_s), which can be used as an indicator of selective pressure acting on a protein-coding gene

Mate pairs Sequences known to be in the 3′ and 5′ of a contig from a single clone

Metadata Definitional data that provide information about or documentation of other data

Metagenome The DNA obtained from uncultured microorganisms

Metagenomics The study of genomic DNA obtained from uncultured microorganisms

Metaproteomics The study of protein molecular data obtained from environmental samples using proteomics techniques

Metatranscriptomics The study of transcription sequence data obtained from environmental samples

ORFan An ORF that has no (or few, depending on definition) homologs in other organisms

OTU Operational taxonomic unit, species distinction in microbiology. Typically using rDNA and a percent similarity threshold for classifying microbes within the same, or different, OTUs

Ontology A formal representation of a set of concepts and the relationships between them. Ontologies are used to create a consensual unambiguous controlled vocabulary

Polony Discrete clonal amplifications of a single DNA molecule, grown in a gel matrix. The clusters can then be individually sequenced, producing short reads. Polony-based sequencing is the basis of most second generation sequencers

Rarefaction curve A curve describing the growth of a number of species discovered as a function of individuals sampled

Ribotype A phylotypic classification based on rDNA sequences

Scaffold A series of contigs that are in the right order but not necessarily connected in one contiguous stretch

Shadow ORF An incorrectly identified ORF that overlaps the coding region of the true ORF.

ACKNOWLEDGMENTS

We would like to thank Linda Amaral-Zettler, Yanay Ofran, Daniel Huson, Jack Gilbert, Mufit Ozden, Frank Oliver Glöckner, and Renzo Kottmann for their critique and input. We would also like to thank the referees for contributing their time and effort towards making this a better article. IF dedicates his contribution to the memory of Ilan Friedberg.

REFERENCES

1. Whitman WB, Coleman DC, Wiebe WJ (1998) Prokaryotes: the unseen majority. Proc Natl Acad Sci U S A 95: 6578–6583.
2. Savage DC (1977) Microbial ecology of the gastrointestinal tract. Annu Rev Microbiol 31: 107–133.
3. Berg R (1996) The indigenous gastrointestinal microflora. Trends Microbiol 4: 430–435.
4. Collins FS, McKusick VA (2001) Implications of the human genome project for medical science. JAMA 285: 540–544.
5. Kaput J, Cotton RGHG, Hardman L, Watson M, Al Aqeel AII, et al. (2009) Planning the human variome project: the spain report. Hum Mut 30: 496–510.
6. O'Hara AM, Shanahan F (2006) The gut flora as a forgotten organ. EMBO Rep 7: 688–693.
7. Fiers W, Contreras R, Duerinck F, Haegeman G, Iserentant D, et al. (1976) Complete nucleotide sequence of bacteriophage ms2 RNA: primary and secondary structure of the replicase gene. Nature 260: 500–507.
8. Sanger F, Coulson AR, Friedmann T, Air GM, Barrell BG, et al. (1978) The nucleotide sequence of bacteriophage phix174. J Mol Biol 125: 225–246.
9. Fleischmann RD, Adams MD, White O, Clayton RA, Kirkness EF, et al. (1995) Whole-genome random sequencing and assembly of haemophilus influenzae rd. Science 269: 496–512.
10. Amann RI, Ludwig W, Schleifer KH (1995) Phylogenetic identification and in situ detection of individual microbial cells without cultivation. Microbiol Rev 59: 143–169.
11. Pace NR (1997) A molecular view of microbial diversity and the biosphere. Science 276: 734–740.
12. Rappé MS, Giovannoni SJ (2003) The uncultured microbial majority. Annu Rev Microbiol 57: 369–394.
13. Handelsman J, Rondon MR, Brady SF, Clardy J, Goodman RM (1998) Molecular biological access to the chemistry of unknown soil microbes: a new frontier for natural products. Chem Biol 5: R245–R249.
14. Rondon MR, August PR, Bettermann AD, Brady SF, Grossman TH, et al. (2000) Cloning the soil metagenome: a strategy for accessing the genetic and functional diversity of uncultured microorganisms. Appl Environ Microbiol 66: 2541–2547.
15. Field D, Garrity G, Gray T, Morrison N, Selengut J, et al. (2008) The minimum information about a genome sequence (migs) specification. Nat Biotech 26: 541–547.
16. Kottmann R, Gray T, Murphy S, Kagan L, Kravitz S, et al. (2008) A standard MIGS/MIMS compliant xml schema: toward the development of the genomic contextual data markup language (gcdml). OMICS 12: 115–121.
17. Brazma A, Hingamp P, Quackenbush J, Sherlock G, Spellman P, et al. (2001) Minimum information about a microarray experiment (MIAME)–toward standards for microarray data. Nat Genet 29: 365–371.
18. Westbrook JD, Fitzgerald PM (2003) The PDB format, mmCIF, and other data formats. Methods Biochem Anal 44: 161–179.
19. .Sanger F, Coulson AR (1975) A rapid method for determining sequences in DNA by primed synthesis with DNA polymerase. J Mol Biol 94: 441–448.
20. Sanger F, Nicklen S, Coulson AR (1977) DNA sequencing with chain-terminating inhibitors. Proc Natl Acad Sci U S A 74: 5463–5467.
21. Sorek R, Zhu Y, Creevey CJ, Francino MP, Bork P, et al. (2007) Genome-wide experimental determination of barriers to horizontal gene transfer. Science 318: 1449–1452.
22. Pedros-Alio C (2007) Ecology: Dipping into the rare biosphere. Science 315: 192–193.
23. Sogin ML, Morrison HG, Huber JA, Mark Welch D, Huse SM, et al. (2006) Microbial diversity in the deep sea and the underexplored "rare biosphere". Proc Natl Acad Sci U S A 103: 12115–12120.
24. Hamp TJ, Jones WJ, Fodor AA (2009) Effects of experimental choices and analysis noise on surveys of the "rare biosphere". Appl Environ Microbiol 75: 3263–3270.
25. Neufeld JD, Li J, Mohn WW (2008) Scratching the surface of the rare biosphere with ribosomal sequence tag primers. FEMS Microbiol Lett 283: 146–153.
26. Mitra RD, Church GM (1999) In situ localized amplification and contact replication of many individual DNA molecules. Nucl Acids Res 27: e34.
27. Porreca GJ, Shendure J, Church GM (2006) Polony DNA sequencing. Hoboken (New Jersey): John Wiley and Sons, Inc. Current protocols in molecular biology. Frederick M. Ausubel, et al., editors. Chapter 7.
28. Nyrén P, Pettersson B, Uhlén M (1993) Solid phase DNA minisequencing by an enzymatic luminometric inorganic pyrophosphate detection assay. Anal Biochem 208: 171–175.
29. Ronaghi M, Uhlén M, Nyrén P (1998) A sequencing method based on real-time pyrophosphate. Science 281: 363–365.
30. Margulies M, Egholm M, Altman WE, Attiya S, Bader JS, et al. (2005) Genome sequencing in microfabricated high-density picolitre reactors. Nature 7057: 376–380.
31. Holt RA, Jones SJM (2008) The new paradigm of flow cell sequencing. Genome Res 18: 839–846.
32. Shendure J, Ji H (2008) Next-generation DNA sequencing. Nat Biotechnol 26: 1135–1145.
33. Harismendy O, Ng P, Strausberg R, Wang X, Stockwell T, et al. (2009) Evaluation of next generation sequencing platforms for population targeted sequencing studies. Genome Biol 10: R32.
34. McPherson JD (2009) Next-generation gap. Nat Methods

6: S2–S5.

35. Clarke J, Wu HC, Jayasinghe L, Patel A, Reid S, et al. (2009) Continuous base identification for single-molecule nanopore DNA sequencing. Nat Nano 4: 265–270.

36. Eid J, Fehr A, Gray J, Luong K, Lyle J, et al. (2008) Real-time DNA sequencing from single polymerase molecules. Science 323: 133–138.

37. Branton D, Deamer DW, Marziali A, Bayley H, Benner SA, et al. (2008) The potential and challenges of nanopore sequencing. Nat Biotech 26: 1146–1153.

38. Lander ES, Waterman MS (1988) Genomic mapping by fingerprinting random clones: a mathematical analysis. Genomics 2: 231–239.

39. Torsvik V, Goksoyr J, Daae FL (1990) High diversity in DNA of soil bacteria. Appl Environ Microbiol 56: 782–787.

40. Youssef NH, Elshahed MS (2008) Species richness in soil bacterial communities: a proposed approach to overcome sample size bias. J of Microb Meth 75: 86–91.

41. Fierer N, Jackson RB (2006) The diversity and biogeography of soil bacterial communities. Proc Natl Acad Sci U S A 103: 626–631.

42. Countway PD, Gast RJ, Pratik Sava I, Caron DA (2005) Protistan diversity estimates based on 18s rDNA from seawater incubations in the western north atlantic. J Euk Micriobiol 52: 95–106.

43. Raes J, Korbel JO, Lercher MJ, von Mering C, Bork P (2007) Prediction of effective genome size in metagenomic samples. Genome Biol 8: R10.

44. Richter DC, Ott F, Auch AF, Schmid R, Huson DH (2008) Metasima sequencing simulator for genomics and metagenomics. PLoS ONE 3: e3373. doi:10.1371/journal.pone.0003373.

45. Batzoglou S, Jaffe DB, Stanley K, Butler J, Gnerre S, et al. (2002) ARACHNE: a whole-genome shotgun assembler. Genome Res 12: 177–189.

46. Aparicio S, Chapman J, Stupka E, Putnam N, ming Chia J, et al. (2002) Whole-genome shotgun assembly and analysis of the genome of fugu rubripes. Science 297: 1301–1310.

47. Myers EW, Sutton GG, Delcher AL, Dew IM, Fasulo DP, et al. (2000) A whole-genome assembly of drosophila. Science 287: 2196–2204.

48. Mavromatis K, Ivanova N, Barry K, Shapiro H, Goltsman E, et al. (2007) Use of simulated data sets to evaluate the fidelity of metagenomic processing methods. Nat Methods 4: 495–500.

49. Pevzner PA, Tang H, Waterman MS (2001) An eulerian path approach to DNA fragment assembly. Proc Natl Acad Sci U S A 98: 9748–9753.

50. Chaisson MJ, Pevzner PA (2008) Short read fragment assembly of bacterial genomes. Genome Res 18: 324–330.

51. Myers EW (2005) The fragment assembly string graph. Bioinformatics 21: Suppl 2ii79–ii85.

52. Chaisson M, Pevzner P, Tang H (2004) Fragment assembly with short reads. Bioinformatics 20: 2067–2074.

53. Zerbino DR, Birney E (2008) Velvet: algorithms for de novo short read assembly using de bruijn graphs. Genome Res 18: 821–829.

54. Sundquist A, Ronaghi M, Tang H, Pevzner P, Batzoglou S (2007) Whole-genome sequencing and assembly with high-throughput, short-read technologies. PLoS ONE 2: e484. doi:10.1371/journal.pone.0000484.

55. Warren RL, Sutton GG, Jones SJ, Holt RA (2007) Assembling millions of short DNA sequences using SSAKE. Bioinformatics 23: 500–501.

56. Ye Y, Tang H (2009) An orfome assembly approach to metagenomics sequences analysis. J Bioinform Comput Biol 7: 455–471.

57. Pop M (2009) Genome assembly reborn: recent computational challenges. Brief Bioinform 4: 354–366.

58. Flicek P, Birney E (2009) Sense from sequence reads: methods for alignment and assembly. Nat Methods 6: S6–S12.

59. Yooseph S, Sutton G, Rusch DB, Halpern AL, Williamson SJ, et al. (2007) The Sorcerer II global ocean sampling expedition: expanding the universe of protein families. PLoS Biol 5: e16. doi:10.1371/journal.pbio.0050016.

60. Altschul SF, Gish W, Miller W, Myers EW, Lipman DJ (1990) Basic local alignment search tool. J Mol Biol 215: 403–410.

61. Altschul SF, Madden TL, Schäffer AA, Zhang J, Zhang Z, et al. (1997) Gapped blast and psi-blast: a new generation of protein database search programs. Nucleic Acids Res 25: 3389–3402.

62. Azad RK, Borodovsky M (2004) Probabilistic methods of identifying genes in prokaryotic genomes: connections to the HMM theory. Brief Bioinform 5: 118–130.

63. Yooseph S, Li W, Sutton G (2008) Gene identification and protein classification in microbial metagenomic sequence data via incremental clustering. BMC Bioinformatics 9: 182.

64. Hoff KJ, Tech M, Lingner T, Daniel R, Morgenstern B, et al. (2008) Gene prediction in metagenomic fragments: a large scale machine learning approach. BMC Bioinformatics 9: 217.

65. Schouls LM, Schot CS, Jacobs JA (2003) Horizontal transfer of segments of the 16s rRNA genes between species of the streptococcus anginosus group. J Bacteriol 185: 7241–7246.

66. DeSantis TZ, Hugenholtz P, Larsen N, Rojas M, Brodie EL, et al. (2006) Greengenes, a chimera-checked 16s rRNA gene database and workbench compatible with arb. Appl Environ Microbiol 72: 5069–5072.

67. Case RJ, Boucher Y, Dahllof I, Holmstrom C, Doolittle FW, et al. (2007) Use of 16s rRNA and rpob genes as molecular markers for microbial ecology studies. Appl Environ Microbiol 73: 278–288.

68. Klappenbach JA, Saxman PR, Cole JR, Schmidt TM (2001) rrndb: the Ribosomal RNA Operon Copy Number Database. Nucl Acids Res 29: 181–184.

69. Walsh DA, Bapteste E, Kamekura M, Doolittle FW (2004) Evolution of the RNA polymerase b' subunit gene (rpob') in halobacteriales: a complementary molecular marker to the ssu rRNA gene. Mol Biol Evol 21: 2340–2351.

70. Achenbach LA, Carey J, Madigan MT (2001) Photosynthetic and phylogenetic primers for detection of anoxygenic phototrophs in natural environments. Appl Environ Microbiol 67: 2922–2926.

71. von Mering C, Hugenholtz P, Raes J, Tringe SG, Doerks T, et al. (2007) Quantitative phylogenetic assessment of microbial communities in diverse environments. Science 315: 1126–1130.

72. Enright MC, Spratt BG (1999) Multilocus sequence typing. Trends Microbiol 7: 482–487.

73. Maiden MCJ, Bygraves JA, Feil E, Morelli G, Russell JE, et al. (1998) Multilocus sequence typing: A portable approach to the identification of clones within populations of pathogenic microorganisms. Proc Natl Acad Sci U S A 95: 3140–3145.

74. Mahenthiralingam E, Baldwin A, Drevinek P, Vanlaere E, Vandamme P, et al. (2006) Multilocus sequence typing breathes life into a microbial metagenome. PLoS ONE 1: e17. doi:10.1371/journal.pone.0000017.

75. Zhu F, Massana R, Not F, Marie D, Vaulot D (2005) Mapping of picoeucaryotes in marine ecosystems with quantitative pcr of the 18s rRNA gene. FEMS Microbiol Ecol 52: 79–92.

76. Loram JE, Boonham N, O'Toole P, Trapido-Rosenthal HG, Douglas AE (2007) Molecular quantification of symbiotic dinoflagellate algae of the genus symbiodinium. Biol Bull 212: 259–268.

77. Colwell RK (2005) EstimateS - statistical estimation of species richness and shared species from samples.

78. Schloss PK (2008) Mothur - the one-stop source for your computational microbial ecology needs.

79. Knight R, Maxwell P, Birmingham A, Carnes J, Caporaso JG, et al. (2007) PyCogent: a toolkit for making sense from sequence. Genome Biol 8: R171.

80. Angly F, Rodriguez-Brito B, Bangor D, McNairnie P, Breitbart M, et al. (2005) Phaccs, an online tool for estimating the structure and diversity of uncultured viral communities using metagenomic information. BMC Bioinformatics 6: 41.

81. 1988) (1988) International committee on systematic bacteriology announcement of the report of the ad hoc committee on reconciliation of approaches to bacterial systematics. J Appl Bacteriol 64: 283–284.

82. Schbath S, Prum B, de Turckheim E (1995) Exceptional motifs in different markov chain models for a statistical analysis of DNA sequences. J Comput Biol 2: 417–437.

83. Teeling H, Waldmann J, Lombardot T, Bauer M, Glöckner FO (2004) Tetra: a web-service and a stand-alone program for the analysis and comparison of tetranucleotide usage patterns in DNA sequences. BMC Bioinformatics 5: 163.

84. Mchardy AC, Martín HG, Tsirigos A, Hugenholtz P, Rigoutsos I (2006) Accurate phylogenetic classification of variable-length DNA fragments. Nat Methods 4: 63–72.

85. Chan CKKK, Hsu AL, Tang SLL, Halgamuge SK (2008) Using growing self-organising maps to improve the binning process in environmental whole-genome shotgun sequencing. J Biomed Biotech 513701.

86. Chan CKK, Hsu AL, Halgamuge SK, Tang SL (2008) Binning sequences using very sparse labels within a metagenome. BMC Bioinformatics 9: 215.

87. Tzahor S, Aharonovich DM, Kirkup B, Yogev T, Frank IB, et al. (2009) A supervised learning approach for taxonomic classification of core-photosystem-ii genes and transcripts in the marine environment. BMC Genomics 10: 229.

88. Huson DH, Auch AF, Qi J, Schuster SC (2007) MEGAN analysis of metagenomic data. Genome Res 17: 377–386.

89. Krause L, Diaz NN, Goesmann A, Kelley S, Nattkemper TW, et al. (2008) Phylogenetic classification of short environmental DNA fragments. Nucl Acids Res 36: 2230–2239.

90. Finn RD, Tate J, Mistry J, Coggill PC, Sammut SJJ, et al. (2008) The pfam protein families database. Nucleic Acids Res 36: D281–D288.

91. Brady A, Salzberg SL (2009) Phymm and phymmbl: metagenomic phylogenetic classification with interpolated markov models. Nat Methods 6: 673–676.

92. Friedberg I (2006) Automated protein function prediction-the genomic challenge. Brief Bioinform 7: 225–242.

93. Kunik V, Meroz Y, Solan Z, Sandbank B, Weingart U, et al. (2007) Functional representation of enzymes by specific peptides. PLoS Comput Biol 3: e167. doi:10.1371/journal.pcbi.0030167.

94. Sharon I, Tzahor S, Williamson S, Shmoish M, Man-Aharonovich D, et al. (2007) Viral photosynthetic reaction center genes and transcripts in the marine environment. ISME J 1: 492–501.

95. Meroz Y, Horn D (2008) Biological roles of specific peptides in enzymes. Proteins 72: 606–612.

96. Meyer F, Paarmann D, D'Souza M, Olson RD, Glass EM, et al. (2008) The metagenomics rast server - a public resource for the automatic phylogenetic and functional analysis of metagenomes. BMC Bioinformatics 9: 386.

97. Li W (2009) Analysis and comparison of very large metagenomes with fast clustering and functional annotation. BMC Bioinformatics 10: 359.

98. Li W, Godzik A (2006) CD-HIT: a fast program for clustering and comparing large sets of protein or nucleotide sequences. Bioinformatics 22: 1658–1659.

99. Haft DH, Selengut JD, White O (2003) The TIGRFAMs database of protein families. Nucl Acids Res 31: 371–373.

100. Turnbaugh PJ, Ley RE, Mahowald MA, Magrini V, Mardis ER, et al. (2006) An obesity-associated gut microbiome with increased capacity for energy harvest. Nature 444: 1027–1031.

101. Brulc JM, Antonopoulos DA, Berg Miller ME, Wilson MK, Yannarell AC, et al. (2009) Gene-centric metagenomics of the fiber-adherent bovine rumen

microbiome reveals forage specific glycoside hydrolases. Proc Natl Acad Sci U S A 106: 1948–1953.

102. Willner D, Furlan M, Haynes M, Schmieder R, Angly FE, et al. (2009) Metagenomic analysis of respiratory tract DNA viral communities in cystic fibrosis and non-cystic fibrosis individuals. PLoS ONE 4: e7370. doi:10.1371/journal.pone.0007370.

103. Mitra S, Klar B, Huson DH (2009) Visual and statistical comparison of metagenomes. Bioinformatics btp341.

104. Huson D, Richter D, Mitra S, Auch A, Schuster S (2009) Methods for comparative metagenomics. BMC Bioinformatics 10: S1–S12.

105. Markowitz VM, Ivanova NN, Szeto E, Palaniappan K, Chu K, et al. (2008) IMG/M: a data management and analysis system for metagenomes. Nucleic Acids Res 36: D534–D538.

106. Lozupone C, Knight R (2005) UniFrac: a new phylogenetic method for comparing microbial communities. Appl Environ Microbiol 71: 8228–8235.

107. White JR, Nagarajan N, Pop M (2009) Statistical methods for detecting differentially abundant features in clinical metagenomic samples. PLoS Comput Biol 5: e1000352. doi:10.1371/journal.pcbi.1000352.

108. Giardine B, Riemer C, Hardison RC, Burhans R, Elnitski L, et al. (2005) Galaxy: a platform for interactive large-scale genome analysis. Genome Res 15: 1451–1455.

109. Kristiansson E, Hugenholtz P, Dalevi D (2009) ShotgunFunctionalizeR: an R-package for functional comparison of metagenomes. Bioinformatics 25: 2737–2738.

110. R Development Core Team (2009) R: a language and environment for statistical computing. Vienna: R Foundation for Statistical Computing. Available: http://www.R-project.org. ISBN 3-900051-07-0.

111. Bohnebeck U, Lombardot T, Kottmann R, Glockner F (2008) Metamine - a tool to detect and analyse gene patterns in their environmental context. BMC Bioinformatics 9: 459.

112. Lombardot T, Kottmann R, Giuliani G, de Bono A, Addor N, et al. (2007) Metalook: a 3d visualisation software for marine ecological genomics. BMC Bioinformatics 8: 406.

113. Egerton FN (2006) A history of the ecological sciences, part 19: Leeuwenhoek's microscopic natural history. Ecol Appl 87: 47–58.

114. Gianoulis TA, Raes J, Patel PV, Bjornson R, Korbel JO, et al. (2009) Quantifying environmental adaptation of metabolic pathways in metagenomics. Proc Natl Acad Sci U S A 106: 1374–1379.

115. Wu D, Daugherty SC, Van Aken SE, Pai GH, Watkins KL, et al. (2006) Metabolic complementarity and genomics of the dual bacterial symbiosis of sharpshooters. PLoS Biol 4: e188. doi:10.1371/journal.pbio.0040188.

116. Woyke T, Teeling H, Ivanova NN, Huntemann M, Richter M, et al. (2006) Symbiosis insights through metagenomic analysis of a microbial consortium. Nature 443: 950–955.

117. Kannan N, Taylor SSS, Zhai Y, Venter JCC, Manning G (2007) Structural and functional diversity of the microbial kinome. PLoS Biol 5: e17. doi:10.1371/journal.pbio.0050017.

118. Brussow H, Hendrix RW (2002) Phage genomics: small is beautiful. Cell 108: 13–16.

119. Edwards RA, Rohwer F (2005) Viral metagenomics. Nat Rev Microbiol 3: 504–510.

120. Mann NH, Cook A, Millard A, Bailey S, Clokie M (2003) Marine ecosystems: bacterial photosynthesis genes in a virus. Nature 424: 741.

121. Millard A, Clokie MRJ, Shub DA, Mann NH (2004) Genetic organization of the psbad region in phages infecting marine synechococcus strains. Proc Natl Acad Sci U S A 101: 11007–11012.

122. Sharon I, Alperovitch A, Rohwer F, Haynes M, Glaser F, et al. (2009) Photosystem i gene cassettes are present in marine virus genomes. Nature 461: 258–262.

123. Delwart EL (2007) Viral metagenomics. Rev Med Virol 17: 115–131.

124. Nishizawa T, Okamoto H, Konishi K, Yoshizawa H, Miyakawa Y, et al. (1997) A novel dna virus (ttv) associated with elevated transaminase levels in posttransfusion hepatitis of unknown etiology. Biochem Biophys Res Commun 241: 92–97.

125. Simons JN, Leary TP, Dawson GJ, Pilot-Matias TJ, Muerhoff AS, et al. (1995) Isolation of novel virus-like sequences associated with human hepatitis. Nat Med 1: 564–569.

126. Yin Y, Fischer D (2006) On the origin of microbial orfans: quantifying the strength of the evidence for viral lateral transfer. BMC Evol Biol 6: 63.

127. Hambly E, Suttle C (2005) The viriosphere, diversity, and genetic exchange within phage communities. Curr Opin Microbiol 8: 444–450.

128. Boyer HW (1971) Dna restriction and modification mechanisms in bacteria. Annu Rev Microbiol 25: 153–176.

129. Kass S (1997) How does DNA methylation repress transcription? Trends Genet 13: 444–449.

130. Tost J, Gut IG (2007) Dna methylation analysis by pyrosequencing. Nat Protoc 2: 2265–2275.

131. Batley J, Edwards D (2009) Genome sequence data: management, storage, and visualization. BioTechniques 46: 333–336.

132. Richter BG, Sexton DP (2009) Managing and analyzing next-generation sequence data. PLoS Comput Biol 5: e1000369. doi:10.1371/journal.pcbi.1000369.

133. Bailly J, Fraissinet-Tachet L, Verner MC, Debaud JC, Lemaire M, et al. (2007) Soil eukaryotic functional diversity, a metatranscriptomic approach. ISME J 1: 632–642.

134. Wilmes P, Bond PL (2004) The application of two-dimensional polyacrylamide gel electrophoresis and downstream analyses to a mixed community of prokaryotic microorganisms. Environ Microbiol 6: 911–920.

135. Wilmes P, Bond PL (2006) Metaproteomics: studying functional gene expression in microbial ecosystems. Trends Microbiol 14: 92–97.

CHAPTER 8

GENE PREDICTION IN METAGENOMIC FRAGMENTS: A LARGE SCALE MACHINE LEARNING APPROACH

Katharina J Hoff, Maike Tech, Thomas Lingner, Rolf Daniel, Burkhard Morgenstern and Peter Meinicke

BACKGROUND

Communities of natural microorganisms often encompass a bewildering range of physiological, metabolic, and genomic diversity. The microbial diversity in most environments exceeds the biodiversity of plants and animals by orders of magnitude. Phylogenetic surveys of complex ecosystems such as soils and sediments have demonstrated that the multitude of discrete prokaryotic species represented in a single sample goes far beyond the number and phenotypes of known cultured microorganisms [1, 2]. Direct cultivation or indirect molecular approaches have been used to explore and to exploit this enormous microbial diversity. Cultivation and isolation of microorganisms are the traditional methods. It has been estimated that less than 1 % of environmental microorganisms are culturable using standard cultivation methods. Thus, only a tiny portion of the gene pool of natural microbial communities has been analyzed so far [2, 3, 4].

To circumvent some of the limitations of cultivation approaches, indirect molecular methods, such as metagenomics have been developed. Metagenomics is based on the direct isolation, cloning, and subsequent analysis of microbial DNA from environmental samples without prior cultivation [5, 6, 7]. Function- and sequence-based analysis of metagenomic DNA fragments have resulted in the identification of a variety of novel genes and gene products [6, 8, 9]. In addition, partial sequencing of metagenomes, such as those from the acid mine biofilm (75 Mbp) [10], Minnesota farm soil (100 Mbp) [11], and Sargasso Sea (1,600 Mbp) [12], have provided a better understanding of the structure and genomic potential of microbial communities.

A major goal of metagenomic sequencing projects is the identification of protein coding genes. Most genes in metagenomic fragments are currently identified by homology to known genes by employing other methods, e.g. BLAST [13]. The disadvantage of such an approach is obvious: it is impossible to find novel genes that way. Particularly in cases where metagenomic studies aim to discover new proteins, homology search is an inadequate tool for gene prediction.

The computational ab initio prediction of genes from microbial DNA has a long history, and a number of tools have been developed and employed for gene prediction and annotation of genomic sequences from single prokaryotic species (e.g. GLIMMER [14] and GeneMark.hmm [15]). A minor restriction in the application of some conventional approaches to metagenomes is that they are based on the identification of open reading frames (ORFs), which begin with a start codon and end with an in-frame stop codon. Sequenced metagenomes comprise a collection of numerous short sequencing reads of varying length depending on the employed sequencing technique. A typical metagenomic fragment derived by Sanger sequencing [16] is approximately 700 bp long and contains two or fewer genes. The majority of these genes are incomplete, meaning one or both gene ends extend beyond fragment end(s). Therefore, most ORFs in metagenomic sequencing reads will be overlooked by ORF-based gene finders. A more profound problem is that most gene finders for prokaryotic genomes rely on statistical sequence models that are estimated from the analyzed or a closely related genome. Most metagenomic fragments do not bear sufficient sequence information for building statistical models able to distinguish coding from non-coding ORFs. One might consider to derive models from a complete metagenome but the resulting gene prediction quality in fragments from underrepresented species in the metagenome is questionable.

Up to now, there are three approaches for predicting genes from metagenomic DNA fragments. One of these methods is based on BLAST search, where the search is not only applied against databases of known proteins but also against a library constructed from the metagenomic sample itself [17]. In principle, this computationally expensive approach is able to find novel genes, provided that homologues of these genes are contained in the sample. However, it is not clear whether

Citation: Katharina J Hoff, Maike Tech, Thomas Lingner, Rolf Daniel, Burkhard Morgenstern and Peter Meinicke, Gene Prediction in Metagenomic Fragments: A Large Scale Machine Learning Approach, DOI: 10.1186/1471-2105-9-217.

interesting genes will always be conserved in a metagenomic sample. The first method that was developed for ab initio gene prediction in short and anonymous DNA sequences is a heuristic approach of GeneMark.hmm that derives an adapted monocodon usage model from the GC-content of an input sequence [18]. Another method that was developed for ab initio gene prediction in metagenomic DNA fragments is MetaGene [19]. Similar to GeneMark.hmm, MetaGene employs GC-content specific monocodon and dicodon models for predicting genes. The time-efficient two step gene prediction algorithm first extracts ORFs and scores them on the basis of statistical models estimated from fully sequenced and annotated genomes. Subsequently, a dynamic program calculates the final ORF combination from different scores. Additionally, MetaGene utilizes ORF length, the distance from the annotated start codon to the left-most start codon, and distances to neighboring ORFs. Two separate models were estimated from bacterial and archaeal genomes, respectively. The domain specific models are simultaneously applied to each fragment and the higher scoring model is selected for final gene prediction. Results in randomly sampled fragments from annotated genomes indicate that MetaGene provides a high sensitivity in finding genes in fragmented DNA, while the specificity of the predictions is slightly lower. In addition, the performance of GeneMark.hmm in 700 bp fragments and for complete genomes was investigated (supplementary table S3 and table 1 of [19]). Comparable performance results were obtained for both methods for both types of input sequences.

Here, we present a novel approach for gene prediction in single fragments, which is based entirely on machine learning techniques. In bioinformatics, state-of-the-art machine learning methods are usually applied to problems where, at most, several thousands of examples exist for training and evaluation. In our application, learning has to be performed on large data sets with millions of examples. This requires the use of a learning architecture that is capable of large-scale training and testing. Here, we propose a combination of neural networks and linear discriminants. While linear discriminants are used for the extraction of features from high-dimensional data which characterize codon usage and potential gene starts, a small neural network is used for non-linear combination of these features with additional information on length and GC-content of gene candidates. Neural networks in combination with linear discriminants or positional weight matrices have also been applied to other gene prediction problems, for instance in promoter recognition [20].

To provide comparability in our experimental evaluation, we use a setup that is similar to the one used for the initial evaluation of MetaGene. We test our program on fragments from thirteen species. However, we provide some important extensions: We use a higher number of fragments which are randomly sampled from the test genomes to avoid any bias that may result from a particular fragmentation technique. The higher number of fragments is used to cope with the variance across different (repeated) sampling experiments. In addition, we provide a detailed analysis of the translation initiation site (TIS) prediction performance and we also investigate the ability to discriminate between complete and incomplete genes.

METHODS

Most prokaryotic protein coding genes consist of a start codon, followed by a variable number of consecutive in-frame codons and are terminated by a stop codon. This particular arrangement of codons is commonly referred to as open reading frame (ORF). The sole identification of ORFs is not sufficient for prokaryotic gene prediction because the majority of ORFs in a genome are, in fact, non-coding. In DNA fragments, ORFs frequently exceed the fragment ends. We therefore extend the ORF definition toincomplete ORFs.

The fact that start codons are identical to some regular codons results in a high number of related ORFs that share a stop codon but have different start codons. We term such a set of related ORFs an ORF-set and we name the possible start codons of an ORF-set translation initiation site (TIS) candidates. Figure 1 illustrates possible cases of ORF occurrence in a DNA fragment: In case 1, the complete ORF-set is located in the fragment. Additional TIS candidates for this ORF-set cannot occur because of an upstream in-frame stop codon. Predicted genes from this ORF-set will always be complete. In case 2, only TIS candidates are located inside the fragment. The range for upstream TIS is again limited by an in-frame stop codon. This candidate, if classified as coding, would result in the prediction of an incomplete gene. In case 3, the stop is located in the fragment. Some TIS candidates are contained in the fragment but there might exist TIS candidates outside the fragment. An ORF-set of this type may result either in a complete or in an incomplete gene. Case 4 is complementary to case 2. Only a stop codon is located inside the fragment. Case 5 and 6 are fragment-spanning ORF-sets, where 5 also includes TIS candidates inside the fragment. Predictions from case 5 will be incomplete but may have a start codon. Case 5 and 6 can both result in the prediction of incomplete genes without start and stop codons.

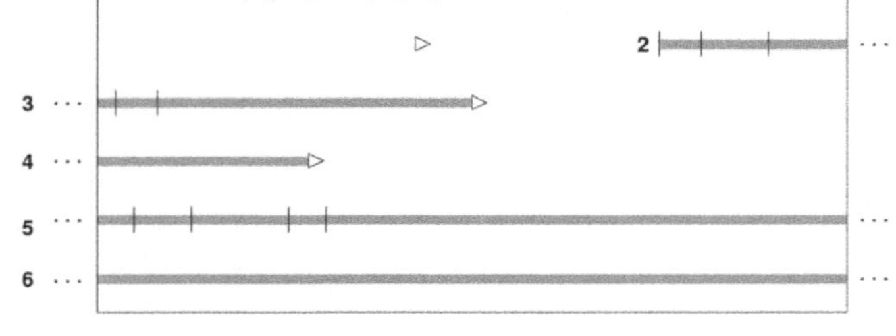

Figure 1: The figure illustrates possible localizations of open reading frames (ORFs) in a fragment (shown only for the forward strand). ORFs are shown as grey bars, »▷«denotes stop codons, »|« indicates the position of translation initiation site candidates. ORFs that are related by

a common stop codon are grouped and we refer to them as ORF-sets. The box symbolizes the fragment range. Everything that might be located outside the box is invisible to gene prediction algorithms. Further explanations are given in section »Methods«.

Our gene prediction algorithm is designed for the discrimination of coding from non-coding ORFs. After the identification of all ORFs in a fragment, we extract features from those ORFs using linear discriminants. Subsequently, we use a neural network that has been particularly trained for the classification of ORFs as coding or non-coding. Classification is based on a gene probability that the neural network assigns to every ORF. Because gene-containing ORF-sets usually comprise of more than one candidate, several ORFs of such an ORF-set may be assigned a high probability by the neural network. The final gene prediction is achieved by a »greedy« method that selects the most probable ORFs that overlap by, at most, 60 bases.

Machine Learning Techniques

To predict whether a particular ORF actually corresponds to a protein coding region or to a non-coding region, we use a neural network for binary ORF classification. In the following sections, we will first describe the features utilized as inputs for the neural network. Subsequently, we will depict the neural network architecture and the methods we used for large scale training and validation from labeled ORFs in artificial fragments.

Features

For realization of the neural network, we use seven features based on sequence characteristics of ORFs. As network inputs, these sequence features are subject to a separate preprocessing step. Below, we explain the methods for computation of these features in detail.

Codon and Dicodon Usage

The perhaps most important features for the discrimination between coding and non-coding ORFs can be derived from codon usage, in particular from 4^3 monocodon and 4^6 dicodon frequencies. These frequencies represent the occurrences of successive trinucleotides (non-overlapping) and hexanucleotides (half-overlapping), respectively. For the characterization of monocodon and dicodon usage, we compute two features based on linear discriminant scores.

Linear discriminants were obtained from training with annotated sequence data. We used coding and non-coding regions from annotated genomes as positive and negative examples, respectively (see section »Training Data for Feature Preprocessing«). Examples are represented by vectors of frequencies of 4^3 and 4^6possible monocodons and dicodons, respectively. In the following, we describe discriminant training for the monocodon case. The same training procedure was applied to the dicodon case.

For the i-th example, we denote a monocodon frequency vector as $\mathbf{x}^i_M \in \mathbb{R}^{64}$, which is the i-th column of the data matrix $\mathbf{X}_M$, containing all training vectors. To remove length information from these data, all training vectors are normalized to unit Euclidean norm. The corresponding label $y^i_M \in \{-1, 1\}$, which is the i-th element of the label vector $\mathbf{y}_M$, indicates whether the example represents a coding ($y^i_M = 1$) or non-coding ($y^i_M = -1$) region. For training of the discriminant weight vector $\mathbf{w}_M$, we use a regularized least squares approach, i.e. we minimize the following regularized error:

$$E(\mathbf{w}_M) = \frac{1}{n}\sum_{i=1}^{n}(\mathbf{w}_M \cdot \mathbf{x}^i_M - y_i)^2 + \lambda \mathbf{w}_M \cdot \mathbf{w}_M \qquad (1)$$

where » · « denotes the dot product. The minimizer of E is obtained by [21]:

$$\mathbf{w}_M = (\mathbf{X}_M \mathbf{X}_M^T + \lambda\, n\mathbf{I})^{-1} \mathbf{X}_M \mathbf{y}_M \qquad (2)$$

with d × d identity matrix **I** and with upper T and -1 indicating matrix transposition and inversion, respectively. The computational cost scale linearly with the number of examples, which makes the approach well suited for large scale learning. Doing the same for the dicodon frequency discriminant vector $\mathbf{w}_D$, we obtain two discriminant scores that serve as the first two input features of the neural network:

$$x_1 = \mathbf{w}_M \cdot \mathbf{x}_M,\ x_2 = \mathbf{w}_D \cdot \mathbf{x}_D. \qquad (3)$$

To adjust the regularization parameter λ, we measure the discriminative power of the respective classifier by means of the area under precision recall curve (»auPRC«) as explained in section »Measures of Performance«. Thereby, we choose a $\lambda \in \{10^m | m = -8, -7, ..., 6\}$ to maximize the auPRC on an independent validation set (see section »Training Data for Feature Preprocessing«).

Translation Initiation Site

A discriminant is derived from up- and downstream regions of translation initiation site (TIS) examples. Here we use a 60 basepair (bp) window centered on a potential start codon at window position 31 (see section »Training Data«). We encode the trinucleotide occurrences in that window to yield binary indicator vectors. In each of its 3712 dimensions (64 trinucleotides × 58 positions), a vector indicates whether a certain trinucleotide occurs at a particular window position. Training of the discriminant proceeds in the same way as for the previous two discriminants based on codon usage. Again, we select the regularization parameter $\lambda \in \{10^m | m = -8, -7, ..., 6\}$ by maximization of the auPRC on an independent validation set.

Because not all genes have a potential TIS region we do not use the TIS score $s = \mathbf{w}_T \cdot \mathbf{x}_T$ directly, but instead we take the posterior probabilities of being a TIS or not. For computation of the posterior probabilities, we use Gaussian probability density functions of the score:

$$p(s \mid \mu, \sigma) = \frac{1}{\sqrt{2\pi}\sigma}\exp(-\frac{1}{2\sigma^2}(s-\mu)^2) \qquad (4)$$

where μ stands for mean and σ for standard deviation.

The features x_3 and x_4 were obtained from a mixture of two Gaussians

$$p(s) = \pi^+ p(s|\mu^+, \sigma^+) + \pi^- p(s|\mu^-, \sigma^-) \qquad (5)$$

with parameters estimated from scores of positive and negative training examples, respectively (π^+ and π^- are the a priori probabilities of the two classes):

$$x_3 = \frac{\pi^+ p(s \mid \mu^+, \sigma^+)}{p(s)}, \quad x_4 = \frac{\pi^- p(s \mid \mu^-, \sigma^-)}{p(s)}. \tag{6}$$

If no TIS candidate is present, both probabilities are set to zero for that ORF. Note that this case is different from the case of missing values, which can be solved by assigning a priori probabilities for true and false TIS. Here we encounter the possible case where we know that none of the two categories is adequate.

Length features

Another feature for discrimination between coding and non-coding ORFs is the sequence length of the ORF. Here, it is important to distinguish between complete and incomplete ORFs. For incomplete ORFs, the observable »incomplete length« is merely a lower bound for the unobservable »complete length« of that ORF and therefore should be treated in a different way. Consequently, we use one »incomplete« and one »complete length« feature. For a particular ORF, only the feature that corresponds to the type of ORF has non-zero value. The value is simply the observed length divided by the maximal length l_{max}. In our evaluation, we set l_{max} to 700 bp. In this way we obtain two more features x_5, $x_6 \geq 0$ for complete and incomplete length.

GC-content

As a last feature $x_7 \in [0, 1]$, we use, for each ORF, the GC-content estimated from the whole fragment in which this ORF occurs.

Neural Network

We use standard multilayer perceptrons with one layer of k hidden nodes and with a single logistic output function. Within a binary classification setup with labels $y_i = 1$ (»true«) or $y_i = 0$ (»false«) the output of the neural network can be viewed as an approximation of the posterior probability of the »true« class [22]. In our case, the »true« class represents coding ORFs and therefore the network output can be interpreted in terms of a gene probability. For an input feature vector **x**, the k hidden layer activations z_i based on input weight vectors w_l^i and bias parameters b_l^i are

$$z_i = \tanh(\mathbf{w}_l^i \cdot \mathbf{x} + b_l^i). \tag{7}$$

Putting the z_i into a vector **z**, the output of the network, i.e. its prediction function based on weight vector $\mathbf{w}_O$ and bias b_O, is

$$g(z) = \frac{1}{1+\exp(-w_O \cdot z - b_O)}. \tag{8}$$

Given a training set $\mathbf{x}_1, \ldots, \mathbf{x}_N$ and a network with weight and bias parameters collected in the vector $\boldsymbol{\theta}$, we now write the corresponding network output as $f(\mathbf{x}_1; \boldsymbol{\theta}), \ldots, f(\mathbf{x}_N; \boldsymbol{\theta})$. With diagonal matrix **A** containing the regularization parameters, the training objective is to minimize the regularized error:

$$E(\;) = \sum_{i=1}^{N} (f(\mathbf{x}_i;\;) - y_i)^2 + \;^T\mathbf{A}\;. \tag{9}$$

The diagonal matrix $\mathbf{A} = \mathrm{diag}(\alpha_1, \ldots, \alpha_1, \alpha_2, \ldots, \alpha_2, \alpha_3, \ldots, \alpha_3, \alpha_4)$ of the regularization term involves four hyperparameters α_1, α_2, α_3, $\alpha_4 > 0$ for separate scaling of the parameters w_l^i, b_l^i, w_O, b_O. Note that the regularization term penalizes the squared magnitude of the weights. For the adaptation of hyperparameters, we utilize the evidence framework [23] based on a Gaussian approximation of the posterior distribution of network weights. The evidence-based adaptation of hyperparameters can be incorporated into the network training procedure and does not require additional validation data. For the minimization of (9) with respect to weight and bias parameters, we use a scaled conjugate gradient scheme, as implemented in the NETLAB toolbox [24]. While weight and bias parameters were initialized randomly according to a standard normal distribution, the hyperparameters were initially set to $\alpha_1 = \alpha_2 = \alpha_3 = \alpha_4 = 0.01$. The complete training scheme performs 50 iterations where each iteration comprises 50 gradient steps and two successive hyperparameter adaptation steps.

Final Candidate Selection

Application of the neural network to a certain fragment results in a list of potential gene candidates with a predicted gene probability above 0.5. Most of these predictions are mutually exclusive in terms of overlap. Many predictions even belong to the same ORF-set, differing only in the position of the start codon. In order to obtain a list G of final genes for a particular fragment, predictions with maximal probability are iteratively selected from the list of candidates C, which is successively reduced according to a maximum overlap constraint. Starting with an empty list G and an initial list C containing all fragment-specific ORFs i with gene probability $P_i = f(\mathbf{x}_i; \boldsymbol{\theta}) > 0.5$, we apply the following »greedy« selection scheme:

While C is nonempty do

- Determine $i_{max} = \arg\max_i P_i$ with respect to all ORFs i in C
- remove ORF i_{max} from C and add it to G
- remove all ORFs from C that overlap with ORF i_{max} by more than o_{max} bp

In our evaluation, we set o_{max} to 60 bp, which corresponds to the minimal gene length we consider for prediction.

Training Data

Our machine learning approach for gene prediction in metagenomic DNA fragments is based on learning the characteristics of coding and non-coding regions from 131 fully sequenced prokaryotic genomes [see Additional file 1] and their GenBank [25] annotation for protein coding genes. The training genomes correspond to the ones that were used for building the statistical models of MetaGene except that we excluded Pseudomonas aeruginosa from the training set because a subset of reliably annotated genes that is valuable for the determination of TIS correctness is available for this species. All training and test data sets described in this article are based on the initial extraction of ORFs with a minimal length of 60 bp. Two types of ORFs are distinguishable: Complete ORFs begin with a start codon (ATG, CTG, GTG or TTG), and are followed by a flexible number of subsequent codons and conclude with a stop codon (TAG, TGA or TAA). Incomplete ORFs stretch from one fragment end to a stop or start codon or to the other fragment end without being interrupted by another in-frame stop codon (compare Figure 1).

In the following paragraphs, we first describe the preparation of training data sets for feature preprocessing and for training of the neural network. Subsequently, we specify the compilation of a test data set for performance evaluation.

Training Data for Feature Preprocessing

Monocodon, dicodon and TIS feature extraction from ORFs require a preprocessing step that is based on the separate training procedure described in section «Codon and Dicodon Usage». Training examples for feature preprocessing were randomly sampled from complete genomes to a coverage of 50 %. Two separate training sets were compiled. For the mono- and dicodon frequencies training set, DNA sequences of genes defined by their exact start and stop codon position served as positive examples ($\approx 1.9 \times 10^5$). The longest candidate out of each non-coding ORF-set was selected for the composition of negative examples ($\approx 2.8 \times 10^6$).

Training of the TIS discriminant was carried out on symmetric 60 bp sequence windows around start codons. The sequence windows of annotated start codons served as positive examples (1.9×10^5) while the windows around other possible start codons of the same ORF-sets were used as negative examples (5.6×10^6).

The examples for both training data sets were randomly split into 50 % for discriminant training and 50 % for validation of the regularization parameter.

Training Data for the Neural Network

The neural network was trained with the extracted features from ORFs in 700 bp fragments that were randomly excised to a 1-fold genome coverage from each training genome. We define an n-fold coverage as the amount of sampled DNA that is in total length (bp) n times longer than the original genome sequence. Annotated genes in these fragments were used as positive examples for coding regions ($\approx 2.6 \times 10^6$) while one candidate out of each non-coding ORF-set was randomly selected for the negative examples ($\approx 4.5 \times 10^6$). The data sets were randomly split into 50% for neural network training and 50 % for validation of the network size (see section »Neural Network«).

Test Data and Experimental Evaluation

The performance of our gene prediction algorithm was evaluated on artificial DNA fragments from three archaeal and ten bacterial species (see Table 1) whose genera were not used for training. Fragments of the lengths 100 to 2000 bp (in intervals of 100 bp) were randomly sampled from each genome to a 5-fold genome coverage for each length. We used the fragments of all lengths to investigate gene prediction performance of our method, which was trained on fragments with the length 700 bp.

Table 1: Genomes of microbial species that were used for the evaluation of our method. The upper three species are archaea while the lower ten species belong to the bacterial domain. The table shows GenBank accession numbers (GenBank Acc.), and genome sizes (Size)

Species	GenBank Acc.	Size (Mbp)
Archaeoglobus fulgidus	NC_000917	2.2
Methanococcus jannaschii	NC_000909	1.7
Natronomonas pharaonis	NC_007426	2.6
Buchnera aphidicola	NC_002528	0.6
Burkholderia pseudomallei	NC_006350, NC_006351	7.2
Bacillus subtilis	NC_000964	4.2
Corynebacterium jeikeium	NC_007164	2.5
Chlorobium tepidum	NC_002932	2.2
Escherichia coli	NC_000913	4.6
Helicobacter pylori	NC_000921	1.6
Pseudomonas aeruginosa	NC_002516	6.3
Prochlorococcus marinus	NC_007577	1.7
Wolbachia endosymbiont	NC_006833	1.1

A more detailed analysis was carried out on 700 bp fragments (also sampled to a 5-fold coverage), including a comparison to MetaGene. In order to determine statistical significance, we used 10 replicates of each randomly sampled fragment stack.

Gene prediction performance was evaluated by comparing predictions of our method to known annotated genes in fragments. The GenBank annotation for protein coding genes was used to measure general gene prediction performance. However, the GenBank gene start annotation has previously been suspected to be inaccurate [26]. Therefore, we used »reliable gene annotation subsets« [27] for the evaluation of translation initiation site (TIS) prediction performance: all genes with an experimentally verified TIS from »EcoGene« forEscherichia coli [28], experimentally verified genes of the Bacillus subtilis GenBank annotation (all non-y genes) and the »PseudoCAP« (Pseudomonas community annotation project) annotation of Pseudomonas aeruginosa[29].

Measures of Performance

The capability of detecting annotated genes (and genes including their annotated TIS) was measured as sensitivity:

$$\text{Sens} = \frac{TP}{TP+FN} \tag{10}$$

For gene prediction sensitivity, TP_{gene} (true positives) denotes correct matches and FN_{gene} (false negatives) indicate overlooked genes. We counted all predictions as TP_{gene} that match at least 60 bp in the same reading-frame to an annotated gene.

In one experiment, we compared gene predictions to a subset of genes that have a reliably annotated gene start in the fragment. For this subset, we measured TIS prediction sensitivity. Here, TP_{TIS} are genes with correctly predicted TIS and FN_{TIS} are genes whose correct start codons were not predicted.

The reliability of gene predictions was measured by specificity:

$$\text{Spec} = \frac{TP_{gene}}{TP_{gene}+FP_{gene}} \tag{11}$$

Gene prediction specificity was calculated with predicted genes that do not correspond to any gene in the annotation as FP_{gene} (false positives).

To provide a suitable composite measure of sensitivity and specificity we use the harmonic mean, which corresponds to a particular realization of the F-measure [30]:

$$\text{HarmonicMean} = \frac{2^{*}\text{Sens}^{*}\text{Spec}}{\text{Sens} + \text{Spec}}. \tag{12}$$

To measure the discriminative power of the codon usage and TIS discriminants for feature extraction (see »Features«), we use the area under precision recall curve (auPRC). The precision recall curve shows for each possible score threshold the relation of sensitivity (on the x-axes) and specificity (on the y-axes). Sensitivity and specificity are not sufficient for measuring TIS prediction performance. When applied to TIS prediction, these measures rather reflect general gene prediction performance than accuracy of TIS prediction. ‹TIS correctness› was therefore measured by the percentage of correctly predicted TIS within a subset of true positive gene predictions TP_{gene} that have an annotated start codon within the fragment (TP_{gene} that have annotated start codon wihin the fragment (TP_{gene}*):

$$\text{TIS correctness} = \frac{TP_{TIS}}{TP_{gene^{*}}}. \tag{13}$$

Accuracy of complete/incomplete gene type prediction was calculated on the basis of correctly predicted genes with an existing true TIS:

$$\text{GeneTypeAccuracy} = \frac{TP_{complete}+TN_{complete}}{TP_{gene}} \tag{14}$$

where $TP_{complete}$ and $TN_{complete}$ account for the number of genes within TP_{gene} that have correctly been predicted as complete and incomplete.

RESULTS AND DISCUSSION

In the following sections, we first describe and discuss the results of discriminant and neural network validation which led to the choice of a hyperparameter λ and a suitable number of nodes for the neural net. Subsequently, we show and discuss gene prediction performance results of the neural network on several fragment lengths and in 700 bp fragments.

Discriminant Validation

Training of the linear discriminants for monocodon, dicodon and TIS features requires the validation of the regularization parameter λ . For each of the three discriminants, we chose λ from the set of values $\{10^{m} | m = -8, -7, ..., 6\}$ by maximizing the area under precision recall curve (auPRC) on separate validation data. While for the TIS discriminant, a well-defined maximum was achieved for an intermediate $\lambda = 10^{-2}$, for the monocodon and dicodon case the maximum was achieved for the smallest value $\lambda = 10^{-8}$. However, as shown in Additional file 2, for small λ values the auPRC performance in these cases reaches a plateau and therefore we did not try smaller values. The resulting discriminant weights for the 64 monocodons are shown in Additional file 1. The high negative weights for the three stop codons TAA, TAG, TGA are due to the large fraction of negative examples. Because negative examples are, by a factor 10, more frequent than positive examples in the training set, a negative shift of the discriminant score is induced by codons that, like stop codons, are present in any example in any of the two classes.

Network Validation

In principle, the evidence-based hyperparameter adaptation (see section »Neural Network«) obviates the search for an adequate size of the network, i.e. to find a suitable number k of hidden nodes. The network size has just to be large enough to provide maximum performance, while larger nets would automatically be subject to stronger regularization in terms of larger regularization parameters. Nevertheless, network size is crucial in terms of computational cost for training and testing.

In order to find a small network with sufficient performance, we started to train networks of increasing size. Trying networks with k = 5, 10, ..., 25 nodes, we found the performance to reach a nearly flat plateau within that k-range, with only very slight increase above k = 15 [see Additional file 2]. Performance was measured in terms of the harmonic mean criterion (see section »Measures of Performance«), computed on an independent validation set (see section »Training Data«). For the final predictions on the test data, we used the largest network with k = 25 nodes.

While training of neural networks is a time consuming process, computing predictions with a trained network on new data is very fast. In our case, training a network with k = 25 hidden nodes from $\cong 3.6 \times 10^{6}$ examples took about 190 cpu hours (AMD Opteron, 2 GHz). The training scheme described above was applied in parallel to 5 networks with different (random) initialization of parameters to avoid weak local minima of the regularized error (9). According to the lowest error, the best resulting network was selected for the final predictions within the test setup. In contrast, testing of the same number of examples, i.e. prediction on more than three million candidates, only took ≅ 16.5 seconds on the same machine.

The parameters of the neural network with 25 nodes that we used for further evaluation are given in Additional file 1.

Gene Prediction Performance in DNA fragments

To evaluate the performance of our machine learning approach, we tested the method on artificially fragmented genomes. In the following section, we present the results in general gene prediction performance on various fragment lengths. Subsequently, we analyze gene prediction performance, TIS prediction correctness and complete/incomplete gene type prediction accuracy in detail for fragments of length 700 bp, which corresponds to the fragment length on which the neural network was trained.

Performance in Fragments of Different Lengths

The predictions of our method in DNA fragments with lengths ranging from 100 to 2000 bp from thirteen species were compared to the GenBank [25] annotation for protein coding genes. Note that on average 15 % of 100 bp fragments do not contain any annotated gene matching the 60 bp minimal length criterion (complete or incomplete), for 700 bp fragments, this fraction of fragments accounts 3 %, for 2000 bp fragments 0.8 %. The average percentage of complete genes within all annotated genes in our test fragments is 0 % for 100 bp fragments, 8 % for 700 bp fragments and 40 % for 2000 bp fragments [see Additional file 2]. The mean of gene prediction sensitivity and specificity for all fragment lengths is shown in Figure 2. On 700 bp fragments, our method has an average gene prediction sensitivity of 89 % and an average specificity of 93 %. Sensitivity and specificity slightly increase with growing fragment size. This can be explained by the fact that ORFs carrying distinct mono-/dicodon and TIS signals occur more often in longer fragments. Gene prediction performance decreases with length and sharply drops for fragments shorter than 200 bp.

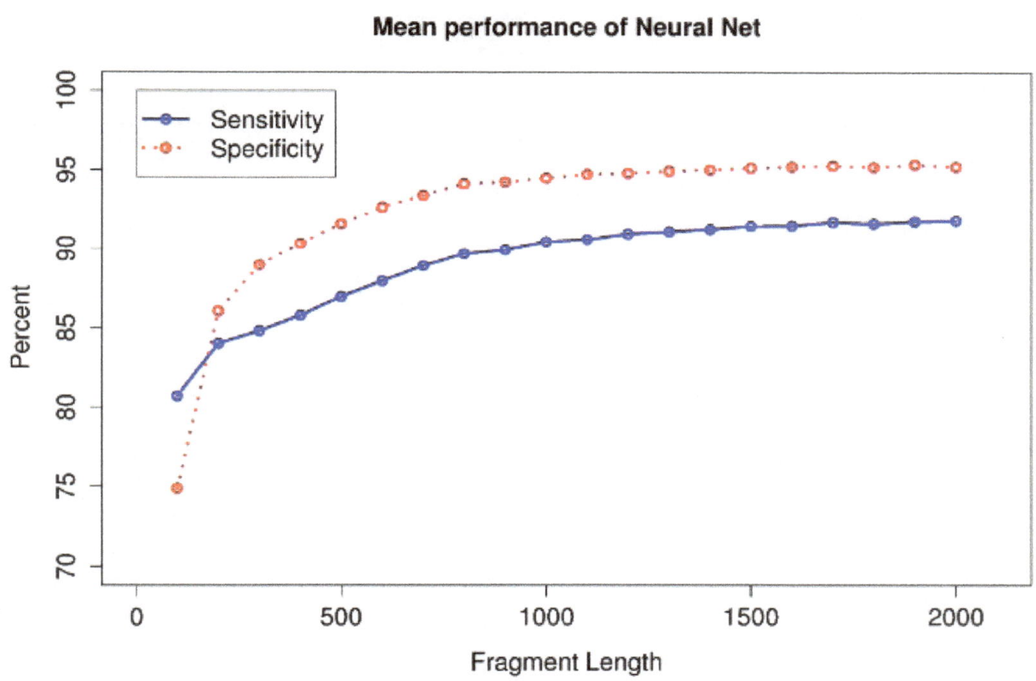

Figure 2: Average gene prediction performance of the neural network in fragments of the lengths 100 to 2000 bp. The performance values from thirteen test species were averaged by arithmetic mean.

Before application of our method on real metagenomes, the neural network should be trained and thoroughly evaluated on fragments of a length that corresponds to the real fragments of interest. Metagenomic sequencing projects differ in their aims and in the applied sequencing and annotation strategies. Improved Sanger sequencing from one or both vector insert ends is applied in many metagenomic projects (examples are listed in review [31]) and yields sequencing reads roughly ranging from 500 to 1000 bp. Based on the current results, our method might be particularly useful for improving gene annotation and discovery on Sanger sequencing reads. However, pyrosequencing [32] has also been introduced to metagenomics [33]. The pyrosequencing approach does not involve any cloning step. With recent improvements, pyrosequencing now yields a read length between 200 and 300 bp [34]. In principle, it should be possible to predict genes in such short fragments with our fragment-based techniques but environmental pyrosequencing projects may rather be focused on phylogenetic studies and habitat comparison than on the discovery of new genes. In some metagenomic sequencing projects, long metagenomic inserts (up to 40 kbp) are fully sequenced [35]. Although gene prediction performance of the neural network does not decrease on longer fragments [see Additional file2], other methods like MetaGene [19] or GeneMark.hmm [18], which also consider the context of putative genes (e.g. operons), may be more suitable for fragments of this size.

Performance in 700 bp Fragments

Predicted genes in fragments with a length of 700 bp from three archaeal and ten bacterial species were compared to the GenBank annotation for protein coding genes. The mean and standard deviation for sensitivity, specificity and the harmonic mean of 10 repetitions per species are shown in Table 2. The neural network has high sensitivity (ranging from 82 to 92 %) and specificity (ranging from 85 to 97 %) in fragments from all species. We could not observe a major performance difference for sensitivity and specificity between archaeal and bacterial fragments but the variation between different species in general is large.

Table 2: Mean and standard deviation for gene prediction performance of our method (Neural Net) and MetaGene. Performance was measured on 700 bp fragments that were randomly excised from each test genome to 5-fold coverage (ten replications per species). The harmonic mean is a measure that combines sensitivity and specificity

	SENSITIVITY		SPECIFICITY		HARMONIC MEAN	
Species	Neural Net	MetaGene	Neural Net	MetaGene	Neural Net	MetaGene
Archaeoglobus fulgidus	87.2 ± 0.21	**93.7 ± 0.15**	**93.4 ± 0.16**	92.7 ± 0.16	90.2 ± 0.17	**93.2 ± 0.14**
Methanococcus jannaschii	91.7 ± 0.17	**95.8 ± 0.14**	**96.2 ± 0.13**	92.7 ± 0.19	93.9 ± 0.10	**94.3 ± 0.15**
Natronomonas pharaonis	87.9 ± 0.22	**95.1 ± 0.09**	**93.9 ± 0.10**	92.7 ± 0.17	90.8 ± 0.16	**93.9 ± 0.12**
Buchnera aphidicola	90.6 ± 0.37	**96.7 ± 0.24**	**95.3 ± 0.31**	91.1 ± 0.29	92.9 ± 0.28	**93.8 ± 0.21**
Burkholderia pseudomallei	87.9 ± 0.11	**94.1 ± 0.11**	**90.1 ± 0.09**	85.1 ± 0.13	89.0± 0.08	**89.4 ± 0.10**
Bacillus subtilis	**91.4 ± 0.16**	89.8 ± 0.14	**95.3 ± 0.09**	89.3 ± 0.19	**93.3 ± 0.10**	89.5 ± 0.14
Corynebacterium jeikeium	89.7 ± 0.24	**91.9 ± 0.12**	**93.8 ± 0.19**	89.2 ± 0.21	**91.7 ± 0.19**	90.5 ± 0.13
Chlorobium tepidum	82.1 ± 0.25	**85.7 ± 0.27**	**91.2 ± 0.17**	88.4 ± 0.26	86.4 ± 0.19	**87.0 ± 0.22**
Escherichia coli	91.7 ± 0.16	**93.3 ± 0.07**	**95.3 ± 0.09**	90.9 ± 0.10	**93.5 ± 0.12**	92.1 ± 0.07
Helicobacter pylori	**92.1 ± 0.11**	90.2 ± 0.14	**96.6 ± 0.15**	89.6 ± 0.23	**94.3 ± 0.11**	89.9 ± 0.15
Pseudomonas aeruginosa	90.4 ± 0.14	**96.2 ± 0.07**	**92.5 ± 0.11**	91.4 ± 0.09	91.4 ± 0.12	**93.7 ± 0.07**
Prochlorococcus marinus	87.2 ± 0.21	**93.7 ± 0.25**	**95.9 ± 0.14**	90.8 ± 0.20	91.4 ± 0.15	**92.2 ± 0.19**
Wolbachia endosymbiont	87.2 ± 0.27	**90.6 ± 0.42**	**85.2 ± 0.44**	71.2 ± 0.54	**86.2 ± 0.29**	79.7 ± 0.45

In comparison to MetaGene, the neural network has a higher specificity in fragments from all test species (on average 4.6 % higher). On the other hand, MetaGene has a higher sensitivity in fragments from most species (on average 3.8 % higher). The neural network only shows a higher sensitivity in Bacillus subtilis andHelicobacter pylori. The overall performance of both methods calculated by harmonic mean is very similar. For some species, the neural network yields a better overall gene prediction performance while MetaGene performs better on other species. In particular, MetaGene performs better in all tested archaea. All local pairwise differences in sensitivity, specificity and harmonic mean between the neural network and MetaGene are significant to a confidence level of 95 % according to Wilcoxon's signed rank test [36] (R-packageexactRankTests [37]).

A precise TIS prediction is very important in metagenomics since the aim of many environmental sequencing projects is the identification and subsequent experimental investigation of novel genes. For example, the expression of a metagenomic protein in a host organism may fail or yield incorrect results if the predicted start codon is incorrect. Accurate TIS prediction is a difficult task, even for conventional gene finders on complete genomes [38, 39, 40, 41, 42]. This is because ATG, CTG, GTG and TTG also occur inside genes.

One gene may for example contain several ATGs but only one corresponds to a TIS. Our approach includes a TIS-model that is based on a linear discriminant. We measured TIS prediction performance of our algorithm for all correctly predicted genes that have annotated start codons within a fragment. First, we investigated TIS performance on our complete set of test species fragments according to GenBank annotation. The results are shown in Table 3. TIS correctness of our algorithm varies remarkably between different test species. On some bacterial species, our algorithm reaches a TIS correctness of 87 % (e.g. in Helicobacter pylori). The lowest TIS performance can be observed in fragments from the bacterium Chlorobium tepidum (68 %). The average TIS correctness of our algorithm is around 78 %. In comparison to this, the highest performance of MetaGene can be observed for fragments of Prochlorococcus marinus (89 %), the lowest for fragments of Bacillus subtilis (66 %). Note that TIS correctness depends on the number of correctly predicted genes with an annotated TIS. Therefore, TIS correctness of our algorithm is not directly comparable to the one obtained by MetaGene, which detects a higher number of genes. However, the variation in TIS correctness of both methods is large.

Table 3: Translation initiation site prediction correctness (TIS correctness) and complete/incomplete classifi-cation accuracy (Gene Type Accuracy) of the Neural Net and MetaGene according to GenBank annotation. Performance was measured on 700 bp fragments that were randomly excised from each test genome to 5-fold coverage (mean and standard deviation for 10 replicates per species are given)

	TIS CORRECTNESS		GENE TYPE ACCURACY	
Species	Neural Net	MetaGene	Neural Net	MetaGene
Archaeoglobus fulgidus	69.8 ± 0.32	73.6 ± 0.32	98.1 ± 0.05	97.2 ± 0.07
Methanococcus jannaschii	69.4 ± 0.52	73.3 ± 0.52	99.0 ± 0.09	97.6 ± 0.12
Natronomonas pharaonis	75.2 ± 0.58	82.9 ± 0.28	96.9 ± 0.16	97.6 ± 0.09
Buchnera aphidicola	86.5 ± 0.40	88.6 ± 0.64	99.1 ± 0.09	98.3 ± 0.21
Burkholderia pseudomallei	70.1 ± 0.45	73.0 ± 0.28	97.6 ± 0.08	96.9 ± 0.09
Bacillus subtilis	79.7 ± 0.32	66.1 ± 0.42	98.6 ± 0.05	97.0 ± 0.08
Corynebacterium jeikeium	78.2 ± 0.49	73.4 ± 0.68	98.1 ± 0.08	96.6 ± 0.11
Chlorobium tepidum	68.1 ± 0.46	71.9 ± 0.45	98.1 ± 0.08	96.7 ± 0.13
Echerichia coli	84.5 ± 0.31	78.2 ± 0.15	98.7 ± 0.06	97.0 ± 0.08
Helicobacter pylori	87.3 ± 0.40	77.1 ± 0.33	99.2 ± 0.09	96.4 ± 0.16
Pseudomonas aeruginosa	78.4 ± 0.22	81.0 ± 0.36	97.7 ± 0.03	97.2 ± 0.07
Prochlorococcus marinus	86.6 ± 0.40	88.6 ± 0.47	99.0 ± 0.07	97.8 ± 0.10
Wolbachia endosymbiont	79.3 ± 0.77	79.9 ± 0.42	98.7 ± 0.13	96.9 ± 0.17

A reason for this variation might be that the GenBank gene annotation contains many hypothetical and not experimentally verified genes [26]. Therefore, we also evaluated TIS prediction performance on »reliable annotation subsets« of the bacteria Escherichia coli, Bacillus subtilis and Pseudomonas aeruginosa (see section »Test Data and Experimental Evaluation«). Evaluating gene prediction performance in fragments according to these annotation subsets, our algorithm achieves a highly consistent TIS prediction performance between 81 and 87 % in fragments from all three test species. The TIS prediction sensitivity varies from 68 % to 80 % (see Table 4). In comparison, MetaGene›s TIS performance shows a higher variation, ranging from 70 to 84 % while the TIS sensitivity ranges from 62 to 80 %.

Table 4: Translation initiation site prediction performance of the new gene prediction algorithm (Neural Net) and MetaGene according to »reliable annotation subsets« (A subset of »verified genes« from »EcoGene« for Escherichia coli [28], all non-y genes of theBacillus subtilis GenBank annotation and the »PseudoCAP« annotation of Pseudomonas aeruginosa [29]). TIS prediction sensitivity and correctness were measured on artificial 700 bp fragments that were randomly excised from each test genome to 5-fold coverage. Mean and standard deviation over 10 replicates per species are shown

	SENSITIVITY TIS		TIS CORRECTNESS	
Species	Neural Net	MetaGene	Neural Net	MetaGene
Bacillus subtilis	**73.4 ± 1.79**	62.1 ± 1.43	84.1 ± 0.51	70.2 ± 0.64
Escherichia coli	**80.0 ± 0.68**	75.1 ± 0.61	86.6 ± 0.57	77.5 ± 0.67
Pseudomonas aeruginosa	68.0 ± 0.22	**79.7 ± 0.44**	80.7 ± 0.20	83.7 ± 0.36

The nature of fragmented DNA results in the occurrence of complete and incomplete genes. A gene may be incorrectly predicted as complete or incomplete if it has several TIS candidates of which at least one is located outside the fragment. Due to the short fragment length of 700 bp, the vast majority of annotated genes (≈90 %) in our test fragments is incomplete. The experimental strategy of many metagenomic projects relies on sequencing the fragment from one or both ends of the vector insert. Although the insert is not always sequenced completely, sequencing of the entire fragment is possible in case the biologist is interested in further analysis of an incomplete gene. Therefore, it is important to know whether a gene is contained in a sequencing read completely or incompletely.

We evaluated the percentage of genes that were correctly classified as complete or incomplete within the correctly identified genes according to GenBank annotation. Our method achieves an average accuracy of 98 % with little variation (see Table 3). It can be noted that MetaGene slightly more often misclassifies genes concerning their completeness. Note here that the performance indices of the neural network and MetaGene in Table 3 are not directly comparable because they rely on different numbers of correctly identified genes.

Remarks on the Experimental Setup

The evaluation of computational methods for metagenomic gene prediction is troubled by the fact that reliably annotated metagenomes are not available. Some metagenomes have been subject to annotation for several years by now, but their gene annotation is far from complete. Particularly, the exact location of gene starts on metagenomes has been verified experimentally only in rare cases. Currently, the only way to reliably investigate gene prediction accuracy is the evaluation on DNA fragments from complete microbial genomes. For the evaluation of our method, we used an experimental setup similar to the one proposed by the authors of MetaGene in order to keep both methods comparable. MetaGene relies on statistical models built from 116 bacterial and 15 archaeal genomes. These species were selected to represent every genus from GenBank in the year 2006. By now, species belonging to many additional genera have been fully sequenced and annotated. Members of these genera should be included in the training set of a future gene prediction tool version in order to collect as much information about the characteristics of coding and non-coding ORFs as possible.

It remains an open question, which criteria are most suitable for the selection of training species. In general, taxonomy does not reflect phylogeny properly. Some species of different genera for example exhibit highly similar codon usage patterns. Particularly for the identification of novel genes in metagenomes whose biological diversity is yet unknown, the transfer of the GenBank bias toward single species should be avoided in the training data set. To reduce this bias, training genomes could be selected according to other criteria, e.g. GC-content, oligonucleotide frequencies or monocodon/dicodon frequencies in protein coding regions.

The experimental setup chosen here also differs from real metagenomes with respect to sequencing errors. The effect of sequencing errors in terms of base-changes on gene prediction performance of our method would depend on the frequency of such kind of error. The effect of a usually small number of base-errors (less than one error per 10 000 bp after routine fragment end removal [19]) can be neglected. As for other alignment-free methods, like MetaGene, our method is susceptible to frame shifts. Only certain alignment-based methods can be expected to be more robust with regard to this kind of error [17].

CONCLUSION

Large scale machine learning is well suitable for gene prediction in metagenomic DNA fragments. Due to performance results obtained with the current experimental setup, we suggest that our machine learning approach, with its high gene prediction specificity, TIS correctness and complete/incomplete prediction capabilities, complements MetaGene with its high gene finding sensitivity well. Thus, a combination of both methods should be considered.

SOFTWARE AVAILABILITY

Linear discriminants and the trained neural network are available as MATLAB files for download at [43]. A command line tool for gene prediction in DNA fragments (Linux, 64-bit architecture) is available from the authors on request.

ACKNOWLEDGEMENTS

We thank Dr. Heiko Liesegang for supporting the development of our method. We are grateful to Andrew Woehler for proofreading the final version of our manuscript. K.J.H was supported by a Georg-Christoph-Lichtenberg stipend granted by the state of Lower Saxony. T.L. was supported by BMBF project MediGrid (01AK803G).

AUTHORS' CONTRIBUTIONS

KJH developed and implemented the combination of ORF extraction and machine learning modules, assembled training and test data sets and performed the evaluation of the new algorithm and MetaGene. MT developed and implemented ORF-set identification and extraction. KJH and MT contributed biological expertise, T.L. implemented fast versions of the discriminant scoring. RD contributed biological expertise on metagenomics and drafted parts of the manuscript. PM contributed machine learning expertise and designed and implemented the machine learning architecture. BM and TL supported the project and contributed conceptually. KJH and PM wrote the manuscript. All authors read and approved the manuscript.

REFERENCES

1. Hugenholtz P: Exploring prokaryotic diversity in the genomic era. Genome Biol 2002, 3(2):reviews0003.1–0003.8. 10.1186/gb-2002-3-2-reviews0003
2. Torsvik V, ∃vreås L: Microbial diversity and function in soil: from genes to ecosystems. Curr Opin Microbiol 2002, 5: 240–245. 10.1016/S1369-5274(02)00324-7
3. Amann R, Ludwig W, Schleifer K: Phylogenetic identification and in situ detection of individual microbial cells without cultivation. Microbiol Rev 1995, 59: 143–169.
4. Rappe MS, Giovannoni SL: The uncultured microbial majority. Annu Rev Microbiol 2003, 57: 369–394. 10.1146/annurev.micro.57.030502.090759
5. Riesenfeld CS, Schloss PD, Handelsman J: Metagenomics: Genomic analysis of microbial communities. Annu Rev Genet2004, 38: 525–552. 10.1146/annurev.genet.38.072902.091216
6. Handelsman J: Metagenomics: Application of genomics to uncultured microorganisms. Microbiol Mol Biol Rev 2004, 64(4):669–685. 10.1128/MMBR.68.4.669-685.2004
7. Daniel R: The metagenomics of soil. Nature Rev Microbiol 2005, 3: 470–478. 10.1038/nrmicro1160
8. Daniel R: The soil metagenome – a rich resource for the discovery of novel natural products. Curr Opin Biotechnol 2004, 15: 199–204. 10.1016/j.copbio.2004.04.005
9. Streit W, Daniel R, Jaeger KE: Prospecting for biocatalysts and drugs in the genomes of non-cultured microorganisms.Curr Opin Biotechnol 2004, 15: 285–290. 10.1016/j.copbio.2004.05.006
10. Tyson GW, Chapman J, Hugenholtz P, Allen EE, Ram RJ, Richardson PM, Solovyev VV, Rubin EM, Rokhsar DS, Banfield JF: Community structure and metabolism through reconstruction of microbial genomes from the environment. Nature2004, 428: 37–43. 10.1038/nature02340
11. Tringe SG, von Mering C, Kobayashi A, Salamov AA, Chen K, Chang HW, Podar M, Short JM, Mathur EJ, Detter JC, Bork P, Hugenholtz P, Rubin EM: Comparative metagenomics of microbial communities. Science 2005,

308: 554–557. 10.1126/science.1107851

12. Venter JC, Remington K, Heidelberg JF, Halpern AL, Rusch D, Eisen JA, Wu D, Paulsen I, Nelson KE, Nelson W, Fouts DE, Levy S, Knap AH, Lomas MW, Nealson K, Peterson OWJ, Hoffman J, Parsons R, Baden-Tillson H, Pfannkoch C, Rogers YH, Smith HO: Environmental shotgun sequencing of the Sargasso Sea. Science 2004, 304: 66–74. 10.1126/science.1093857
13. Altschul S, Gish W, Miller W, Myers E, Lipman D: Basic local alignment search tool. J Mol Biol 1990, 215: 403–410.
14. Delcher AL, Harmon D, Kasif S, White O, Salzberg SL: Improved microbial gene identification with GLIMMER. Nucleic Acids Res 1999, 27(23):4636–4641. 10.1093/nar/27.23.4636
15. Lukashin A, Borodovsky M: GeneMark.hmm: new solutions for gene finding. Nucleic Acids Res 1998, 26(4):1107–1115. 10.1093/nar/26.4.1107
16. Sanger F, Nicklen S, Coulson AR: DNA sequencing with chain-terminating inhibitors. Proc Natl Acad Sci USA 1977, 74(12):5463–5467. 10.1073/pnas.74.12.5463
17. Krause L, Diaz NN, Bartels D, Edwards RA, Pühler A, Rohwer F, Meyer F, Stoye J: Finding novel genes in bacterial communities isolated from the environment. Bioinformatics 2006, 22(14):e281-e289. 10.1093/bioinformatics/btl247
18. Besemer J, Borodovsky M: Heuristic approach to deriving models for gene finding. Nucleic Acids Res 1999, 27(19):3911–3920. 10.1093/nar/27.19.3911
19. Noguchi H, Park J, Takagi T: MetaGene: prokaryotic gene finding from environmental shotgun sequences. Nucleic Acids Res 2006, 34(19):5623–5630. 10.1093/nar/gkl723
20. Bajic VB, Seah SH, Chong A, Zhang G, Koh JLY, Brusic V: Dragon Promoter Finder: recognition of vertebrate RNA polymerase II promoters. Bioinformatics 2002, 18: 198–199. 10.1093/bioinformatics/18.1.198
21. Hastie T, Tibshirani R, Friedman J: The Elements of Statistical Learning. Berlin: Springer-Verlag; 2001.
22. Bishop CM: Neural Networks for Pattern Recognition. Oxford: Clarendon Press; 1995.
23. MacKay DJC: A Practical Bayesian Framework for Backpropagation Networks. Neural Comput 1992, 4(3):448–472. 10.1162/neco.1992.4.3.448
24. Nabney IT: Netlab: Algorithms for Pattern Recognition. New York: Springer-Verlag; 2001.
25. Benson DA, Karsch-Mizrachi I, Lipman DJ, Ortell J, Wheeler DL: GenBank. Nucleic Acids Res 2007, 35: D21-D25. 10.1093/nar/gkl986
26. Nielson P, Krogh A: Large-scale prokaryotic gene prediction and comparison to genome annotation. Bioinformatics2005, 21(24):4322–4329. 10.1093/bioinformatics/bti701
27. Tech M, Meinicke P: An unsupervised classification scheme for improving predictions of prokaryotic TIS. BMC Bioinformatics 2006., 7(121):
28. Rudd KE: EcoGene: a genome sequence database for Escherichia coli K-12. Nucleic Acids Res 2000, 28: 60–64. 10.1093/nar/28.1.60
29. PseudoCAP Pseudomonas aeruginosa Community Annotation Project[http://pseudomonas.com/]
30. van Rijsbergen CJ: Information retrieval. 2nd edition. London: Butterworths; 1979.
31. Chen K, Pachter L: Bioinformatics for whole-genome shotgun sequences of microbial communities. PloS Comput Biol2005, 1(2):106–112. 10.1371/journal.pcbi.0010024
32. Ronaghi M, Uhlén M, Nyreén P: A sequencing method based on real-time pyrophosphate. Science 1998, 281(5375):363–365. 10.1126/science.281.5375.363
33. Edwards RA, Rodriguez-Britol B, Wegley L, Haynes M, Breitbart M, Peterson DM, Saar MO, Alexander S, Alexander EC, Rohwer F: Using pyrosequencing to shed light on deep mine microbial ecology. BMC Genomics 2006, 7: 57. 10.1186/1471-2164-7-57
34. Jarvie T, Harkins T: Metagenomics Analysis Using the Genome Sequencer FLX System. Biochemica 2007, 3: 4–6.
35. Voget S, Leggewie C, Uesbeck A, Raasch C, Jaeger KE, Streit WR: Prospecting for Novel Biocatalysts in a Soil Metagenome.Appl Env Microbiol 2003, 69(10):6235–6242. 10.1128/AEM.69.10.6235-6242.2003
36. Wilcoxon F: Individual comparisons by ranking methods. Biometrics 1945, 1(6):80–83. 10.2307/3001968
37. R Development Core Team: R: A language and environment for statistical computing. R Foundation for Statistical Computing, Vienna, Austria; 2004. ISBN 3–900051–00–3
38. Suzek BE, Ermolaeva MD, Schreiber M, Salzberg SL: A probabilistic method for identifying start codons in bacterial genomes. Bioinformatics 2001, 17(12):1123–1130. 10.1093/bioinformatics/17.12.1123
39. Frishman D, Mironov A, Gelfand M: Starts of bacterial genes: estimating the reliability of computer predictions. Gene1999, 234(2):257–265. 10.1016/S0378-1119(99)00200-0
40. Ou HY, Guo FB, Zhang CT: GS-Finder: a program to find bacterial gene start sites with a self-training method. Int J Biochem Cell Biol 2004, 36(3):535–544. 10.1016/j.biocel.2003.08.013
41. Tech M, Pfeifer N, Morgenstern B, Meinicke P: TICO: a tool for improving predictions of prokaryotic translation initiation sites. Bioinformatics 2005, 17(21):3568–3569. 10.1093/bioinformatics/bti563
42. Tech M, Morgenstern B, Meinicke P: TICO: a tool for postprocessing the predictions of prokaryotic translation initiation sites. Nucleic Acids Res 2006, 34: 588–590. 10.1093/nar/gkl313
43. sets D[http://orphelia.gobics.de]

CHAPTER 9 METAGENOMICS: A GUIDE FROM SAMPLING TO DATA ANALYSIS

Torsten Thomas, Jack Gilbert and Folker Meyer

Keywords: Sampling, sequencing, Assembly, binning, annotation, data storage, data sharing, DNA extraction, microbial ecology, microbial diversity.

INTRODUCTION

Arguably, one of the most remarkable events in the field of microbial ecology in the past decade has been the advent and development of metagenomics. Metagenomics is defined as the direct genetic analysis of genomes contained with an environmental sample. The field initially started with the cloning of environmental DNA, followed by functional expression screening [1], and was then quickly complemented by direct random shotgun sequencing of environmental DNA [2,3]. These initial projects not only showed proof of principle of the metagenomic approach, but also uncovered an enormous functional gene diversity in the microbial world around us [4].

Metagenomics provides access to the functional gene composition of microbial communities and thus gives a much broader description than phylogenetic surveys, which are often based only on the diversity of one gene, for instance the 16S rRNA gene. On its own, metagenomics gives genetic information on potentially novel biocatalysts or enzymes, genomic linkages between function and phylogeny for uncultured organisms, and evolutionary profiles of community function and structure. It can also be complemented with metatranscriptomic or metaproteomic approaches to describe expressed activities [5,6]. Metagenomics is also a powerful tool for generating novel hypotheses of microbial function; the remarkable discoveries of proteorhodopsin-based photoheterotrophy or ammonia-oxidizing Archaea attest to this fact [7,8].

The rapid and substantial cost reduction in next-generation sequencing has dramatically accelerated the development of sequence-based metagenomics. In fact, the number of metagenome shotgun sequence datasets has exploded in the past few years. In the future, metagenomics will be used in the same manner as 16S rRNA gene fingerprinting methods to describe microbial community profiles. It will therefore become a standard tool for many laboratories and scientists working in the field of microbial ecology.

This review gives an overview of the field of metagenomics, with particular emphasis on the steps involved in a typical sequence-based metagenome project (Figure.1. We describe and discuss sample processing, sequencing technology, assembly, binning, annotation, experimental design, statistical analysis, and data storage and sharing. Clearly, any kind of metagenomic dataset will benefit from the rich information available from other metagenome projects, and it is hoped that common, yet flexible, standards and interactions among scientists in the field will facilitate this sharing of information. This review article summarizes the current thinking in the field and introduces current practices and key issues that those scientists new to the field need to consider for a successful metagenome project.

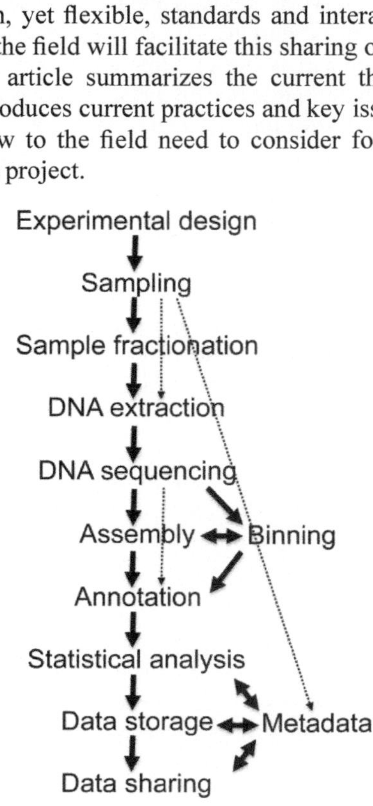

Figure 1: Flow diagram of a typical metagenome projects. Dashed arrows indicate steps that can be omitted.

Citation: Torsten Thomas, Jack Gilbert and Folker Meyer, Metagenomics: A Guide From Sampling To Data Analysis, doi: 10.1186/2042-5783-2-3.

SAMPLING AND PROCESSING

Sample processing is the first and most crucial step in any metagenomics project. The DNA extracted should be representative of all cells present in the sample and sufficient amounts of high-quality nucleic acids must be obtained for subsequent library production and sequencing. Processing requires specific protocols for each sample type, and various robust methods for DNA extraction are available (e.g. [3,9,10]). Initiatives are also under way to explore the microbial biodiversity from tens of thousands of ecosystems using a single DNA extraction technology to ensure comparability [11].

If the target community is associated with a host (e.g. an invertebrate or plant), then either fractionation or selective lysis might be suitable to ensure that minimal host DNA is obtained (e.g. [9,12]). This is particularly important when the host genome is large and hence might "overwhelm" the sequences of the microbial community in the subsequent sequencing effort. Physical fractionation is also applicable when only a certain part of the community is the target of analysis, for example, in viruses seawater samples. Here a range of selective filtration or centrifugation steps, or even flow cytometry, can be used to enrich the target fraction [3,13,14]. Fractionation steps should be checked to ensure that sufficient enrichment of the target is achieved and that minimal contamination of non-target material occurs.

Physical separation and isolation of cells from the samples might also be important to maximize DNA yield or avoid coextraction of enzymatic inhibitors (such as humic acids) that might interfere with subsequent processing. This situation is particularly relevant for soil metagenome projects, and substantial work has been done in this field to address the issue ([10] and references therein). Direct lysis of cells in the soil matrix versus indirect lysis (i.e. after separation of cells from the soil) has a quantifiable bias in terms of microbial diversity, DNA yield, and resulting sequence fragment length [10]. The extensive work on soil highlights the need to ensure that extraction procedures are well benchmarked and that multiple methods are compared to ensure representative extraction of DNA.

Certain types of samples (such as biopsies or groundwater) often yield only very small amounts of DNA [15]. Library production for most sequencing technologies require high nanograms or micrograms amounts of DNA (see below), and hence amplification of starting material might be required. Multiple displacement amplification (MDA) using random hexamers and phage phi29 polymerase is one option employed to increase DNA yields. This method can amplify femtograms of DNA to produce micrograms of product and thus has been widely used in single-cell genomics and to a certain extent in metagenomics [16,17]. As with any amplification method, there are potential problems associated with reagent contaminations, chimera formation and sequence bias in the amplification, and their impact will depend on the amount and type of starting material and the required number of amplification rounds to produce sufficient amounts of nucleic acids. These issues can have significant impact on subsequent metagenomic community analysis [15], and so it will be necessary to consider whether amplification is permissible.

SEQUENCING TECHNOLOGY

Over the past 10 years metagenomic shotgun sequencing has gradually shifted from classical Sanger sequencing technology to next-generation sequencing (NGS). Sanger sequencing, however, is still considered the gold standard for sequencing, because of its low error rate, long read length (> 700 bp) and large insert sizes (e.g. > 30 Kb for fosmids or bacterial artificial chromosomes (BACs)). All of these aspects will improve assembly outcomes for shotgun data, and hence Sanger sequencing might still be applicable if generating close-to-complete genomes in low-diversity environments is the objective [18]. A drawback of Sanger sequencing is the labor-intensive cloning process in its associated bias against genes toxic for the cloning host [19] and the overall cost per gigabase (appr. USD 400,000).

Of the NGS technologies, both the 454/Roche and the Illumina/Solexa systems have now been extensively applied to metagenomic samples. Excellent reviews of these technologies are available [20,21], but a brief summary is given here with particular attention to metagenomic applications.

The 454/Roche system applies emulsion polymerase chain reaction (ePCR) to clonally amplify random DNA fragments, which are attached to microscopic beads. Beads are deposited into the wells of a picotitre plate and then individually and in parallel pyrosequenced. The pyrosequencing process involves the sequential addition of all four deoxynucleoside triphosphates, which, if complementary to the template strand, are incorporated by a DNA polymerase. This polymerization reaction releases pyrophosphate, which is converted via two enzymatic reactions to produce light. Light production of ~ 1.2 million reactions is detected in parallel via a charge-coupled device (CCD) camera and converted to the actual sequence of the template. Two aspects are important in this process with respect to metagenomic applications. First, the ePCR has been shown to produce artificial replicate sequences, which will impact any estimates of gene abundance. Understanding the amount of replicate sequences is crucial for the data quality of sequencing runs, and replicates can be identified and filtered out with bioinformatics tools [22,23]. Second, the intensity of light produced when the polymerase runs through a homopolymer is often difficult to correlate to the actual number of nucleotide positions. Typically, this results in insertion or deletion errors in homopolymers and can hence cause reading frameshifts, if protein coding sequences (CDSs) are called on a single read. This type of error can however be incorporated into models of CDS prediction thus resulting in high, albeit not perfect, accuracy [24]. Despite these disadvantages, the much cheaper cost of ~ USD 20,000 per gigabase pair has made 454/Roche pyrosequencing a popular choice for shotgun-sequencing metagenomics. In addition, the 454/Roche technology produces an average read length between 600-800 bp, which is long enough to cause only minor loss in the number of reads that can be annotated [25]. Sample preparation has also been optimized so that tens of nanograms of DNA are sufficient for sequencing single-end libraries [26,27], although pair-end sequencing might still require micrograms quantities. Moreover, the 454/Roche sequencing platform offers multiplexing allowing for up to 12 samples to be analyzed in a single run of ~500 Mbp.

The Illumina/Solexa technology immobilizes random DNA fragments on a surface and then performs solid-surface PCR amplification, resulting in clusters of identical DNA fragments. These are then sequenced with reversible terminators in a sequencing-by-synthesis process [28]. The cluster density is enormous, with hundreds of millions of reads per surface channel and 16 channels per run on the HiSeq2000 instrument. Read length is now approaching 150 bp, and clustered fragments can be sequenced from both ends. Continuous sequence information of nearly 300 bp can be obtained from two overlapping 150 bp paired-reads from a single insert. Yields of ~60 Gbp can therefore be typically expected in a single channel. While Illumina/Solexa has limited systematic errors, some datasets have shown high error rates at the tail ends of reads [29]. In general, clipping reads has proven to be a good strategy for eliminating the error in "bad" datasets, however, sequence quality values should also be used to detect "bad" sequences. The lower costs of this technology (~ USD 50 per Gbp) and recent success in its application to metagenomics, and even the generation of draft genomes from complex dataset [30,31], are currently making the Illumina technology an increasingly popular choice. As with 454/Roche sequencing, starting material can be as low as a 20 nanograms, but larger amounts (500-1000 ng) are required when matepair-libraries for longer insert libraries are made. The limited read length of the Illumina/Solexa technology means that a greater proportion of unassembled reads might be too short for functional annotation than are with 454/Roche technology [25]. While assembly might be advisable in such a case, potential bias, such as the suppression of low-abundance species (which can not be assembled) should be considered, as should the fact that some current software packages (e.g. MG-RAST) are capable of analyzing unassembled Illumina reads of 75 bp and longer. Multiplexing of samples is also available for individual sequencing channels, with more than 500 samples multiplexed per lane. Another important factor to consider is run time, with a 2 × 100 bp paired-end sequencing analysis taking approx. 10 days HiSeq2000 instrument time, in contrast to 1 day for the 454/ Roche technology. However, faster runtime (albeit at higher cost per Gbp of approx. USD 600) can be achieved with the new Illumina MiSeq instrument. This smaller version of Illumina/Solexa technology can also be used to test-run sequencing libraries, before analysis on HiSeq instrument for deeper sequencing.

A few additional sequencing technologies are available that might prove useful for metagenomic applications, now or in the near future. The Applied Biosystems SOLiD sequencer has been extensively used, for example, in genome resequencing [32]. SOLiD arguably provides the lowest error rate of any current NGS sequencing technology, however it does not achieve reliable read length beyond 50 nucleotides. This will limit its applicability for direct gene annotation of unassembled reads or for assembly of large contigs. Nevertheless, for assembly or mapping of metagenomic data against a reference genome, recent work showed encouraging outcomes [33]. Roche is also marketing a smaller-scale sequencer based on pyrosequencing with about 100 Mbp output and low per run costs. This system might be useful, because relatively low coverage of metagenomes can establish meaningful gene profiles [34]. Ion Torrent (and more recently Ion Proton) is another emerging technology and is based on the principle that protons released during DNA polymerization can detect nucleotide incorporation. This system promises read lengths of > 100 bp and throughput on the order of magnitude of the 454/Roche sequencing systems. Pacific Biosciences (PacBio) has released a sequencing technology based on single-molecule, real-time detection in zero-mode waveguide wells. Theoretically, this technology on its RS1 platform should provide much greater read lengths than the other technologies mentioned, which would facilitate annotation and assembly. In addition, a process called strobing will mimic pair-end reads. However, accuracy of single reads with PacBio is currently only at 85%, and random reads are «dropped,» making the instrument unusable in its current form for metagenomic sequencing [35]. Complete Genomics is offering a technology based on sequencing DNA nanoballs with combinatorial probe-anchor ligation [36]. Its read length of 35 nucleotides is rather limited and so might be its utility for *de novo* assemblies. While none of the emerging sequencing technologies have been thoroughly applied and tested with metagenomics samples, they offer promising alternatives and even further cost reduction.

ASSEMBLY

If the research aims at recovering the genome of uncultured organisms or obtain full-length CDS for subsequent characterization rather than a functional description of the community, then assembly of short read fragments will be performed to obtain longer genomic contigs. The majority of current assembly programs were designed to assemble single, clonal genomes and their utility for complex pan-genomic mixtures should be approached with caution and critical evaluation.

Two strategies can be employed for metagenomics samples: reference-based assembly (co-assembly) and *de novo* assembly.

Reference-based assembly can be done with software packages such as Newbler (Roche), AMOShttp://sourceforge.net/projects/amos/, or MIRA [37]. These software packages include algorithms that are fast and memory-efficient and hence can often be performed on laptop-sized machines in a couple of hours. Reference-based assembly works well, if the metagenomic dataset contains sequences where closely related reference genomes are available. However, differences in the true genome of the sample to the reference, such as a large insertion, deletion, or polymorphisms, can mean that the assembly is fragmented or that divergent regions are not covered.

De novo assembly typically requires larger computational resources. Thus, a whole class of assembly tools based on the de Bruijn graphs was specifically created to handle very large amounts of data [38,39]. Machine requirements for the de Bruijn assemblers Velvet [40] or SOAP [41] are still significantly higher than for reference-based assembly (co-assembly), often requiring hundreds of gigabytes of memory in a single machine and run times frequently being days.

The fact that most (if not all) microbial communities include significant variation on a strain and species level makes the use of assembly algorithms that assume clonal genomes less suitable for metagenomics. The "clonal" assumptions built into many assemblers might lead to suppression of contig formation for certain heterogeneous taxa at specific parameter settings. Recently, two de Bruijn-type assemblers, MetaVelvet and Meta-IDBA [42] have been released that deal explicitly with the non-clonality of natural populations. Both assemblers aim to identify within the entire de Bruijn graph a sub-graph that represents related genomes. Alternatively, the metagenomic sequence mix can be partition into "species bins" via k-mer binning (Titus Brown, personal communications). Those subgraphs or subsets are then resolved to build a consensus sequence of the genomes. For Meta-IDBA a improvement in terms of N50 and maximum contig length has been observed when compared to "classical" de Bruijn assembler (e.g. Velvet or SOAP; results from the personal experience of the authors; data not shown here). The development of "metagenomic assemblers" is however still at an early stage, and it is difficult to access their accuracy for real metagenomic data as typically no references exist to compare the results to. A true gold standard (i.e. a real dataset for a diverse microbial community with known reference sequences) that assemblers can be evaluated against is thus urgently required.

Several factors need to be considered when exploring the reasons for assembling metagenomic data; these can be condensed to two important questions. First, what is the length of the sequencing reads used to generate the metagenomic dataset, and are longer sequences required for annotation? Some approaches, e.g. IMG/M, prefer assembled contigs, other pipelines such as MG-RAST [43] require only 75 bp or longer for gene prediction or similarity analysis that provides taxonomic binning and functional classification. On the whole, however, the longer the sequence information, the better is the ability to obtain accurate information. One obvious impact is on annotation: the longer the sequence, the more information provided, making it easier to compare with known genetic data (e.g. via homology searches [25]). Annotation issues will be discussed in the next section. Binning and classification of DNA fragments for phylogenetic or taxonomic assignment also benefits from long, contiguous sequences and certain tools (e.g. Phylopythia) work reliably only over a specific cut-off point (e.g. 1 Kb) [44]. Second, is the dataset assembled to reduce data-processing requirements? Here, as an alternative to assembling reads into contigs, clustering near-identical reads with cd-hit [45] or uclust [46] will provide clear benefits in data reduction. The MG-RAST pipeline also uses clustering as a data reduction strategy.

Fundamentally, assembly is also driven by the specific problem that single reads have generally lower quality and hence lower confidence in accuracy than do multiple reads that cover the same segment of genetic information. Therefore, merging reads increases the quality of information. Obviously in a complex community with low sequencing depth or coverage, it is unlikely to actually get many reads that cover the same fragment of DNA. Hence assembly may be of limited value for metagenomics.

Unfortunately, without assembly, longer and more complex genetic elements (e.g., CRISPRS) cannot be analyzed. Hence there is a need for metagenomic assembly to obtain high-confidence contigs that enable the study of, for example, major repeat classes. However, none of the current assembly tools is bias-free. Several strategies have been proposed to increase assembly accuracy [38], but strategies such as removal of rare k-mers are no longer considered adequate, since rare k-mers do not represent sequence errors (as initially assumed), but instead represent reads from less abundant pan-genomes in the metagenomic mix.

BINNING

Binning refers to the process of sorting DNA sequences into groups that might represent an individual genome or genomes from closely related organisms. Several algorithms have been developed, which employ two types of information contained within a given DNA sequence. Firstly, compositional binning makes use of the fact that genomes have conserved nucleotide composition (e.g. a certain GC or the particular abundance distribution of k-mers) and this will be also reflected in sequence fragments of the genomes. Secondly, the unknown DNA fragment might encode for a gene and the similarity of this gene with known genes in a reference database can be used to classify and hence bin the sequence.

Compositional-based binning algorithms include Phylopythia [44], S-GSOM [47], PCAHIER [48,49] and TACAO [49], while examples of purely similarity-based binning software include IMG/M [50], MG-RAST [43], MEGAN [51], CARMA [52], SOrt-ITEMS [53] and MetaPhyler [54]. There is also number of binning algorithms that consider both composition and similarity, including the programs PhymmBL [55] and MetaCluster [56]. All these tools employ different methods of grouping sequences, including self-organising maps (SOMs) or hierarchical clustering, and are operated in either an unsupervised manner or with input from the user (supervised) to define bins.

Important considerations for using any binning algorithm are the type of input data available and the existence of a suitable training datasets or reference genomes. In general, composition-based binning is not reliable for short reads, as they do not contain enough information. For example, a 100 bp read can at best possess only less than half of all 256 possible 4-mers and this is not sufficient to determine a 4-mer distribution that will reliably relate this read to any other read. Compositional assignment can however be improved, if training datasets (e.g. a long DNA fragment of known origin) exist that can be used to define a compositional classifier [44]. These "training" fragments can either be derived from assembled data or from sequenced fosmids and should ideally contain a phylogenetic marker (such as a rRNA gene) that can be used for high-resolution, taxonomic assignment of the binned fragments [57].

Short reads may contain similarity to a known gene and this information can be used to putatively assign the read to a specific taxon. This taxonomic assignment obviously requires the availability of reference data. If the query sequence is only distantly related to known reference genomes, only a taxonomic assignment at a very high level (e.g. phylum) is possible. If the metagenomic dataset, however, contains two or more genomes that would fall into this high taxon assignment, then "chimeric" bins might be produced. In this case, the two genomes might be separated by additional binning based on compositional features. In general, however this might again require that the unknown fragments have a certain length.

Binning algorithm will obviously in the future benefit from the availability of a greater number and phylogenetic breadth of reference genomes, in particular for similarity-based assignment to low taxonomic levels. Post-assembly the binning of contigs can lead to the generation of partial genomes of yet-uncultured or unknown organisms, which in turn can be used to perform similarity-based binning of other metagenomic datasets. Caution should however been taken to ensure the validity of any newly created genome bin, as "contaminating" fragments can rapidly propagate into false assignments in subsequent binning efforts. Prior to assembly with clonal assemblers binning can be used to reduce the complexity of an assembly effort and might reduce computational requirement.

As major annotation pipelines like IMG/M or MG-RAST also perform taxonomic assignments of reads, one needs to carefully weigh the additional computational demands of the particular binning algorithm chosen against the added value they provide.

ANNOTATION

For the annotation of metagenomes two different initial pathways can be taken. First, if reconstructed genomes are the objective of the study and assembly has produced large contigs, it is preferable to use existing pipelines for genome annotation, such as RAST [58] or IMG [59]. For this approach to be successful, minimal contigs length of 30,000 bp or longer are required. Second, annotation can be performed on the entire community and relies on unassembled reads or short contigs. Here the tools for genome annotation are significantly less useful than those specifically developed for metagenomic analyses. Annotation of metagenomic sequence data has in general two steps. First, features of interest (genes) are identified (feature prediction) and, second, putative gene functions and taxonomic neighbors are assigned (functional annotation).

Feature prediction is the process of labeling sequences as genes or genomic elements. For completed genome sequences a number of algorithms have been developed [60,61] that identify CDS with more than 95% accuracy and a low false negative ratio. A number of tools were specifically designed to handle metagenomic prediction of CDS, including FragGeneScan [24], MetaGeneMark [62], MetaGeneAnnotator (MGA)/ Metagene [63] and Orphelia [64,65]. All of these tools use internal information (e.g. codon usage) to classify sequence stretches as either coding or non-coding, however they distinguish themselves from each other by the quality of the training sets used and their usefulness for short or error-prone sequences. FragGeneScan is currently the only algorithm known to the authors that explicitly models sequencing errors and thus results in gene prediction errors of only 1-2%. True positive rates of FragGeneScan are around 70% (better than most other methods), which means that even this tool still misses a significant subset of genes. These missing genes can potentially be identified by BLAST-based searches, however the size of current metagenomic datasets makes this computational expensive step often prohibitive.

There exists also a number of tools for the prediction of non-protein coding genes such as tRNAs [66,67], signal peptides [68] or CRISPRs [69,70], however they might require significant computational resources or long contiguous sequences. Clearly subsequent analysis depends on the initial identification of features and users of annotation pipelines need to be aware of the specific prediction approaches used. MG-RAST uses a two-step approach for feature identification, FGS and a similarity search for ribosomal RNAs against a non-redundant integration of the SILVA [71], Greengenes [72] and RDP [73] databases. CAMERA's RAMCAPP pipeline [74] uses FGA and MGA, while IMG/M employs a combination of tools, including FGS and MGA [58,59].

Functional annotation represents a major computational challenge for most metagenomic projects and therefore deserves much attention now and over the next years. Current estimates are that only 20 to 50% of a metagenomic sequences can be annotated [75], leaving the immediate question of importance and function of the remaining genes. We note that annotation is not done *de novo*, but via mapping to gene or protein libraries with existing knowledge (i.e., a non-redundant database). Any sequences that cannot be mapped to the known sequence space are referred to as ORFans. These ORFans are responsible for the seemingly never-ending genetic novelty in microbial metagenomics (e.g. [76]. Three hypotheses exist for existence of this unknown fraction. First, ORFans might simply reflect erroneous CDS calls caused by imperfect detection algorithms. Secondly, these ORFans are real genes, but encode for unknown biochemical functions. Third, ORFan genes have no sequence homology with known genes, but might have structural homology with known proteins, thus representing known protein families or folds. Future work will likely reveal that the truth lies somewhere between these hypotheses [77]. For improving the annotation of ORFan genes, we will rely on the challenging and labor-intensive task of protein structure analysis (e.g. via NMR and x-ray crystallography) and on biochemical characterization.

Currently, metagenomic annotation relies on classifying sequences to known functions or taxonomic units based on homology searches against available "annotated" data. Conceptually, the annotation is relatively simple and for small datasets (< 10,000 sequences) manual curation can be used increase the accuracy of any automated annotation. Metagenomic datasets are typically very large, so manual annotation is not possible. Automated annotation therefore has to become more accurate and computationally inexpensive. Currently, running a BLASTX similarity search is computationally expensive; as much as ten times the cost of sequencing [78]. Unfortunately, computationally less demanding methods involving detecting feature composition in genes [44] have limited success for short reads. With growing dataset sizes, faster algorithms are urgently needed, and several programs for similarity searches have been developed to resolve this issue [46,79-81].

Many reference databases are available to give functional context to metagenomic datasets, such as KEGG [82], eggNOG [83], COG/KOG [84], PFAM [85], and TIGRFAM [86]. However, since no reference database covers all biological functions, the ability to visualize and merge the interpretations of all database searches within a single framework is important, as implemented in the most recent versions of MG-RAST and IMG/M. It is essential that metagenome analysis platforms be able to share data in ways that map and visualize data in the framework of other platforms. These metagenomic exchange languages should also reduce the burden associated with re-processing large datasets, minimizing, the redundancy of searching and enabling the sharing of annotations that can be mapped to different ontologies and nomenclatures, thereby allowing multifaceted interpretations. The Genomic

Standards Consortium (GSC) with the M5 project is providing a prototypical standard for exchange of computed metagenome analysis results, one cornerstone of these exchange languages.

Several large-scale databases are available that process and deposit metagenomic datasets. MG-RAST, IMG/M, and CAMERA are three prominent systems [43,50,74]. MG-RAST is a data repository, an analysis pipeline and a comparative genomics environment. Its fully automated pipeline provides quality control, feature prediction and functional annotation and has been optimized for achieving a trade-off between accuracy and computational efficiency for short reads using BLAT {Kent, 2002 #64}. Results are expressed in the form of abundance profiles for specific taxa or functional annotations. Supported are the comparison of NCBI taxonomies derived from 16S rRNA gene or whole genome shotgun data and the comparison of relative abundance for KEGG, eggNOG, COG and SEED subsystems on multiple levels of resolution. Users can also download all data products generated by MG-RAST, share them and publish within the portal. The MG-RAST web interface allows comparison using a number of statistical techniques and allows for the incorporation of metadata into the statistics. MG-RAST has more than 7000 users, > 38,000 uploaded and analyzed metagenomes (of which 7000 are publicly accessible) and 9 Terabases analyzed as of December 2011. These statistics demonstrate a move by the scientific community to centralize resources and standardize annotation.

IMG/M also provides a standardized pipeline, but with "higher" sensitivity as it performs, for example, hidden Markov model (HMM) and BLASTX searches at substantial computational cost. In contrast to MG-RAST, comparisons in IMG/M are not performed on an abundance table level, but are based on an all vs. all genes comparison. Therefore IMG/M is the only system that integrates all datasets into a single protein level abstraction. Both IMG/M and MG-RAST provide the ability to use stored computational results for comparison, enabling comparison of novel metagenomes with a rich body of other datasets without requiring the end-user to provide the computational means for reanalysis of all datasets involved in their study. Other systems, such as CAMERA [74], offer more flexible annotation schema but require that individual researchers understand the annotation of data and analytical pipelines well enough to be confident in their interpretation. Also for comparison, all datasets need to be analyzed using the same workflow, thus adding additional computational requirements. CAMERA allows the publication of datasets and was the first to support the Genomic Standards Consortium's Minimal Information checklists for metadata in their web interface [87].

MEGAN is another tool used for visualizing annotation results derived from BLAST searches in a functional or taxonomic dendrogram [51]. The use of dendrograms to display metagenomic data provides a collapsible network of interpretation, which makes analysis of particular functional or taxonomic groups visually easy.

EXPERIMENTAL DESIGN AND STATISTICAL ANALYSIS

Owing to the high costs, many of the early metagenomic shotgun-sequencing projects were not replicated or were focused on targeted exploration of specific organisms (e.g. uncultured organisms in low-diversity acid mine drainage [2]). Reduction of sequencing cost (see above) and a much wider appreciation of the utility of metagenomics to address fundamental questions in microbial ecology now require proper experimental designs with appropriate replication and statistical analysis. These design and statistical aspects, while obvious, are often not properly implemented in the field of microbial ecology [88]. However, many suitable approaches and strategies are readily available from the decades of research in quantitative ecology of higher organisms (e.g. animals, plants). In a simplistic way, the data from multiple metagenomic shotgun-sequencing projects can be reduced to tables, where the columns represent samples and the rows indicate either a taxonomic group or a gene function (or groups thereof) and the fields containing abundance or presence/absence data. This is analogous to species-sample matrices in ecology of higher organisms, and hence many for the statistical tools available to identify correlations and statistically significant patterns are transferable. As metagenomic data however often contain many more species or gene functions then the number of samples taken, appropriate corrections for multiple hypothesis testing have to be implemented (e.g. Bonferroni correction for t-test based analyses).

The Primer-E package [89] is a well-established tool, allowing for a range of multivariate statistical analyses, including the generation of multidimensional scaling (MDS) plots, analysis of similarities (ANOSIM), and identification of the species or functions that contribute to the difference between two samples (SIMPER). Recently, multivariate statistics was also incorporated in a web-based tools called Metastats [90], which revealed with high confidence discriminatory functions between the replicated metagenome dataset of the gut microbiota of lean and obese mice [91]. In addition, the ShotgunFunctionalizeR package provides several statistical procedures for assessing functional differences between samples, both for individual genes and for entire pathways using the popular R statistical package [92].

Ideally, and in general, experimental design should be driven by the question asked (rather than technical or operational restriction). For example, if a project aims to identify unique taxa or functions in a particular habitat, then suitable reference samples for comparison should be taken and processed in consistent manner. In addition, variation between sample types can be due to true biological variation, (something biologist would be most interested in) and technical variation and this should be carefully considered when planning the experiment. One should also be aware that many microbial systems are highly dynamic, so temporal aspects of sampling can have a substantial impact on data analysis and interpretation. While the question of the number of replicates is often difficult to predict prior to the final statistical analysis, small-scale experiments are often useful to understand the magnitude of variation inherent in a system. For example, a small number of samples could be selected and sequenced to shallower depth, then analyzed to determine if a larger sampling size or greater sequencing effort are required to obtain statistically meaningful results [88]. Also, the level at which replication takes place is something that should not lead to false interpretation of the data. For example, if one is interested in the level of functional variation of the microbial community in habitat A, then multiple samples from this habitat should be taken and processed completely separately, but in the same manner. Taking just one sample and splitting it up prior to processing will provide information only about technical, but not biological, variation in habitat A. Taking multiple samples and then pooling them will lose all information on variability and hence will be of little use for statistical purposes. Ultimately, good experimental design of metagenomic projects will facilitate integration of datasets into new or existing ecological theories [93].

As metagenomics gradually moves through a range of explorative biodiversity surveys, it will also prove itself extremely valuable for manipulative experiments. These will allow for observation of treatment impact on the functional and phylogenetic composition of microbial communities. Initial experiments already showed promising results [94]. However, careful experimental planning and interpretations should be paramount in this field.

One of the ultimate aims of metagenomics is to link functional and phylogenetic information to the chemical, physical, and other biological parameters that characterize an environment. While measuring all these parameters can be time-consuming and cost-intensive, it allows retrospective correlation analysis of metagenomic data that was perhaps not part of the initial aim of the project or might be of interest for other research questions. The value of such metadata cannot be overstated and, in fact, has become mandatory or optional for deposition of metagenomic data into some databases [50,74].

SHARING AND STORAGE OF DATA

Data sharing has a long tradition in the field of genome research, but for metagenomic data this will require a whole new level of organization and collaboration to provide metadata and centralized services (e.g., IMG/M, CAMERA and MG-RAST) as well as sharing of both data and computational results. In order to enable sharing of computed results, some aspects of the various analytical pipelines mentioned above will need to be coordinated - a process currently under way under the auspices of the GSC. Once this has been achieved, researchers will be able to download intermediate and processed results from any one of the major repositories for local analysis or comparison.

A suite of standard languages for metadata is currently provided by the Minimum Information about any (x)Sequence checklists (MIxS) [95]. MIxS is an umbrella term to describe MIGS (the Minimum Information about a Genome Sequence), MIMS (the Minimum Information about a Metagenome Sequence) and MIMARKS (Minimum Information about a MARKer Sequence)[87] and contains standard formats for recording environmental and experimental data. The latest of these checklists, MIMARKS builds on the foundation of the MIGS and MIMS checklists, by including an expansion of the rich contextual information about each environmental sample.

The question of centralized versus decentralized storage is also one of "who pays for the storage," which is a matter with no simple answer. The US National Center for Biotechnology Information (NCBI) is mandated to store all metagenomic data, however, the sheer volume of data being generated means there is an urgent need for appropriate ways of storing vast amounts of sequences. As the cost of sequencing continues to drop while the cost for analysis and storing remains more or less constant, selection of data storage in either biological (i.e. the sample that was sequenced) or digital form in (de-) centralized archives might be required. Ongoing work and successes in compression of (meta-) genomic data [96], however, might mean that digital information can still be stored cost-efficiently in the near future.

CONCLUSION

Metagenomics has benefited in the past few years from many visionary investments in both financial and intellectual terms. To ensure that those investments are utilized in the best possible way, the scientific community should aim to share, compare, and critically evaluate the outcomes of metagenomic studies. As datasets become increasingly more complex and comprehensive, novel tools for analysis, storage, and visualization will be required. These will ensure the best use of the metagenomics as a tool to address fundamental question of microbial ecology, evolution and diversity and to derive and test new hypotheses. Metagenomics will be employed as commonly and frequently as any other laboratory method, and "metagenomizing" a sample might become as colloquial as "PCRing." It is therefore also important that metagenomics be taught to students and young scientists in the same way that other techniques and approaches have been in the past.

AUTHORS' CONTRIBUTIONS

All authors contributed to the conception and writing of the review article. All authors have read and approved the final manuscript.

ACKNOWLEDGEMENTS

This work was supported by the Australian Research Council and the U.S. Dept. of Energy under Contract DE-AC02-06CH11357.

The submitted manuscript has been created by UChicago Argonne, LLC, Operator of Argonne National Laboratory ("Argonne"). Argonne, a U.S. Department of Energy Office of Science laboratory, is operated under Contract No. DE-AC02-06CH11357. The U.S. Government retains for itself, and others acting on its behalf, a paid-up nonexclusive, irrevocable worldwide license in said article to reproduce, prepare derivative works, distribute copies to the public, and perform publicly and display publicly, by or on behalf of the Government.

REFERENCES

1. Handelsman J, Rondon MR, Brady SF, Clardy J, Goodman RM. Molecular biological access to the chemistry of unknown soil microbes: a new frontier for natural products. Chem Biol.1998;5(10):R245–249. doi: 10.1016/S1074-5521(98)90108-9.
2. Tyson GW, Chapman J, Hugenholtz P, Allen EE, Ram RJ, Richardson PM, Solovyev VV, Rubin EM, Rokhsar DS, Banfield JF. Community structure and metabolism through reconstruction of microbial genomes from the environment. Nature. 2004;428(6978):37–43. doi: 10.1038/nature02340.
3. Venter JC, Remington K, Heidelberg JF, Halpern AL, Rusch D, Eisen JA, Wu D, Paulsen I, Nelson KE,

Nelson W, Fouts DE, Levy S, Knap AH, Lomas MW, Nealson K, White O, Peterson J, Hoffman J, Parsons R, Baden-Tillson H, Pfannkoch C, Rogers YH, Smith HO. Environmental genome shotgun sequencing of the Sargasso Sea. Science. 2004;304(5667):66–74. doi: 10.1126/science.1093857.

4. Simon C, Daniel R. Metagenomic analyses: past and future trends. Appl Environ Microbiol.2011;77(4):1153–1161. doi: 10.1128/AEM.02345-10.

5. Wilmes P, Bond PL. Metaproteomics: studying functional gene expression in microbial ecosystems. Trends Microbiol. 2006;14(2):92–97. doi: 10.1016/j.tim.2005.12.006.

6. Gilbert JA, Field D, Huang Y, Edwards R, Li W, Gilna P, Joint I. Detection of large numbers of novel sequences in the metatranscriptomes of complex marine microbial communities. PLoS One.2008;3(8):e3042. doi: 10.1371/journal.pone.0003042.

7. Beja O, Aravind L, Koonin EV, Suzuki MT, Hadd A, Nguyen LP, Jovanovich SB, Gates CM, Feldman RA, Spudich JL, Spudich EN, DeLong EF. Bacterial rhodopsin: evidence for a new type of phototrophy in the sea. Science. 2000;289(5486):1902–1906. doi: 10.1126/science.289.5486.1902.

8. Nicol GW, Schleper C. Ammonia-oxidising Crenarchaeota: important players in the nitrogen cycle?Trends Microbiol. 2006;14(5):207–212. doi: 10.1016/j.tim.2006.03.004.

9. Burke C, Kjelleberg S, Thomas T. Selective extraction of bacterial DNA from the surfaces of macroalgae. Appl Environ Microbiol. 2009;75(1):252–256. doi: 10.1128/AEM.01630-08.

10. Delmont TO, Robe P, Clark I, Simonet P, Vogel TM. Metagenomic comparison of direct and indirect soil DNA extraction approaches. J Microbiol Methods. 2011;86(3):397–400. doi: 10.1016/j.mimet.2011.06.013.

11. Knight R, Desai N, Field D, Fierer N, Fuhrman J, Gordon J, Hu B, Hugenholtz P, Jansson J, Meyer F, Stevens R, Bailey M, Kowalchuk G, Gilbert J. Designing Better Metagenomic Surveys: The role of experimental design and metadata capture in making useful metagenomic datasets for ecology and biotechnology. Nature Biotechnology. in review.

12. Thomas T, Rusch D, DeMaere MZ, Yung PY, Lewis M, Halpern A, Heidelberg KB, Egan S, Steinberg PD, Kjelleberg S. Functional genomic signatures of sponge bacteria reveal unique and shared features of symbiosis. ISME J. 2010;4(12):1557–1567. doi: 10.1038/ismej.2010.74.

13. Palenik B, Ren Q, Tai V, Paulsen IT. Coastal *Synechococcus* metagenome reveals major roles for horizontal gene transfer and plasmids in population diversity. Environ Microbiol. 2009;11(2):349–359. doi: 10.1111/j.1462-2920.2008.01772.x.

14. Angly FE, Felts B, Breitbart M, Salamon P, Edwards RA, Carlson C, Chan AM, Haynes M, Kelley S, Liu H, Mahaffy JM, Mueller JE, Nulton J, Olson R, Parsons R, Rayhawk S, Suttle CA, Rohwer F. The marine viromes of four oceanic regions. PLoS Biol. 2006;4(11):e368. doi: 10.1371/journal.pbio.0040368.

15. Abbai NS, Govender A, Shaik R, Pillay B. Pyrosequence analysis of unamplified and whole genome amplified DNA from hydrocarbon-contaminated groundwater. Mol Biotechnol. 2011.

16. Lasken RS. Genomic DNA amplification by the multiple displacement amplification (MDA) method.Biochem Soc Trans. 2009;37(Pt 2):450–453.

17. Ishoey T, Woyke T, Stepanauskas R, Novotny M, Lasken RS. Genomic sequencing of single microbial cells from environmental samples. Curr Opin Microbiol. 2008;11(3):198–204. doi: 10.1016/j.mib.2008.05.006.

18. Goltsman DS, Denef VJ, Singer SW, VerBerkmoes NC, Lefsrud M, Mueller RS, Dick GJ, Sun CL, Wheeler KE, Zemla A, Baker BJ, Hauser L, Land M, Shah MB, Thelen MP, Hettich RL, Banfield JF. Community genomic and proteomic analyses of chemoautotrophic iron-oxidizing "*Leptospirillum rubarum*" (Group II) and "*Leptospirillum ferrodiazotrophum*" (Group III) bacteria in acid mine drainage biofilms. Appl Environ Microbiol. 2009;75(13):4599–4615. doi: 10.1128/AEM.02943-08.

19. Sorek R, Zhu Y, Creevey CJ, Francino MP, Bork P, Rubin EM. Genome-wide experimental determination of barriers to horizontal gene transfer. Science. 2007;318(5855):1449–1452. doi: 10.1126/science.1147112.

20. Metzker ML. Sequencing technologies - the next generation. Nat Rev Genet. 2010;11(1):31–46. doi: 10.1038/nrg2626.

21. Mardis ER. The impact of next-generation sequencing technology on genetics. Trends Genet.2008;24(3):133–141. doi: 10.1016/j.tig.2007.12.007.

22. Niu B, Fu L, Sun S, Li W. Artificial and natural duplicates in pyrosequencing reads of metagenomic data. BMC Bioinformatics. 2010;11:187. doi: 10.1186/1471-2105-11-187.

23. Teal TK, Schmidt TM. Identifying and removing artificial replicates from 454 pyrosequencing data.Cold Spring Harb Protoc. 2010;2010(4):pdb prot5409.

24. Rho M, Tang H, Ye Y. FragGeneScan: predicting genes in short and error-prone reads. Nucleic Acids Res. 2010;38(20):e191. doi: 10.1093/nar/gkq747.]

25. Wommack KE, Bhavsar J, Ravel J. Metagenomics: read length matters. Appl Environ Microbiol.2008;74(5):1453–1463. doi: 10.1128/AEM.02181-07.

26. White RA, Blainey PC, Fan HC, Quake SR. Digital PCR provides sensitive and absolute calibration for high throughput sequencing. BMC Genomics. 2009;10:116. doi: 10.1186/1471-2164-10-116.

27. Adey A, Morrison HG, Asan Xun X, Kitzman JO, Turner EH, Stackhouse B, MacKenzie AP, Caruccio NC, Zhang X, Shendure J. Rapid, low-input, low-bias construction of shotgun fragment libraries by high-density in vitro transposition. Genome Biol. 2010;11(12):R119. doi: 10.1186/gb-2010-11-12-r119.

28. Bentley DR, Balasubramanian S, Swerdlow HP, Smith GP, Milton J, Brown CG, Hall KP, Evers DJ, Barnes CL, Bignell HR, Boutell JM, Bryant J, Carter RJ, Keira Cheetham R, Cox AJ, Ellis DJ, Flatbush MR, Gormley NA, Humphray SJ, Irving LJ, Karbelashvili MS, Kirk SM, Li H, Liu X, Maisinger KS, Murray LJ, Obradovic B, Ost T, Parkinson ML, Pratt MR. et al. Accurate whole human genome sequencing using reversible terminator chemistry. Nature. 2008;456(7218):53–59. doi: 10.1038/nature07517.

29. Nakamura K, Oshima T, Morimoto T, Ikeda S, Yoshikawa H, Shiwa Y, Ishikawa S, Linak MC, Hirai A, Takahashi H, Altaf-Ul-Amin M, Ogasawara N, Kanaya S. Sequence-specific error profile of Illumina sequencers. Nucleic Acids Res. 2011;39(13):e90. doi: 10.1093/nar/gkr344.

30. Hess M, Sczyrba A, Egan R, Kim TW, Chokhawala H, Schroth G, Luo S, Clark DS, Chen F, Zhang T, Mackie RI, Pennacchio LA, Tringe SG, Visel A, Woyke T, Wang Z, Rubin EM. Metagenomic discovery of biomass-degrading genes and genomes from cow rumen. Science. 2011;331(6016):463–467. doi: 10.1126/science.1200387.

31. Qin J, Li R, Raes J, Arumugam M, Burgdorf KS, Manichanh C, Nielsen T, Pons N, Levenez F, Yamada T, Mende DR, Li J, Xu J, Li S, Li D, Cao J, Wang B, Liang H, Zheng H, Xie Y, Tap J, Lepage P, Bertalan M, Batto JM, Hansen T, Le Paslier D, Linneberg A, Nielsen HB, Pelletier E, Renault P. et al. A human gut microbial gene catalogue established by metagenomic sequencing.Nature. 2010;464(7285):59–65. doi: 10.1038/nature08821.

32. Gulig PA, de Crecy-Lagard V, Wright AC, Walts B, Telonis-Scott M, McIntyre LM. SOLiD sequencing of four *Vibrio vulnificus* genomes enables comparative genomic analysis and identification of candidate clade-specific virulence genes. BMC Genomics. 2010;11:512. doi: 10.1186/1471-2164-11-512.

33. Tyler HL, Roesch LF, Gowda S, Dawson WO, Triplett EW. Confirmation of the sequence of 'Candidatus Liberibacter asiaticus' and assessment of microbial diversity in Huanglongbing-infected citrus phloem using a metagenomic approach. Mol Plant Microbe Interact. 2009;22(12):1624–1634. doi: 10.1094/MPMI-22-12-1624.

34. Kunin V, Raes J, Harris JK, Spear JR, Walker JJ, Ivanova N, von Mering C, Bebout BM, Pace NR, Bork P, Hugenholtz P. Millimeter-scale genetic gradients and community-level molecular convergence in a hypersaline microbial mat. Mol Syst Biol. 2008;4:198.

35. Rasko DA, Webster DR, Sahl JW, Bashir A, Boisen N, Scheutz F, Paxinos EE, Sebra R, Chin CS, Iliopoulos D, Klammer A, Peluso P, Lee L, Kislyuk AO, Bullard J, Kasarskis A, Wang S, Eid J, Rank D, Redman JC, Steyert SR, Frimodt-Moller J, Struve C, Petersen AM, Krogfelt KA, Nataro JP, Schadt EE, Waldor MK. Origins of the *E. coli* strain causing an outbreak of hemolytic-uremic syndrome in Germany. N Engl J Med. 2011;365(8):709–717. doi: 10.1056/NEJMoa1106920.

36. Drmanac R, Sparks AB, Callow MJ, Halpern AL, Burns NL, Kermani BG, Carnevali P, Nazarenko I, Nilsen GB, Yeung G, Dahl F, Fernandez A, Staker B, Pant KP, Baccash J, Borcherding AP, Brownley A, Cedeno R, Chen L, Chernikoff D, Cheung A, Chirita R, Curson B, Ebert JC, Hacker CR, Hartlage R, Hauser B, Huang S, Jiang Y, Karpinchyk V. et al. Human genome sequencing using unchained base reads on self-assembling DNA nanoarrays. Science. 2010;327(5961):78–81. doi: 10.1126/science.1181498.

37. Chevreux B, Wetter T, Suhai S. Genome Sequence Assembly Using Trace Signals and Additional Sequence Information Computer Science and Biology. Proceedings of the German Conference on Bioinformatics. 1999;99:45–56.

38. Miller JR, Koren S, Sutton G. Assembly algorithms for next-generation sequencing data. Genomics.2010;95(6):315–327. doi: 10.1016/j.ygeno.2010.03.001.

39. Pevzner PA, Tang H, Waterman MS. An Eulerian path approach to DNA fragment assembly. Proc Natl Acad Sci USA. 2001;98(17):9748–9753. doi: 10.1073/pnas.171285098.

40. Zerbino DR, Birney E. Velvet: algorithms for de novo short read assembly using de Bruijn graphs.Genome Res. 2008;18(5):821–829. doi: 10.1101/gr.074492.107.

41. Li R, Li Y, Kristiansen K, Wang J. SOAP: short oligonucleotide alignment program. Bioinformatics.2008;24(5):713–714. doi: 10.1093/bioinformatics/btn025.

42. Peng Y, Leung HC, Yiu SM, Chin FY. Meta-IDBA: a de Novo assembler for metagenomic data.Bioinformatics. 2011;27(13):i94–101. doi: 10.1093/bioinformatics/btr216.

43. Glass EM, Wilkening J, Wilke A, Antonopoulos D, Meyer F. Using the metagenomics RAST server (MG-RAST) for analyzing shotgun metagenomes. Cold Spring Harb Protoc. 2010;2010(1) pdb prot5368.

44. McHardy AC, Martin HG, Tsirigos A, Hugenholtz P, Rigoutsos I. Accurate phylogenetic classification of variable-length DNA fragments. Nat Methods. 2007;4(1):63–72. doi: 10.1038/nmeth976.

45. Li W, Godzik A. Cd-hit: a fast program for clustering and comparing large sets of protein or nucleotide sequences. Bioinformatics. 2006;22(13):1658–1659. doi: 10.1093/bioinformatics/btl158.

46. Edgar RC. Search and clustering orders of magnitude faster than BLAST. Bioinformatics.2010;26(19):2460–2461. doi: 10.1093/bioinformatics/btq461.

47. Chan CK, Hsu AL, Halgamuge SK, Tang SL. Binning sequences using very sparse labels within a metagenome. BMC Bioinformatics. 2008;9:215. doi: 10.1186/1471-2105-9-215.

48. Zheng H, Wu H. Short prokaryotic DNA fragment binning using a hierarchical classifier based on linear discriminant analysis and principal component analysis. J Bioinform Comput Biol.2010;8(6):995–1011. doi: 10.1142/S0219720010005051.

49. Diaz NN, Krause L, Goesmann A, Niehaus K, Nattkemper TW. TACOA: taxonomic classification of environmental

genomic fragments using a kernelized nearest neighbor approach. BMC Bioinformatics.2009;10:56. doi: 10.1186/1471-2105-10-56.

50. Markowitz VM, Ivanova NN, Szeto E, Palaniappan K, Chu K, Dalevi D, Chen IM, Grechkin Y, Dubchak I, Anderson I, Lykidis A, Mavromatis K, Hugenholtz P, Kyrpides NC. IMG/M: a data management and analysis system for metagenomes. Nucleic Acids Res. 2008. pp. D534–538.

51. Huson DH, Auch AF, Qi J, Schuster SC. MEGAN analysis of metagenomic data. Genome Res.2007;17(3):377–386. doi: 10.1101/gr.5969107.

52. Krause L, Diaz NN, Goesmann A, Kelley S, Nattkemper TW, Rohwer F, Edwards RA, Stoye J. Phylogenetic classification of short environmental DNA fragments. Nucleic Acids Res.2008;36(7):2230–2239. doi: 10.1093/nar/gkn038.

53. Monzoorul Haque M, Ghosh TS, Komanduri D, Mande SS. SOrt-ITEMS: Sequence orthology based approach for improved taxonomic estimation of metagenomic sequences. Bioinformatics.2009;25(14):1722–1730. doi: 10.1093/bioinformatics/btp317.

54. Liu B, Gibbons T, Ghodsi M, Treangen T, Pop M. Accurate and fast estimation of taxonomic profiles from metagenomic shotgun sequences. BMC Genomics. 2011;12(Suppl 2):S4. doi: 10.1186/1471-2164-12-S2-S4.

55. Brady A, Salzberg SL. Phymm and PhymmBL: metagenomic phylogenetic classification with interpolated Markov models. Nat Methods. 2009;6(9):673–676. doi: 10.1038/nmeth.1358.

56. Leung HC, Yiu SM, Yang B, Peng Y, Wang Y, Liu Z, Chen J, Qin J, Li R, Chin FY. A robust and accurate binning algorithm for metagenomic sequences with arbitrary species abundance ratio.Bioinformatics. 2011;27(11):1489–1495. doi: 10.1093/bioinformatics/btr186.

57. Yung PY, Burke C, Lewis M, Egan S, Kjelleberg S, Thomas T. Phylogenetic screening of a bacterial, metagenomic library using homing endonuclease restriction and marker insertion. Nucleic Acids Res.2009;37(21):e144. doi: 10.1093/nar/gkp746.

58. Aziz RK, Bartels D, Best AA, DeJongh M, Disz T, Edwards RA, Formsma K, Gerdes S, Glass EM, Kubal M, Meyer F, Olsen GJ, Olson R, Osterman AL, Overbeek RA, McNeil LK, Paarmann D, Paczian T, Parrello B, Pusch GD, Reich C, Stevens R, Vassieva O, Vonstein V, Wilke A, Zagnitko O. The RAST Server: rapid annotations using subsystems technology. BMC Genomics. 2008;9:75. doi: 10.1186/1471-2164-9-75.

59. Markowitz VM, Mavromatis K, Ivanova NN, Chen IM, Chu K, Kyrpides NC. IMG ER: a system for microbial genome annotation expert review and curation. Bioinformatics. 2009;25(17):2271–2278. doi: 10.1093/bioinformatics/btp393.

60. Lukashin AV, Borodovsky M. GeneMark.hmm: new solutions for gene finding. Nucleic Acids Res.1998;26(4):1107–1115. doi: 10.1093/nar/26.4.1107.

61. Delcher AL, Harmon D, Kasif S, White O, Salzberg SL. Improved microbial gene identification with GLIMMER. Nucleic Acids Res. 1999;27(23):4636–4641. doi: 10.1093/nar/27.23.4636.

62. McHardy ACZ, Wenhan Martin HGL, Alexandre Tsirigos A, Hugenholtz P, Rigoutsos IB, Mark. Accurate phylogenetic classification of variable-length DNA fragments. Nat Methods. 2007;4(1):63–72. doi: 10.1038/nmeth976.

63. Noguchi H, Taniguchi T, Itoh T. MetaGeneAnnotator: detecting species-specific patterns of ribosomal binding site for precise gene prediction in anonymous prokaryotic and phage genomes. DNA Res.2008;15(6):387–396. doi: 10.1093/dnares/dsn027.

64. Hoff KJ, Lingner T, Meinicke P, Tech M. Orphelia: predicting genes in metagenomic sequencing reads. Nucleic Acids Res. 2009. pp. W101–105.

65. Yok NG, Rosen GL. Combining gene prediction methods to improve metagenomic gene annotation.BMC Bioinformatics. 2011;12:20. doi: 10.1186/1471-2105-12-20.

66. Gardner PP, Daub J, Tate JG, Nawrocki EP, Kolbe DL, Lindgreen S, Wilkinson AC, Finn RD, Griffiths-Jones S, Eddy SR, Bateman A. Rfam: updates to the RNA families database. Nucleic Acids Res. 2009. pp. D136–140.

67. Lowe TM, Eddy SR. tRNAscan-SE: a program for improved detection of transfer RNA genes in genomic sequence. Nucleic Acids Res. 1997;25(5):955–964. doi: 10.1093/nar/25.5.955.

68. Bendtsen JD, Nielsen H, von Heijne G, Brunak S. Improved prediction of signal peptides: SignalP 3.0. J Molec Biol. 2004;340(4):783–795. doi: 10.1016/j.jmb.2004.05.028.

69. Bland C, Ramsey TL, Sabree F, Lowe M, Brown K, Kyrpides NC, Hugenholtz P. CRISPR recognition tool (CRT): a tool for automatic detection of clustered regularly interspaced palindromic repeats. BMC Bioinformatics. 2007;8:209. doi: 10.1186/1471-2105-8-209.

70. Grissa I, Vergnaud G, Pourcel C. CRISPRFinder: a web tool to identify clustered regularly interspaced short palindromic repeats. Nucleic Acids Res. 2007. pp. W52–57.

71. Pruesse E, Quast C, Knittel K, Fuchs BM, Ludwig W, Peplies J, Glöckner FO. SILVA: a comprehensive online resource for quality checked and aligned ribosomal RNA sequence data compatible with ARB. Nucleic Acids Res. 2007;35(21):7188–7196. doi: 10.1093/nar/gkm864.

72. DeSantis TZ, Hugenholtz P, Larsen N, Rojas M, Brodie EL, Keller K, Huber T, Dalevi D, Hu P, Andersen GL. Greengenes, a chimera-checked 16S rRNA gene database and workbench compatible with ARB. Appl Environ Microbiol. 2006;72(7):5069–5072. doi: 10.1128/AEM.03006-05.

73. Cole JR, Wang Q, Cardenas E, Fish J, Chai B, Farris RJ, Kulam-Syed-Mohideen AS, McGarrell DM, Marsh T, Garrity GM, Tiedje JM. The Ribosomal Database Project: improved alignments and new tools for rRNA analysis. Nucleic Acids Res. 2009. pp. D141–145.

74. Sun S, Chen J, Li W, Altintas I, Lin A, Peltier S, Stocks K, Allen EE, Ellisman M, Grethe J, Wooley J. Community cyberinfrastructure for Advanced Microbial Ecology Research and Analysis: the CAMERA resource. Nucleic Acids Res. 2011. pp. D546–551.

75. Gilbert JA, Field D, Swift P, Thomas S, Cummings D, Temperton B, Weynberg K, Huse S, Hughes M, Joint I, Somerfield PJ, Muhling M. The taxonomic and functional diversity of microbes at a temperate coastal site: a 'multi-omic' study of seasonal and diel temporal variation. PLoS One.2010;5(11):e15545. doi: 10.1371/journal.pone.0015545.

76. Yooseph S, Sutton G, Rusch DB, Halpern AL, Williamson SJ, Remington K, Eisen JA, Heidelberg KB, Manning G, Li W, Jaroszewski L, Cieplak P, Miller CS, Li H, Mashiyama ST, Joachimiak MP, van Belle C, Chandonia JM, Soergel DA, Zhai Y, Natarajan K, Lee S, Raphael BJ, Bafna V, Friedman R, Brenner SE, Godzik A, Eisenberg D, Dixon JE, Taylor SS. et al. The Sorcerer II Global Ocean Sampling expedition: expanding the universe of protein families. PLoS Biol. 2007;5(3):e16. doi: 10.1371/journal.pbio.0050016.

77. Godzik A. Metagenomics and the protein universe. Curr Opin Struct Biol. 2011;21(3):398–403. doi: 10.1016/j.sbi.2011.03.010.

78. Wilkening J, Desai N, Meyer F, A W. Using clouds for metagenomics - case study. IEEE Cluster.2009.

79. Ye Y, Choi JH, Tang H. RAPSearch: a fast protein similarity search tool for short reads. BMC Bioinformatics. 2011;12:159. doi: 10.1186/1471-2105-12-159.

80. Kent WJ. BLAT-the BLAST-like alignment tool. Genome Res. 2002;12(4):656–664.

81. Wang W, Zhang P, Liu X. Short read DNA fragment anchoring algorithm. BMC Bioinformatics.2009;10(Suppl 1):S17. doi: 10.1186/1471-2105-10-S1-S17.

82. Kanehisa M, Goto S, Kawashima S, Okuno Y, Hattori M. The KEGG resource for deciphering the genome. Nucleic Acids Res. 2004. pp. D277–280.

83. Muller J, Szklarczyk D, Julien P, Letunic I, Roth A, Kuhn M, Powell S, von Mering C, Doerks T, Jensen LJ, Bork P. eggNOG v2.0: extending the evolutionary genealogy of genes with enhanced non-supervised orthologous groups, species and functional annotations. Nucleic Acids Res. 2010. pp. D190–195.

84. Tatusov RL, Fedorova ND, Jackson JD, Jacobs AR, Kiryutin B, Koonin EV, Krylov DM, Mazumder R, Mekhedov SL, Nikolskaya AN, Rao BS, Smirnov S, Sverdlov AV, Vasudevan S, Wolf YI, Yin JJ, Natale DA. The COG database: an updated version includes eukaryotes. BMC Bioinformatics.2003;4:41. doi: 10.1186/1471-2105-4-41.

85. Finn RD, Mistry J, Tate J, Coggill P, Heger A, Pollington JE, Gavin OL, Gunasekaran P, Ceric G, Forslund K, Holm L, Sonnhammer EL, Eddy SR, Bateman A. The Pfam protein families database.Nucleic Acids Res. 2010. pp. D211–222.

86. Selengut JD, Haft DH, Davidsen T, Ganapathy A, Gwinn-Giglio M, Nelson WC, Richter AR, White O. TIGRFAMs and Genome Properties: tools for the assignment of molecular function and biological process in prokaryotic genomes. Nucleic Acids Res. 2007. pp. D260–264.

87. Field D, Amaral-Zettler L, Cochrane G, Cole JR, Dawyndt P, Garrity GM, Gilbert J, Glockner FO, Hirschman L, Karsch-Mizrachi I, Klenk HP, Knight R, Kottmann R, Kyrpides N, Meyer F, San Gil I, Sansone SA, Schriml LM, Sterk P, Tatusova T, Ussery DW, White O, Wooley J, Yilmaz P, Gilbert JA, Johnston A, Vaughan R, Hunter C, Park J, Morrison N. et al. The Genomic Standards Consortium: Minimum information about a marker gene sequence (MIMARKS) and minimum information about any (x) sequence (MIxS) specifications. PLoS Biol. 2011;9(6):e1001088. doi: 10.1371/journal.pbio.1001088.

88. Prosser JI. Replicate or lie. Environ Microbiol. 2010;12(7):1806–1810. doi: 10.1111/j.1462-2920.2010.02201.x.

89. Clarke KR. Non-parametric multivariate analyses of changes in community structure. Australian J Ecology. 1993. pp. 117–143.

90. White JR, Nagarajan N, Pop M. Statistical methods for detecting differentially abundant features in clinical metagenomic samples. PLoS Comput Biol. 2009;5(4):e1000352. doi: 10.1371/journal.pcbi.1000352.

91. Turnbaugh PJ, Hamady M, Yatsunenko T, Cantarel BL, Duncan A, Ley RE, Sogin ML, Jones WJ, Roe BA, Affourtit JP, Egholm M, Henrissat B, Heath AC, Knight R, Gordon JI. A core gut microbiome in obese and lean twins. Nature. 2009;457(7228):480–484. doi: 10.1038/nature07540.

92. Kristiansson E, Hugenholtz P, Dalevi D. ShotgunFunctionalizeR: an R-package for functional comparison of metagenomes. Bioinformatics. 2009;25(20):2737–2738. doi: 10.1093/bioinformatics/btp508.

93. Burke C, Steinberg P, Rusch D, Kjelleberg S, Thomas T. Bacterial community assembly based on functional genes rather than species. Proc Natl Acad Sci USA. 2011;108(34):14288–14293. doi: 10.1073/pnas.1101591108.

94. Mou X, Sun S, Edwards RA, Hodson RE, Moran MA. Bacterial carbon processing by generalist species in the coastal ocean. Nature. 2008;451(7179):708–711. doi: 10.1038/nature06513.

95. Yilmaz P, Kottmann R, Field D, Knight R, Cole JR, Amaral-Zettler L, Gilbert JA, Karsch-Mizrachi I, Johnston A, Cochrane G, Vaughan R, Hunter C, Park J, Morrison N, Rocca-Serra P, Sterk P, Arumugam M, Bailey M, Baumgartner L, Birren BW, Blaser MJ, Bonazzi V, Booth T, Bork P, Bushman FD, Buttigieg PL, Chain PS, Charlson E, Costello EK, Huot-Creasy H. et al. Minimum information about a marker gene sequence (MIMARKS) and minimum information about any (x) sequence (MIxS) specifications. Nat Biotechnol. 2011;29(5):415–420. doi: 10.1038/nbt.1823.

96. Hsi-Yang Fritz M, Leinonen R, Cochrane G, Birney E. Efficient storage of high throughput DNA sequencing data using reference-based compression. Genome Res. 2011;21(5):734–740. doi: 10.1101/gr.114819.110.

CHAPTER 10

RECONSTRUCTION OF RIBOSOMAL RNA GENES FROM METAGENOMIC DATA

Lu Fan, Kerensa McElroy, Torsten Thomas

INTRODUCTION

Microorganisms are vital components of our planet's ecosystems. PCR amplification and sequencing of 16S ribosomal RNA (16S rRNA) genes directly from environmental samples has over the last two decades revealed an astonishing amount of new microbial diversity [1], [2]. However, as the 'universal' primers used in PCR are designed based on already known groups of organisms, a skewed picture of community composition is likely obtained, especially for environmental samples containing divergent microbial lineages [3].

Direct sequencing of total environmental DNA (metagenomics) has the potential to assess the true diversity of the environment without primer bias [4], [5]. Metagenomic sequences can be assigned to taxa using their similarity to reference genomes based on either sequence similarity [6]–[9] or genomic composition [10]–[13]. However, these types of assignments are only informative when the genomes of closely related taxa are present in the reference set. As reference genomes are only available for a limited part of the phylogenetic tree of life [14], these taxonomic predictions are generally of low resolution (e.g. phyla or order) and hence often give only an unsatisfactory description of community composition.

In contrast, several comprehensive databases exist for the 16S rRNA gene that provide detailed phylogenetic trees [15] and allow for taxonomic resolution down to the species level[16]. Shotgun metagenomic datasets obviously also contain fragmented 16S rRNA genes and these have been directly assigned to taxa through BLAST-based comparisons [4] or phylogenetic distance-based clustering [17]. However, the short and random nature of metagenomic sequences may not contain the phylogenetically most informative regions of the 16S rRNA genes, thus diminishing the efficiency of taxonomic assignments. Sequence assembly can potentially increase the length of the 16S rRNA gene sequences recovered [18], but low sequence coverage may limit assembly success for 16S rRNA genes and low-stringency assemblies may result in chimeric sequences [19], [20]. The recently released EMIRGE software uses iterative mapping of short Illumina reads against reference sequences to reconstruct 16S rRNA genes [19]. Although this approach has an explicit accuracy to single nucleotide difference, its potential to avoid chimeras is strongly dependent on the quality of the reference database. Further, EMIRGE's algorithm is currently not designed for pyrosequencing reads, which contain high rates of insertion and deletions errors (e.g. in homopolymers) [21]. There is thus a need for an approach that reconstructs 16S rRNA genes with high accuracy from pyrosequencing data.

In the present study, we describe a strategy to reconstruct nearly full-length 16S rRNA sequences from metagenomicpyrosequencing data. Through simulation of communities with different diversities we developed a process of stringent assembly and data filtering that generates 16S rRNAcontigs with minimal chimera rates. We then applied our process to assess the microbial symbiont communities from two marine sponges species and compared the outcome to PCR-based assessments of the community structure (pyro-tag-sequencing). We show that about 30% of the abundant phylotypes reconstructed from metagenomic reads failed to be amplified by PCR, which is most likely due to primer mismatches.

MATERIALS AND METHODS

Simulated metagenomes and metagenomic samples

Ninety completed genomes were selected as references, including 76 bacteria and 14 archaea and combined using established profiles of community diversity with high- (HC), median- (MC), and low- (LC) complexity [22] . Genomic sequences, 16S rRNA gene sequences and gene copy number per genome were obtained from the Integrated Microbial Genomes website (http://img.jgi.doe.gov/cgi-bin/w/main.

Citation: Fan L, McElroy K, Thomas T (2012) Reconstruction of Ribosomal RNA Genes from Metagenomic Data. PLoS ONE 7(6): e39948. doi:10.1371/journal.pone.0039948.

cgi). Heterogenous 16S rRNA genes within a genome were considered separately. For each metagenome complexity, three read data set (1,000,000 reads each, 350 nt) were simulated using empirically derived and context-based error models (Gem SIM software [23]).

Three environmental DNA samples for each of the two sponges *Cymbastelaconcentrica* and *C. coralliophila* were obtained as described in ref. [24]. Shotgun pyrosequencing (454 Titanium) was conducted at the J. Craig Venter Institute, Rockville, USA and the resulting average read length corresponded to the simulated datasets above. The shotgun sequencing is available through the Community Cyberinfrastructure for Advanced Microbial Ecology Research and Analysis website (http://camera.calit2.net/) under project accession 'CAM_PROJ_BotanyBay'.

Reconstruction of 16S rRNA gene sequences

The metagenomic reads of the simulated communities and the sponge microbial communities were pre-processed with PrinSeq [25] using the settings '("minlen":"60","maxlen":"700","minqualm":"20","nsmaxp":"1","complval":"50", "noniupac":"true","derep0":"true","derep1":"true","complmethod":"2","trimtails":"6","trimns":"1","trimscore":"15","trimwindow":"2","trimstep":"1","tailsite":"1","trimsite":"3","trimtype":"2","trimrule":"1")'. Metaxa (version 1.0.2) [26] was then used to identify reads containing 16S rRNA sequences. Reads (>300 nt) from triplicates were then pooled and assembled with the GS De Novo Assembler 2.3 (454 Life Sciences, Branford, CT) using the 'cDNA' option, which is optimized for the uneven and high coverage typically expected in RNA assemblies. Default settings were used except 'overlap identity', which was set to 99%. Additionally, 'reads limited to one contig' and 'extending low depth overlaps' were selected. The 99% cut-off was chosen to allow overlap of reads with a 1% error, which is typical seen towards the end of pyrosequencing reads [23]. Lower stringency (e.g. 97% as used by Radax*et al.* during the assembly of 16S rRNA gene [27]) resulted in unacceptable rates of chimera formation (data not shown). After aligning contigs to the SILVA 1.08 database by SINA [28], flanking regions that were not part of the 16S rRNA gene sequences were removed. Resulting contigs were then examined for chimerism. If a contig constituted reads from more than one strain and any of these strains was less than 99% sequence identity to the other strains, it was considered a chimera.

Pyrosequencing of 16S rRNA genes amplified by PCR

Amplification of the 16S rRNA gene was performed on the same DNA sample as used for shotgun sequencing. Primers 28F 'GAGTTTGATCNTGGCTCAG' and 519R 'GTNTTACNGCGGCKGCTG' were used for amplification of the variable regions V1-3. PCR and subsequent sequencing are described in Dowd *et al.* 2008 [29] and were performed at the Research and Testing Laboratory (Lubbock, USA). Trace data was deposited at the NCBI Sequencing Read Archive database with the project accession SRP011939.

Analysis of the 16S rRNA tag-sequencing data was performed using Mothur v1.23.1 [30]. Specifically, 'shhh.flows' was used for de-noising, 'trim.seqs (pdiffs=2, bdiffs=1, maxhomop=8, minlength=200)' was used for barcode removal and quality filtering, SINA was used for sequence alignment with the SILVA 1.08 database [28], 'screen.seqs(start=1048, minlength=245)' and 'filter.seqs (vertical=T, trump=.)' were used for alignment quality filtering, 'pre.cluster(diffs=2)' was used for further error reduction, 'chimera.uchime' was used for *de novo* removal of chimeric reads, and Metaxa (version 1.0.2) [26] was used to remove mitochondrial and chloroplast sequences.

Operational taxonomic unit (OTU) analysis

For simulated data, filtered 16S rRNAcontigs (with coverage of more than 10 reads and length greater than 700 nt; see below) and 16S rRNA reads not in contigs were pooled with the 16S rRNA sequences of the reference genomes used for simulation. Redundancy within these pools was removed with CD-hit (99% identify cut-off). PhylOTU [17] was then used to generate OTUs with 0.01, 0.03 and 0.05 phylogenetic distance cut-off. OTUs containing both reference sequences and simulated shotgun sequences (filtered contigs or reads) were assigned as 'recovered'. OTUs containing only reference sequences were termed as 'missed', while those containing only shotgun sequences were assigned as 'artificial'. OTU coverage was defined as the number of reads contained in each OTU. For the sponge samples, filtered 16S Rrn Acontigs (with coverage of more than 10 reads and length greater than 700 nt) and 16S RNA reads not in contigs were pooled with PCR-amplified tag-sequences and then processed as above to generate OTUs. Diversity analysis was performed with QIIME [31] and phylogenetic distance-based rarefaction was based on the tree of non-redundant sequences generated during the PhylOTU process.

Taxonomic classification and phylogenetic analysis

16S rRNA classification was performed with the RDP Classifier 2.3 [32], except for the classification of the abundant OTUs in sponge samples, which was performed with the Greengenes Classifier (March 6, 2012) [33] followed by manual examination. Single-copy gene based analysis was performed using MLTreeMap (version 2.05, 'minimal sequence length after Gblocks' set to 35) [7]. For phylogenetic analysis, Maximum-Likelihood trees of the 16S rRNA gene contigs were constructed using RAxML [34] after alignment by SINA and removal of ambiguous positions by Gblocks (–t=d –b4=5 –b5=h) [35].

RESULTS

16S rRNA gene assembly with minimal chimera formation

As chimera formation was a major issue in previous assembly approaches [18], [19], [27], we first examined the occurrence of chimeric 16S rRNAcontigs in our assembly strategy on simulated datasets (see Materials and Methods). 9,931 (0.11%) reads containing 16S rRNA gene information were detected from 8,997,875 shotgun reads after quality filtering (Table 1). After applying our assembly strategy we recovered between 125–130 contigs containing full or partial 16S rRNA genes (Table 1).

Table 1: Reads, 16S rRNAcontigs, OTUs and chimera examination of the simulated communities. http://dx.doi.org/10.1371/journal.pone.0039948.t001

Sample	HC-A	HC-B	HC-C	MC-A	MC-B	MC-C	LC-A	LC-B	LC-C
Reads after quality filtering	999913	999909	999912	999703	999775	999769	999603	999606	999685
16S rRNA gene - containing reads	1303	1353	1376	984	1112	1153	874	916	860
16S rRNAcontigs> 350 nt (chimera, chimera containing >1 contaminating read)	130 (3, 1)			126 (7, 1)			125 (4, 3)		
Reads in 16S rRNAcontigs>350 nt (chimera, chimera containing >1 contaminating read)	3733 (85, 15)			3005 (365, 8)			2386 (374, 150)		
Filtered 16SrRNAcontigs (chimera, chimera containing >1 contaminating read)	73 (0, 0)			53 (3, 0)			54 (3, 2)		
Reads in filtered 16SrRNAcontigs (chimera, chimera containing >1 contaminating read)	3257 (0, 0)			2610 (330, 0)			2004 (364, 140)		
Length of filtered 16S rRNAcontigs (min, max, mean) (nt)	458, 1548, 1262			574, 1529, 1127			515, 1532, 1174		
Recovered, missed, artificial OTUs (0.01)	81, 0, 0	81, 0, 0	81, 0, 0	75, 1, 1	77, 1, 1	77, 1, 1	80, 0, 0	79, 0, 0	80, 0, 0
Reads in recovered, missed, artificial OTUs (0.01)	1303, 0, 0	1353, 0, 0	1376, 0, 0	978, 2, 4	1106, 2, 2	1148, 4, 4	870, 0, 0	915, 0, 0	857, 0, 0
Recovered, missed, artificial OTUs (0.03)	74, 0, 0	74, 0, 0	74, 0, 0	69, 0, 0	72, 0, 0	72, 0, 0	72, 0, 0	71, 0, 0	72, 0, 0
Reads in recovered, missed, artificial OTUs (0.03)	1303, 0, 0	1353, 0, 0	1376, 0, 0	982, 0, 0	1108, 0, 0	1150, 0, 0	870, 0, 0	915, 0, 0	857, 0, 0
Recovered, missed, artificial OTUs (0.05)	52, 0, 0	53, 0, 0	52, 0, 0	49, 0, 0	50, 0, 0	49, 0, 0	49, 0, 0	48, 0, 0	48, 0, 0
Reads in recovered, missed, artificial OTUs (0.05)	1303, 0, 0	1353, 0, 0	1376, 0, 0	982, 0, 0	1108, 0, 0	1150, 0, 0	870, 0, 0	915, 0, 0	857, 0, 0

doi:10.1371/journal.pone.0039948.t001

16S rRNAcontigs larger than 350 nt were plotted by their length and read coverage (Figure 1). Fourteen chimeric contigs (3.6%) were detected in all 381 contigs generated from the nine datasets (solid circle and triangles in Figure 1). Four of these contigs could be readily detected using UChime [36] (arrows in Figure 1). Eight chimeras contain only one 'contaminating' read (solid circles in Figure 1), which were mostly aligned to highly conserved regions of the 16S rRNA gene (data not shown). To examine whether these chimeras would affect the accuracy of community structure prediction, we generated OTUs with different phylogenetic distance cut-offs (0.01, 0.03 and 0.05). In nearly all case, all reference OTUs were recovered and no artificial OTUs were generated. The only exception was for MC communities at a 0.01 OTU level where one artificial OTU was generated and one OTU present in the reference was missed (Table 1). This result shows that our assembly strategy recovers effectively the true microbial community structure, and especially OTU groupings of greater than 0.03 phylogenetic distance.

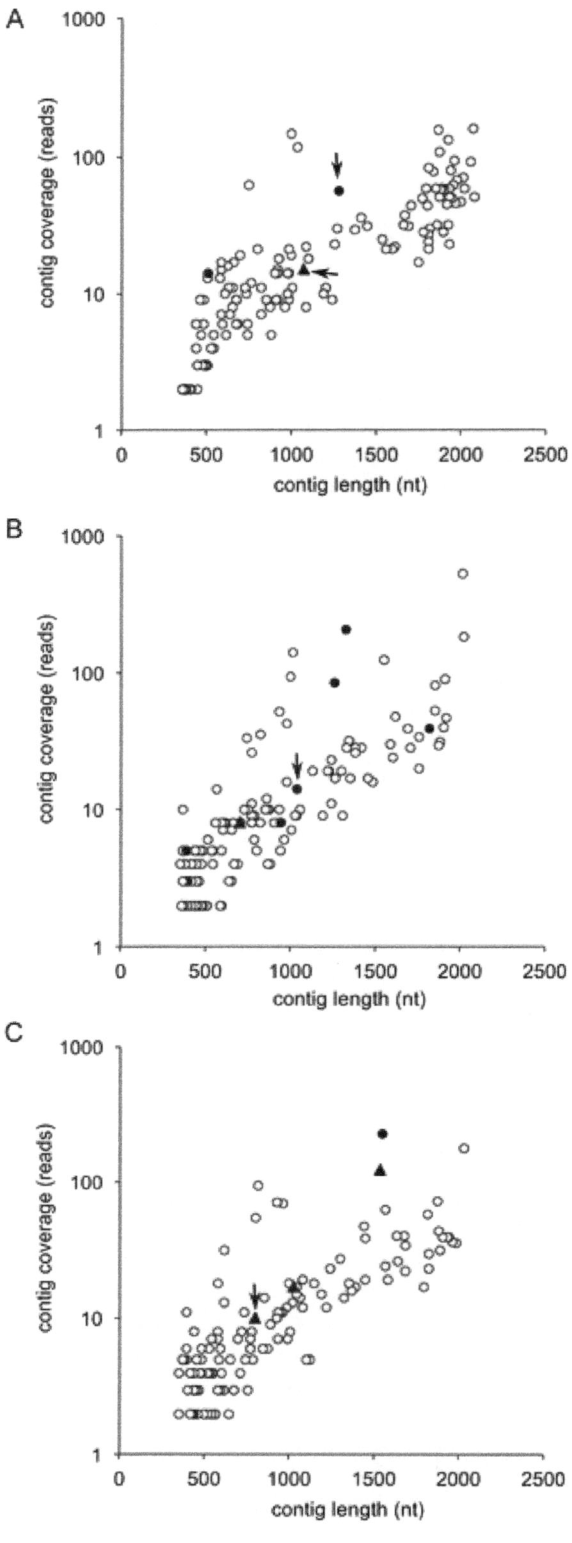

Figure 1: 16S rRNA gene contigs and chimeric contigs for simulated datasets. Open circle: non-chimeric contigs; solid circle: chimeric contigs containing one contaminating read; solid triangles: chimeric contigs containing more than one contaminating read. Arrow: chimera detected by UChime. (A) HC. (B) MC. (C) LC.

With the aim of recovering long 16S rRNA sequences for phylogenetic analysis and to minimize the effects of potential chimeric assembly, we filtered contigs for length of greater than 700 nt and for a coverage of more than 10 reads (Figure 1). In addition we used UChime for chimera removal. Sequences flanking the 16S rRNA gene were removed. This resulted in 180 contigs (mean length: 1,174–1,262 nt) in the nine samples with only two (1.1%) of them containing more than one contaminating read (Table 1). This value is below the chimeric amplification rate generally reported for PCR-based assessment of 16S rRNA gene diversity (5 to 45%) [5], [37]–[40].

Assembly of 16S rRNA sequences improves taxonomic classification

With the assumption that longer 16S rRNA gene sequences can improve the taxonomic description of a community, we compared the proportion of reads before and after assembly that could be confidently assigned using the RDP Classifier (80% confidence). Despite all strains in the simulated datasets being deposited in the RDP database, a steady decline of classification success was observed with between 60–70% of unassembled reads being assigned at the genus level. In contrast, assembled data showed generally higher classification success and at genus level more than 80% could be confidently assigned (Figure 2). This shows a clear benefit of 16S rRNA gene assembly for taxonomic classification and will also improve phylogenetic analysis (see below).

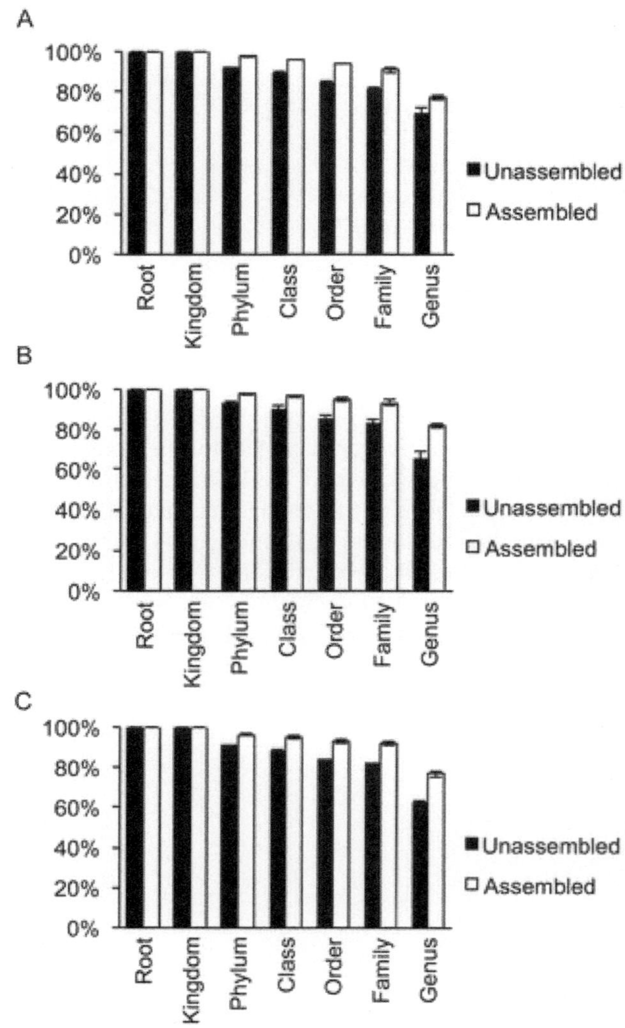

Figure 2: Taxonomic classification of assembled and unassembled shotgun 16S rRNA gene reads for simulated datasets. (A) HC. (B) MC. (C) LC.

16S rRNA gene reconstruction reveals community diversity that is missed by PCR-based approaches

Sponges (phylum Porifera) host complex communities of microbial symbionts, which are essential for the host's function [41]. Over the last decade substantial efforts have been made to describe the phylogenetic diversity and biogeography of sponge-associated microorganisms [41], [42]. However, the vast majority of sponge microbiome surveys are based on PCR-amplification of the 16S rRNA gene. Only recently has one study generated 16S rRNAcontigs from a shotgun-sequenced transcriptome of a sponge microbial community [27]. However, this study generated relatively short contigs (729 nt on average) despite extremely high sequencing coverage (66,743 reads containing 16S rRNA gene sequences) and the loose stringency during assembly could have created many chimeras (see above) [27].

To evaluate the phylogenetic diversity generated by our 16S rRNA gene reconstruction method, we analyzed six shotgun metagenomes from the two spongesC. *concentrica* and *C. coralliophila*. From 5,322,385 quality-filtered pyrosequencingreads, we could identify 1,942 reads containing 16S rRNA genes (0.04%) and generated 25 filteredcontigs (Table 2). The majority of contigs were full or near-full length (Table 2). Community composition of the six sponge DNA samples was also assessed by PCR-amplifying and pyrosequencing the variable region V1-3 of the 16S rRNA gene (pyro-tag-sequencing). 22,392 16S rRNA gene sequences were obtained and 1,366 were unique sequences after quality filtering and pre-clustering (Materials and Methods, Table 3).

Table 2: The sponge metagenomic datasets

Sample	Cyr-A shotgun	Cyr-B shotgun	Cyr-C shotgun	Cyn-A shotgun	Cyn-B shotgun	Cyn-C shotgun
Sponge host	*C. coralliophila*			*C. concentrica*		
Raw reads	897408	971976	888127	678283	1168872	1523499
Average read size (nt)	387.6	151.2	376.8	358.0	408.1	390.6
Reads after quality filtering	819525	898161	780652	668589	1804075	1111085
16S rRNA gene – containing reads	282	385	95	337	530	412
16S rRNA gene contigs(>350 nt (reads))	46 (537)			66 (808)		
Filtered 16S rRNA gene contigs (reads)	13 (919)			12 (717)		
Length of filtered 16S rRNA gene contigs (min, max, mean) (nt)	1216, 1535, 1418			490, 1517, 1251		

doi:10.1371/journal.pone.0059948.t002

Table 3: The sponge tag-sequencing data sets

Sample	Cyr-A PCR	Cyr-B PCR	Cyr-C PCR	Cyn-A PCR	Cyn-B PCR	Cyn-C PCR
Sponge host	*C. coralliophila*			*C. concentrica*		
Raw reads	4888	3906	13861	8257	5364	12809
Average read size (nt)	281.3	303.5	305.7	306.6	317.2	304.1
Reads after quality filtering	2942	3035	4985	5754	2140	6050
Unique sequences	212	179	311	265	155	244
Average size of unique sequences (nt)	269.8	268.9	272.2	267.2	271	269.2

doi:10.1371/journal.pone.0059948.t003

We first compared community composition derived from the pyro-tag-sequencing data, the shotgun reads with and without assembly and single-copy genes (Material and Methods) at the phylum level (Figure 3). In general, more phyla were detected in shotgun sequencing reads compared to pyro-tag-sequencing data. Specifically, the PCR-based approach using the 28F/519R primer set recovered predominately phylotypes belonging to cyanobacteria and proteobacteria, while the shotgun data also detected sequences in Actinobacteria, Nitrospira, Chloroflexi, and Verrucomicrobia (Figure 3A, B). This may be not only due to potential primer bias (see below), but also the short sequences (~250nt after quality processing) (Materials and Methods, Table 3) that are difficult to classify. The presence of these 'missed' phyla (e.g. Chloroflexi) was also confirmed by single-copy gene based search (Figure 3D). However, this single-copy gene approach also failed to detect some taxa (e.g. Nitrospira and Verrucomicrobia), which is likely due to the low number of reference genomes available for these phyla. Overall, these results show that 16S rRNA gene analysis from metagenomic datasets has superior capacity to detect a broad range of phylogenetic diversity.

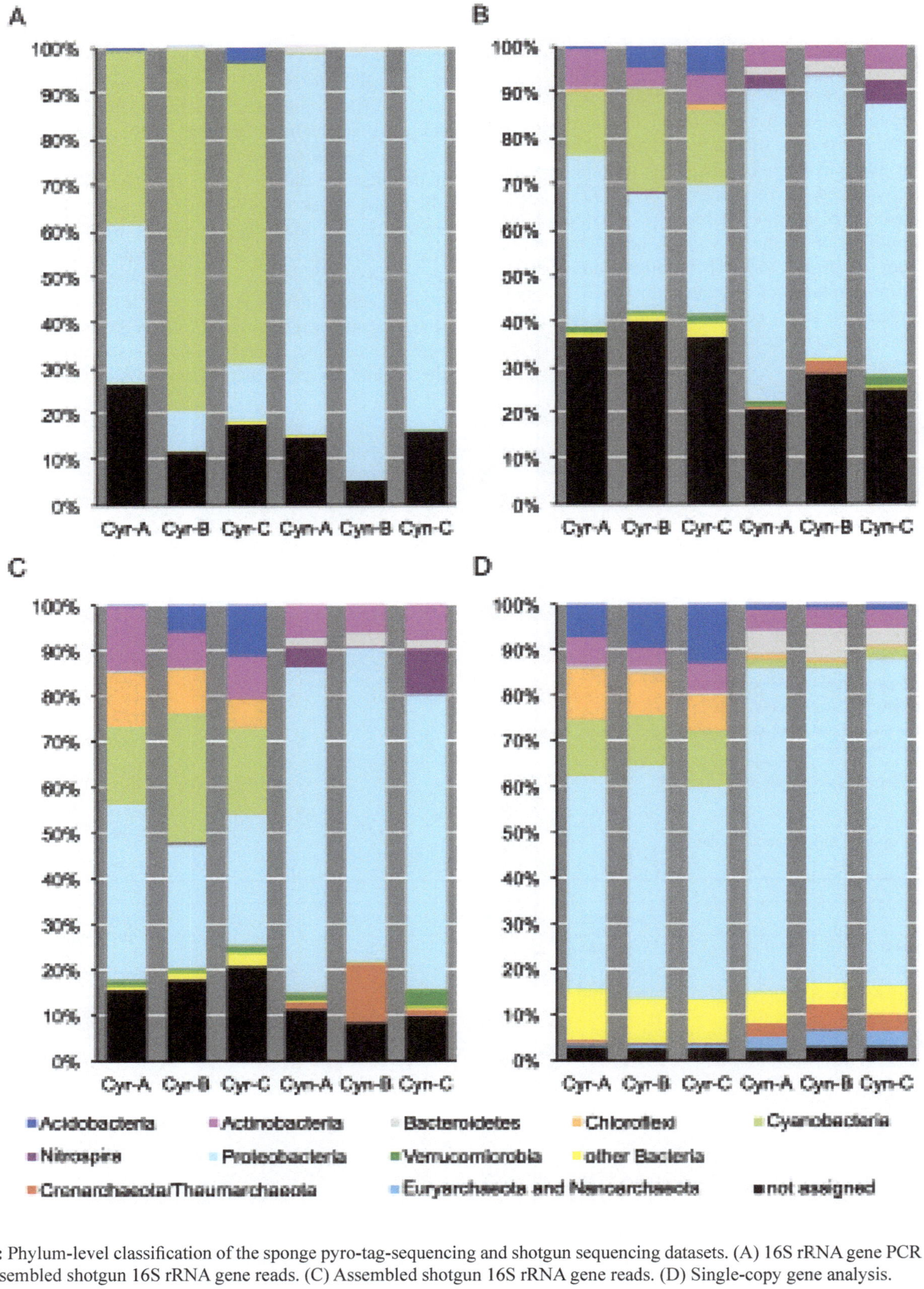

Figure 3: Phylum-level classification of the sponge pyro-tag-sequencing and shotgun sequencing datasets. (A) 16S rRNA gene PCR approach. (B) Unassembled shotgun 16S rRNA gene reads. (C) Assembled shotgun 16S rRNA gene reads. (D) Single-copy gene analysis.

We then compared the pyro-tag-sequencing data and the 16S rRNA gene reconstruction approach by generating OTUs at different phylogenetic distance cut-offs (Materials and Methods). In general, the PCR-based approach produced more OTUs than the metagenome-based approach, except at the 0.05 OTU-level for *C. concentrica* (Figure 4). This is obviously because of the much higher sequencing depth for the 16S rRNA gene in the pyro-tag samples (Table 2, 3). A relative low number of common OTUs were observed between the two approaches. However, the OTUs unique to the PCR-based approach only present a low proportion (2.5–8.3%) of all pyro-tag reads at OTU-levels of 0.03 and 0.05. This result shows that the majority of pyro-tag reads come from phylotypes that are also contained in the metagenomic data set and that the unique OTUs of the PCR-based approach either constitute low abundance phylotypes (e.g. are part of the rare biosphere) [43] or are undetected chimeras[44]. In contrast, a high proportion of reads (~30%) belong to unique OTUs generated from the 16S rRNA gene reconstruction, which indicates that they come from abundant organisms that were missed by PCR-based approaches. Different levels of diversity of the PCR analysis and metagenomic reconstruction are also reflected in rarefaction plots . Although the sampling depths of the shotgun samples were relatively low, the trends reflected in their rarefaction plots compared to the plots of the PCR samples clearly suggests a higher community diversity.

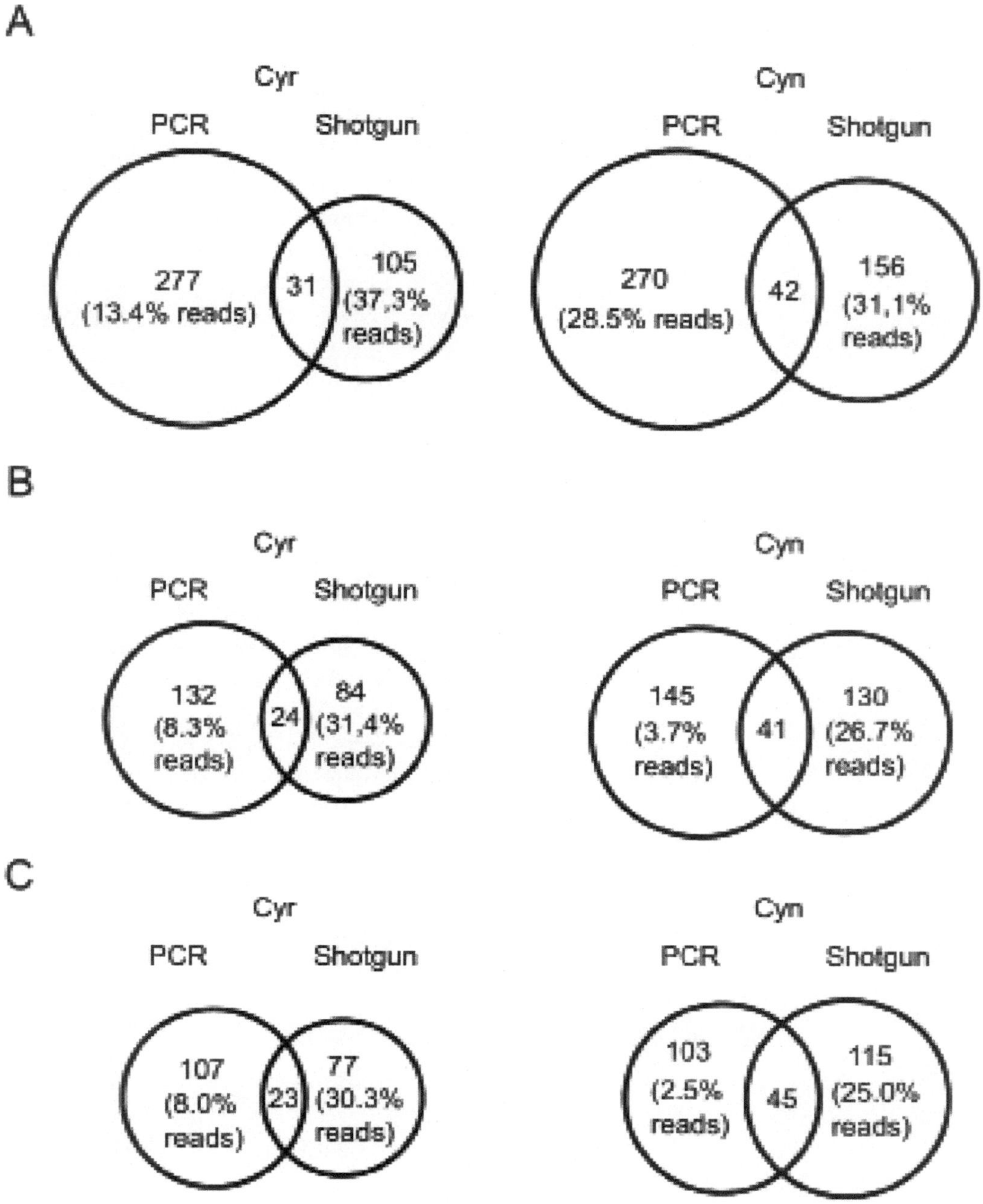

Figure 4: Shared and unique OTUs of the PCR-based and shotgun-based sponge datasets. Circle sizes are proportional to OTU number. (A) 0.01 phylogenetic distance OTU. (B) 0.03 phylogenetic distance OTU. (C) 0.05 phylogenetic distance OTU.

Primer bias can explain the lack of OTU detection

To further investigate how PCR-amplification failed to detect certain groups of bacteria (see above), we taxonomically classified the most abundant 0.01-level OTUs (>2% in any of the 12 samples) (Figure 5). OTUs assigned to the bacterial groups of *Robiginitomaculum*, *Phyllobacteriaceae_*4, OCS116, *Rhodobacteraceae*, *Rhodospirillaceae*, *Acinetobacter*,*Oceanospirillacea e*, *Thiotrichaceae*, *Vibrionaceae*, PAUC26f, Sva0996 and*Verrucomicrobiaceae* were consistently missed or poorly recovered by PCR. Among them, eight 16S rRNA gene contigs belonging to seven 0.01 OTUs (i.e. *Robiginitomaculum*,*Rhodobactera ceae*, *Acinetobacter*, *Oceanospirillaceae*, PAUC26f, Sva0996, and *Verrucomicrobiaceae*, including two contigs belonging to Sva0996) covered the entire V1-3 region of the 16S rRNA gene. Alignment of these eight contigs to the degenerate primers

28F/519R found seven of them had mismatches (either one or both primers) (asterisks in Figure 5). This suggests that primer bias is one of the major causes for the PCR-based approach missing certain OTUs (Figure 4).

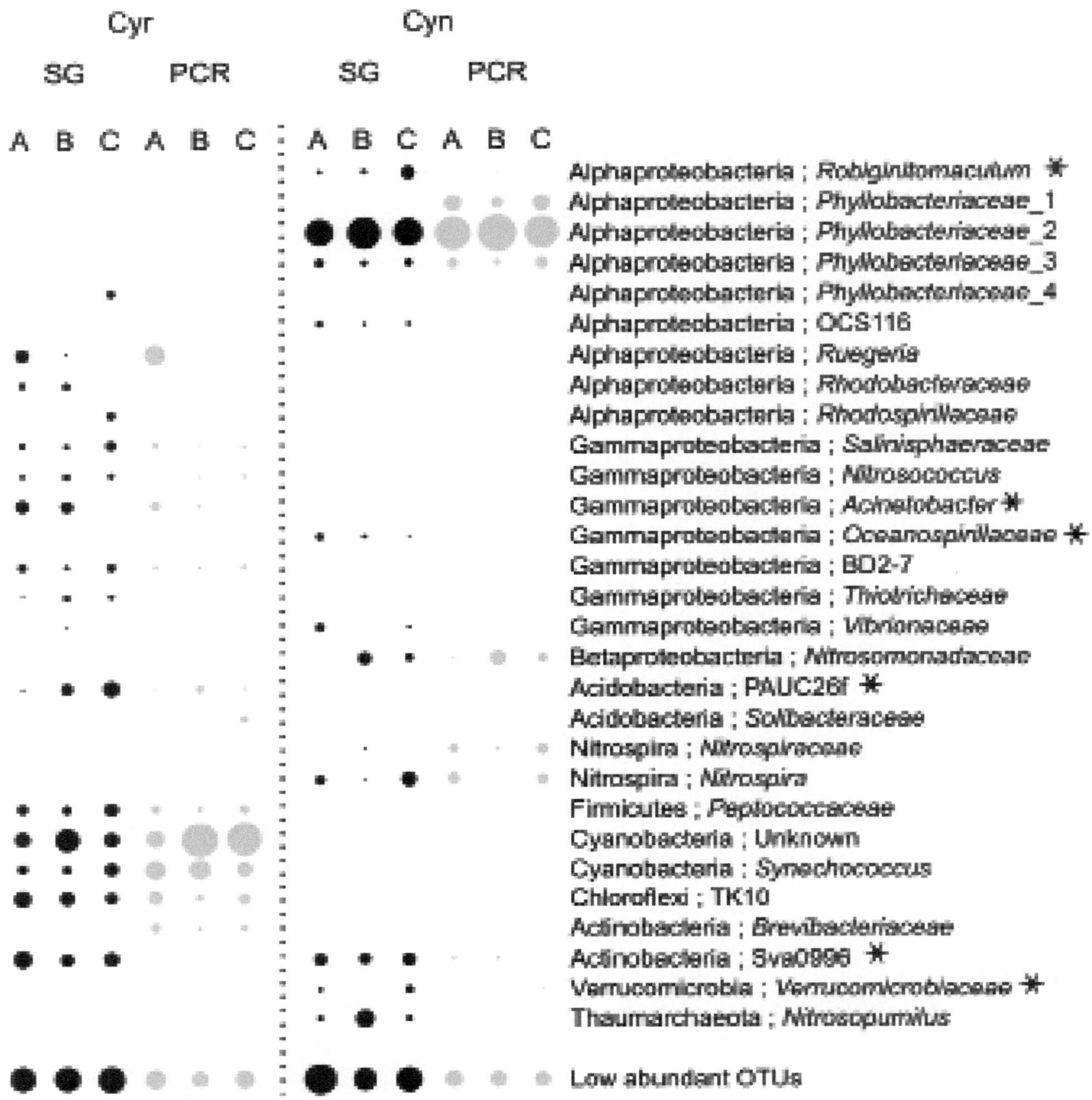

Figure 5: Abundance and primer-mismatches in the top OTUs at the 0.01 phylogenetic distance level for the sponge datasets. Asterisk, primer-mis-match event.

Phylogenetic analysis of the novel 16S rRNA sequences detected by the shotgunapproach

To examine how many of the 25 16S rRNA gene contigs reconstructed from shotgun sequencing data have so far not been detected by PCR-based approaches in these two sponges, we performed searches against the NCBI nt database (7 April 2012) and the full-length 16S rRNA genes (primes 27F and 1492R) previously amplified from *C. concentrica* by Thomas *et al.* [45]. Any match with a BlastN identity of >99% was considered as an amplicon counterpart to the contigs. While none of the 13 contigs from *C. coralliophila* found amplicon counterparts, 10 of the 12 contigs from *C. concentrica* had been previously detected.

Among the 15 undetected sequences, ten were amplified by the primers used in the present study (Figure 5). Of the five remaining contigs, the archaeon *Nitrosopumilus* has been previously detected from the functional metaproteogenomic study of *C. concentrica* [46]. The four bacterial contigs were classified as Sva0996, *Rhodobacteraceae*, BD2-11 and*Oceanospirillaceae* and then further phylogenetically analyzed . The Acidimicrobiales- and the Gemmatimonadetes-phylotypes are part of sponge/coral specific clades in the Sva0996 group and the BD2-11 group, respectively. The*Rhodobacteraceae*-phylotype branches distantly from the most closely related free-living neighbors. The *Oceanospirillaceae*-phylotype has a closely related free-living strain. This phylotype in the sponge *C. concentrica* has been consistently missed by PCR-based approaches despite current and previous extensive sequencing efforts using different protocols and primers [45], [47]–[49].

DISCUSSION

In the present study, we describe how stringent assemblies and filtering can recover nearly full-length 16S rRNA gene sequences from metagenomicpyrosequencing datasets. Through simulation of communities with various complexities, we show that chimera formation is minimal and will not impact on prediction of community composition. These properties make the described approach readily applicable to existing and future metagenomic datasets. Advances in next generation sequencing technology have in recent years led to a surge of metagenomic studies and thousands of datasets are currently available [50], [51]. Our approach will thus prove itself useful in defining the phylogenetic diversity and community composition harbored in these metagenomic resources. We are also expecting that this will lead to the discovery of new phylotypes that have previously eluded PCR-based detection and our analysis of sponge symbiont communities has provided examples of this.

Pyro-tag-sequencing has been become a standard approach for defining community composition and has thus been extensively applied in, for example, the Human Microbiome Project [52] and clinical diagnosis [53]. We show here that PCR can cause a substantial impact on the assessment of communities in terms of diversity, composition and abundance. It might therefore be worthwhile to benchmark primer choice based on 16S rRNA genes reconstructed from metagenomic data before establishing routine assays based on PCR methods.

ACKNOWLEDGMENTS

We acknowledge the J. Craig Venter Institute's Joint Technology Center under the leadership of Yu-Hui Rogers and the assistance of Matt Lewis for producing the sequencing data.

AUTHOR CONTRIBUTIONS

Conceived and designed the experiments: LF TT. Performed the experiments: LF. Analyzed the data: LF TT. Contributed reagents/materials/analysis tools: LF KM. Wrote the paper: LF KM TT.

REFERENCES

1. Pace NR (1997) A molecular view of microbial diversity and the biosphere. Science 276: 734–740.
2. Tringe SG, Hugenholtz P (2008) A renaissance for the pioneering 16S Rrna gene.CurrOpinMicrobiol 11: 442–446.
3. Hong S, Bunge J, Leslin C, Jeon S, Epstein SS (2009) Polymerase chain reaction primers miss half of rRNA microbial diversity. ISME J 3: 1365–1373.
4. Venter JC, Remington K, Heidelberg JF, Halpern AL, Rusch D, et al (2004) Environmental genome shotgun sequencing of the Sargasso Sea. Science 304: 66–74.
5. Haas BJ, Gevers D, Earl AM, Feldgarden M, Ward DV, et al (2011) Chimeric 16S rRNA sequence formation and detection in Sanger and 454-pyrosequenced PCR amplicons. Genome Res 21: 494–504.
6. Huson DH, Auch AF, Qi J, Schuster SC (2007) MEGAN analysis of metagenomic data. Genome Res 17: 377–386.
7. Stark M, Berger SA, Stamatakis A, von Mering C (2010) MLTreeMap – accurate Maximum Likelihood placement of environmental DNA sequences into taxonomic and functional reference phylogenies. BMC Genomics 11: 461.
8. Wu M, Eisen JA (2008) A simple, fast, and accurate method of phylogenomic inference. Genome Biol 9: R151.
9. Liu B, Gibbons T, Ghodsi M, Treangen T, Pop M (2011) Accurate and fast estimation of taxonomic profiles from metagenomic shotgun sequences. BMC Genomics 12 Suppl 2: S4.
10. Teeling H, Waldmann J, Lombardot T, Bauer M, Glo¨ckner FO (2004) TETRA: a web-service and a stand-alone program for the analysis and comparison of tetranucleotide usage patterns in DNA sequences. BMC Bioinformatics 5: 163.
11. McHardy AC, Martı´n HG, Tsirigos A, Hugenholtz P, Rigoutsos I (2007) Accurate phylogenetic classification of variable-length DNA fragments. Nat Methods 4: 63–72.
12. Brady A, Salzberg SL (2009) Phymm and PhymmBL: metagenomic phylogenetic classification with interpolated Markov models. Nat Methods 6: 673–676.
13. Saeed I, Tang S-L, Halgamuge SK (2011) Unsupervised discovery of microbial population structure within metagenomes using nucleotide base composition. Nucleic Acids Res 40: e34.
14. Wu D, Hugenholtz P, Mavromatis K, Pukall R, Dalin E, et al (2009) A phylogeny-driven genomic encyclopaedia of Bacteria and Archaea. Nature 462: 1056–1060.
15. Schloss PD, Handelsman J (2004) Status of the microbial census.MicrobiolMolBiol Rev 68: 686–691.
16. Pedro´s-Alio´ C (2006) Marine microbial diversity: can it be determined? Trends Microbiol 14: 257–263.
17. Sharpton TJ, Riesenfeld SJ, Kembel SW, Ladau J, O'Dwyer JP, et al (2011) PhylOTU: A High-Throughput Procedure Quantifies Microbial Community Diversity and Resolves Novel Taxa from Metagenomic Data. PLoSComputBiol 7: e1001061.
18. Rusch DB, Halpern AL, Sutton G, Heidelberg KB, Williamson S, et al (2007) The Sorcerer II Global Ocean Sampling expedition: northwest Atlantic through eastern tropical Pacific. PLoSBiol 5: e77.
19. Miller CS, Baker BJ, Thomas BC, Singer SW, Banfield JF (2011) EMIRGE: Reconstruction of full length ribosomal genes from microbial community short read sequencing data. Genome Biol 12: R44.
20. Schloss PD, Handelsman J (2005) Metagenomics for studying unculturable microorganisms: cutting the Gordian knot. Genome Biol 6: 229.
21. Margulies M, Egholm M, Altman WE, Attiya S, Bader JS, et al (2005) Genome sequencing in microfabricated

high-density picolitre reactors. Nature 437: 376– 380.

22. Mavromatis K, Ivanova N, Barry K, Shapiro H, Goltsman E, et al (2007) Use of simulated data sets to evaluate the fidelity of metagenomic processing methods. Nat Methods 4: 495–500.

23. McElroy KE, Luciani F, Thomas T (2012) GemSIM: General, Error-Model based SIMulator of next-generation sequencing data. BMC Genomics 13: 74.

24. Fan L, Reynolds D, Liu M, Stark M, Kjelleberg S, et al (2012) Functional equivalence and evolutionary convergence in complex communities of microbial sponge symbionts. ProcNatlAcadSci U S A (In Press).

25. Schmieder R, Edwards R (2011) Quality control and preprocessing of metagenomic datasets. Bioinformatics 27: 863–864.

26. Bengtsson J, Eriksson KM, Hartmann M, Wang Z, Shenoy BD, et al (2011) Metaxa: a software tool for automated detection and discrimination among ribosomal small subunit (12S/16S/18S) sequences of archaea, bacteria, eukaryotes, mitochondria, and chloroplasts in metagenomes and environmental sequencing datasets. Antonie Van Leeuwenhoek 100: 471–475.

27. Radax R, Rattei T, Lanzen A, Bayer C, Rapp HT, et al (2012) Metatranscriptomics of the marine sponge Geodiabarretti: tackling phylogeny and function of its microbial community. Environ Microbiol 14: 1308–1324.

28. Pruesse E, Quast C, Knittel K, Fuchs BM, Ludwig W, et al (2007) SILVA: a comprehensive online resource for quality checked and aligned ribosomal RNA sequence data compatible with ARB. Nucleic Acids Res 35: 7188–7196.

29. Dowd SE, Callaway TR, Wolcott RD, Sun Y, McKeehan T, et al (2008) Evaluation of the bacterial diversity in the feces of cattle using 16S rDNA bacterial tag-encoded FLX ampliconpyrosequencing (bTEFAP). BMC Microbiol 8: 125.

30. Schloss, Gevers PDA, Westcott DA, Sarah S (2011) Reducing the Effects of PCR Amplification and Sequencing Artifacts on 16S rRNA-Based Studies. PLoS One 6: e27310.

31. Caporaso JG, Kuczynski J, Stombaugh J, Bittinger K, Bushman FD, et al (2010) QIIME allows analysis of high-throughput community sequencing data. Nat Methods 7: 335–336.

32. Wang Q, Garrity GM, Tiedje JM, Cole JR (2007) Naive Bayesian classifier for rapid assignment of rRNA sequences into the new bacterial taxonomy. Appl Environ Microbiol 73: 5261–5267.

33. McDonald D, Price MN, Goodrich J, Nawrocki EP, Desantis TZ, et al (2011) An improved Greengenes taxonomy with explicit ranks for ecological and evolutionary analyses of bacteria and archaea. ISME J 6: 610–618.

34. Stamatakis A (2006) RAxML-VI-HPC: maximum likelihood-based phylogenetic analyses with thousands of taxa and mixed models. Bioinformatics 22: 2688–2690.

35. Talavera G, Castresana J (2007) Improvement of phylogenies after removing divergent and ambiguously aligned blocks from protein sequence alignments.-SystBiol 56: 564–577.

36. Edgar RC, Haas BJ, Clemente JC, Quince C, Knight R (2011) UCHIME improves sensitivity and speed of chimera detection. Bioinformatics 27: 2194– 2200.

37. Ashelford KE, Chuzhanova NA, Fry JC, Jones AJ, Weightman AJ (2005) At least 1 in 20 16S rRNA sequence records currently held in public repositories is estimated to contain substantial anomalies. Appl Environ Microbiol 71: 7724– 7736.

38. Huber T, Faulkner G, Hugenholtz P (2004) Bellerophon: a program to detect chimeric sequences in multiple sequence alignments. Bioinformatics 20: 2317– 2319.

39. Ashelford KE, Chuzhanova NA, Fry JC, Jones AJ, Weightman AJ (2006) New screening software shows that most recent large 16S rRNA gene clone libraries contain chimeras. Appl Environ Microbiol 72: 5734–5741.

40. Quince C, Lanze´n A, Curtis TP, Davenport RJ, Hall N, et al (2009) Accurate determination of microbial diversity from 454 pyrosequencing data. Nat Methods 6: 639–641.

41. Taylor MW, Radax R, Steger D, Wagner M (2007) Sponge-associated microorganisms: evolution, ecology, and biotechnological potential.MicrobiolMolBiol Rev 71: 295–347.

42. Schmitt S, Tsai P, Bell J, Fromont J, Ilan M, et al (2012) Assessing the complex sponge microbiota: core, variable and species-specific bacterial communities in marine sponges. ISME J 6: 564–576.

43. Sogin ML, Morrison HG, Huber JA, Mark Welch D, Huse SM, et al (2006) Microbial diversity in the deep sea and the underexplored "rare biosphere". ProcNatlAcadSci U S A 103: 12115–12120.

44. Quince C, Lanzen A, Davenport RJ, Turnbaugh PJ (2011) Removing noise from pyrosequencedamplicons. BMC Bioinformatics 12: 38.

45. Thomas T, Rusch D, Demaere MZ, Yung PY, Lewis M, et al (2010) Functional genomic signatures of sponge bacteria reveal unique and shared features of symbiosis. ISME J 4: 1557–1567.

46. Liu M, Fan L, Zhong L, Kjelleberg S, Thomas T (2012) Metaproteogenomic analysis of a community of sponge symbionts. ISME J (doi:10.1038/ ismej.2012.1).

47. Taylor MW, Schupp PJ, Dahllo¨f I, Kjelleberg S, Steinberg PD (2004) Host specificity in marine sponge-associated bacteria, and potential implications for marine microbial diversity. Environ Microbiol 6: 121–130.

48. Taylor MW, Schupp PJ, de Nys R, Kjelleberg S, Steinberg PD (2005) Biogeography of bacteria associated with the marine sponge Cymbastelaconcentrica. Environ Microbiol 7: 419–433.

49. Yung PY, Burke C, Lewis M, Egan S, Kjelleberg S, et al (2009) Phylogenetic screening of a bacterial, metagenomic library using homing endonuclease

restriction and marker insertion. Nucleic Acids Res 37: e144.

50. Simon C, Daniel R (2011) Metagenomic analyses: past and future trends. Appl Environ Microbiol 77: 1153–1161.

51. Thomas T, Gilbert J, Meyer F (2012) Metagenomics-a guide from sampling to data analysis. Microbial Informatics and Experimentation 2: 3.

52. Peterson J, Garges S, Giovanni M, McInnes P, Wang L, et al (2009) The NIH Human Microbiome Project. Genome Res 19: 2317–2323.

53. Siqueira JF, Fouad AF, Roˆc¸as IN (2012) Pyrosequencing as a tool for better understanding of human microbiomes. J Oral Microbiol 4: 10743.

SECTION-IV CONSORTIA AND DATABASES

CHAPTER 11

THE METAGENOMICS RAST SERVER- A PUBLIC RESOURCE FOR THE AUTOMATIC PHYLOGENETIC AND FUNCTIONAL ANALYSIS OF METAGENOMES

F Meyer D Paarmann, M D'Souza, R Olson, EM Glass, M Kubal, T Paczian, A Rodriguez, R Stevens, A Wilke, J Wilkening and RA Edwards

BACKGROUND

The genomic revolution of the 1990s has yielded almost a thousand sequenced microbial genomes. More recently, the explosion of random community genomics, or metagenomics, where DNA is sequenced directly from environmental samples has provided insights into microbial communities. Currently, two approaches to sequencing metagenome samples are commonly used. In the traditional approach, DNA is cloned into BACs, or small plasmids, and dideoxy chain termination sequencing ("Sanger sequencing") is used to determine the sequences [1, 2]. In the alternative approach, DNA is sequenced without cloning, using one of the so-called next-generation sequencing techniques, usually pyrosequencing. Both approaches have advantages and disadvantages. For example, Sanger sequencing generates longer sequence reads but has inherent biases due to the cloning. Pyrosequencing has much higher throughput and a lower error rate per base sequenced compared to Sanger sequencing, but those errors are biased toward certain mistakes [3].

Regardless of the sequencing approach used to generate the data, the first steps in analysis of any metagenome involve comparing those sequences to known sequence databases. This computationally intensive task provides the basic data types for many subsequent analyses, including phylogenetic comparisons, functional annotations, binning of sequences, phylogenomic profiling, and metabolic reconstructions.

Here we describe the development of a freely available, fully automated open source system for processing metagenome sequence data to generate these basic elements. A public implementation of this system has been provided for all researchers to analyze their metagenomes. Our service, the metagenomics RAST server (mg-RAST for short), is available over the web to all researchers, and access is not limited to specific groups or data types. Almost 500 metagenomes have been processed through the beta version of the pipeline so far.

IMPLEMENTATION AND RESULTS

The MG-RAST server is an open source system based on the SEED framework for comparative genomics [4, 5]. Users can upload raw sequence data in fasta format; the sequences will be normalized and processed and summaries automatically generated. Genome annotation systems are ever evolving; therefore, in order to accommodate new methods that may be developed, the pipeline was designed with a modular framework that allows the rapid addition of new analysis steps or comparative data at any stage of the analysis. The server provides several methods to access the different data types, including phylogenetic and metabolic reconstructions, and the ability to compare the metabolism and annotations of one or more metagenomes and genomes. In addition, the server offers a comprehensive search capability. Access to the data is password protected, and all data generated by the automated pipeline is available for download and analysis in variety of common formats. Here we describe the key components of the pipeline, which are summarized in Figure 1.

Citation: F Meyer, D Paarmann, M D'Souza, R Olson, EM Glass, M Kubal, T Paczian, A Rodriguez, R Stevens, A Wilke, J Wilkening and RA Edwards, The metagenomics RAST server – a public resource for the automatic phylogenetic and functional analysis of metagenomes, doi:10.1186/1471-2105-9-386.

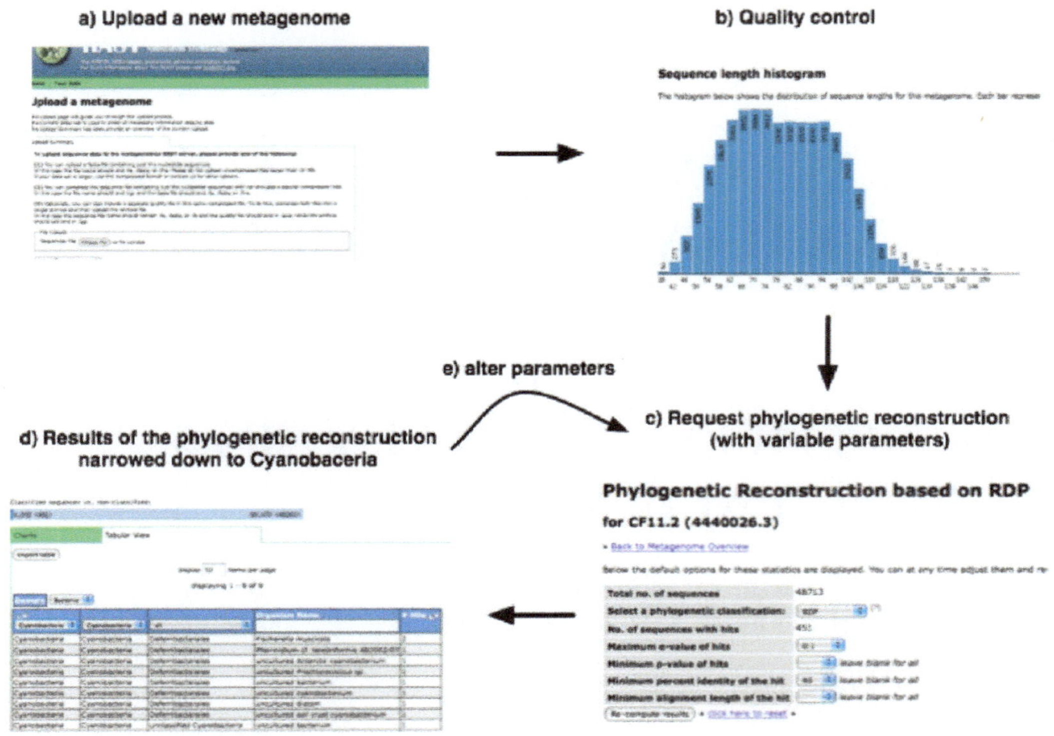

Figure 1: After uploading a dataset (a), the system computes initial quality control (b) and allows the user to set the parameters for phylogenetic analysis (c). The system then displays the results (d) and allows the user to alter the parameters (e). Data shown in this example is from the dataset CF11.2 (ID: 4440026.3) that is publicly available in the MG-RAST server.

User Registration and Management

The user registration serves two functions: to limit access to each data set to the user and their colleagues and to secure a valid email address in case correspondence is required, for example if a data-processing problem occurs. Once logged in, users can view their own metagenomes, those to which the owner has granted them rights, and the default set of publicly available metagenomes. The system supports delegation of authorization so that users can allow others to access one or more of their metagenomes. In addition, data owners can release their metagenomes to the public at any point, allowing all users of the system to view their data.

Data Types

The pipeline accepts data in a number of formats: 454 reads may be uploaded directly in the format delivered by 454 [6], and fasta files typical of Sanger-sequences and used by other platforms may also be uploaded. The pipeline will also accept assembled sequences in fasta format. Sequence data may be compressed by one of several common computer programs to speed upload.

Users may choose to upload raw unassembled reads or assembled contigs. As discussed below, each approach has advantages and disadvantages. Users with a limited number of larger contigs, where the average contig length exceeds 40 kb, should consider using the RAST server for the analysis of complete Bacterial and Archaeal genomes [7].

The Genomics Standards Consortium has proposed a minimal set of data, called the Minimum Information about a Genome Sequence (MIGS) [8], that should be collected with every metagenome sequence. Although this is an evolving standard, the metagenomics-RAST server is MIGS-compliant. Metadata, accessory data about the metagenome (e.g., date and location where the sample was collected), is requested from the user at the time of sequence submission. This data is stored with the user's data and can be provided to the GSC genome catalogue, and other archives, when the sequence data is ready for public release.

Implementation and Core Analyses

The pipeline is implemented in Perl by using a number of open source components, including the SEED framework [4], NCBI BLAST [9], SQLite, and Sun Grid Engine [10] as components. The system also uses the publicly available SEED subsystems, SEED nr, and FIGfam protein families (see http://www.theSEED.org).

The distinct steps are implemented to provide a flexible, extensible processing pipeline. The steps incrementally add data to a self-contained "job directory" that contains all job-relevant data in flat file and SQLite [11] format. Relational database technology is used to efficiently provide a mapping of sequences in a metagenome to both organisms and metabolic functions and at the same time allow the user to change the parameters for the underlying sequence matches. The user interface enables the download of the user's job directories, and a future version of the software will allow uploading of user-created directories into the server.

After uploading the data, a normalization step (see Figure 2) is executed, generating unique internal IDs and removing exactly duplicate sequences from 454 data sets. (These sequences are an artefact of the sequencing technique and are not scientifically meaningful [12].)

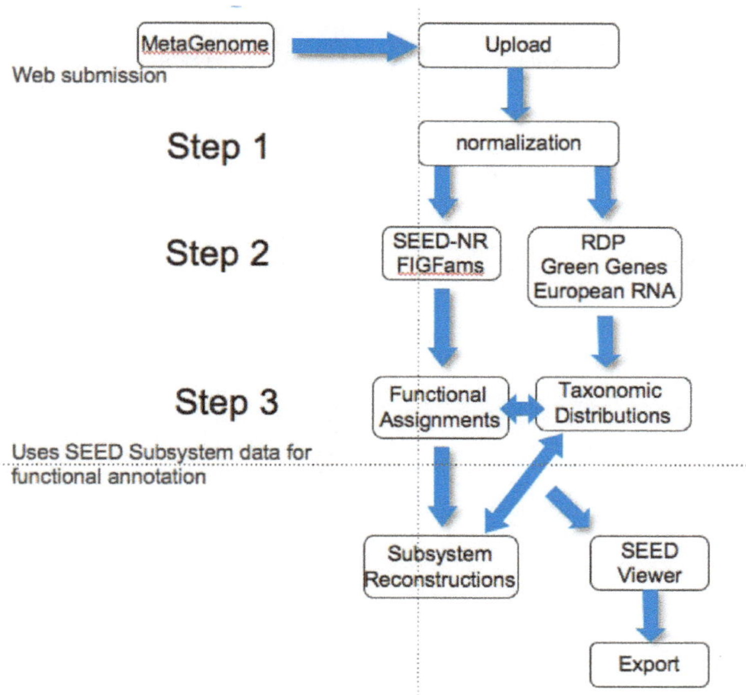

Figure 2: Overview of the workflow implemented in the metagenomics RAST pipeline. Three distinct stages of processing are executed, each adding data to a single directory, and ultimately enabling web-based browsing of results.

In the second step, the sequences are screened for potential protein encoding genes (PEGs) via a BLASTX [9] search against the SEED comprehensive nonredundant database sourced from the INSDC databases, sequencing centers, and other sources [4]. An expect value (E) cut-off of 0.01 is used to pick up potentially coding elements. (This was chosen empirically to increase the number of potentially coding elements while not being overwhelming for data analysis.) In parallel with the BLASTX searches, the sequence data is compared to all accessory databases by using the appropriate algorithms and significance selection criteria. These databases include several rDNA databases, including GREENGENES [13], RDP-II [14], and the European 16S RNA database [15], and boutique databases such as the chloroplast database, mitochondrial database, and ACLAME database of mobile elements [16]. The search criteria are specific for each database. For example, screens for ribosomal RNA genes are performed by using BLASTN against the rDNA databases, but much more stringent selection criteria are used to identify candidate RNA genes than for identifying protein-encoding genes (by default, the similarity must exceed 50 bp in length and have an expect value less than 1×10^{-5}).

In the third step, these matches to external databases are used to compute the derived data. First, a phylogenomic reconstruction of the sample is computed by using both the phylogenetic information contained in the SEED nr database and the similarities to the ribosomal RNA database. Functional classifications of the PEGs are computed by projecting against SEED FIGfams [17] and subsystems based on these similarity searches [4]. These functional assignments become the raw input to an automatically generated initial metabolic reconstruction of the sample, providing suggestions for metabolic fluxes and flows, reactions, and enzymes.

One of the design goals of this server was easy accessibility via a web-based interface. The interface provides views for browsing and analysis of the data, as well as a means to download all result tables and the sequences for every subset displayed. Figure 3 provides an overview of the various elements of the user interface and highlights the options for downloading various subsets. The user interface provides a means to alter some of the parameters used to compute the functional, metabolic, and phylogenetic reconstruction. This allows more stringent match criteria (e.g., expectation value, score, overall percent identity, length of match, and number of mismatches); and, by restricting the matches, the derived data is dynamically changed. The default parameters have been chosen by empirical testing and represent a tradeoff between accuracy and specificity.

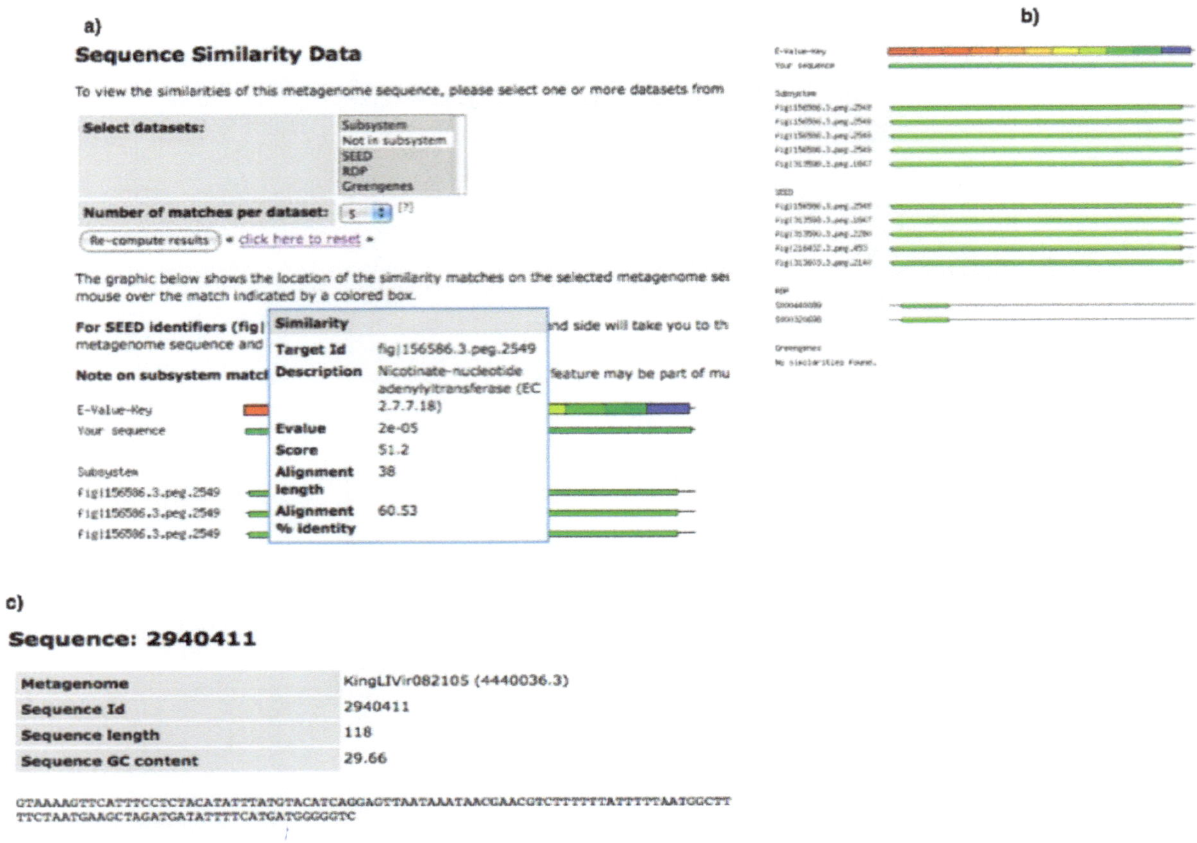

Figure 3: We emphasize data accessibility, (a) sequence analysis results (e.g. BLAST matches) and all sequences in a metagenome are visible and can be downloaded. In addition the server provides an overview (b) of the sequence analysis results per fragment in a metagenome (c).

Comparative Metagenomics

The abundance of comparative metagenomics tools is central to the utility of the mg-RAST platform. Various tools have been built into the framework, allowing users to compare their data against other metagenomes or complete genomes taken from the SEED [4] environment. The subsystems heat map and the taxonomic heat map provide comparative metagenomics summaries that encapsulate the differences between samples.

The subsystem comparison tools identify the number of pegs in each metagenome that are connected to a subsystem via protein level similarity. Based on these connections, each subsystem present in a sample is scored by counting the number of sequences that are similar to a protein in each subsystem. This score is divided by the total number of sequences from the sample that are similar to any protein in a subsystem, to give a fraction of sequences in subsystems that are in a given subsystem. This approach allows comparisons between samples that have different numbers of sequences. Since the fractions tend to be small (a few sequences hit each subsystem, but there are now over 600 subsystems in the SEED), the scores can be factored for display purposes. Furthermore, a nonquantitative approach is provided to group the subsystem scores, emphasizing those subsystems that are most different between the samples. Moreover, the display can be limited to specific areas of metabolism, or other subsystem groups, as desired by the user.

The taxonomic heat map works in an analogous fashion but highlights the different taxonomic profiles in each sample, as determined by the phylogenetic or phylogenomic approaches selected by the end user (e.g., 16S comparisons, phylogenomics from BLAST results). Again, samples may be grouped in a nonquantitative fashion to rapidly highlight particular phylogenetic groups that predominate in different samples.

Often a metagenome comprises a few dominant organisms, and many of the pathways in the metagenome can be predicted. The automatically generated metabolic reconstructions can be compared to any given metagenome or complete microbial genome. This approach highlights subsystems that are unique to a metagenome, a comparative genome, or the subsystems common to both. With these tools, users can identify shared metabolism present in their samples.

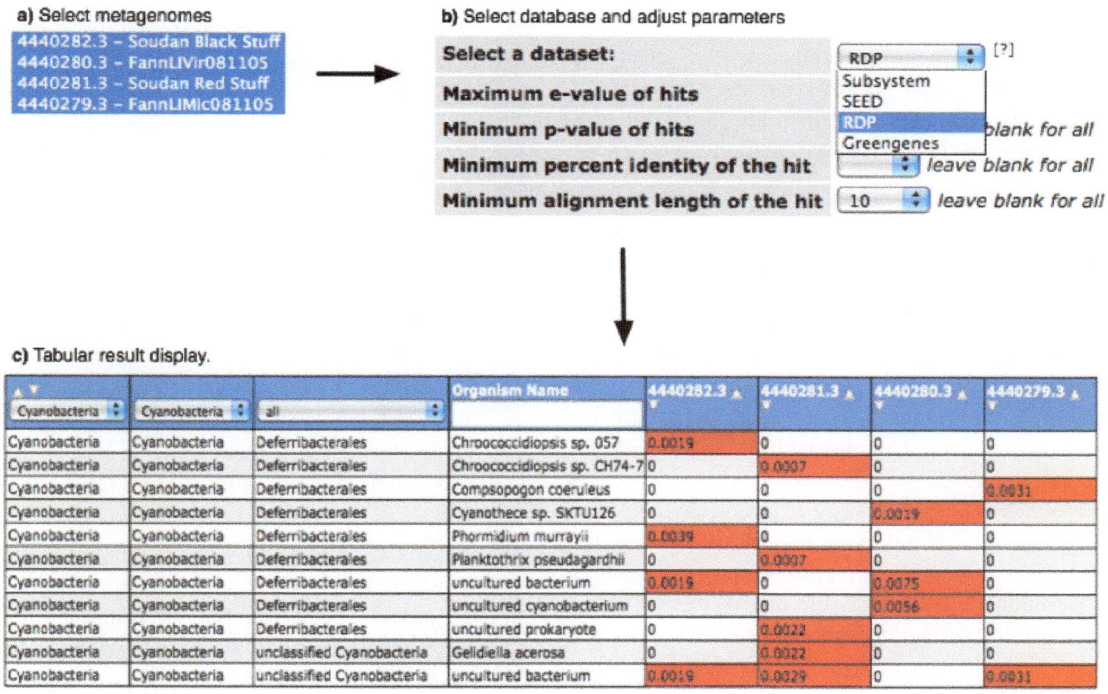

Cyanobacteria	Cyanobacteria	all	Organism Name	4440282.3	4440281.3	4440280.3	4440279.3
Cyanobacteria	Cyanobacteria	Deferribacterales	Chroococcidiopsis sp. 057	0.0019	0	0	0
Cyanobacteria	Cyanobacteria	Deferribacterales	Chroococcidiopsis sp. CH74-7	0	0.0007	0	0
Cyanobacteria	Cyanobacteria	Deferribacterales	Compsopogon coeruleus	0	0	0	0.0031
Cyanobacteria	Cyanobacteria	Deferribacterales	Cyanothece sp. SKTU126	0	0	0.0019	0
Cyanobacteria	Cyanobacteria	Deferribacterales	Phormidium murrayii	0.0039	0	0	0
Cyanobacteria	Cyanobacteria	Deferribacterales	Planktothrix pseudagardhii	0	0.0007	0	0
Cyanobacteria	Cyanobacteria	Deferribacterales	uncultured bacterium	0.0019	0	0.0075	0
Cyanobacteria	Cyanobacteria	Deferribacterales	uncultured cyanobacterium	0	0	0.0056	0
Cyanobacteria	Cyanobacteria	Deferribacterales	uncultured prokaryote	0	0.0022	0	0
Cyanobacteria	Cyanobacteria	unclassified Cyanobacteria	Gelidiella acerosa	0	0.0022	0	0
Cyanobacteria	Cyanobacteria	unclassified Cyanobacteria	uncultured bacterium	0.0019	0.0029	0	0.0031

Figure 4: Comparing the phylogenetic composition of four metagenomes. Initially (a) the user selects a subset of metagenomes or genomes (here we selected 2 Soudan mine samples and 2 marine samples). The next step (b) allows selecting the basis for the comparison (protein-based-only SEED subsystems or all SEED proteins vs. RNA based RDP or Greengenes) and the parameters for the matches. The parameters include e-value, minimal alignment length, p-value, and percent identity. Finally, the result (c) is displayed in tabular format, in which a heatmap-style color coding is used to highlight differences. The resulting table can be downloaded as a spreadsheet.

DISCUSSION

A completely new public metagenome annotation system has been developed and released. The process is the result of several years of planning and engineering. Designed to leverage the SEED microbial genome annotation platform, the mg-RAST platform provides seamless integration of metagenome data, microbial genomics, and manually curated annotations. Each metagenome project has its own requirements for stringency, datasets to be analyzed, and output format for results. The metagenomics SEED pipeline was designed to allow alterations to the parameters for the sequence matches underlying both the phylogenetic and metabolic reconstructions to restrict matches. It has been built by using an extensible format allowing the integration of new datasets and algorithms without a need for recomputation of existing results.

The mg-RAST service handles both assembled and unassembled data. Each approach has advantages that should be considered when comparing metagenomes. For example, if one is carrying out comparative metagenomics or if statistics are being used to compare samples [18, 19], the sequences cannot be assembled, since the assembly process loses the frequency information critical for determining differences between samples. In contrast, assembled sequences tend to be longer and therefore more likely to accurately identify gene function or phylogenetic source from binning [20].

The analytical methods integrated into the pipeline provide core annotation and analysis tools to compare and contrast a diverse set of metagenomes [21, 22, 23, 24]. The approach underlying the subsystems-based functional analysis of metagenomes has been validated with 90 different samples from nine major biomes. The analysis demonstrated that the biomes could clearly be separated by their functional composition [25]. All of the metagenomes present in that study are included in the publicly available datasets visible in the mg-RAST server.

Although the service contains core functionality for the annotation and analysis of metagenomes, many of the techniques traditionally used for genome analysis (e.g., approaches for the prediction of coding sequences) either do not work with metagenomes or show a significant performance degradation [26]. Many of the differences between complete genome annotation and metagenome annotation are reminiscent of those encountered previously with the analysis of expressed sequenced tags [27]. Therefore, new analytical methods are needed to fully understand metagenomics data. The most obvious problem is with the large number of unknown sequences in any sample. Depending on the specific sample processed, as few as 10% of the sequences or as many as 98% of the sequences may have no known similarity to anything in the database [28]. We and others are developing new binning, clustering, and coding region prediction tools to handle these unknown sequences, and effective tools will be incorporated into the pipeline when available. Another problem is that the rapid pace with which sequence data is being generated outpaces increases in computational speed, and therefore improvements in common search algorithms are required to ensure that sequence space can be accurately and efficiently searched. A third problem, common to all annotation platforms, is that metabolic reconstructions and analyses are dependent on the underlying quality of the data. The SEED has the most consistent and accurate microbial genome annotations of any publicly available source because of the subsystems approach to annotation. However, the SEED subsystems are necessarily focused on core metabolism and pathogenesis of a select few organisms. Comprehensive subsystem coverage of secondary metabolism, and especially of metabolism specific to diverse environments, is required to truly comprehend those data sets.

CONCLUSION

We have provided a free, public resource for the analysis of metagenome sequence data. Our service does not require a specific type of sequence data and has no requirement for release or control of the data. All sequence data remains protected by a password mechanism and is visible only to permitted users. This metagenomics annotation pipeline was specifically developed to handle pyrosequencing data and accommodate some of the nuances associated with that data. However, the tools and approaches we have developed are applicable and available for any metagenome project, regardless of sequence type. This service has removed one of the primary bottlenecks in metagenome sequence analysis – the availability of high-performance computing for annotating the data.

ACKNOWLEDGEMENTS

Part of this project has been funded with federal funds from the National Institute of Allergy and Infectious Diseases, National Institutes of Health, Department of Health and Human Services, under Contract No. HHSN266200400042C. Argonne National Laboratory's work was supported under U.S. Department of Energy contract DE-AC02-06CH11357. We thank Forest Rohwer and all members of his group for helpful advice while developing the pipeline. We thank Veronika Vonstein and Ross Overbeek and the members of their group for advice and comments in using subsystem technology.

The submitted manuscript has been created by UChicago Argonne, LLC, operator of Argonne National Laboratory ("Argonne"). Argonne, a U.S. Department of Energy Office of Science laboratory, is operated under Contract No. DE-AC02-06CH11357. The U.S. Government retains for itself, and others acting on its behalf, a paid-up, nonexclusive, irrevocable worldwide license in said article to reproduce, prepare derivative works, distribute copies to the public, and perform publicly and display publicly, by or on behalf of the Government.

REFERENCES

1. Venter JC, Remington K, Heidelberg JF, Halpern AL, Rusch D, Eisen JA, Wu D, Paulsen I, Nelson KE, Nelson W, *et al.*: Environmental genome shotgun sequencing of the Sargasso Sea. *Science* 2004, 304(5667):66–74. 10.1126/science.1093857
2. Tyson GW, Chapman J, Hugenholtz P, Allen EE, Ram RJ, Richardson PM, Solovyev VV, Rubin EM, Rokhsar DS, Banfield JF: Community structure and metabolism through reconstruction of microbial genomes from the environment. *Nature*2004, 428(6978):37–43. 10.1038/nature02340
3. Huse SM, Huber JA, Morrison HG, Sogin ML, Welch DM: Accuracy and quality of massively parallel DNA pyrosequencing.*Genome biology* 2007, 8(7):R143. 10.1186/gb-2007-8-7-r143
4. Overbeek R, Begley T, Butler RM, Choudhuri JV, Diaz N, Chuang H-Y, Cohoon M, de Crécy-Lagard V, Disz T, Edwards R, *et al.*: The Subsystems Approach to Genome Annotation and its Use in the Project to Annotate 1000 Genomes. *Nucleic Acids Res* 2005., 33(17):
5. McNeil LK, Reich C, Aziz RK, Bartels D, Cohoon M, Disz T, Edwards RA, Gerdes S, Hwang K, Kubal M, *et al.*: The National Microbial Pathogen Database Resource (NMPDR): a genomics platform based on subsystem annotation. *Nucleic Acids Res* 2007, (35 Database):D347–353. 10.1093/nar/gkl947
6. Margulies M, Egholm M, Altman WE, Attiya S, Bader JS, Bemben LA, Berka J, Braverman MS, Chen YJ, Chen Z, *et al.*: Genome sequencing in microfabricated high-density picolitre reactors. *Nature* 2005, 437(7057):376–380.
7. Aziz RK, Bartels D, Best AA, DeJongh M, Disz T, Edwards RA, Formsma K, Gerdes S, Glass EM, Kubal M, *et al.*: The RAST Server: Rapid Annotations using Subsystems Technology. *BMC Genomics* 2008, in press.
8. Field D, Morrison N, Selengut J, Sterk P: Meeting report: eGenomics: Cataloguing our Complete Genome Collection II.*Omics* 2006, 10(2):100–104. 10.1089/omi.2006.10.100
9. Altschul SFl, Madden TL, Schaffer AA, Zhang J, Zhang Z, Miller W, Lipman DJ: Gapped BLAST and PSI-BLAST: a new generation of protein database search programs. *Nucleic Acids Res* 1997, 25: 3389–3402. 10.1093/nar/25.17.3389
10. gridengine – Project home[http://gridengine.sunsource.net/]
11. SQLite[http://www.sqlite.org/]
12. Jarvie T: personal communication. 2006.r
13. DeSantis TZ, Hugenholtz P, Larsen N, Rojas M, Brodie EL, Keller K, Huber T, Dalevil D, Hu P, Andersen GL: Greengenes, a chimera-checked 16S rRNA gene database and workbench compatible with ARB. *Appl Environ Microbiol* 2006, 72(7):5069–5072. 10.1128/AEM.03006-05
14. Cole JR, Chai B, Farris RJ, Wang Q, Kulam-Syed-Mohideen AS, McGarrell DM, Bandela AM, Cardenas E, Garrity GM, Tiedje JM: The ribosomal database project (RDP-II): introducing myRDP space and quality controlled public data. *Nucleic Acids Res* 2007, (35 Database):D169–172. 10.1093/nar/gkl889
15. Wuyts J, Peer Y, Winkelmans T, De Wachter R: The European database on small subunit ribosomal RNA. *Nucleic Acids Res* 2002, 30(1):183–185. 10.1093/nar/30.1.183
16. Leplae R, Hebrant A, Wodak SJ, Toussaint A: ACLAME: a CLAssification of Mobile genetic Elements. *Nucleic Acids Res*2004, (32 Database):D45–49. 10.1093/nar/gkh084
17. Meyer F, Overbeek R, Rodriquez A: FIGfams – Yet another protein family collection. 2008, in press.
18. Tringe SG, von Mering C, Kobayashi A, Salamov AA, Chen K, Chang HW, Podar M, Short JM, Mathur

EJ, Detter JC, *et al.*: Comparative metagenomics of microbial communities. *Science* 2005, 308(5721):554–557. 10.1126/science.1107851

19. Rodriguez-Brito B, Rohwer F, Edwards RA: An application of statistics to comparative metagenomics. *BMC bioinformatics [electronic resource]* 2006, 7: 162. 10.1186/1471-2105-7-162
20. McHardy AC, Martin HG, Tsirigos A, Hugenholtz P, Rigoutsos I: Accurate phylogenetic classification of variable-length DNA fragments. *Nature methods* 2007, 4(1):63–72. 10.1038/nmeth976
21. Edwards RA, Rodriguez-Brito B, Wegley L, Haynes M, Breitbart M, Peterson DM, Saar MO, Alexander S, Alexander EC Jr, Rohwer F: Using pyrosequencing to shed light on deep mine microbial ecology. *BMC Genomics* 2006, 7: 57. 10.1186/1471-2164-7-57
22. Fierer N, Breitbart M, Nulton J, Salamon P, Lozupone C, Jones R, Robeson M, Edwards RA, Felts B, Rayhawk S, *et al.*: Metagenomic and small-subunit rRNA analyses reveal the genetic diversity of bacteria archaea, fungi, and viruses in soil.*Appl Environ Microbiol* 2007, 73(21):7059–7066. 10.1128/AEM.00358-07
23. Wegley L, Edwards R, Rodriguez-Brito B, Liu H, Rohwer F: Metagenomic analysis of the microbial community associated with the coral Porites astreoides. *Environmental microbiology* 2007, 9(11):2707–2719. 10.1111/j.1462-2920.2007.01383.x
24. Mou XSS, Edwards RA, Hodson RE, Moran MA: Generalist Species Dominate Bacterial Carbon Processing in the Coastal Ocean. *Nature* 2008.
25. Dinsdale EA, Edwards RA, Hall D, Angly F, Breitbart M, Brulc JM, Furlan M, Desnues C, Haynes M, Li LM, *et al.*: Functional Metagenomic Profiling of Nine Biomes. *Nature* 2008.
26. Krause L, Diaz NN, Bartels D, Edwards RA, Puhler A, Rohwer F, Meyer F, Stoye J: Finding novel genes in bacterial communities isolated from the environment. *Bioinformatics* 2006, 22(14):e281–289. 10.1093/bioinformatics/btl247
27. Liang F, Holt I, Pertea G, Karamycheva S, Salzberg SL, Quackenbush J: An optimized protocol for analysis of EST sequences. *Nucleic Acids Res* 2000, 28(18):3657–3665. 10.1093/nar/28.18.3657
28. Rohwer F: personal communication. 2007.
29. Paarmann D, Paczian T, Meyer F: SEED-Viewer – a web user interface for studying genomics datasets. *unpublished*

CHAPTER 12 MOCAT: A METAGENOMICS ASSEMBLY AND GENE PREDICTION TOOLKIT

Jens Roat Kultima, Shinichi Sunagawa, Junhua Li, Weineng Chen, Hua Chen, Daniel R. Mende, Manimozhiyan Arumugam, Qi Pan , Binghang Liu, Junjie Qin, Jun Wang, Peer Bork

INTRODUCTION

The emerging field of metagenomics has enabled researchers to study the structure, dynamics, and functionality of uncultured microbial communities. Processing the vast amounts of metagenomics data usually involves quality-controlling raw sequence reads, aligning them to reference databases, and assembling them into longer contigs prior to predicting genes. Several packages are available for processing and analyzing metagenomics data, either as web- and cloud-based services or stand-alone computational pipelines [1]–[7]. But currently none of them supports the assembly and gene prediction of metagenomics data produced by the Illumina platform.

As exemplified by recent clinical, large-scale, and ongoing studies (e.g., the Human and Earth Microbiome Projects), the usage of high throughput sequencing (HTS) data can be anticipated to further increase considerably in both terms of data volume and scope of application [8]–[11]. Thus, there is an imminent need for applications providing standardized methods for processing of HTS data in the form of pipelines [12] to facilitate comparative downstream analyses.

To address these issues, we have developed MOCAT, a metagenomics assembly and gene prediction toolkit for both small and large-scale processing of metagenomic data produced by the Illumina sequencing technology.

RESULTS AND DISCUSSION

The main pipeline is divided into five major steps: (i) quality trimming and filtering of raw reads, (ii) optional mapping to remove, extract, and/or quantify reads matching a reference database, (iii) assembly, (iv) assembly revision, and (v) gene prediction (Figure 1). Statistics from each step are summarized into multi-sheet Excel documents, as well as queryable SQLite databases. Full details of output files and statistics produced in each processing step are given

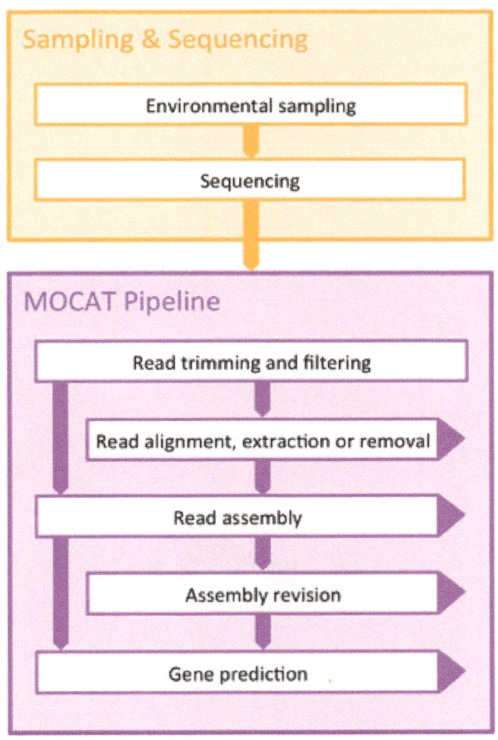

Figure 1: The MOCAT data processing pipeline. Metagenomic samples are collected and sequenced. The raw sequence reads are given as input to the pipeline, which are processed by modular steps resulting in metagenome assemblies and predicted genes. Arrows extending to the right from boxes, indicate input to various downstream analyses. Statistics from each step are summarized into multi-sheet Excel documents, as well as queryable SQLite databases.

Citation: Kultima JR, Sunagawa S, Li J, Chen W, Chen H, Mende DR, et al. (2012) MOCAT: A Metagenomics Assembly and Gene Prediction Toolkit. PLoS ONE 7(10): e47656. doi:10.1371/journal.pone.0047656.

The individual processing steps in MOCAT were benchmarked using three different data sets: 124 published human gut metagenomic samples [8], a mock community produced by the Human Microbiome Project (HMP) with 22 species from 19 genera [9], and a simulated metagenome with 100 strains from 85 species [13]. By using this combination of host associated, artificial, and simulated metagenomes with different taxonomical resolution, we show that MOCAT can efficiently process a variety of metagenomic samples, ranging in both size (0.5–16.6 Gbp), origin and composition owing to new developments in each of the five major steps.

i) Quality Trimming and Filtering of Raw Reads

Read quality trimming and filtering can greatly improve the length and accuracy of metagenomic assemblies [13]. Therefore, in the first processing step, raw reads below specified quality and length cutoffs are trimmed or removed using either the FastX program (http://hannonlab.cshl.edu/fastx_toolkit/) or the DynamicTrim algorithm in the SolexaQA package [14]. The supported FastX program removes bases from the 3′ end below a user-defined threshold, whereas the DynamicTrim algorithm in the SolexaQA package keeps the longest contiguous read segment in which all quality scores are above the user-defined threshold [14]. After quality trimming and filtering our three test datasets, 57–79% of the reads remained as high quality reads.

Additionally, to reduce base composition-biases that commonly occur in HTS data [15], the frequency of each base at each position over all reads is calculated, and bases that exceed two standard deviations of the average base frequency within a sample are trimmed from the 5′ end of all reads. Using our test data set of 124 published human gut microbial samples, on average, the fraction of reads that could be mapped to assemblies was 1% higher when using 5′ trimmed reads, compared to non-trimmed reads.

MOCAT also supports the FastQC package, for evaluating raw read quality statistics (http://www.bioinformatics.bbsrc.ac.uk/projects/fastqc).

ii) Mapping, and Removal or Extraction of Reads Matching a Reference Database

In the second step, reads can be mapped to reference sequences in order to extract or remove reads from the original data set as well as to calculate base or read coverages. For example, reads from a human fecal metagenomic sample can be mapped to the provided human genome database (hg19, Genome Reference Consortium Human Reference 37) to remove reads of human origin using SOAPAligner2 [16], or reads containing adapters used for sequencing library construction can be removed using Usearch [17]. Reads can also be mapped to any other custom reference database to calculate base and insert coverage of reference sequences to estimate taxonomic and/or functional composition of a sample, for example.

Here, we estimated the taxonomic composition of the simulated metagenome by mapping reads to the set of original reference genomes and calculating genome size-normalized base and read coverages. The Pearson and Spearman correlation coefficients between the observed and expected composition of the simulated metagenome were 0.95 and 0.90, respectively, for both base and read counts (Figure 2), and only 80 out of more than 30 Million reads were not aligned. However, the observed abundances of genomes with very high sequence identity may deviate from the expected abundances due to reads mapping to both the genome of origin and other highly similar genomes.

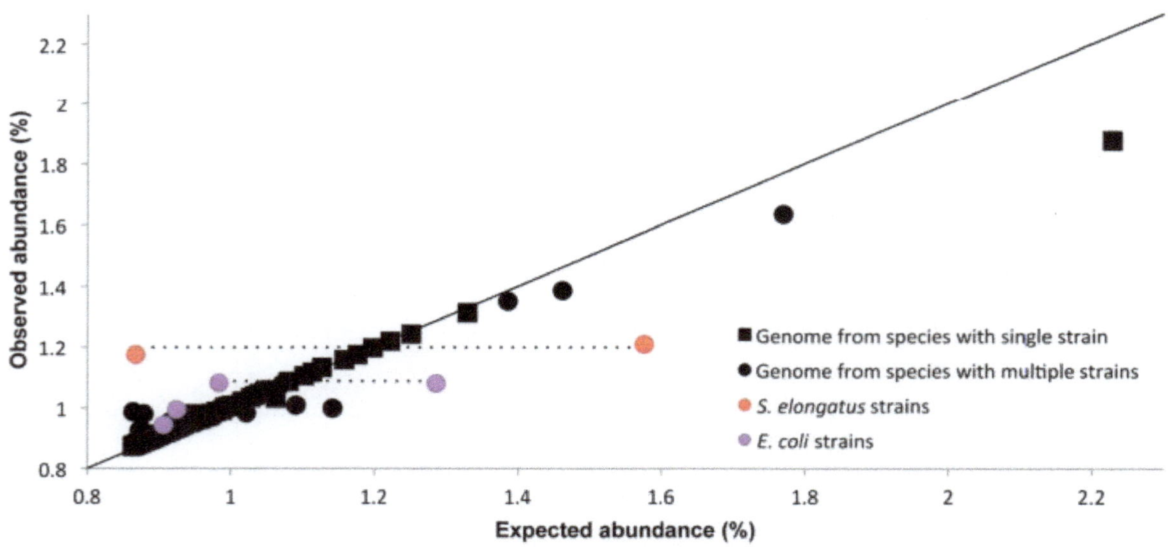

Figure 2: Relative abundance of each reference genome present in the simulated metagenome. The observed abundances by mapping reads to reference genomes and the expected abundance correlate with a Pearson correlation coefficient of 0.95 (base and read counts). Circles represent genomes with multiple strains from one species and squares represent genomes with only one strain within the species. All, but one, of the observations deviating from the diagonal are strains from the same species. These strains are either over- or under represented because reads are mapped to other closely related strains in addition to the strain of origin. Highlighted by dashed lines, are two examples where a high sequence similarity between strains (99.9% and 98.7% for the *Synechococcus elongatus* and *Escherichia coli* strains, respectively) can result in deviations from expected abundances.

When estimating taxonomic composition of the HMP mock community, reads were mapped to reference sequences of the community. By first removing quality filtered and trimmed reads matching known Illumina adapter sequences, the percentage of bases and reads mapping to the reference genomes increased from 94.3% to 97.3%, and 95.0% to 97.6%, respectively, indicating the usefulness of a pre-screening step. The taxonomic composition estimated here is similar to the values calculated by the HMP consortium (Pearson correlation coefficient of 0.75 and 0.83 for bases and reads mapping, respectively, Figure 3), and also to estimates of 16S sequences using 454 sequencing presented in ([18]). Experimental errors, not applicable to estimates of computationally simulated metagenomes, may explain the lower correlation in the mock community, compared to the simulated metagenome.

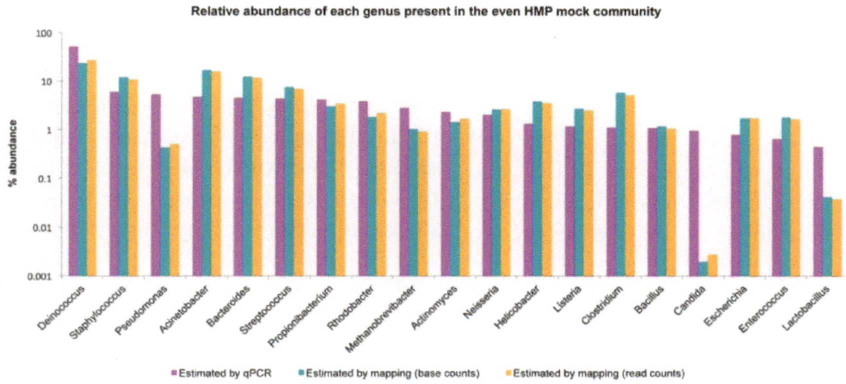

Figure 3: Relative abundance of each genus present in the even HMP mock community. The estimated abundances using qPCR and by mapping reads to reference genomes correlate with a Pearson correlation coefficient of 0.75 (base counts) and 0.83 (read counts).

iii) Assembly

In the assembly step, a new version (1.06) of SOAPdenovo [19] is used. For paired-end sequences, the insert size of each sequencing library is estimated at run-time by mapping reads to either reference marker genes [20] prior to assembly, or assembled contigs prior to scaffolding. Similarly, Kmer sizes used for assemblies are calculated at run-time for each individual metagenome. Empirical tests on a large number of samples show that estimating a Kmer size for each sample as the closest odd number larger or equal to half the average read length may not yield the best possible assembly, but balances assembly throughput and accuracy.

The accuracy of metagenomic assemblies was assessed using data from the simulated metagenome and the mock community. We used the percentage of predicted complete genes aligning to the reference sequences of origin, as a proxy for correctly assembled scaftigs (contigs that were extended and linked using the paired-end information of sequencing reads). For the simulated metagenome this value was 95.2% (12,385 complete genes predicted), and for the mock community 89.3% of the complete genes aligned (1,042 complete genes predicted). The lower number of predicted complete genes in the mock community may be explained by the relatively low number of high quality reads used in the assembly for this metagenome.

The effect of using variable Kmer sizes, rather than a fixed kmer, in the assembly step, was evaluated using the 124 gut metagenomes. Estimating Kmer sizes at run-time for each individual metagenome, rather than using a fixed Kmer size across all samples, improved the number and frequency of complete gene calls as well as overall average gene length (column 1 in Table 1).

Table 1: Progressive improvement of gene prediction metrics in 124 human gut metagenomes

Quality metric	Improvement compared to fixed kmer = 23 (%)	
	No assembly revision	Revised assembly
Number of complete genes	8.1	10.2
Number of complete genes/Mbp	4.6	18.5
Average gene length	1.7	1.8

Gene prediction metrics are improved when using an automated kmer size in SOAPdenovo and with assembly revision (correction of base errors, short indels, and chimeric contigs), compared to a fixed kmer size of = 23 in SOAPdenovo and no assembly revision. The Kmer size is estimated as the closest odd number greater than half the average read length for a sample. Numbers reported are in percent improvement of the respective quality metric. The calculated Kmer for each sample is given in Table S8.

doi:10.1371/journal.pone.0047656.t001

iv) Assembly Revision

In the assembly revision step, a feature independent of the utilized assembly packages, MOCAT can revise existing paired-end read assemblies by aligning the reads to assembled scaftigs using the gap-tolerant BWA aligner [21] to correct for base errors and short indels, and the fast SOAPaligner2 to resolve chimeric regions. Performing assembly revision on the 124 human fecal metagenomes further improved gene prediction metrics (column 2 in Table 1).

v) Gene Prediction

Finally, protein coding genes on the metagenomes are predicted using either the default component Prodigal [22] or MetaGeneMark [23]. An in depth comparison of the gene prediction software is beyond the scope of this article. However, each software have been benchmarked by the respective authors (http://prodigal.ornl.gov/results.php and [23]). An independent comparison determined that MetaGeneMark had a higher precision and Prodigal a higher recall rate (http://genome.jgi.doe.gov/programs/metagenomes/benchmarks.jsf).

Conclusions

The functionality and versatility of the pipeline has been demonstrated using an artificial mock community metagenome, a simulated metagenome with 100 species, and 124 human gut metagenomes. Based on parameter exploration and data driven parameter optimization at run-time, the MOCAT pipeline can process metagenomes in a standardized and automated way while improving the quality of assembly and gene prediction compared to using default parameters for the supported programs. To date, MOCAT has additionally been used to process and assemble hundreds of host-associated and ocean metagenomes within the scope of the MetaHIT [8] and TARA Oceans projects [24].

Implementation, Availability, and Requirements

MOCAT is implemented in Perl and installed by extracting the package and executing one script, which downloads the default external software used by the pipeline and sets up the software. This reduces the otherwise tedious process of downloading all the individual components, a common drawback of in-house pipelines [12]. Optional components requiring a license, such as MetaGeneMark [23] for gene prediction, and Usearch [17] for removal or extraction of reads by alignment to a FASTA-formatted sequence file, require a manual download.

A new project is quickly setup requiring only single- or paired-end FastQ formatted sequencing reads files [25] for each sample in a separate directory. The use of a project-specific configuration file, with suggested default settings, offers users to run all processing steps up to gene prediction without additional setup, while allowing experienced users to modify parameters and programs used in MOCAT. All of the settings are described in the MOCAT documentation.

A queuing system enables processing of a large number of samples in parallel. If present, MOCAT seamlessly integrates all processing steps with the SGE and PBS queuing systems. However, if no queuing system is available, MOCAT processes samples serially on the machine it was executed.

MOCAT runs on 64-bit UNIX systems and can be freely downloaded athttp://www.bork.embl.de/mocat/. Perl version 5.8.8 or above is required. MOCAT is also available in a Virtual Machine package, which could be used to run MOCAT on a PC or a cloud based system. The open source code and modular architecture allow users to modify or exchange the programs that are utilized in the various processing steps. There are no minimum hardware requirements for the pipeline itself to run, however, requirements for analyzing metagenomic datasets vary depending on the number of samples to process in parallel and the sequencing depth of each sample. To aid in determining whether local computational resources are adequate, we provide the maximum resources required to process the datasets in this article. We recommend at least 16 GB of RAM to process smaller metagenomes and 64 GB of RAM to process medium sized metagenomes, but these requirements may vary depending on project settings and systems used. The hard disk space requirements depend on the size and number of metagenomes to analyze, but we recommend at least 500 GB of hard disk space.

METHODS

Data Sources

Data for the simulated metagenome is publically available athttp://www.bork.embl.de/~mende/simulated_data/[13]. This dataset consisted of simulated paired-end raw reads and 193 reference sequences (chromosomes and plasmids) from 100 genomes used to simulate this metagenome. Metagenomic data for the even HMP mock community were downloaded from http://www.ncbi.nlm.nih.gov/bioproject/48475, and the references sequences were downloaded from the NCBI database, with the exception of *Candida albicans*, which was downloaded fromhttp://www.candidagenome.org/download/sequence/C_albicans_SC5314/Assembly21/current/. Metadata for the mock community was downloaded from http://www.hmpdacc.org/HMMC/. Datasets for the simulated metagenome and the mock community can optionally be downloaded automatically when installing the MOCAT pipeline.

Raw reads for the 124 human gut microbiomes were downloaded from the EBI homepage (accession number ERA000116, http://ftp.sra.ebi.ac.uk/vol1/ERA000/ERA000116/fastq/).

Data Processing and Software Settings

The three datasets were processed by the *read_trim_filter* step in MOCAT with length cut off set to *30* and quality cut off set to *20*, using *solexaqa* for the mock community and the simulated metagenome, and *fastx* for the 124 gut metagenomes.

Estimated taxonomic compositions for the simulated metagenome and the mock community were calculated in three steps. First, quality trimmed and filtered reads from the mock community were screened against a FASTA-file with Illumina adapter sequences, using the *screen_fastafile* option and e-value set to *0.01*. Second, screened reads from the mock community and quality trimmed and filtered reads from the simulated metagenome were mapped and filtered against the custom-made reference databases with chromosome and plasmid sequences from the 22 mock genomes and 100 genomes from the simulated metagenome ([13]), respectively. This was done by executing the *screen* and *filter* commands with length cutoff set to *30*, percentage identity set to *90 and* paired_end_filtering set to *yes* for the simulated metagenome and set to *no* for the mock community. Finally, the taxonomic composition was estimated using the *calculate_coverage* command.

Assembly and gene prediction, on the simulated metagenome and mock community, were performed using the *assembly* (SOAPdenovo version *1.06*) and *gene_prediction* (*MetaGeneMark*) options. Quality trimmed and filtered reads from the simulated metagenome, and adapter-screened reads from the mock community, were assembled into scaftigs *60* bp or longer. Predicted complete genes were aligned to their respective metagenomes using blastall v2.2.26 [26] (program blastn, 95% sequence identity, alignment length >=90%, and e-value 0.1) and only the best hit selected.

The 124 human gut microbiomes were processed with and without 5′ trimming. 5′ trimmed reads were assembled using SOAPdenovo *1.05*, using both the Kmer determined by MOCAT and a fixed Kmer size set to 23. These assemblies were revised using SOAPdenovo *1.06*using the *assembly_revision* options, and genes were predicted, with *MetaGeneMark* as selected software, on scaftigs from both assemblies and revised assemblies. The non 5′ trimmed and 5′ trimmed reads were mapped to the assembled scaftigs using the *screen* option using length cutoff *30* and quality cutoff *15*.

Complete commands for processing the simulated metagenome and mock community in MOCAT are bundled with the installation of the pipeline.

ACKNOWLEDGMENTS

We wish to thank the MetaHIT consortium and members of the Bork group, especially Siegfried Schloissnig, for fruitful discussions and code improvements.

AUTHOR CONTRIBUTIONS

Conceived and designed the experiments: JRK SS PB. Performed the experiments: JRK SS. Analyzed the data: JRK SS. Contributed reagents/materials/analysis tools: DRM MA JL WC HC QP BL JQ JW. Wrote the paper: JRK SS PB.

REFERENCES

1. Goecks J, Nekrutenko A, Taylor J (2010) Galaxy: a comprehensive approach for supporting accessible, reproducible, and transparent computational research in the life sciences. Genome biology 11: R86 doi:10.1186/gb-2010-11-8-r86. doi: 10.1186/gb-2010-11-8-r86
2. Meyer F, Paarmann D, D'Souza M, Olson R, Glass EM, et al. (2008) The metagenomics RAST server - a public resource for the automatic phylogenetic and functional analysis of metagenomes. BMC bioinformatics 9: 386 doi:10.1186/1471-2105-9-386. doi: 10.1186/1471-2105-9-386
3. Sun S, Chen J, Li W, Altintas I, Lin A, et al. (2011) Community cyberinfrastructure for Advanced Microbial Ecology Research and Analysis: the CAMERA resource. Nucleic acids research 39: D546–51 doi:10.1093/nar/gkq1102. doi: 10.1093/nar/gkq1102
4. Markowitz VM, Chen I-MA, Chu K, Szeto E, Palaniappan K, et al. (2012) IMG/M: the integrated metagenome data management and comparative analysis system. Nucleic acids research 40: D123–9 doi:10.1093/nar/gkr975. doi: 10.1093/nar/gkr975
5. Angiuoli SV, Matalka M, Gussman A, Galens K, Vangala M, et al. (2011) CloVR: A virtual machine for automated and portable sequence analysis from the desktop using cloud computing. BMC Bioinformatics 12: 356 doi:10.1186/1471-2105-12-356. doi: 10.1186/1471-2105-12-356
6. Arumugam M, Harrington ED, Foerstner KU, Raes J, Bork P (2010) SmashCommunity: A metagenomic annotation and analysis tool. Bioinformatics (Oxford, England) 26: 2977–2978 doi:10.1093/bioinformatics/btq536. doi: 10.1093/bioinformatics/btq536
7. Huson DH, Mitra S, Ruscheweyh H-J, Weber N, Schuster SC (2011) Integrative analysis of environmental sequences using MEGAN4. Genome research 21: 1552–1560 doi:10.1101/gr.120618.111. doi: 10.1101/gr.120618.111
8. Qin J, Li R, Raes J, Arumugam M, Burgdorf KS, et al. (2010) A human gut microbial gene catalogue established by metagenomic sequencing. Nature 464: 59–65 doi:10.1038/nature08821. doi: 10.1038/nature08821
9. .Peterson J, Garges S, Giovanni M, McInnes P, Wang L, et al. (2009) The NIH Human Microbiome Project. Genome research 19: 2317–2323 doi:10.1101/gr.096651.109. doi: 10.1101/gr.096651.109
10. Gilbert JA, Meyer F, Antonopoulos D, Balaji P, Brown CT, et al. (2010) Meeting report: the terabase metagenomics workshop and the vision of an Earth microbiome project. Standards in genomic sciences 3: 243–248 doi:10.4056/sigs.1433550. doi: 10.4056/sigs.1433550
11. Simpson DA, Clark GR, Alexander S, Silvestri G, Willoughby CE (2011) Molecular diagnosis for heterogeneous genetic diseases with targeted high-throughput DNA sequencing applied to retinitis pigmentosa. Journal of medical genetics 48: 145–151 doi:10.1136/jmg.2010.083568. doi: 10.1136/jmg.2010.083568
12. Gonzalez A, Knight R (2012) Advancing analytical algorithms and pipelines for billions of microbial sequences. Current Opinion in Biotechnology 23: 64–71 doi:10.1016/j.copbio.2011.11.028. doi: 10.1016/j.copbio.2011.11.028
13. Mende DR, Waller AS, Sunagawa S, Järvelin AI, Chan MM, et al. (2012) Assessment of metagenomic assembly using simulated next generation sequencing data. PloS one 7: e31386 doi:10.1371/journal.pone.0031386. doi: 10.1371/journal.pone.0031386
14. Cox MP, Peterson DA, Biggs PJ (2010) SolexaQA: At-a-glance quality assessment of Illumina second-generation sequencing data. BMC bioinformatics 11: 485 doi:10.1186/1471-2105-11-485. doi: 10.1186/1471-2105-11-485
15. Schwartz S, Oren R, Ast G (2011) Detection and removal of biases in the analysis of next-generation sequencing

reads. PloS one 6: e16685 doi:10.1371/journal.pone.0016685. doi: 10.1371/journal.pone.0016685

16. Li R, Yu C, Li Y, Lam T-W, Yiu S-M, et al. (2009) SOAP2: an improved ultrafast tool for short read alignment. Bioinformatics (Oxford, England) 25: 1966–1967 doi:10.1093/bioinformatics/btp336. doi: 10.1093/bioinformatics/btp336
17. Edgar CRC (2012) USEARCH 5.1. Clinical infectious diseases?: an official publication of the Infectious Diseases Society of America 54: NP. doi:10.1093/cid/cir977.
18. Haas BJ, Gevers D, Earl AM, Feldgarden M, Ward DV, et al. (2011) Chimeric 16S rRNA sequence formation and detection in Sanger and 454-pyrosequenced PCR amplicons. Genome research 21: 494–504 doi:10.1101/gr.112730.110. doi: 10.1101/gr.112730.110
19. Li R, Zhu H, Ruan J, Qian W, Fang X, et al. (2010) De novo assembly of human genomes with massively parallel short read sequencing. Genome research 20: 265–272 doi:10.1101/gr.097261.109. doi: 10.1101/gr.097261.109
20. Ciccarelli FD, Doerks T, von Mering C, Creevey CJ, Snel B, et al. (2006) Toward automatic reconstruction of a highly resolved tree of life. Science (New York, NY) 311: 1283–1287 doi:10.1126/science.1123061. doi: 10.1126/science.1123061
21. Li H, Durbin R (2009) Fast and accurate short read alignment with Burrows-Wheeler transform. Bioinformatics (Oxford, England) 25: 1754–1760 doi:10.1093/bioinformatics/btp324. doi: 10.1093/bioinformatics/btp324
22. Hyatt D, Chen G-L, Locascio PF, Land ML, Larimer FW, et al. (2010) Prodigal: prokaryotic gene recognition and translation initiation site identification. BMC bioinformatics 11: 119 doi:10.1186/1471-2105-11-119. doi: 10.1186/1471-2105-11-119
23. Zhu W, Lomsadze A, Borodovsky M (2010) Ab initio gene identification in metagenomic sequences. Nucleic acids research 38: 1–15 doi:10.1093/nar/gkq275. doi: 10.1093/nar/gkq275
24. Karsenti E, Acinas SG, Bork P, Bowler C, De Vargas C, et al. (2011) A holistic approach to marine eco-systems biology. PLoS biology 9: e1001177 doi:10.1371/journal.pbio.1001177. doi: 10.1371/journal.pbio.1001177
25. Cock PJA, Fields CJ, Goto N, Heuer ML, Rice PM (2010) The Sanger FASTQ file format for sequences with quality scores, and the Solexa/Illumina FASTQ variants. Nucleic acids research 38: 1767–1771 doi:10.1093/nar/gkp1137. doi: 10.1093/nar/gkp1137
26. Altschul SF, Gish W, Miller W, Myers EW, Lipman DJ (1990) Basic local alignment search tool. Journal of molecular biology 215: 403–410 doi:10.1016/S0022-2836(05)80360-2. doi: 10.1016/s0022-2836(05)80360-2

SECTION-V COMPUTER ASSISTED ANALYSIS

CHAPTER 13

THE EFFECTS OF ALIGNMENT QUALITY, DISTANCE CALCULATION METHOD, SEQUENCE FILTERING, AND REGION ON THE ANALYSIS OF 16S RRNA GENE-BASED STUDIES

Patrick D. Schloss

INTRODUCTION

The recent advent of massively-parallelized pyrosequencing platforms has enabled a renaissance in the field of microbial ecology [1], [2]. Pyrosequencing has engendered much enthusiasm since it is now possible to obtain nearly 100-times as many sequences by pyrosequencing for the same cost as using traditional Sanger sequencing technology. Although pyrosequencing is capable of generating 10^5–10^6 sequences per run, the sequences are between 100 and 400 bp in length. This method has become widely used among microbial ecologists to sequence PCR amplicons from variable regions within the ca. 1,500-bp 16S rRNA gene.

These massive datasets have been analyzed through the generation of phylogenetic trees[e.g. 3], assignment of sequences to operational taxonomic units (OTUs) for based on distance thresholds [e.g. 4], and classification of sequences to phylogenentic bins based on similarity to reference sequences [e.g. 5]. Each approach has received some level of evaluation using pyrotag sequencing. Liu et al. [6] asserted that phylogenies generated using pyrotags were as good as full-length sequences based on similarity of UniFrac test statistics. Several studies have evaluated various regions and methods for assigning sequences to phylotypes [7]–[9]. Finally, a recent study emphasized differences in α-diversity metrics using different regions within the 16S rRNA gene and OTU definitions [10].

Each of these studies have focused on a limited range of phylogenetic groups found in a particular environment (e.g. soil, mouse cecum, human feces) and have glossed over more fundamental questions related to how alignment quality, methods of calculating pairwise genetic distances, sequence filtering, and region affects downstream analysis and their relationship to full-length sequences. Alignment quality is expected to significantly affect pairwise distances. Investigators have either used reference alignments to align sequences that implicitly incorporate the secondary structure of the 16S rRNA molecule [11]–[14] or they have used methods that do not consider the secondary structure [15], [16]. Previous results have shown that the manually-curated SILVA reference alignment provides superior complementary base-pairing within the secondary structure compared to the greengenes alignment, which appears haphazard; the RDP alignment does not align the variable regions[11]. Considering the focus of these studies is on the variable region, there is the added complication that these areas are difficult to align accurately. To overcome limitations in alignment of variable regions, many studies have employed the use of masks to filter the troublesome regions [e.g. 3]. Yet, these filters remove a considerable amount of information from already information-sparse data (Table 1). The actual method of calculating distances is also typically taken for granted. Practically every 16S rRNA survey has made use of substitution models that assume that an alignment gap represents missing data instead of a mutation [e.g. 17]. The decision to use such a model seems motivated more by a sense of phylogenetic guilt than by biology. It is also unknown how distances calculated between partial sequences predict distances between full-length sequences. To make data analysis more tractable, some have employed heuristics based on correlations between kmer- and sequence-based pairwise distances to select which pairs of sequences to align and group within OTUs[16]. It is unclear how these correlations vary across regions within the 16S rRNA gene or what the level of risk is for falsely ignoring pairs of similar sequences. Finally, most studies make the implicit assumption that distances between partial sequences are not significantly different from those of full-length sequences; however, this is a questionable

Citation: Schloss PD (2010) The Effects of Alignment Quality, Distance Calculation Method, Sequence Filtering, and Region on the Analysis of 16S rRNA Gene-Based Studies. PLoS Comput Biol 6(7): e1000844. doi:10.1371/journal.pcbi.1000844.

assumption as it is well-established that the 16S rRNA gene does not evolve uniformly along its length. This is apparent in the choice a 3% distance cutoff, which is used as a proxy species definition for full-length sequences, to define species using sequences from variable regions [e.g. 2], [4]. Each of these factors is expected to have a significant effect on the analysis, interpretation, and generalizability of 16S rRNA gene surveys.

Table 1: Summary of the 13 regions within the 16S rRNA gene that were used in this study

Region	*E. coli* numbering	Platform	Example Ref.	Masking	Average number of bases[a]
V19	2–1491	Sanger[b]	[32]	None	1454 (1399–1490)
				Lane	1255 (1244–1256)
V12	28–337	Titanium[c]		None	306 (276–332)
				Lane	239 (238–239)
V13	28–514	Titanium	HMP[f]	None	480 (428–508)
				Lane	386 (384–386)
V14	28–784	Sanger	[17]	None	750 (698–779)
				Lane	656 (653–656)
V2	100–337	FLX[d]	[3]	None	240 (223–257)
				Lane	198 (197–198)
V23	100–514	Titanium		None	415 (378–437)
				Lane	345 (343–345)
V3	357–514	FLX/Illumina[e]	[33]	None	158 (135–161)
				Lane	128 (127–128)
V35	357–906	Titanium	HMP	None	546 (523–552)
				Lane	507 (504–507)
V4	578–784	FLX	[8]	None	207 (206–208)
				Lane	207 (206–207)
V6	986–1045	FLX/Illumina	[5]	None	60 (57–66)
				Lane	27 (27–27)
V69	986–1491	Titanium	HMP	None	507 (489–516)
				Lane	411 (407–412)
V89	1100–1491	Titanium		None	392 (373–403)
				Lane	330 (326–331)
V9	1300–1491	FLX	[4]	None	192 (182–197)
				Lane	170 (146–147)

Each sub-region was generated from the sequences in the V19 database.
[a]The numbers in the parentheses represent the 95% confidence interval.
[b]Sanger reads are estimated to be up to 800 bp and can be used to sequence the same molecule multiple times.
[c]GS FLX Titanium reads average ~400 bp (amplicon kit released 11/2009).
[d]GS FLX reads average ~240 bp.
[e]Illumina reads average ~200 bp.
[f]The NIH Human Microbiome Project is considering these regions for their cross-sectional studies.
doi:10.1371/journal.pcbi.1000844.t001

Here, I used a collection of full-length 16S rRNA gene sequences representing 43 bacterial phyla to quantify how alignment quality, distance calculation methods, masking, and region within the 16S rRNA gene affect out ability to assess α- and β-diversity. The results of these analyses urge greater caution in how surveys are designed and interpreted.

RESULTS

The effect of alignment on genetic distances

For each of the 13 regions I used various alignment methods to calculate 91,131,750 pairwise distances assuming that a series of consecutive gaps represented one insertion or deletion. The SILVA, greengenes, and RDP alignments represent a gradation in the level of attention given to aligning the variable regions and are each guided by the secondary structure of the 16S rRNA gene. In contrast, the MUSCLE and pairwise alignments are attempts to optimize the alignment between sequences based on a limited number of parameters that are set *a priori*. To compare the pairwise distances calculated for the same pairs of sequences across alignments, I calculated the regression coefficients describing the relationship between the distances for the greengenes, RDP, MUSCLE, and pairwise alignments and the SILVA alignment for each region (Table 2). Distance calculations for this analysis assumed that consecutive gap positions were the product of a single insertion or deletion mutation (i.e. one gap). With the exception of the V3 and V4 regions, the RDP alignment for each of the regions predicted greater genetic diversity than that of the SILVA alignment. Interestingly, the greengenes alignment, which does a poor job of aligning the variable regions, predicted between 9 and 33% more genetic diversity for each region than the RDP alignment, which does not attempt to align the variable regions. Visual inspection of the greengenes alignment suggests that in many instances the variable region alignments are somewhat random [11]. I observed that the MUSCLE-generated alignments described considerably greater genetic diversity than any of the other methods for the V3, V6, and V9 regions (Table 2); however, the use of pairwise alignments yielded smaller distances than those calculated with the other alignment methods because pairwise alignment methods optimize the alignment without the constraint of preserving positional homology across multiple sequences (Table 2). Perhaps most worrisome is the observation that with the exception of the distances calculated from pairwise alignments, regressions of the other alignment methods to the SILVA-based alignment typically did a poor job of accounting for the variation in the distances (Table 2). These data make it clear that variation in alignment quality can have a significant impact on the genetic diversity that is calculated between the same pairs of sequences.

Table 2: Slope coefficients and R^2 values for the comparison of one gap distances calculated for SINA-aligned sequences extracted from different regions within the 16S rRNA gene sequence to one gap distances calculated using sequences aligned by different methods

Region	Statistic	RDP	greengenes	MUSCLE	Needleman
V19	Slope	1.06	1.17	1.04	0.93
	R^2	0.97	0.77	0.98	0.99
V12	Slope	1.13	1.25	1.11	0.93
	R^2	0.80	0.52	0.77	0.91
V13	Slope	1.08	1.20	1.07	0.93
	R^2	0.88	0.62	0.92	0.93
V14	Slope	1.06	1.16	1.05	0.94
	R^2	0.94	0.74	0.96	0.97
V2	Slope	1.04	1.21	1.16	0.97
	R^2	0.94	0.67	0.64	0.95
V23	Slope	1.04	1.18	1.09	0.95
	R^2	0.96	0.74	0.92	0.96
V3	Slope	1.00	1.11	2.07	1.02
	R^2	0.91	0.67	0.14	0.96
V35	Slope	1.01	1.12	1.04	0.95
	R^2	0.98	0.83	0.97	0.98
V4	Slope	1.00	1.09	1.08	0.98
	R^2	0.99	0.77	0.87	0.98
V6	Slope	1.09	1.42	3.04	0.98
	R^2	0.66	0.14	0.30	0.97
V69	Slope	1.07	1.21	1.10	0.94
	R^2	0.92	0.70	0.91	0.97
V89	Slope	1.05	1.18	1.12	0.96
	R^2	0.94	0.78	0.86	0.98
V9	Slope	1.08	1.19	1.58	0.96
	R^2	0.80	0.67	0.36	0.96

doi:10.1371/journal.pcbi.1000844.t002

Effect of alignment on interpretation of α-diversity

Considering the poor correlation between the distances generated from the five alignment methods, it was necessary to determine the effect of this variation on the ability to accurately describe and compare communities. As expected based on the genetic distance analysis, the number of OTUs observed using the green genes alignment was routinely higher than that observed using the other alignment methods and the number of OTUs observed using the pairwise alignment method was routinely the lowest (Fig. 1). Inspection of these lineage through time plots identified a stair-like appearance for many of the regions. This was due to the loss of information as sequence length decreased. The most extreme example of this phenomenon was for the V6 region that had an average sequence length of 60 bp. Each difference between a pair of V6 sequences changed the distance by approximately 0.0167 units, which is the step-length observed for the V6 data in Fig. 1. When the phylogenetic diversity of the datasets was calculated, the greengenes aligned sequences had the highest phylogenetic diversity and the pairwise aligned sequences had the lowest (Fig. 2). One limitation of the phylogenetic diversity metric is that it is difficult to interpret the statistic and so it is unclear how biologically meaningful the level of variation observed is in Fig. 2.

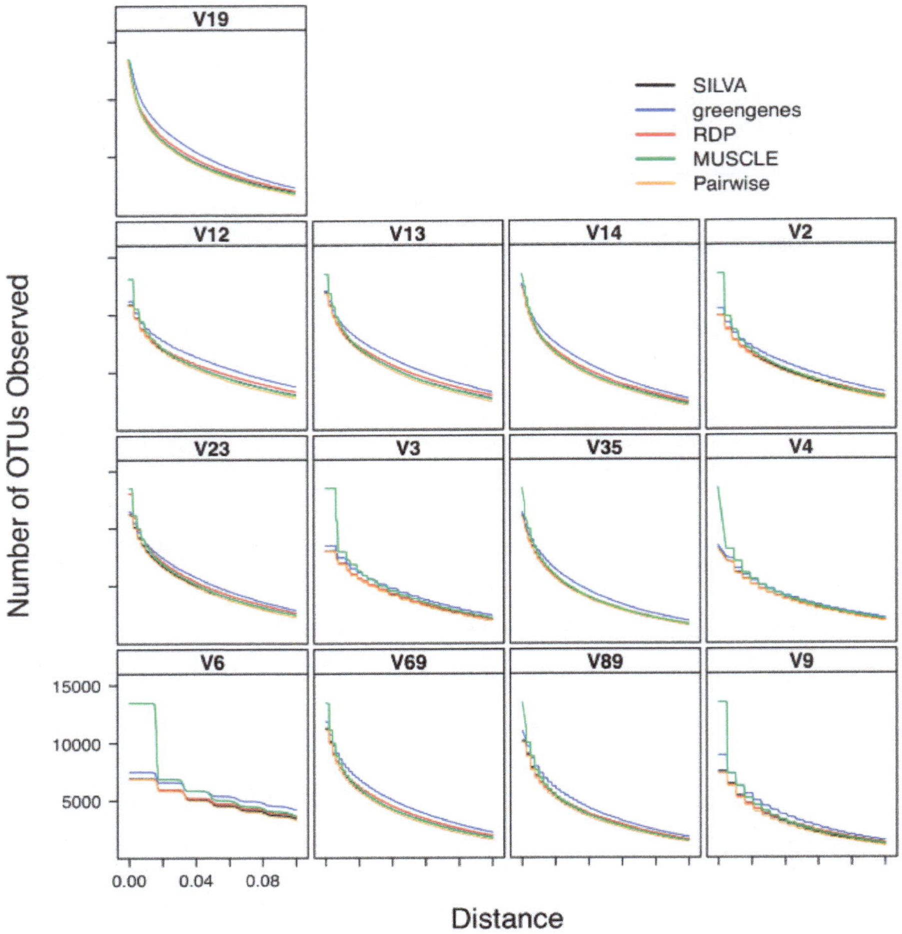

Figure 1: The number of OTUs observed as a function of genetic distance for various regions within the 16S rRNA gene when using different sequence alignments.

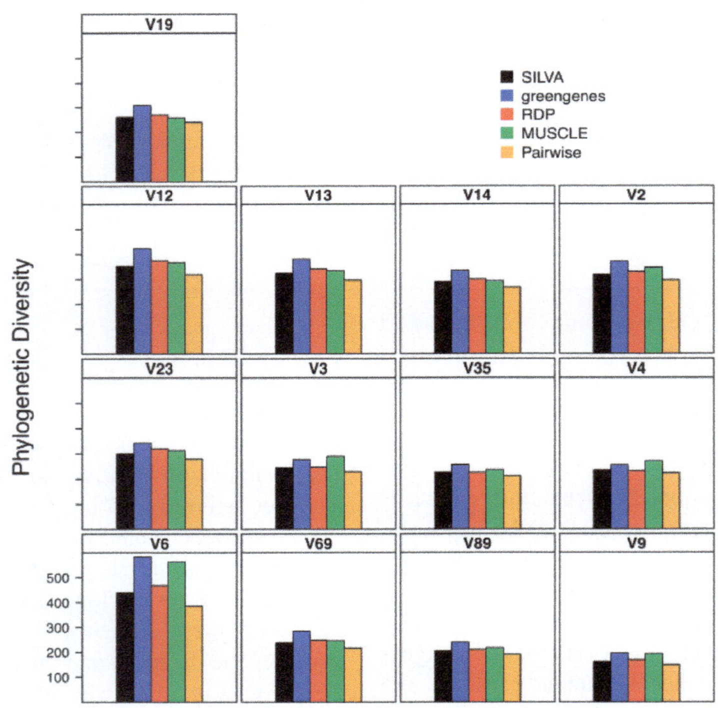

Figure 2: The phylogenetic diversity observed for different regions within the 16S rRNA gene when using different alignments. Phylogenetic diversity was measured by calculating the total branch length for a phylogenetic tree.

Effect of alignment on interpretation of β-diversity

To describe β-diversity, I used two OTU-based metrics (Figs. 3 and 4) and two phylogenetic-based metrics (Fig. 5) to measure the sensitivity of the metrics to alignment quality. Sequences were partitioned so that they would represent two samplings of communities whose Jaccard similarity index was 0.80, but whose Morisita-Horn similarity index was 0.60 with a cutoff of 0.05 when defining OTUs with full-length sequences. Because the sampling of the two simulated communities was limited (ca. 6,750 sequences per community), the Jaccard and unweighted UniFrac statistics did not equal the expected values. Within this simulation framework, the effect of alignment was generally highly statistically significant across metrics of β-diversity ($p \ll 0.001$); however it is unclear how biologically meaningful the observed differences were.

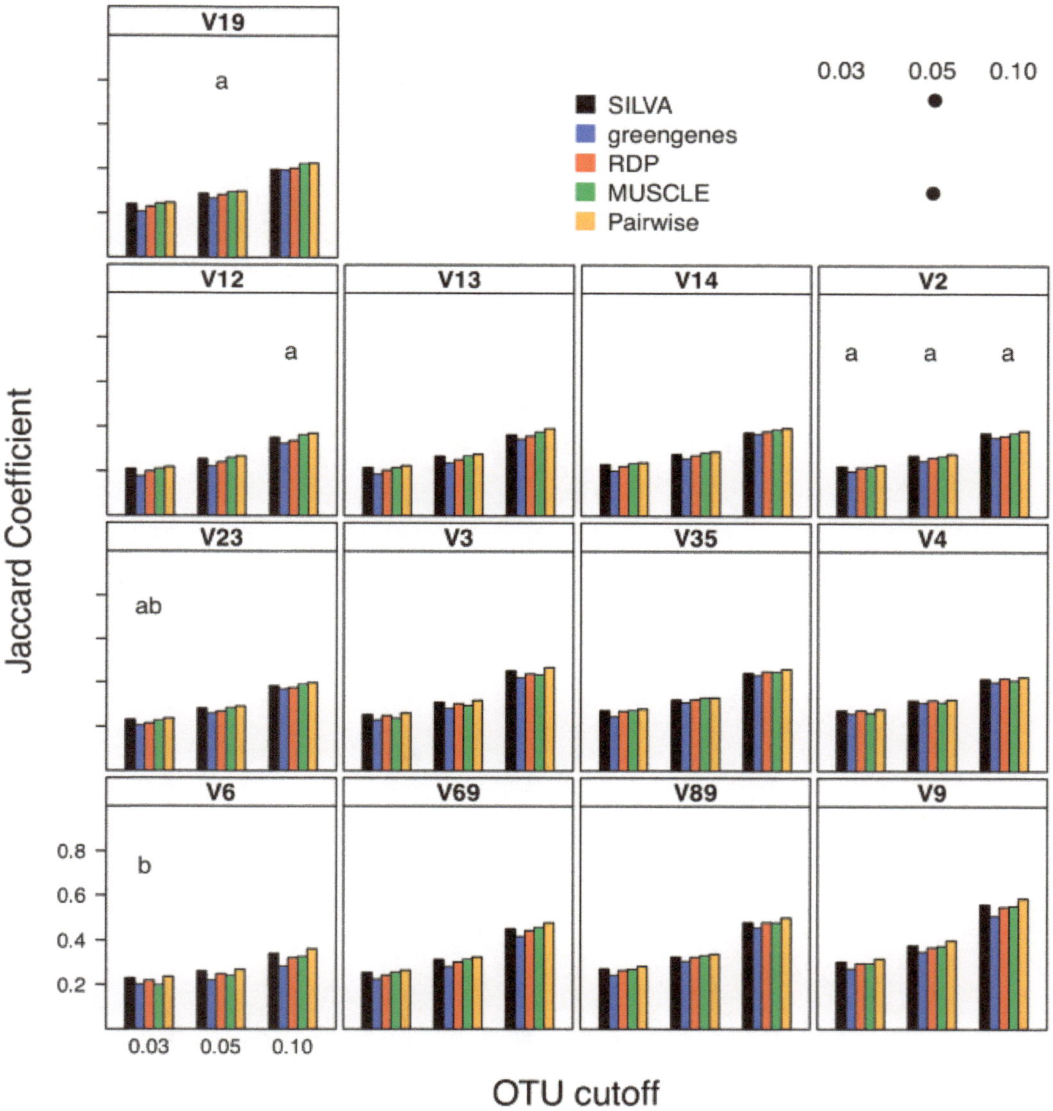

Figure 3: The Jaccard coefficient calculated between two mock communities (described in Materials & Methods) for different OTU definitions and alignments. Each bar represents the average coefficient value for 100 randomized partitionings of the data. Within the same OTU cutoff, alignment strategies with the same symbol and regions with the same letter were not significantly different from each other.

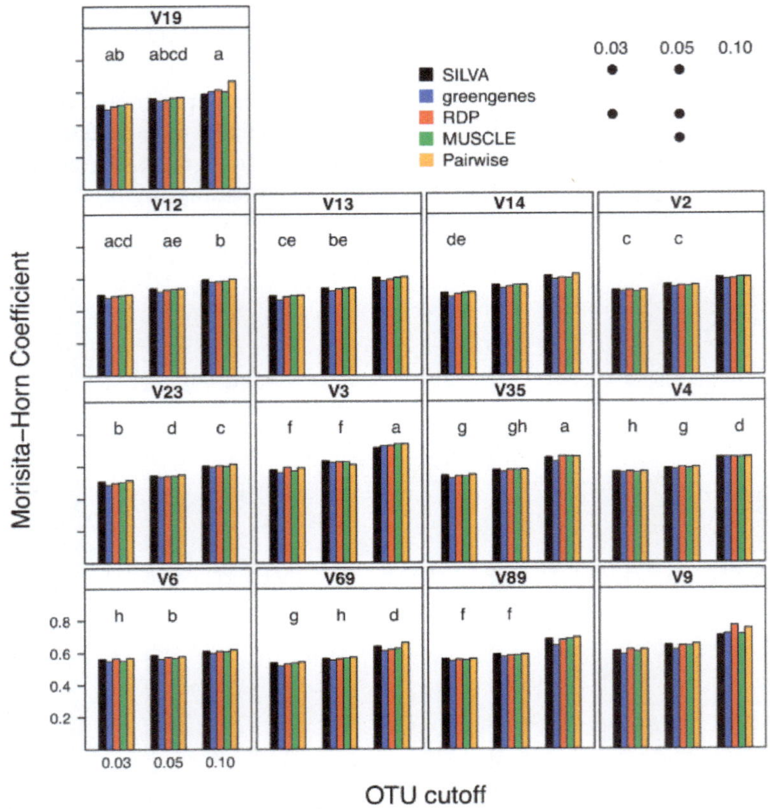

Figure 4: The Morisita-Horn coefficient calculated between two mock communities (described in Materials & Methods) using different OTU definitions and alignments. Each bar represents the average coefficient value for 100 randomized partitionings of the data. Within the same OTU cutoff, alignment strategies with the same symbol and regions with the same letter were not significantly different from each other.

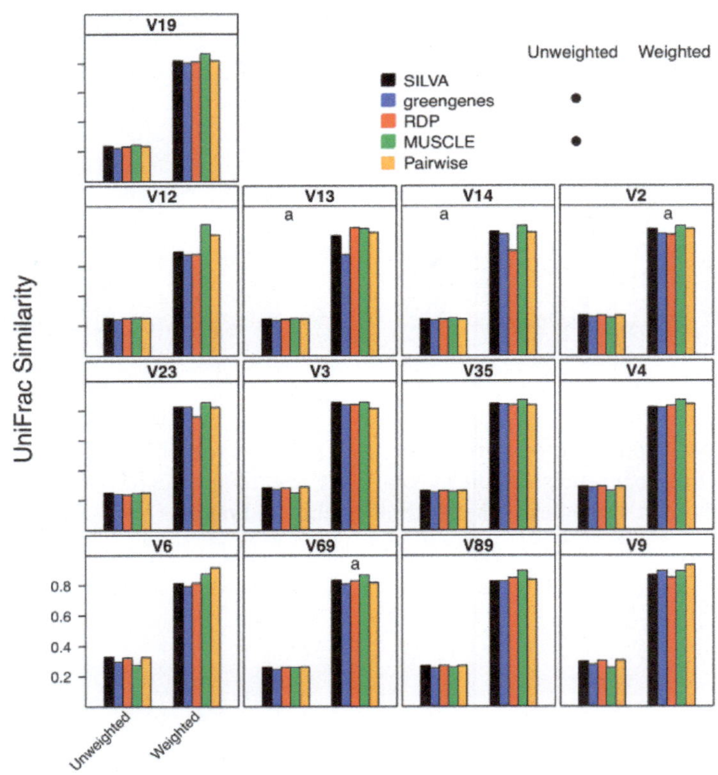

Figure 5: Unweighted and weighted UniFrac similarity values calculated between two mock communities (described in Materials & Methods) using different alignments. Each bar represents the average coefficient value for 100 randomized partitionings of the data. Within the same UniFrac approach, alignment strategies with the same symbol and regions with the same letter were not significantly different from each other.

The effect of distance calculation method on genetic distances

Using the same SILVA-aligned sequences that I analyzed above, I investigated the effect of different distance calculation methods on downstream analyses. Specifically, I considered the one gap calculator (i.e. a gap of any length between two sequences represents a single mutation) and each gap (i.e. gaps length *n*, represent *n* mutations) and ignore gap calculators (i.e. gapped characters are not considered in calculating a distance; Table 3). The slope of lines forced through the origin indicated that the each gap calculator calculated between 0 (V4) and 9% (V3) more genetic diversity than the one gap calculator. With the exception of the V3 region (69%), the regression between the each gap and one gap calculators accounted for more than 87% of the variation in the distances. The differences in the explanatory power of the regression were a function of frequency of gaps longer than 1 nucleotide. The ignore gap calculator calculated between 2 (V4) and 7% (V9) less genetic diversity than the one gap calculator. The regression between the ignore gap and one gap calculators accounted for more than 94% of the variation in the data. Until there is a more well-developed theoretical basis for selecting a method for treating gaps in sequence alignments, these results suggest that treating gaps of any length as a single mutation is a middle ground between ignoring them and treating each of them as a separate evolutionary event.

Table 3: Slope and R^2 values for the regression of one gap pairwise distances onto each gap, ignore gap, Lane mask-filtered one gap, and kmer distances using SINA-aligned sequences over the same region

Region	Statistic	Each gap	Ignore gaps	Filtered	kmer
V19	Slope	1.02	0.94	0.66	3.73
	R^2	0.99	0.99	0.89	0.97
V12	Slope	1.02	0.94	0.56	3.91
	R^2	0.87	0.97	0.47	0.89
V13	Slope	1.02	0.94	0.55	3.80
	R^2	0.92	0.97	0.56	0.91
V14	Slope	1.02	0.95	0.67	3.87
	R^2	0.96	0.98	0.75	0.95
V2	Slope	1.01	0.97	0.71	4.31
	R^2	0.90	0.99	0.58	0.92
V23	Slope	1.01	0.96	0.64	3.99
	R^2	0.92	0.98	0.66	0.93
V3	Slope	1.09	0.95	0.77	4.69
	R^2	0.69	0.94	0.52	0.83
V35	Slope	1.01	0.97	0.79	4.24
	R^2	0.99	0.98	0.84	0.94
V4	Slope	1.00	0.98	0.99	4.70
	R^2	1.00	0.98	0.99	0.93
V6	Slope	1.00	0.96	1.17	4.82
	R^2	1.00	0.95	0.38	0.86
V69	Slope	1.02	0.94	0.73	3.94
	R^2	0.98	0.98	0.79	0.94
V89	Slope	1.03	0.93	0.85	4.35
	R^2	0.94	0.97	0.86	0.95
V9	Slope	1.01	0.93	0.76	4.60
	R^2	0.95	0.94	0.69	0.91

doi:10.1371/journal.pcbi.1000844.t003

Pairwise kmer distances were much larger than the alignment-based calculators and their regression onto the one gap calculated distances accounted for between 83 and 97% of the variation observed between the distances. In order to have no risk of falsely ignoring true one gap pairwise distances smaller than 0.10, it was necessary to keep kmer distances smaller than 0.45 (V19) to 0.73 (V6). This would result in needing to calculate between 3.3- and 9.1-fold more distances than would be needed by alignment-based methods.

Effect of distance calculation method on interpretation of α-diversity

Lacking a theoretical basis for treating gaps as a single evolutionary event, I was curious how much measures of α- and β-diversity are affected by the choice of a distance calculator. I used an OTU-based approach to determine the effect of distance calculation methods on the richness of OTUs within the dataset (Fig. 6) and a phylogeny-based approach using total branch length to measure phylogenetic diversity (Fig. 7). As would be predicted, the number of observed OTUs at any genetic distance was greatest with the each gap and least with the ignore gap calculators; the one gap and each gap calculators generated comparable numbers of OTUs. When I analyzed the effect of region and distance calculation method on the phylogenetic diversity of the datasets, there were qualitative trends between methods and regions that could have been predicted from the regression analysis in Table 2 (Fig. 7). These analyses suggest that the difference observed in α-diversity when using either the one gap or each gap calculator is unlikely to be biologically meaningful.

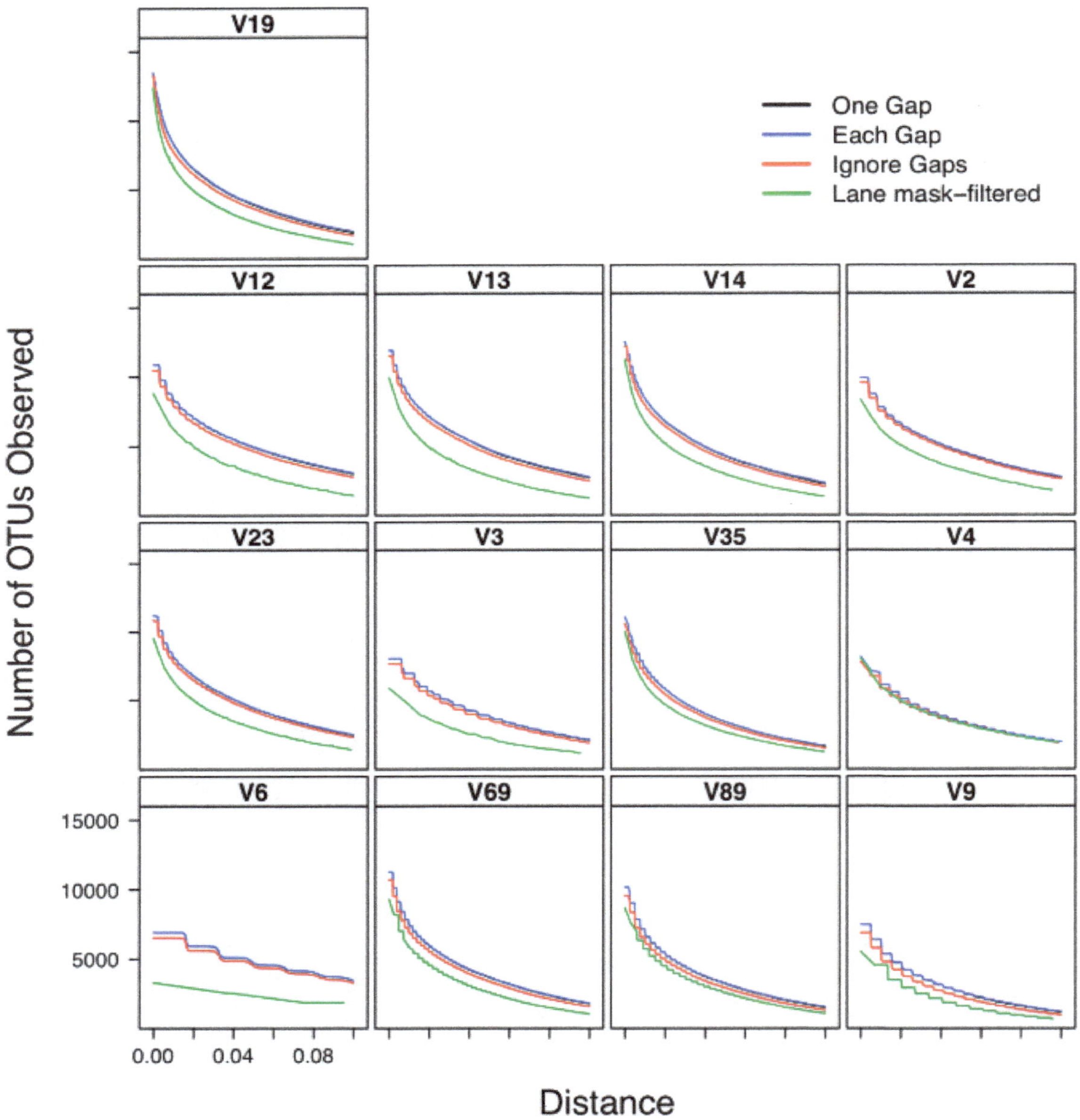

Figure 6: The number of OTUs observed as a function of genetic distance for various regions within the 16S rRNA gene when using different methods of calculating distances and masking sequences.

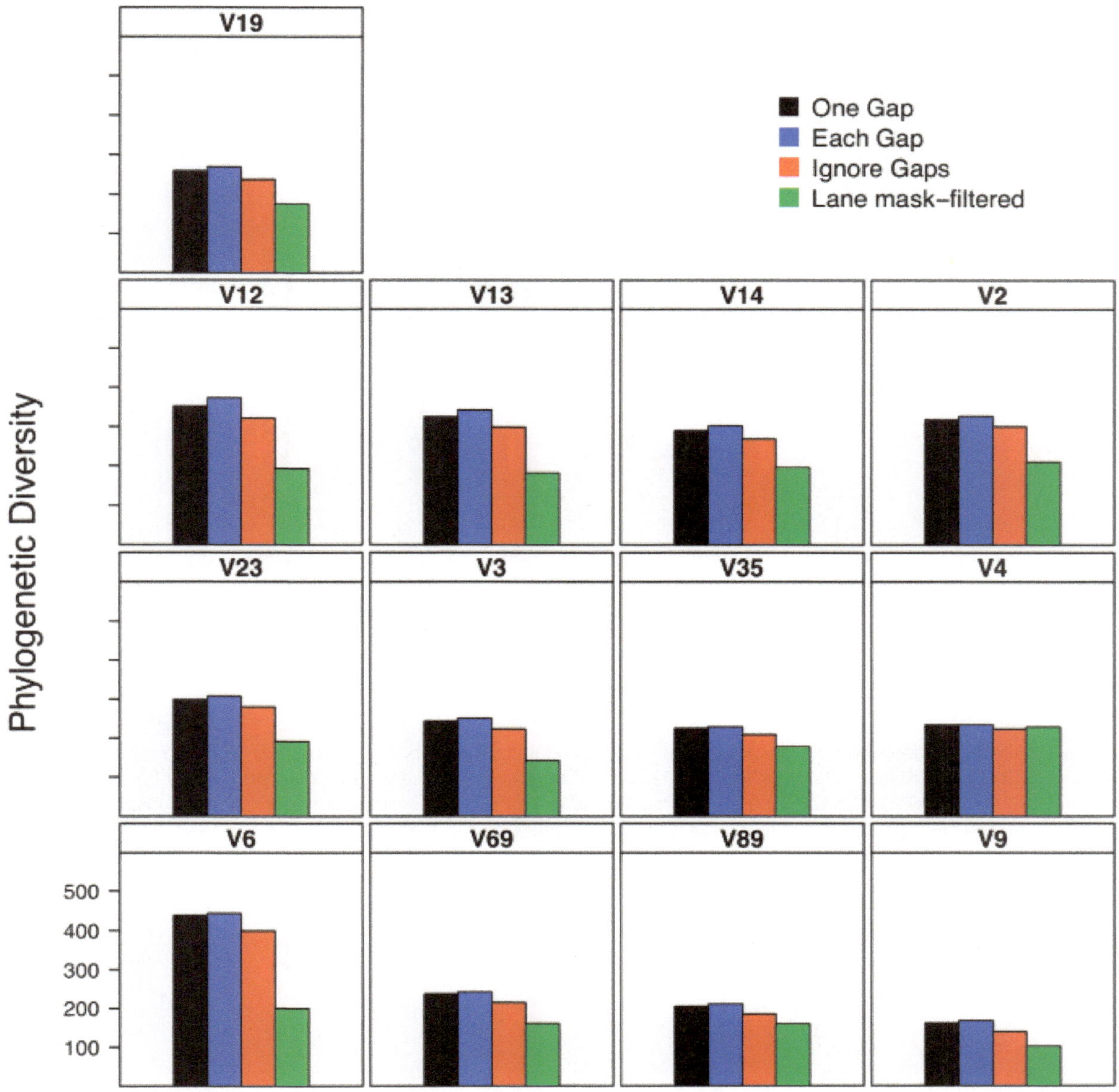

Figure 7: The phylogenetic diversity observed for different regions within the 16S rRNA gene when using different methods of calculating distances and masking sequences. Phylogenetic diversity was measured by calculating the total branch length for a phylogenetic tree.

Effect of distance calculation method on interpretation of β-diversity

I next investigated what effect each calculator method had on two OTU-based (Figs. 8 and 9) and two phylogeny-based β-diversity measures (Fig. 10). For the OTU-based metrics, ignoring gaps resulted in an over-estimate of the similarity between the two communities and counting each gap resulted in an under-estimate. Increasing and decreasing the cutoff used to define the OTUs had a parallel effect on the Jaccard and Morisita-Horn indices (Figs. 8 and 9). These results occurred because ignoring gaps and increasing the threshold each dampen the differences between sequences and pull more sequences into an OTU so that more OTUs are likely to be shared; the same phenomenon was observed when sequences were filtered using the Lane mask (see below). Penalizing each gap or making the OTU definition more stringent had the opposite effect. The calculated Jaccard coefficients were not significantly different between the one gap and each gap distance calculation methods when using the 0.03 and 0.05 OTU cutoffs (Fig. 8); all four distance calculation methods yielded statistically significant differences in Morisita-Horn coefficients, regardless of the OTU cutoff. For the phylogeny-based methods, the observed differences between each of the distance calculation methods were statistically significant (Fig 10). Although the differences between distance calculation methods were highly statistically significant ($p \ll 0.001$), it is unclear how biologically meaningful the differences were.

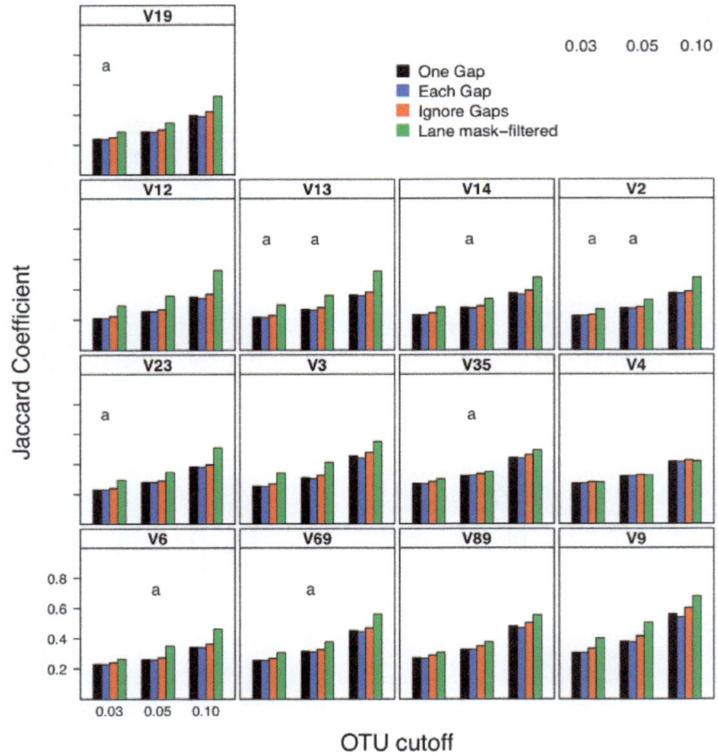

Figure 8: Jaccard similarity values calculated between two mock communities (described in Materials & Methods) for different OTU definitions, methods of calculating distances, and masking sequences. Each bar represents the average coefficient value for 100 randomized partitionings of the data. Within the same OTU cutoff, regions with the same letter were not significantly different from each other; for each OTU cutoff all distance calculation methods were significantly different from each other.

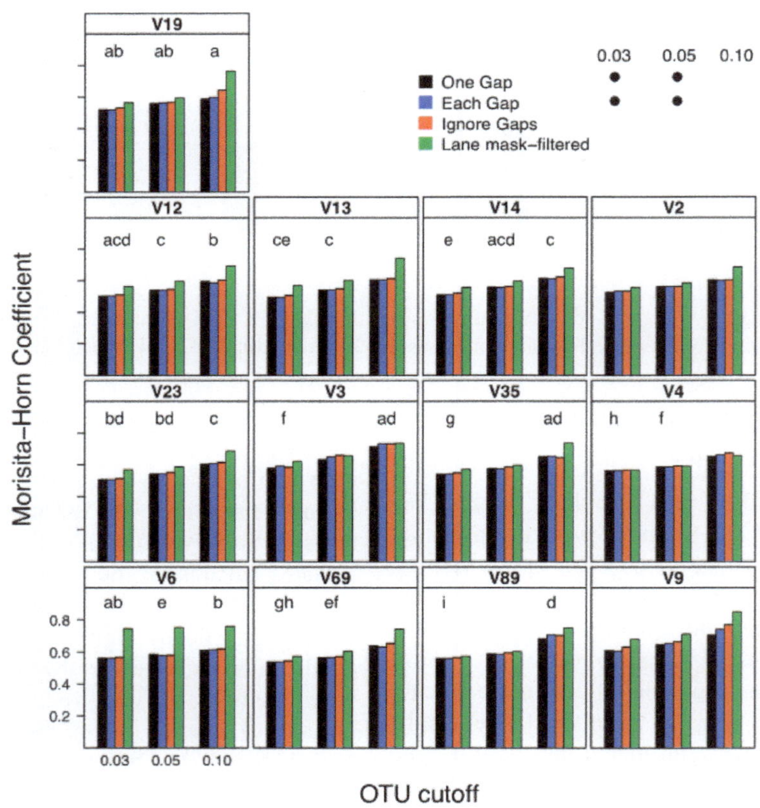

Figure 9: The Morisita-Horn coefficient calculated between two mock communities (described in Materials & Methods) using different OTU definitions, methods of calculating distances, and masking sequences. Each bar represents the average coefficient value for 100 randomized partitionings of the data. Within the same OTU cutoff, distance calculation methods with the same symbol and regions with the same letter were not significantly different from each other.

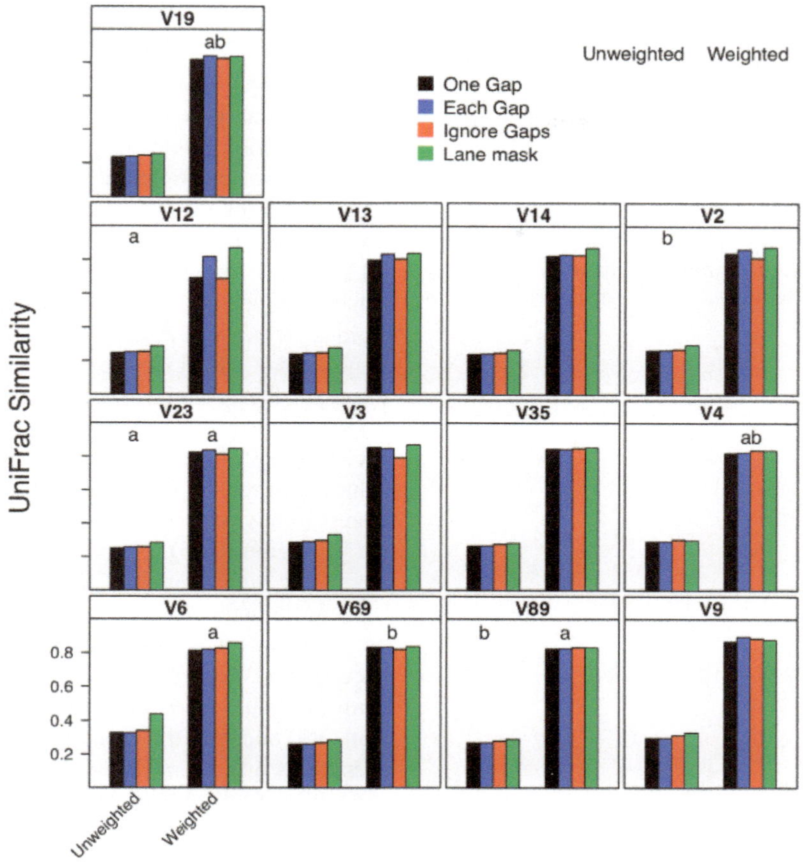

Figure 10: Unweighted and weighted UniFrac similarity values calculated between two mock communities (described in Materials & Methods) using different methods of calculating distances and masking sequences. Each bar represents the average coefficient value for 100 randomized partitionings of the data. Within the same UniFrac method, regions with the same letter were not significantly different from each other; for both UniFrac methods the distance calculation methods were all significantly different from each other.

Effects of filtering sequences using the Lane mask on analysis

To circumvent alignment quality problems, the Lane mask has been used to filter variable regions from 16S rRNA genes. Results of analyses using filtered sequences aligned by any method or when distances were calculated by any method did not vary to a meaningful degree. Comparison of distances calculated using filtered sequences to those calculated using unfiltered sequences showed that filtering significantly reduced the genetic diversity observed between sequences (Table 3). With the exception of the V4 and V6 regions, masking removed between 15 and 45% of the genetic diversity. The V4 region is largely unaffected by the Lane mask and the average length of V6 sequences following the Lane mask treatment was only 27 bp, which made the resulting pairwise distances of dubious value (Table 1). As would be expected, the number of OTUs and phylogenetic diversity observed using Lane mask-filtered sequences was significantly lower than those calculated with the unfiltered sequences. For the four β-diversity measures, when the Lane mask-filtered sequences were analyzed, the communities appeared more similar than for non-filtered SILVA-aligned sequences (Figs. 8–10). One explanation for this observation is that because filtering makes sequences more similar to each other, it also makes communities appear more similar to each other. Although useful for broad-scale phylogenetic analysis at the level of a kingdom or phylum, filters remove the sequence information necessary to differentiate populations within a community. Ultimately, application of such filters is troublesome because it mutes the signals that differentiate communities.

Relationship between the genetic diversity calculated between full-length and regional sequences

I compared the one gap distances calculated for each of the 12 regions from each alignment to the one gap distances calculated from the full-length SILVA alignments (Table 4). The regression of pairwise distances calculated from a sub-region onto distances calculated from full-length sequences was rarely near 1.00. The most extreme case was the V6 region for which distances were nearly 3-fold higher than distances calculated using full-length sequences. Conversely, sequences from the V9 region were 33% less diverse than their full-length counterparts. In general, genetic diversity decreased along the length of the 16S rRNA gene. Although one could use these regression coefficients to relate data collected from one region to that from full-length sequences, the ability of the regression to explain the variation observed between sub-region and full-length sequences was quite poor. As expected, longer regions did the best job of relating the variation between sub-regions and full-length sequences. For example, when using the SILVA alignments, the regression of the V14, V35, and V69 distances onto the full-length distances accounted for 87, 77, and 77% of the variation in distances. Shorter regions such as the V3, V6, and V9 accounted for 26, 36, and 46% of the variation (Table 4). This analysis revealed that all sub-regions are limited in their capacity to serve as surrogates for full-length 16S rRNA gene sequences.

Table 4: Regression coefficients and R^2 values for the comparison of one gap distances calculated for different regions within the 16S rRNA gene sequence and aligned by different methods to the one gap distances calculated using SINA aligned full-length sequences

Region	Statistic	SILVA	greengenes	RDP	MUSCLE	Needleman
V19	Slope	NA	1.17	1.06	1.04	0.93
	R^2	NA	0.77	0.97	0.98	0.99
V12	Slope	1.50	1.79	1.65	1.74	1.36
	R^2	0.70	0.55	0.64	0.57	0.70
V13	Slope	1.31	1.52	1.40	1.40	1.19
	R^2	0.73	0.60	0.66	0.74	0.72
V14	Slope	1.13	1.29	1.19	1.18	1.05
	R^2	0.87	0.72	0.83	0.88	0.87
V2	Slope	1.39	1.61	1.43	1.57	1.31
	R^2	0.70	0.59	0.69	0.62	0.70
V23	Slope	1.21	1.38	1.26	1.29	1.13
	R^2	0.73	0.64	0.71	0.72	0.74
V3	Slope	1.05	1.15	1.06	1.63	0.97
	R^2	0.26	0.26	0.24	0.23	0.27
V35	Slope	0.90	0.99	0.90	0.93	0.85
	R^2	0.77	0.68	0.76	0.75	0.76
V4	Slope	0.97	1.05	0.97	1.05	0.94
	R^2	0.57	0.50	0.56	0.55	0.57
V6	Slope	2.98	3.52	3.30	4.99	2.62
	R^2	0.36	0.31	0.34	0.24	0.37
V69	Slope	0.98	1.16	1.05	1.04	0.90
	R^2	0.77	0.62	0.72	0.77	0.78
V89	Slope	0.79	0.93	0.84	0.86	0.75
	R^2	0.70	0.62	0.67	0.68	0.70
V9	Slope	0.67	0.83	0.76	0.85	0.63
	R^2	0.46	0.43	0.42	0.39	0.46

doi:10.1371/journal.pcbi.1000844.t004

Relationship between sub-region differences and differences in α-diversity

The distance-based analysis clearly showed significant differences between distances calculated from sub-regions and full-length sequences. The OTU-based analysis in Fig. 1demonstrates that there was a clear difference in the number ofOTUs observed across regions for a given genetic distance as well as the level of curvature observe observed in the lineage-through-time plots (Figs. 1 and 6). In the phylogenetic-based analysis those regions that described more genetic diversity than the full-length sequences had greater phylogenetic diversity than the phylogenetic diversity calculated for the full-length sequences whereas the regions that described less genetic diversity yielded greater phylogenetic diversity (Figs. 2 and7).

Relationship between sub-region differences and differences in β-diversity

Using pyrotag data introduces several complexities to β-diversity analyses. Moving across regions, but using the same OTU definition could lead one to overestimate community similarity. For example, the average Morisita-Horn similarity for full-length SILVA-aligned sequences with one gap distances was 0.56. Using similarly treated sequences from the V12, V13, V14, and V23 regions I calculated Morisita-Horn values between 0.57 and 0.60; however those from the other 8 regions yielded values between 0.64 (V2) and 0.79 (V9). For a single region, changing the OTU cutoff also had a significant effect on the Morisita-Horn index. For instance, full-length SILVA-aligned sequences yielded 0.52, 0.56, and 0.86 for cutoffs of 0.03, 0.05, and 0.10. This spread in Morisita-Horn values between the 0.03 and 0.10 OTU cutoffs (0.34) was the largest of any region. The narrowest spread was observed for the V6 region (0.06). In contrast to the Morisita-Horn values, there was little variation in the unweighted or weighted UniFrac statistic when comparing sequences analyzed by the same alignment and distance calculation method. With the exception of the V6 region (0.33), the average unweighted UniFrac values varied between 0.24 (V13, V14, V19) and 0.30 (V9) and with the exception of the V12 region (0.69), the average weighted UniFrac values varied between 0.80 (V13) and 0.87 (V9); the value for the full-length sequence was 0.82. Similar to the α-diversity measure of phylogenetic diversity, an added complication of phylogeny-based methods is the complexity of interpreting the proportion of branch length that is shared between or unique to two communities and how such proportions relate to classical β-diversity measures. Thus, it is difficult to interpret the biological significance of such variation. Regardless, the results of the OTU- and phylogeny-based analyses demonstrate that caution must be taken in extrapolating results from one region to another.

DISCUSSION

The ability to define OTUs and reconstruct phylogenies allows an investigator to approach their problem using the data as they present themselves without being confined to an *a priori*taxonomy. Regardless, the analysis I have presented indicates that comparing results obtained by sequencing one region of the 16S rRNA gene can not be easily compared to those obtained using full-length sequences. Ultimately, the fact that the 16S rRNA gene does not evolve uniformly across its length complicates its analysis. Technical limitations require investigators to select a region based on the availability of conserved PCR primers, fragment length, and the ability to generate high quality sequence. Analytical limitations require investigators to select a region based on the availability of database sequences for that region, the ability to accurately classify sequences, and the level of genetic diversity found in the region. Until there is a standardized approach, individual investigators will continue to select different regions for their analysis. Studies such as this are necessary to inform investigators about the strengths and weaknesses of the various regions within the 16S rRNA gene. Based on this analysis, it is clear that regardless of the region, longer reads will improve one's ability to relate their analysis to full-length sequences. As sequence lengths increase to the point that pyrosequencing full-length 16S rRNA genes is possible, this discussion will be unnecessary. Ultimately, all pyrotag regions represent a marker of a marker of genomic diversity. Even if full-length 16S rRNA gene sequencing is possible, it is still just a marker of genomic diversity. Correlations between the complete genome sequence and full-length 16S rRNA gene sequences are probably just as poor as correlations between full-length 16S rRNA gene sequences and their sub-regions [e.g. 18]. Although one may endeavor to characterize and compare the composition of multiple communities, any cutoffs that are employed are at best empirical and hopefully have some biological meaning.

I have shown that alignment quality has a significant impact on downstream data analysis. Because the 16S rRNA gene sequence follows a well-determined secondary structure, it is possible to objectively state that one alignment is better than another. Furthermore, pairwise and multiple sequence alignments that ignore the secondary structure are unadvisable on theoretical grounds. Such methods are also unadvisable on technical grounds as the time and memory required to complete them typically scales in excess of the number of sequences squared; the time required to perform a profile-based alignment scales linearly with the number of sequences.

A significant factor in the analysis of DNA sequences is the calculation of pairwise distances. The rich literature developed for protein-coding sequences has generated the Jukes-Cantor, Kimura, Hasegawa-Kishino-Yano and other substitution models [reviewed in 19]. Yet these models ignore gapped positions, which I have shown to have a significant impact on downstream analyses. Substitution models for structural RNA molecules such as the 16S rRNA gene are not well developed or widely used [20], [21]. It is underappreciated that use of short sequence or filtering methods such as the Lane mask reduces the precision and information represented by a distance. For instance, if there are fewer than 200 bases being considered, then it is difficult to place much confidence in an OTU threshold of 0.03 (i.e. 6 differences) when one considers the potential impact of PCR, sequencing, and alignment artifacts. Furthermore, reducing the information content of a 1,500 bp molecule to a 200-bp sequence read will affect the confidence placed in the generation of phylogenetic trees and OTU assignments. Althoguh removing non-informative positions can be helpful for reconstructing broad phylogenies, the α- and β-diversity analyses described here are adversely affected by removing this fine level sequence diversity. These are clearly issues that warrant further attention.

This study has ramifications on how analyses are performed. Since it is clear that the 16S rRNA gene does not evolve uniformly across its length, it is critical that sequences fully overlap before they are compared. For example, consider an analysis that includes sequences from the V2 region and those from the V12 region. The V12 sequences will have higher pairwise distances amongst each other than compared to the V2 region because the V1 region is evolving at a faster rate. Thus, the comparison of short and long sequence reads will add artifacts into the analysis, which will overstate the richness within the community. Although not explored here, it is likely that similar problems will be encountered in analyses where a taxonomy hierarchy is used to assign sequences to bins. Thus it is critical that sequences are trimmed to start and end at the same sequence-based landmarks. Because pyrosequencing does not yield a uniform length sequence read, this introduces a conundrum of whether to favor fewer long reads or many short reads. Because it is impossible to compare pyrotags to the full-length sequences accurately, it seems appropriate to increase the power of other statistical analyses by sacrificing sequence length in favor of having more sequence reads.

Next generation sequence analysis of 16S rRNA genes offers the first opportunity to replicate analyses, develop more complex experimental designs, and to increase sampling depth and breadth. The results of this study encourage one to see pyrotags as markers within a metagenome and suggest a different way of considering microbial community analysis. Just as single nucleotide polymorphisms (SNPs) have been used as markers of disease in genome-wide association studies (GWAS), which may have no direct effect on a genes phenotype, pyrotags no doubt will serve as a useful analog to SNPs for the nascent field of metagenome-wide association studies (MWAS).

MATERIALS AND METHODS

Sequence collection

I obtained the SSURef 16S rRNA gene sequence database from the SILVA project (version 98;http://www.arb-silva.de) [22]. From this collection of sequences longer than 1,200 bp, I identified bacterial sequences that had an alignment quality score (ARB database field "align_quality_slv") of 100 and were not chloroplasts, mitochondria, or suspected of being chimeric. The collection was further screened to remove sequences that had more than 5 ambiguous base positions and did not start by *E. coli* position 28 or end after position 1491. Of the remaining sequences, 13,501 sequences were unique and shared between the SILVA [22], greengenes [23], and RDP sequence collections [14]. I then generated 12 datasets from the full-length sequences using the SILVA, greengenes, and RDP alignments by extracting sub-regions of various lengths (Table 1). These regions were selected because they had already been used in publications or are amenable to the available sequencing platforms. Lane masks were generated by mapping the original mask onto the *E. coli* reference sequence and then it was applied to each of the three reference alignments [24]. In addition to the three reference alignments I generated pairwise alignments between all pairs of sequences using the Needleman-Wunsch algorithm [25] and multiple sequence alignments using MUSCLE with two iterations (maxiters=2) and the diags option [15].

Distance calculation methods

I implemented three sequence-based methods for calculating pairwise distances and a kmer-based distance metric. The first sequence-based method ignored any site that contained a gap; this method is implemented in the commonly used DNADIST program from the PHYLIP package [26]. The second sequence-based method counted gaps as a fifth character so that any comparison between a gap and a base was penalized as a mismatch; comparisons between two gaps were ignored. This approach asserts that every gap represents a distinct mutation. The third sequence-based method calculated distances by only penalizing a string of gaps as one mismatch [2]. This approach asserts that a gap, of any length, represents a single mutation. Distances were not corrected for multiple substitutions to simplify analysis of the data. Furthermore, some distances were so large that when they were corrected, they yielded undefined values. Distances were calculated as implemented in the mothur software package with precision to 0.0001 [27]. Finally, kmer-based distances were calculated between pairs of unaligned sequences based on their 7-base kmer profiles [28].

Distance analysis

Pairwise distances were compared using a custom C++-coded program that calculated the linear regression coefficient using the origin as the intercept and the Pearson product-moment correlation coefficient [29]. Because several of the datasets did not demonstrate a linear correlation with the V19 region when the V19 pairwise distances were larger than 0.10, all regression and correlation coefficients are presented for V19 distances smaller than 0.10. Assessments of how much genetic diversity was either gained or lost represent the deviation from a slope of 1.0. The square of the Pearson product-moment correlation coefficient (i.e. R^2) was used to quantify the fraction of the variation that was accounted for by the linear regression.

α-diversity analysis

OTU- and phylogeny-based analyses were performed to assess the intra-sample biodiversity. Sequences were assigned to OTUs using the mothur implementation of the furthest-neighbor clustering algorithm [27]; although parallel analyses using the nearest and average neighbor algorithms yielded different α- and β-diversity values, the overall relationships observed with furthest neighbor algorithm were observed. The observed richness (i.e. the number of OTUs in a sample) of the dataset was calculated using every possible cutoff that the data could describe. Traditional neighbor-joining trees were generated using the clearcut software program and the distance matrices that were used in the OTU-based analyses [30]; however, the relaxed neighbor-joining algorithm was not used. The phylogenetic diversity of the data was calculated by summing the branch length for the entire tree [31]. Both analyses were replicated 50 times to assess the effects of randomization on α-diversity.

β-diversity analysis

The OTU assignments and neighbor-joining trees created to study α-diversity were used to evaluate the effects of each variable on the ability to calculate β-diversity. Towards this end, I segregated the sequences to create two mock communities that shared 80% of their membership but had different structures. To create the mock communities full-length SILVA-aligned sequences were first assigned to OTUs using a furthest neighbor clustering of one gap distances with a cutoff of 0.05. Second, OTUs were randomly ordered. Third, 10% of the OTUs were assigned exclusively to the first community, another 10% were assigned exclusively to the second community, and the remaining OTUs were shared. For half of the shared OTUs, the probability of a sequence being from the first community was 0.375 and for the other half of the shared OTUs, the probability was 0.625. These probabilities were selected to simulate sampling two communities that had a Jaccard similarity index of 0.80 and Morisita-Horn Index value of 0.60. This process was repeated to create 100 simulated communities. Because the mock communities were not exhaustively sampled, it was unlikely that the measures would actually equal 0.80 and 0.60 for the Jaccard and Morisita-Horn indices. All β-diversity calculations were made using the mothur software package [27]. The same 100 partitions were used to analyze all distance calculation methods, alignments, regions, and β-diversity measures. I analyzed the effects of region and the alignment or distance calculation methods using a two-way analysis of variance. Each factor was highly significant ($p \ll 0.001$) and so I used the Tukey's honestly significant difference test for pairwise comparisons. Only those differences, which were non-significant ($p>0.05$) are indicated in figures. All test were performed within an OTU cutoff or UniFrac method.

AUTHOR CONTRIBUTIONS

Conceived and designed the experiments: PDS. Performed the experiments: PDS. Analyzed the data: PDS. Contributed reagents/materials/analysis tools: PDS. Wrote the paper: PDS.

REFERENCES

1. Margulies M, Egholm M, Altman WE, Attiya S, Bader JS, et al. (2005) Genome sequencing in microfabricated high-density picolitre reactors. Nature 437: 376–380.
2. Sogin ML, Morrison HG, Huber JA, Welch DM, Huse SM, et al. (2006) Microbial diversity in the deep sea and the underexplored "rare biosphere". Proc Natl Acad Sci U S A 103: 12115–12120.
3. Costello EK, Lauber CL, Hamady M, Fierer N, Gordon JI, et al. (2009) Bacterial community variation in human body habitats across space and time. Science 326: 1694–1697.
4. Roesch LFW, Fulthorpe RR, Riva A, Casella G, Hadwin AKM, et al. (2007) Pyrosequencing enumerates and contrasts soil microbial diversity. ISME J 1: 283–290.
5. Huber JA, Mark Welch DB, Morrison HG, Huse SM, Neal PR, et al. (2007) Microbial population structures in the deep marine biosphere. Science 318: 97–100.
6. Liu Z, Lozupone C, Hamady M, Bushman FD, Knight R (2007) Short pyrosequencing reads suffice for accurate microbial community analysis. Nucleic Acids Res 35: e120.
7. Huse SM, Dethlefsen L, Huber JA, Welch DM, Relman DA, et al. (2008) Exploring microbial diversity and taxonomy using SSU rRNA hypervariable tag sequencing. PLoS Genet 4: e1000255.
8. Wang Q, Garrity GM, Tiedje JM, Cole JR (2007) Naive Bayesian classifier for rapid assignment of rRNA sequences into the new bacterial taxonomy. Appl Environ Microbiol 73: 5261–5267.
9. Liu Z, DeSantis TZ, Andersen GL, Knight R (2008) Accurate taxonomy assignments from 16S rRNA sequences produced by highly parallel pyrosequencers. Nucleic Acids Res 36: e120.
10. Youssef N, Sheik CS, Krumholz LR, Najar FZ, Roe BA, et al. (2009) Comparison of species richness estimates obtained using nearly complete fragments and simulated pyrosequencing-generated fragments in 16S rRNA gene-based environmental surveys. Appl Environ Microbiol 75: 5227–5236.

11. Schloss PD (2009) A high-throughput DNA sequence aligner for microbial ecology studies. PLoS ONE 4: e8230.
12. DeSantis TZ Jr, Hugenholtz P, Keller K, Brodie EL, Larsen N, et al. (2006) NAST: a multiple sequence alignment server for comparative analysis of 16S rRNA genes. Nucleic Acids Res 34: W394–399.
13. Caporaso JG, Bittinger K, Bushman FD, DeSantis TZ, Andersen GL, et al. (2010) PyNAST: a flexible tool for aligning sequences to a template alignment. Bioinformatics 26: 266–267.
14. Cole JR, Wang Q, Cardenas E, Fish J, Chai B, et al. (2009) The Ribosomal Database Project: improved alignments and new tools for rRNA analysis. Nucleic Acids Res 37: D141–145.
15. Edgar RC (2004) MUSCLE: a multiple sequence alignment method with reduced time and space complexity. BMC Bioinformatics 5: 113.
16. Sun Y, Cai Y, Liu L, Yu F, Farrell ML, et al. (2009) ESPRIT: estimating species richness using large collections of 16S rRNA pyrosequences. Nucleic Acids Research e76–e76.
17. Schloss PD, Handelsman J (2006) Toward a census of bacteria in soil. PLoS Comp Biol 2: e92.
18. Welch RA, Burland V, Plunkett G 3rd, Redford P, Roesch P, et al. (2002) Extensive mosaic structure revealed by the complete genome sequence of uropathogenic*Escherichia coli*. Proc Natl Acad Sci U S A 99: 17020–17024.
19. Felsenstein J (2004) Inferring phylogenies. Sunderland, , Mass.: Sinauer Associates..
20. Smith AD, Lui TW, Tillier ER (2004) Empirical models for substitution in ribosomal RNA. Mol Biol Evol 21: 419–427.
21. Tillier ER, Collins RA (1998) High apparent rate of simultaneous compensatory base-pair substitutions in ribosomal RNA. Genetics 148: 1993–2002.
22. Pruesse E, Quast C, Knittel K, Fuchs BM, Ludwig W, et al. (2007) SILVA: a comprehensive online resource for quality checked and aligned ribosomal RNA sequence data compatible with ARB. Nucleic Acids Res 35: 7188–7196.
23. DeSantis TZ, Hugenholtz P, Larsen N, Rojas M, Brodie EL, et al. (2006) Greengenes, a chimera-checked 16S rRNA gene database and workbench compatible with ARB. Appl Environ Microbiol 72: 5069–5072.
24. Lane DJ (1991) 16S/23S rRNA sequencing. In: Stackebrandt E, Goodfellow M, editors. Nucleic Acid Techniques in Bacterial Systematics. New York: Wiley. pp. 115–175.
25. Needleman SB, Wunsch CD (1970) A general method applicable to the search for similarities in the amino acid sequence of two proteins. J Mol Biol 48: 443–453.
26. Felsenstein J (1989) PHYLIP – Phylogeny Inference Package. Cladistics 5: 164–166.
27. Schloss PD, Westcott SL, Ryabin T, Hall JR, Hartmann M, et al. (2009) Introducing mothur: Open-source, platform-independent, community-supported software for describing and comparing microbial communities. Appl Environ Microbiol 75: 7537–7541.
28. Edgar RC (2004) MUSCLE: multiple sequence alignment with high accuracy and high throughput. Nucleic Acids Res 32: 1792–1797.
29. Sokal RR, Rohlf FJ (1995) Biometry: the principles and practice of statistics in biological research. New York: Freeman..
30. Sheneman L, Evans J, Foster JA (2006) Clearcut: a fast implementation of relaxed neighbor joining. Bioinformatics 22: 2823–2824.
31. Faith DP, Lozupone CA, Nipperess D, Knight R (2009) The cladistic basis for the phylogenetic diversity (PD) measure links evolutionary features to environmental gradients and supports broad applications of microbial ecology's "phylogenetic beta diversity" framework. Int J Mol Sci 10: 4723–4741.
32. Eckburg PB, Bik EM, Bernstein CN, Purdom E, Dethlefsen L, et al. (2005) Diversity of the human intestinal microbial flora. Science 308: 1635–1638.
33. Dethlefsen L, Huse S, Sogin ML, Relman DA (2008) The pervasive effects of an antibiotic on the human gut microbiota, as revealed by deep 16S rRNA sequencing. PLoS Biol 6: e280.

CHAPTER 14 METASIM: A SEQUENCING SIMULATOR FOR GENOMICS AND METAGENOMICS

Daniel C. Richter, Felix Ott, Alexander F. Auch, Ramona Schmid, Daniel H. Huson

INTRODUCTION

Metagenomics is based on the isolation and characterization of DNA from environmental samples without the need for prior cultivation of microorganisms. In contrast to single genome studies, analyses are applied to entire communities of microbes instead of only few isolated organisms. It has already led to exciting insights into the ecology of different habitats such as ocean [1], soil [2], acid mine [3], human and mouse gut [4], [5] and even into ancient DNA [6].

The research field of Metagenomics is spurred by the recent development and improvement of next-generation sequencing technologies like Roche's 454 pyrosequencing [7]. Although these high through-put technologies promise faster and relatively inexpensive generation of reads, Sanger sequencing still has been used in environmental genome projects [5] to avoid the drawbacks of shorter read lengths.

In general, studies show that algorithms developed for single-genome assembly are only suitable for environmental sequences under special conditions, for example in low complexity populations [2], [8]. In particular, it is very difficult to assemble reads from highly diverse ecologic systems [9]. The problem is that the arrangement of reads into contigs fails or is misleading because contigs are put together from reads from many different genomes.

Currently, the primary goals of metagenomic studies are the investigation of the phylogenetic composition of the sample (taxonomical binning, "Who is out there"), the quantitive analysis ("How many are there?") and the prediction of genes and their functions (functional binning, "What are they doing").

Citation: Richter DC, Ott F, Auch AF, Schmid R, Huson DH (2008) MetaSim—A Sequencing Simulator for Genomics and Metagenomics. PLoS ONE 3(10): e3373. doi:10.1371/journal.pone.0003373.

Since the amount of comparable environmental data is rapidly growing, comparative studies of multiple metagenomic data sets are of great interest as wells. As of September 2008, 44 metagenome studies have already been conducted whereas 86 projects still are on-going [10].

Common strategies for taxonomical binning are for example: (1) detecting phylogenetic markers like rRNA, RecA, heat shock protein (HSP70) and elongation factors (EF-Tu, EF-G)[11], (2) comparing reads against a reference database such as NCBI-nr [12] and then analyzing the matches to place the reads in the NCBI taxonomy [13] and (3) measuring the oligonucleotide frequency caused by codon usage or restriction-site frequency [14]–[18].

When it comes to functional binning, sequences are compared to known protein functions, families and pathways provided by several databases, for example COG, KEGG, PFAM, SEED, STRING and TIGRFAM [19]–[24]. A de novo search for (unknown) functional units is only feasible if either long reads or contigs are available for the detection of open reading frames.

Another challenge in metagenomic studies is the development of robust statistical techniques[25]. Particularly with regard to comparative metagenomics dealing with highly variable data, these techniques are considered as indispensable for a well-founded analysis.

Despite the enormous amount of sequence data that was generated and analyzed in the past few years, the number of publicly available software specialized in metagenomic data analysis is surprisingly low. Hence, many studies still make use of classic methods, software or web services that originally were not intended for metagenomic data analysis and have to be adapted or pipelined to produce the desired results [8].

Thus, there is a great demand for specialized metagenomic software supporting the analysis process. Because of the complexity of metagenomic data, it is crucial to benchmark new and existing software with standardized test cases using simulated and verifiable data. A first study [9] provides three data sets with varying complexity by selecting original sequence reads from 113 isolated genomes. In their paper, the authors anticipate that these data sets will be used as standard test cases for software testing.

Some other publications already applied the software ReadSim (pre-version of MetaSim,unpublished) to generate simulated read data sets for testing their software [18], [26].

Description of MetaSim

MetaSim takes as input a set of known genome sequences and an abundance profile. This profile determines which genome sequences are selected for the simulation and the relative abundance of each genome sequence in the dataset.

MetaSim integrates an "induced tree view" of the NCBI taxonomy [27] that can be used to interactively select taxa and inner nodes of the taxonomy to configure their relative abundances. Additionally, the user is able to simulate an "evolved" population of a single genome sequence, using a population simulator. This feature is aimed at simulating the common real world situation that many different, but closely related strains of a lineage coexist in the same habitat.

Finally, for the construction of a realistic read data set, MetaSim includes a versatile read sequencing simulator. The user is able to choose from different (adaptable) error models of current sequencing technologies (e.g. Sanger [28], [29], Roche's 454 [7] and Illumina (former Solexa) [30]).

MetaSim allows one to construct verifiable read data sets, and additionally, metagenomes variable in size, taxonomical composition and abundance to reflect the diverse and complex output of real metagenomic studies. The resulting data sets can be used to plan and design metagenomic studies and for evaluation and improvement of metagenomics software tools, statistical methods or assembly algorithms.

Availability

MetaSim is written in Java and can be run with a graphical user interface or in command line mode. Installers for Linux/Unix, MacOS X and Windows are freely available from our website at: http://www-ab.informatik.uni-tuebingen.de/software/metasim.

METHODS

MetaSim's processing pipeline consists of several phases:

- Selection of source genome sequences from the internal database
- Configuration of the species abundance profile by setting the relative copy number of the genome sequences
- Sampling sequencing of fragments according to the species abundance profiles
- Application of technology-specific error models to the fragments to create sequencing reads

Configuration of Species Abundance Profiles

At the beginning, whole genome sequences available from public database can be stored locally as source sequences in an integrated database. The user specifies the relative abundance of each genome sequence in a text-based profile file. An interesting feature of MetaSim is the possibility of assigning frequency values not only at the species level but also at higher taxonomical levels. For example, if the genus Escherichia is assigned a certain amount of genome copies, this amount is split and applied uniformly to all descendant species whose sequences are available from the internal database.

To facilitate this data composition process in GUI mode, MetaSim provides an interactivetaxonomy editor that visualizes the induced NCBI taxonomy, i.e. the genome sequences listed in the profile file are displayed as nodes in a rooted tree (Figure 1). Node sizes reflect the relative number of genome copies for each given taxon.

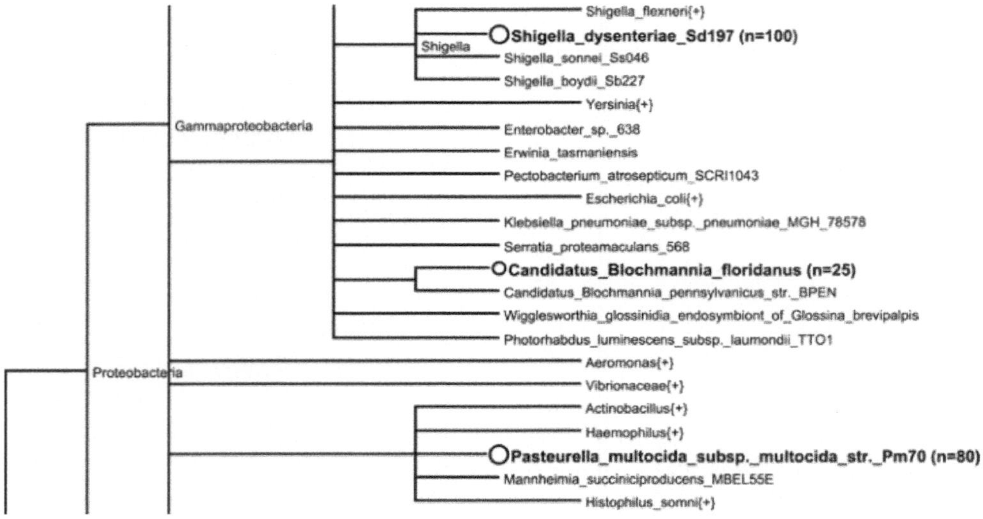

Figure 1: Taxonomy Editor. A clipping of the taxonomy editor view is shown. Three taxa are assigned an abundance value (number in parenthesis). These settings can be either determined in a text-based abundance profile file or directly in the taxonomy editor by right-clicking on a node.

Population sampling

The current genome databases reflect only a small part of earth's still unexplored microbial diversity. Thus, a simulated metagenome only based on known genome sequences does not adequately reflect the complexity of realistic data sets.

MetaSim therefore includes a population sampler that optionally generates a set of evolved (mutated) offsprings derived from single source genomes, using a given evolutionary tree. This tree describes how the offspring sequences descend from the source sequence. By default, a random pyholgenetic tree is generated under the Yule-Harding model [31], [32], but alternatively, user-defined trees can also be loaded. As a simple model of DNA evolution, the Jukes-Cantor formula [33] is applied to estimate a probability of change for each base pair, with a customizable transition rate α (0.001 by default) and time t based on the edge weights. MetaSim then generates the designated number of evolved genomes and then adds them to the internal genome database. As an example, a fragment recruitment plot (according to [1]) shows 10000 sampled Sanger reads of 100 evolved offspring sequences (α=0.004) mapped to the source genome (Escherichia coli K-12 substr. MG1655) using blastn (Figure 2). Read sequences sampled directly from the source genome show a significantly higher identity compared to the mutated sequencing reads.

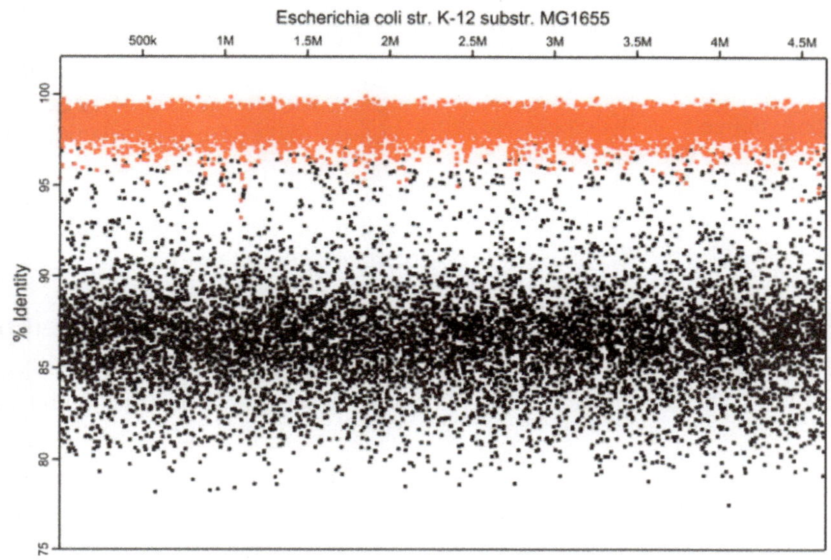

Figure 2: Fragment Recruitment Plot. Black dots represent 10,000 sequencing reads (Sanger technology, ≈800 bp) drawn from 100 evolved offsprings (α=0.004) of the source genome Escherichia coli K-12 substr. MG1655. Their sequence identity is lower compared to the mapped reads sampled directly form the source genome (red dots).

Read Sampling

MetaSim simulates both Sanger sequencing and Roche's 454 (sequencing-by-synthesis) approach. Additionally, it provides a flexible, empirical error model usable to simulate Illumina's ultra-short reads.

For the simulation of read sequences, statistical approaches are adopted to simulate the distribution of read lengths, its frequency rate and the use of error models depending on the chosen sequencing technology.

To be able to model mate-pairs as well, MetaSim first extracts large fragments called clones from the set of genomes with normally or uniformly distributed lengths. For example, clones with a length of 1000 bp and a standard deviation of 100 bp are modelled with a normal distribution N(1000,100) (Figure 3). The overall number of clones is determined by the number of reads or mate-pairs the user desires to generate.

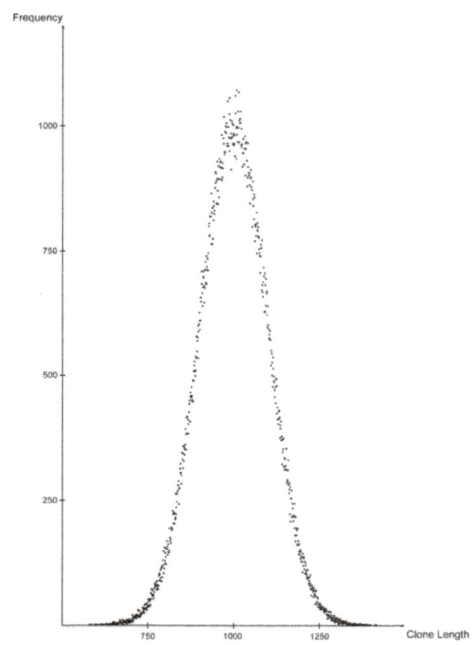

Figure 3: Frequency distribution of clone lengths. As an example, 250,000 clones with mean length 1000 bp and standard deviation of 100 bp were modelled with a normal distribution.

If only one source genome is present in the given profile, the clones are randomly extracted from this single sequence. In contrast, in a typical metagenome simulation, the clones have to be sampled from many genomes of varying length, copy number (e.g. to model the abundance of plasmids versus the organsim genomes) and abundancies.

So, each genome sequence s is assigned a weight

$$w_s = l_s \times c_s \times a_s \quad (1)$$

where l_s is the length, c_s is the copy number and a_s is the specified relative abundance of the genome sequence s as determined in the profile.

For each length of the clone length distribution, the weights of all sequences are summed up to receive the summarized weight w_{sum} that is used to compute a sequence probability $p_s=w_s/w_{sum}$. Considering the overall lengths distribution, a frequency value for each source sequence is then obtained.

After the clone sampling, the ends of the clones are the basis for the subsequent sampling of the reads or mate-pairs, respectively. Again, read lengths can be either uniformly or normally distributed. Finally, read sequences are processed and modified by applying the selected error model.

Simulation of Sanger sequencing

A widely-used approach to sequencing large DNA molecules is Sanger sequencing, using a shotgun approach that involves cloning small pieces (or inserts) of DNA and then determining their sequence using fluorescent dideoxynucleotides for termination and capillary electrophoresis.

To simulate Sanger sequencing, we closely followed the implementation of celsim reported in[34]. Each read is subjected to a linearly increasing error rate. We model fixed percentages of deletion errors, insertion errors and substitutions. Further, the simulator is capable of modeling mate-pairs and one can specify the length distribution of inserts.

Simulation of sequencing-by-synthesis

In pyrosequencing, the intensity of emitted light is used to estimate the length of homopolymers, i.e. runs of identical nucleotides in a sequence. During sequencing, the four DNA composing nucleotides are periodically flowed over the inserts to be sequenced. Within each flow, the intensity of the signal emitted (which is linear up to 8 bp) reflects the number of nucleotides incorporated. Thus, the addition of a single base or even homopolymer stretches of multiple bases in a single flow can be detected.. For chemical and technical reasons, this signal is subject to fluctuations that lead to sequencing errors. In [7], an error rate of about 3% is reported.

Let r denote the length of a given homopolymer. We model the emitted light intensity using a normal distribution $N(\mu, \sigma)$, with mean μ=r and standard deviation $\sigma = k \cdot \sqrt{r}$, where k is a fixed proportionality factor. Following [7], by default we use k=0.15. Although basic statistics implies that the standard deviation should grow with the square root of r, in [7] the standard deviation of the light intensity emitted during 454-sequencing is reported to be σ=k·r. Both variants of the calculation are implemented in our software.

A negative flow is a flow of nucleotides in which the sequence to synthesize is not elongated. Light intensities of negative flows follow a lognormal distribution, with mean μ=0.23, and standard deviation σ=0.15, see [7]. A random variable X is said to be lognormally distributed, if the random variable ln(X) is normally distributed.

We simulate base-calling intensities of negative flows and model the misinterpretation of null-mers as homopolymers of length=1 (insertion). Our algorithm takes the order of the sequencing flows into account. Since the nucleotides are cyclically flowed in the order T,A,C,G, after a given base only two specific negative flows in a specific order are allowed.

During base-calling the algorithm looks at each homopolymer, generates a $N(\mu, \sigma)$ distributed random variable and, according to this variable, decides which length to set for the observed homopolymer.

The signal intensity space is separated into disjoint intervals by probability density functions of all homopolymer signals. The intersections of the density functions $f_{N(r_1,\sigma)}(x)$, and $f_{N(r_2,\sigma)}(x)$ of the normal distributions for different homopolymer lengths r_1 and r_2 are calculated and stored. In each interval one probability density function is maximized. These values are used to decide which homopolymer length is called given a certain signal intensity.

Simulation of reads with empirical models

As an additional feature, MetaSim includes an empirical error model that allows the incorporation of user-defined error statistics.

The probability of an occurrence of a sequencing error often depends on the position of the erroneous base and its surrounding bases. The program GenFrag [35] originally developed an error model which incorporates different error types (deletion, insertion, substitution) at certain positions with empirical error probabilities.

MetaSim adopts this general approach. The error model is based on mappings (error curves) that assign error rates to specific base positions. Each mapping has three parameters (the last two are optional):

- type of error,
- base at the position where the error occurs and
- base preceding the position where the error occurs.

In this way, 48 independent mappings can be individually determined. In addition, in case of a substitution error the user can specify the probability of integrating a particular base depending on the type of the base at the error position and the preceding base.

Using empirical data, MetaSim provides an error model for the short reads of the Illumina technology.

RESULTS

MetaSim can be controlled either by a graphical user interface (Figure 4) or by command line. The command line mode provides access to most of the functions needed for automatic simulation runs. A simulation run generating 400,000 454 reads of length ≈250 bp (a total of 100 Mbp) takes less than 80 seconds on a 2,13 GHz single processor computer.

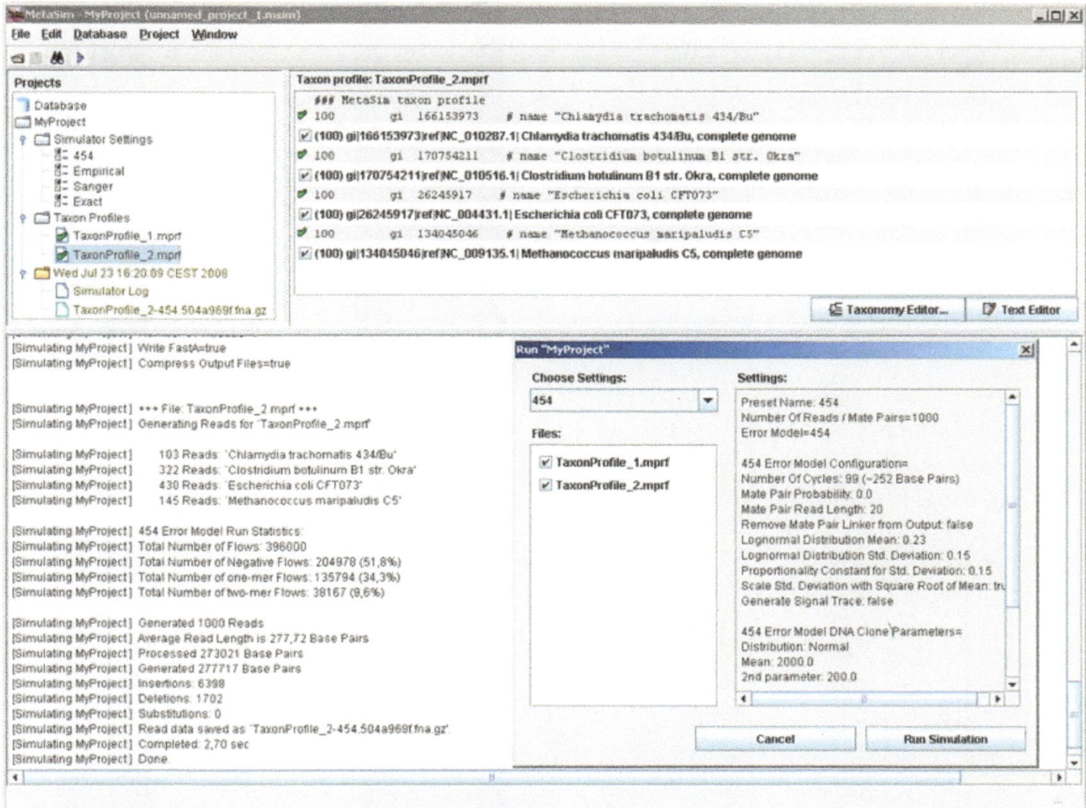

Figure 4: The graphical user interface of MetaSim is divided into three panels: a project tree on the left containing all simulation settings and taxon profiles, an overview and edit panel on the right and a message panel at the bottom. Additionally, a configuration window is shown. http://dx.doi.org/10.1371/journal.pone.0003373.g004

To show the utility of this simulation software, for example, in the benchmarking of new software, we generated 9 data sets using a range of parameters and used them to test how well the MEGAN software succeeded in successfully binning sequences based on taxonomic classification by homology.

Simulation study

MetaSim was used to generate three simulated data sets of different species compositions. These data sets were designed and named simLC, simMC and simHC representing low, medium and high complexity communities, respectively (in correspondence to Mavromatis et al. [9]). For each data set, three runs were conducted with different sequencing error models and read lengths. Also, each resulting read data set comprises of approximately 15 Mbp, so the number of reads differ accordingly: for the 454 sequencing technology with read length ≈100 bp (≈250 bp), 150,000 (60,000) read sequences were generated. The third data set consists of Sanger reads (≈800 bp) and consisted of 18,750 reads. For a list with all simulation parameters of MetaSim see Supplementary Information.

Table 1–3 show the relative abundancies and the resulting number of sampled reads for each selected taxon. The simLC data set (Table 1) consists of two taxonomically distant microbes whose relative abundance values differ significantly. The simMC data set (Table 2) is composed of nine microbial species (all from the phylum γ-Proteobacteria) with two dominant populations. The simHC data set (Table 3) consists of 11 diverse microbial species covering several phyla from the superkingdom bacteria. All species in the simHC data set are sampled with the same relative abundance.

Table 1: Species abundance and percentage of sampled reads of the simLC dataset

Abdce	Species	Mbp	454-100[a]	454-250[b]	S-800[c]
90	Methanoculleus marisnigri JR1	2.5	82,70	82,61	82,71
10	Escherichia coli str. K-12 substr. MG1655	4.6	17,30	17,39	17,29

[a]454 technology, 150000 reads (length: 100 bp).
[b]454 technology, 60000 reads, (length: 250 bp).
[c]Sanger technology, 18750 reads, (length: 800 bp).
doi:10.1371/journal.pone.0003373.t001

Table 2: Species abundance and percentage of sampled reads of the simMC dataset

Abdce	Species	Mbp[a]	454-100[b]	454-250[c]	S-800[d]
100	Pseudomonas fluorescens PfO-1	6.4	38,42	38,39	38,13
100	Shigella dysenteriae Sd197	4.6	27,07	27,47	27,25
80	Pasteurella multocida subsp. multocida str. Pm70	2.3	10,82	10,87	10,81
50	Buchnera aphidicola str. APS	6.6	1,97	1,94	1,78
50	Francisella tularensis subsp. tularensis Schu 4	1.9	5,69	5,60	5,56
25	Alcanivorax borkumensis SK2	3.1	4,68	4,57	4,60
25	Candidatus Blochmannia floridanus	7.1	1,03	1,08	1,21
25	Pseudomonas entomophila L48	5.9	8,89	8,65	9,26
5	Escherichia coli str. K-12 substr. MG1655	4.6	1,43	1,43	1,40

[a]The length of plasmids is considered as well.
[b]454 technology, 150,000 reads (length: 100 bp).
[c]454 technology, 60,000 reads, (length: 250 bp).
[d]Sanger technology, 18,750 reads, (length: 800 bp).
doi:10.1371/journal.pone.0003373.t002

Table 3: Species abundance and percentage of sampled reads of the simHC dataset

Abdce	Species	Mbp[a]	454-100[b]	454-250[c]	S-800[d]
100	Agrobacterium tumefaciens str. C58	5.7	11,7	11,7	11,3
100	Anabaena variabilis ATCC 29413	7.1	14,7	14,9	14,6
100	Archaeoglobus fulgidus DSM 4304	2.2	4,54	4,41	4,55
100	Bdellovibrio bacteriovorus HD100	3.8	7,84	7,79	7,83
100	Campylobacter jejuni subsp. jejuni 81-176	1.7	3,52	3,6	3,57
100	Clostridium acetobutylicum ATCC 824	4.1	8,6	8,56	8,49
100	Lactococcus lactis subsp. cremoris SK11	2.6	5,38	5,32	5,54
100	Nitrosomonas europaea ATCC 19718	2.8	5,81	5,66	5,59
100	Pseudomonas aeruginosa PA7	6.6	13,6	13,6	14
100	Streptomyces coelicolor A3(2)	9.1	18,7	18,9	18,9
100	Sulfolobus tokodaii str. 7	2.7	5,64	5,59	5,6

[a]The length of plasmids is considered as well.
[b]454 technology, 150,000 reads (length: 100 bp).
[c]454 technology, 60,000 reads, (length: 250 bp).
[d]Sanger technology, 18,750 reads, (length: 800 bp).
doi:10.1371/journal.pone.0003373.t003

Obviously, the amount and ratio of sampled reads in each simulation result reflects the configured abundances and genome sizes. For example, in the simHC data set bothCampylobacter jejuni subsp. jejuni 81-176 and Streptomyces coelicolor A3(2) are assigned the same relative abundance of 100. The percentage of sampled reads differs about 15% which can be explained by the difference in genome size (length of contained plasmids are considered as well). Considering all simulations runs, the ratio of sampled reads is almost equal for each species. For example in Table 1, 17,30%, 17,39% and 17,29% reads were sampled for Escherichia coli str. K12 substr. MG1655 when simulating 454 sequencing technology with read length 100 bp and 250 bp and Sanger sequencing with read length of 800 bp, respectively.

Taxonomical classification using MEGAN

Following the generation of the nine data sets, a taxonomical assignment of the reads with the MEGAN software was conducted to test its binning functionality. Therefore, all generated read sets were blasted against the NCBI-nr database (downloaded March 2008). MEGAN then assigns these reads to taxa in such a way that the taxonomical level of the assigned taxon reflects the level of conservation of the sampled sequence (Parameter settings of MEGAN: minscore: 0.0, toppercent: 1.0, minsupport: 2, winscore: 0.0).

The result for each simulation run is shown in Figure 5A–C. (For a list of all results see Supplementary Information.) Summarizing all simulations, one observation can be made: The amount of total assignments of reads to taxa correlates with the read length. The longer the read sequence, the more assignments for a taxon are found. This is also true for the correct assignments (true positives) (Supplementary Information).

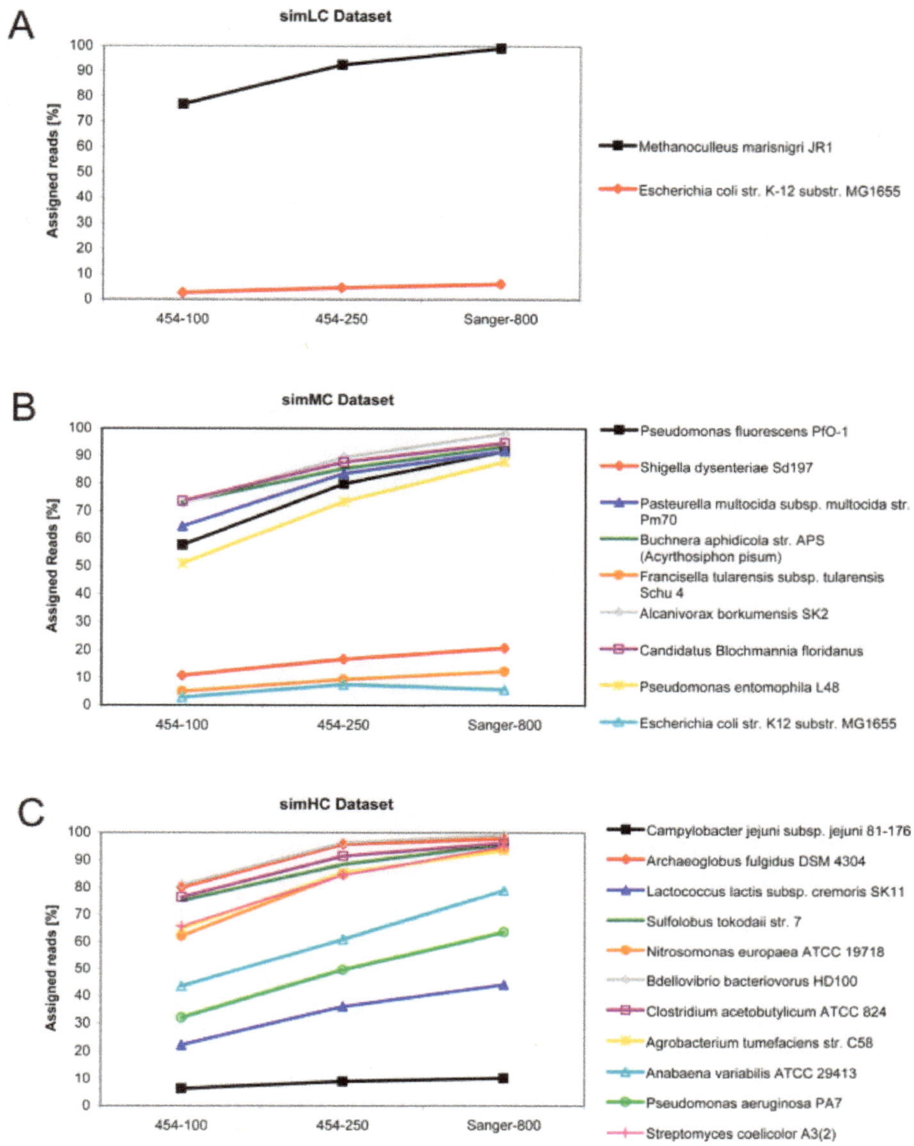

Figure 5: Assignment curves of reads taxonomically classified by MEGAN. The precentage values refer to the number of sampled reads generated for each organism. (A) The simLC dataset consists of only two organisms. The number of assigned reads to M. marisnigri JR1 almost equals the number of its sampled reads whereas E. coli str. K-12 substr. MG1655 has only few assignmentss. (B) In the simMC dataset, the number of assigned reads increases significantly with longer read lengths (except for Shigella dysenteriae Sd197, Francisella tularensis subsp. tularensis Schu 4 and E. coli str. K12 substr. MG1655). (C) In the simHC dataset, the fraction of assigned reads to Campylobacter jejuni subsp. jejuni 81-176, Lactococcus lactis subsp. cremoris SK11 and Pseudomonas aeruginosa PA7 is rather low compared to the other organisms.

Accordingly, the fraction of reads that did not match anything in the NCBI-nr database (% no hits) (Table 4) decreases significantly in case of longer read sequences. These findings were expected since longer read sequences generally give rise to more significant high scoring pairs using BLAST.

Table 4: Percentage of assigned, unassigned and "No Hits" reads for all simulation runs

	Total Reads	%Assigned Reads	%Unassigned Reads	%No Hits[a]
simLC-454-100	150000	83,14	0,46	16,40
simLC-454-250	60000	98,58	0,85	0,57
simLC-S-800	18750	99,45	0,55	0,00
simMC-454-100	150000	81,71	0,52	17,76
simMC-454-250	60000	98,08	1,02	0,91
simMC-S-800	18750	99,28	0,71	0,01
simHC-454-100	150000	81,68	0,51	17,81
simHC-454-250	60000	97,55	0,93	1,52
simHC-S-800	18750	99,08	0,87	0,05

[a]Reads that did not match anything in the NCBI-nr database.
doi:10.1371/journal.pone.0003373.t004

Independent of the read length, MEGAN is able to assign reads with a high true positive rate. Except for E. coli str. K12 substr. MG1655 and Shigella dysenteriae Sd197 in simMC, virtually all reads (98–100%) are correctly classified.

It is important to mention that the composition of the simulated data sets was explicitly intended as an example for a benchmarking study, not for modelling real ecological environments. Thus, simLC consists of only two microbial species that derive from two distinct superkingdoms of the taxonomy (Bacteria and Archae) (Table 1 and Figure 5A). The majority of the sampled reads of E.coli str. K12 substr. MG1655 were assigned to other taxa and clades in the subtree of Bacteria (Figure 6) leading to false positive hits in MEGAN. However, many closely related strains in the subtree of genus Escherichia were hit supporting the evidence that a high number of genetic functions are shared among them.

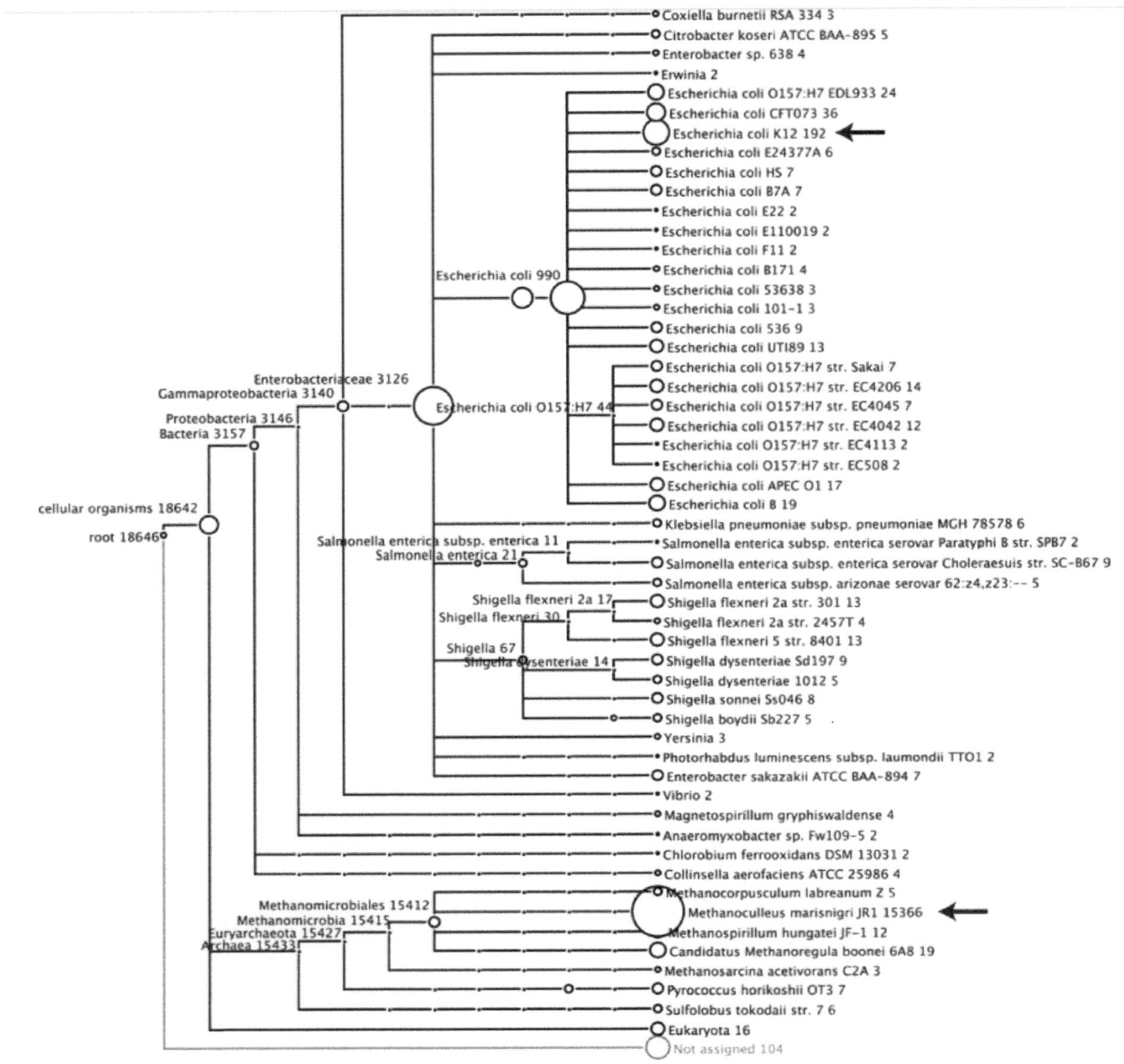

Figure 6: MEGAN visualization of the simLC data set (Sanger technology, read length ≈800 bp). Two arrows point out the two source genomes of the simulation run. The number of assigned reads to E.coli K12 (192) is rather small compared to the number of sampled reads from the genome of E. coli str. K12 substr. MG1655 (192 assigned vs. 3214 sampled reads). Many reads have BLAST hits in multiple strains and clades, so that MEGAN assigns them to an high-order level in the tree e.g. node Bacteria (3157 reads). M. marisnigri JR1 has only few related strains. In this case, the assignment of reads is more specific (15,366 assigned vs. 15,509 sampled reads).

A similar problem can be observed in the simMC data set (Figure 5B). To complicate the distinct classification of the reads, all species were taken from the class γ-Proteobacteria. In this example, reads of S. dysenteriae Sd197, Francisella tularensis subsp. tularensis Schu 4and E. coli str. K12 substr. MG1655 could not be completely assigned to the correct taxa. However, the high assignment accuracy (correct hits, %TP (Supplementary Information)) means that practically all assigned reads were correctly classified by MEGAN.

Data set simHC contains sequences of many phyla assigned with the same relative abundance of 100 (Table 3). This implies that the amount of sampled reads depends mainly on the length of the source genome sequences. At first sight, the fraction of assigned reads for the three genomes Campylobacter jejuni subsp. jejuni 81-176, Lactococcus lactis subsp. cremoris SK11 and Pseudomonas aeruginosa PA7 is quite low (10.15%, 43.93% and 63.62% for 800 bp Sanger reads, respectively)

(Figure 5C). Though, in the taxonomic tree at genus level (Campylobacter, Lactococcus and Pseudomonas), the number of assigned reads virtually equals the number of sampled reads (Figure 7). This is due to the fact that closely related organisms from the same genus share a lot of genes. Again, the number of correct assignments is almost optimal.

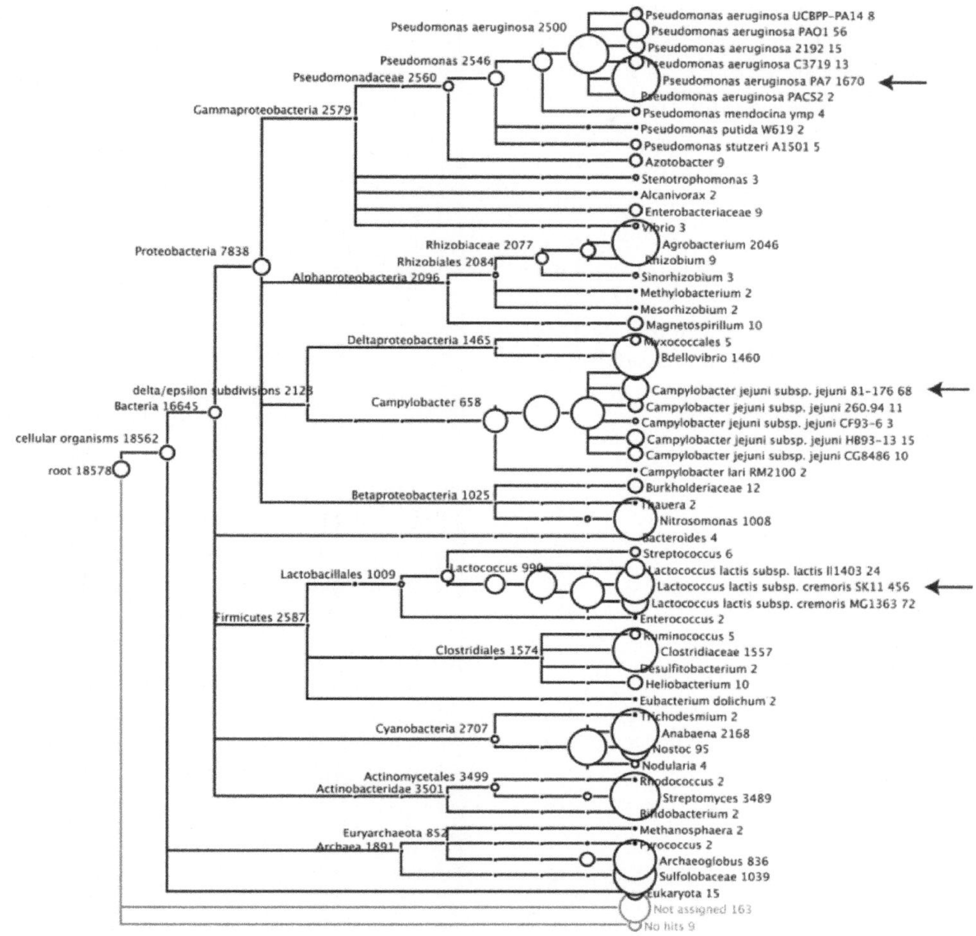

Figure 7: MEGAN visualization of the simHC data set (Sanger technology, read length ≈800 bp). Arrows point out three of the 11 source genomes of the simulation run that show only few assigned reads at species level compared to the number of originally sampled reads. Due to the fact that C. jejuni subsp. jejuni 81-176, L. lactis subsp. cremoris SK11 and P. aeruginosa PA7 share genes with many closely related strains, most of the sampled reads were assigned by MEGAN to an high-order level in the tree (e.g. genus).

Summary of MEGAN results

The analysis of the nine artifical data sets help to reveal the pros and cons of taxonomical binning based on homology as done by MEGAN. All simulation runs indicated that MEGAN generally is capable of binning the majority of all generated reads correctly. Additionally, the number of unassigned reads i.e. reads that actually hit sequences in the database but could not be assigned due to MEGAN's parameter settings, is very low. However, reads having many homologous sequences in the database due to high conservation among microbial families and lineages, lead to a rather "diffuse", but still correct read assignment: due to gene sharing, MEGAN assigns these reads to high-order taxa closer to the root (e.g. at genus level), thus avoiding probable false-positive assignments (MEGAN uses an LCA-based algorithm[13].). This means that MEGAN classifies those reads rather "generally" but the assignment still is adequate for interpreting metagenomic data in sufficiency.

Another observation is that the amount of assigned reads correlates strongly with the read length. Obviously, with reads produced by Sanger sequencing (800 bp), taxonomical binning becomes easier compared to short reads (100–250 bp). This oberservation confirms the findings in [13]. Moreover, the positive effect of better assignment and classification of long reads comes with higher costs and workload in the sequencing phase.

The expense for sequencing with Sanger technology is about $500/Mbp (800 bp reads), whereas e.g. pyrosequencing with Roche's 454 technology is only about $100/Mbp (250 bp reads) yielding many more base pairs per run at the same time. On top of the positive cost factor, compared to the Sanger technology, next-generation sequencers do not suffer from cloning bias.

DISCUSSION

The appearance of next-generation sequencers on the market has boosted the number and scope of (meta-)genomic sequencing projects. A lot of data can be generated in less time demanding fast and precise analysis algorithms and software. However, especially in the field of metagenomics, the problem of producing individual, simulated test cases for benchmarking is open.

We try to fill this gap with MetaSim, a flexible tool for producing simulated read data sets, useful for designing metagenome projects and for testing and comparing metagenomic or assembly software. A lot of parameters can be adapted to generate user defined sequence sets that can serve as verified example data. Currently, the Sanger and Roche's 454 sequencing error model can be selected, as well as the Illumina error model which is based on empirical data. The empirical error model can easily be configured and adapted to other sequencing technologies or error probabilities.

Given this flexibility of the read simulator, it is possible to construct many kinds of individual fragments like, for example reads, contigs or expressed sequence tags (ESTs) derived from existing sequences. We plan to integrate further error models of upcoming sequencers. In addition, future version of MetaSim will extend the population sampler by using more sophisticated models for sequence evolution like HKY [36], as, e.g., already implemented in Seq-Gen [37].

ACKNOWLEDGMENTS

We would like to thank Korbinian Schneeberger and Stephan Ossowski for providing the empirical error models for the Illumina sequencing.

AUTHOR CONTRIBUTIONS

Conceived and designed the experiments: DCR AFA. Wrote the paper: DCR AFA DHH. Contributed to the implementation of the software: FO AFA RS DHH.

REFERENCES

1. Rusch DB, Halpern AL, Sutton G, Heidelberg KB, Williamson S, et al. (2007) The Sorcerer II Global Ocean Sampling expedition: northwest Atlantic through eastern tropical Pacific. PLoS Biol 5: e77.
2. Tringe SG, von Mering C, Kobayashi A, Salamov AA, Chen K, et al. (2005) Comparative Metagenomics of Microbial Communities. Science 308: 554–557.
3. Tyson GW, Chapman J, Hugenholtz P, Allen EE, Ram RJ, et al. (2004) Community structure and metabolism through reconstruction of microbial genomes from the environment. Nature 428: 37–43.
4. Gill SR, Pop M, Deboy RT, Eckburg PB, Turnbaugh PJ, et al. (2006) Metagenomic analysis of the human distal gut microbiome. Science 312: 1355–1359.
5. Turnbaugh PJ, Ley RE, Mahowald MA, Magrini V, Mardis ER, et al. (2006) An obesity-associated gut microbiome with increased capacity for energy harvest. Nature 444: 1027–1031.
6. Poinar HN, Schwarz C, Qi J, Shapiro B, Macphee RDE, et al. (2006) Metagenomics to paleogenomics: large-scale sequencing of mammoth DNA. Science 311: 392–394.
7. Margulies M, Egholm M, Altman WE, Attiya S, Bader JS, et al. (2005) Genome sequencing in microfabricated high-density picolitre reactors. Nature 437: 376–380.
8. Pachter L (2007) Interpreting the unculturable majority. Nat Methods 4: 479–480.
9. Mavromatis K, Ivanova N, Barry K, Shapiro H, Goltsman E, et al. (2007) Use of simulated data sets to evaluate the fidelity of metagenomic processing methods. Nat Methods 4: 495–500.
10. Bernal A, Ear U, Kyrpidesa N (2001) Genomes OnLine Database (GOLD): a monitor of genome projects world-wide. Nucleic Acids Res 29: 126–127.
11. Venter JC, Remington K, Heidelberg JF, Halpern AL, Rusch D, et al. (2004) Environmental Genome Shotgun Sequencing of the Sargasso Sea. Science 304: 66–74.
12. Benson DA, Karsch-Mizrachi I, Lipman DJ, Ostell J, Wheeler DL (2005) GenBank. Nucleic Acids Res 1: D34–38.
13. Huson DH, Auch AF, Qi J, Schuster SC (2007) MEGAN analysis of metagenomic data. Genome Res 17: 377–386.
14. Karlin S, Burge C (1995) Dinucleotide relative abundance extremes: a genomic signature. Trends Genet 11: 283–290.
15. Deschavanne PJ, Giron A, Vilain J, Fagot G, Fertil B (1999) Genomic signature: characterization and classification of species assessed by chaos game representation of sequences. Mol Biol Evol 16: 1391–1399.
16. Teeling H, Waldmann J, Lombardot T, Bauer M, Glšckner FO (2004) TETRA: a web-service and a stand-alone program for the analysis and comparison of tetranucleotide usage patterns in DNA sequences. BMC Bioinformatics 5: 163.
17. McHardy AC, Mart'n HGa, Tsirigos A, Hugenholtz P, Rigoutsos I (2007) Accurate phylogenetic classification of variable-length DNA fragments. Nat Methods 4: 63–72.
18. Chatterji S, Yamazaki I, Bai Z, Eisen JA Vingron M, Wong L, editors. (2008) CompostBin: A DNA composition-based algorithm for binning environmental shotgun reads. Lecture Notes in Computer Science Springer. 17–28.
19. Tatusov RL, Koonin EV, Lipman DJ (1997) A genomic perspective on protein families. Science 278: 631–637.
20. Kanehisa M, Goto S (2000) KEGG: kyoto encyclopedia of genes and genomes. Nucleic Acids Res 28: 27–30.
21. Finn RD, Tate J, Mistry J, Coggill PC, Sammut SJ, et al. (2008) The Pfam protein families database. Nucleic Acids Res 36: D281–D288.
22. Overbeek R, Begley T, Butler RM, Choudhuri JV, Chuang H-Y, et al. (2005) The subsystems approach to genome annotation and its use in the project to annotate 1000 genomes. Nucleic Acids Res 33: 5691–5702.
23. von Mering C, Jensen LJ, Snel B, Hooper SD, Krupp M, et al. (2005) STRING: known and predicted protein-protein associations, integrated and transferred across organisms. Nucleic Acids Res 33: D433–D437.
24. Haft DH, Selengut JD, White O (2003) The TIGRFAMs database of protein families. Nucleic Acids Res 31: 371–373.
25. Rodriguez-Brito B, Rohwer F, Edwards RA (2006) An application of statistics to comparative metagenomics. BMC Bioinformatics 7: 162.

26. Krause L, Diaz NN, Goesmann A, Kelley S, Nattkemper TW, et al. (2008) Phylogenetic classification of short environmental DNA fragments. Nucleic Acids Res 36(7): 2230–2239.
27. Wheeler DL, Barrett T, Benson DA, Bryant SH, Canese K, et al. (2008) Database resources of the National Center for Biotechnology Information. Nucleic Acids Res 36: D13–D21.
28. Meldrum D (2000) Automation for Genomics, Part one: Preparation for Sequencing. Genome Research 10: 1081–1092.
29. Meldrum D (2000) Automation for Genomics, Part two: Sequencers, Microarrays, and Future trends. Genome Research 10: 1288–1303.
30. Bentley DR (2006) Whole-genome re-sequencing. Current Opinion in Genetics & Development 16: 545–552.
31. Yule GU (1925) A mathematical theory of evolution, based on the conclusions of Dr. J.C. Willis. Philosophical Transactions of the Royal Society of London Ser B, Biol Sci 213: 21–87.
32. Harding EF (1971) The Probabilities of Rooted Tree-Shapes Generated by Random Bifurcation. Advances in Applied Probability 3: 44–77.
33. Jukes TH, Cantor CR (1969) Evolution of Protein Molecules. In: Munro HN, editor. Mammalian Protein Metabolism. New York, NY: Academic Press. pp. 21–132.
34. Myers G (1999) A dataset generator for whole genome shotgun sequencing. pp. 202–210.
35. Engle ML, Burks C (1994) GenFrag 2.1: new features for more robust fragment assembly benchmarks. Comput Appl Biosci 10: 567–568.
36. Hasegawa M, Kishino H, Yano T (1985) Dating of the human-ape splitting by a molecular clock of mitochondrial DNA. J Mol Evol 22: 160–174.
37. Rambaut A, Grassly NC (1997) Seq-Gen: an application for the Monte Carlo simulation of DNA sequence evolution along phylogenetic trees. Comput Appl Biosci 13: 235–238.

CHAPTER 15

THE NATURAL PRODUCT DOMAIN SEEKER NAPDOS: A PHYLOGENY BASED BIOINFORMATIC TOOL TO CLASSIFY SECONDARY METABOLITE GENE DIVERSITY

Nadine Ziemert, Sheila Podell, Kevin Penn, Jonathan H. Badger, Eric Allen, Paul R. Jensen

INTRODUCTION

Genome sequencing has revealed that the secondary metabolite potential of even well studied bacteria has been severely underestimated [1], [2]. This revelation has led to an explosion of interest in genome mining as an approach to natural product discovery [3], [4], [5], [6], [7], [8]. Considering that natural products remain one of the primary sources of therapeutic agents [9],[10], sequence analysis provides opportunities to identify strains with the greatest genetic potential to yield novel secondary metabolites prior to chemical analysis and thus increase the rate and efficiency with which new drug leads are discovered. In addition, community or metagenomic analyses can be used to identify environments with the greatest secondary metabolite potential and to address ecological questions related to secondary metabolism. To capitalize on these opportunities, it is critical that new bioinformatics tools be developed to handle the massive influx of sequence data that is being generated from next generation sequencing technologies [11].

Polyketide synthases (PKSs) and non-ribosomal peptide synthetases (NRPSs) are large enzyme families that account for many clinically important pharmaceutical agents. These enzymes employ complimentary strategies to sequentially construct a diverse array of natural products from relatively simple carboxylic acid and amino acid building blocks using an assembly line process [12], [13]. The molecular architectures of PKS and NRPS genes have been reviewed in detail and minimally consist of activation (AT or A), thiolation (ACP or PCP), and condensation (KS or C) domains, respectively [14], [15], [16], [17], [18]. These genes are among the largest found in microbial genomes and can include highly repetitive modules that create considerable challenges to accurate assembly and subsequent bioinformatic analysis[8].

When the challenges associated with PKS and NRPS gene assembly can be overcome, a number of effective bioinformatics tools have been developed for domain parsing [19], [20] and domain string analysis [21], [22]. In cases of modular type I PKSs and NRPSs where domain strings follow the "co-linearity rule" such that substrates are incorporated and processed according to the precise domain organization observed in the pathway, bioinformatics has been used to make accurate structural predictions about the metabolic products of those pathways[23]. However, the increasing number of exceptions to co-linearity, such as module skipping and stuttering [24], create limitations for precise, sequence-based structure prediction. The bioinformatic tools currently available for secondary metabolism have been reviewed [25], [26]and are complemented by the recent release of antiSMASH, which has the capacity to accurately identify and provide detailed sequence analysis of gene clusters associated with all known secondary metabolite chemical classes [27]. While all of these tools have useful applications, NaPDoS employs a phylogeny based classification system that can be used to quantify and distinguish KS and C domain types from a variety of datasets including the incomplete genome assemblies typically obtained using next generation sequencing technologies. These specific domains were selected because they are highly conserved and have proven to be among the most informative in a phylogenetic context [28], [29].

Phylogenomics provides a useful approach to infer gene function based on phylogenetic relationships as opposed to sequence similarities [30], [31]. While the evolutionary histories of PKS and NRPS genes are largely uninformative due to their size and complexity, KS and C domain phylogenies reveal highly supported clustering patterns. These patterns have been

Citation: Ziemert N, Podell S, Penn K, Badger JH, Allen E, Jensen PR (2012) The Natural Product Domain Seeker NaPDoS: A Phylogeny Based Bioinformatic Tool to Classify Secondary Metabolite Gene Diversity. PLoS ONE 7(3): e34064. doi:10.1371/journal.pone.0034064.

used to distinguish type II PKSs associated with spore pigment and antibiotic biosynthesis[32], type I modular and hybrid PKSs [33], and subsequently to identify many different PKSs types [34]. KS phylogeny has also been used to predict pathway associations [26], [35] and, in some cases, the secondary metabolic products of those pathways [28], [36], [37]. Phylogenetics has also been used to successfully identify PKS sequences from complex metagenomic datasets [38]. Likewise, C domain phylogeny clearly delineates functional subtypes as opposed to species relationships [39] and has been used to identify new functional classes, such as the "starter" C domain [29]. Taken together, the established phylogenetic relationships of KS and C domains provide an effective framework within which to assess secondary metabolite gene richness and diversity and to identify new functional classes that may be associated with uncharacterized biosynthetic mechanisms.

Here we introduce the web tool Natural Product Domain Seeker (NaPDoS), which extracts and rapidly classifies KS and C domains from a wide range of sequence data. The results can be used to assess the potential for PKS and NRPS secondary metabolite biosynthesis in organisms or environments and to identify new phylogenetic lineages, which can subsequently be investigated as a source of new mechanistic biochemistry. We tested NaPDoS on four draft bacterial genome sequences and two metagenomic datasets. The results reveal a remarkable level of secondary metabolite gene diversity among closely related strains and provide a mechanism to assess secondary metabolism from poorly assembled genomic data.

MATERIALS AND METHODS

Reference database

KS and C domains were extracted from select PKS and NRPS genes associated with experimentally characterized biosynthetic pathways using the online program NRPS-PKS (http://www.nii.res.in/searchall.html) [19], [21]. The pathways selected include representatives of the currently known enzyme architectures and functions associated with type I and II PKSs and NRPSs and thus this database is not meant to be comprehensive. The biochemical function and enzyme architecture of each domain was manually confirmed by analysis of the associated domain string and secondary metabolic product. Based on these results, each sequence was preliminarily assigned to a domain class. The compound produced by the associated pathway, the literature reference including PubMed ID, and the gene accession number was also recorded for each domain.

Sequence alignment and phylogeny

The amino acid sequences of all reference KS and C domains were aligned using either MUSCLE [40] or ClustalX (version 1.83) [41] with the BLOSUM 62 protein weight matrix. The alignments were manually adjusted using Mesquite [42]. Maximum likelihood, parsimony, and neighbor-joining phylogenetic trees were constructed using the "a la carte" mode at the Phylogeny.fr website (http://www.phylogeny.fr/) [43]. Final maximum likelihood trees were constructed from the reference data set with the program PHYML [44]. Final domain classifications were made based on the phylogenetic relationships observed in these trees.

NaPDoS and Webportal

The NaPDoS web portal identifies candidate KS and C domains through a combination of hidden markov model (HMM) searches and the basic local alignment search tool (BLAST) algorithm [45] optimized for query input type as shown in Figure 1. PCR products or coding sequences (CDS) in nucleotide or amino acid format are analyzed directly by local BLASTX or BLASTP searches against the manually curated reference database of experimentally verified KS and C domains described above. This BLAST-based approach proved more effective than HMM models in detecting the target domains from short query sequences. Genomic sequences (including contigs, incomplete drafts, or complete genomes) and metagenomic nucleotide data sets are first pre-screened to obtain rough coordinates for KS and C domains using the KS domain HMM developed by Yadav and co-workers [21] and the PFAM C domain model PF00668 [46]. The resulting candidate domains are then subjected to BLAST analyses using the same manually curated reference database as described above.

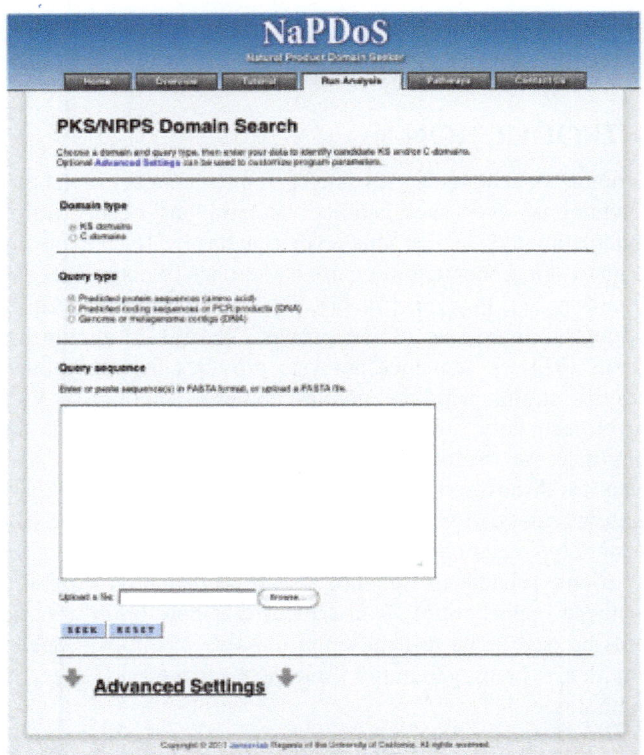

Figure 1: NaPDoS bioinformatic pipeline. The web interface to this pipeline is divided 3 consecutive steps. Nucleic acid sequences are translated into predicted amino acids and genomic sequences are screened using Hidden Markov Models (HMM). For protein and small nucleic acid sequences a BLAST search is performed against curated reference database examples to identify matches to known PKS/NRPS pathways. Selected candidate sequences plus the BLAST results are trimmed and inserted into a manually curated reference alignment, keeping the original reference alignment intact. This alignment is used to build a tree.

BLAST results are linked to a back-end MySQL relational database via CGI-scripting to retrieve and report domain classification and related pathway information. Query sequences are trimmed according to their BLAST match coordinates by a custom Perl script then aligned to each other and their database matches using MUSCLE [40]. Trimmed sequences can be downloaded along with best BLAST matches in FASTA or MSF aligned format. Finally, trimmed and aligned candidate KS and C sequences plus BLAST matches can be inserted into a phylogenetic tree generated from the reference database using FastTree to estimate maximum likelihood [44]. Newick format output from FastTree is converted to SVG format graphic images using the Newick-Utilities program [47]. NaPDoS does not employ any stand-alone software that was created specifically for its operation but instead employs pre-existing and publically available programs as described above.

Draft genomes and metagenomes

Draft genome sequences of *S. arenicola* strain CNH-643 (accession number PRJNA84391), *S. arenicola* strain CNT-088 (accession number PRJNA84269), *"S. pacifica"* strain CNS-143 (accession number PRJNA84389), and *"S. pacifica"* strain CNT-133 (accession number PRJNA84271) were obtained at 8× coverage at the J. Craig Venter Institute using 454 GS FLX pyrosequencing and 0.5× Sanger sequencing as previously described [48] based on an estimated genome size of 5.6 Mb. The sequence data were assembled using the Newbler Assembler with the mapping option [49]. *S. arenicola* strains were mapped onto the complete*S. arenicola* strain CNS-205 genome and the *S. pacifica* strains were mapped to the complete*S. tropica* CNB-440 genome [50] while any unmapped sequence data was assembled de novo. The four draft *Salinispora* genomes were mined for KS and C domains using NaPDoS with default settings. The metagenomic datasets (whale fall, AAFZ00000000, AAFY00000000, AAGA00000000 and Minnesota farm soil, AAFX00000000, [51]) were mined using default HMM settings (e^{-5}) and the resulting sequences further subjected to a loose BLAST analysis with an e-value cut-off of 1 to obtain more precise coordinates and assign initial domain classifications.

RESULTS

The Natural Product Domain Seeker (NaPDoS)

The publically available web tool NaPDoS (http://npdomainseeker.ucsd.edu/) was created to detect and classify KS and C domains in nucleotide and amino acid sequence data. The query data can be PCR amplicons, genes, contigs, genomes, or metagenomes. The current query size limits are <30 MB and <50,000 individual sequences. The website provides a detailed tutorial on the use of this tool, which is implemented using a web interface (Figure 2) that follows the bioinformatic pipeline shown in Figure 1. Query sequences are BLASTed against the reference database, which currently contains 459 KS and 190 C domains derived from 66 PKS, 20 NRPS, 8 PKS/NRPS hybrid, and 5 fatty-acid synthase (FAS) biosynthetic pathways. These sequences can be downloaded from the website and encompass all major classes of type I and II KS and C domains currently described in the literature [12], [29], [52], [53]. This manually curated database will be updated periodically as new modular architectures and biochemical features are discovered for each domain type.

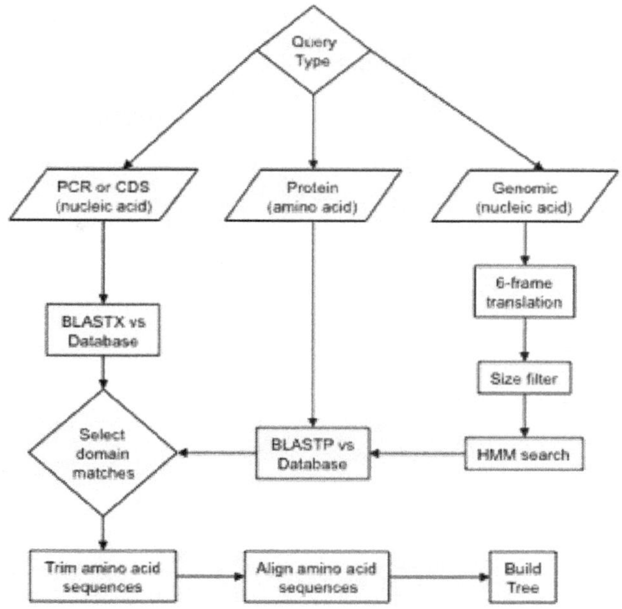

Figure 2: Screen shot of the NaPDoS webpage.

The primary output for all analyses includes the query identification, best database match, percent identity, alignment length, e-value, and product and classification of the biosynthetic pathway associated with the best match. KS and C domain sequences derived from the input data can then be output in raw format or aligned with the best BLAST matches. A NaPDoS independent BLAST of the output domain sequence(s) against the NCBI nr database is also highly recommended to check for matches that do not occur in the reference database.

To generate a final classification for each domain sequence, it is highly recommended to construct a phylogenetic tree, especially in cases where the percent sequence identity to the top database match is low. If that option is chosen, a profile alignment is generated in which the query sequences are incorporated into a carefully curated reference alignment generated from the sequences in the reference database. This alignment is then used to create a phylogenetic tree, which needs to be manually interpreted to establish a final classification for each sequence. Interpreting sequences in the context of a phylogenetic tree is particularly important given that the NaPDoS pipeline is intentionally set to low stringency in an effort to detect all possible KS and C domains. Thus, homologs not involved in secondary metabolism such KSs associated with fatty acid biosynthesis are regularly detected. These sequences can be readily classified based on the phylogenetic tree.

Domain classification

KS and C domain phylogenies form the basis of the NaPDoS classification system (Figure 3). KS domains clade based on biochemical function and enzyme architecture, which are described in Table 1. In some cases, e.g. enediynes, these clades are also predictive of structural motifs associated with the secondary metabolites produced. The KS phylogeny clearly delineates type I and II PKSs (Figure 3A). The shared ancestry reported between type II PKS and FAS sequences [54] is clearly maintained in this tree. The vast majority of the reference sequences fall into the type I PKS clade. This clade can be further resolved into seven classes, which are not always monophyletic. This polyphyly reflects the complex evolutionary histories of the different classes such as the *trans*-AT KSs, which evolved by extensive HGT and exploit considerably greater modular architectures than the *cis*-AT group, which has largely evolved by gene duplication [55]. However, all of these lineages are highly supported in the tree (likelihood values 0.7–1.0) and largely agree with previous phylogenetic studies [29], [56].

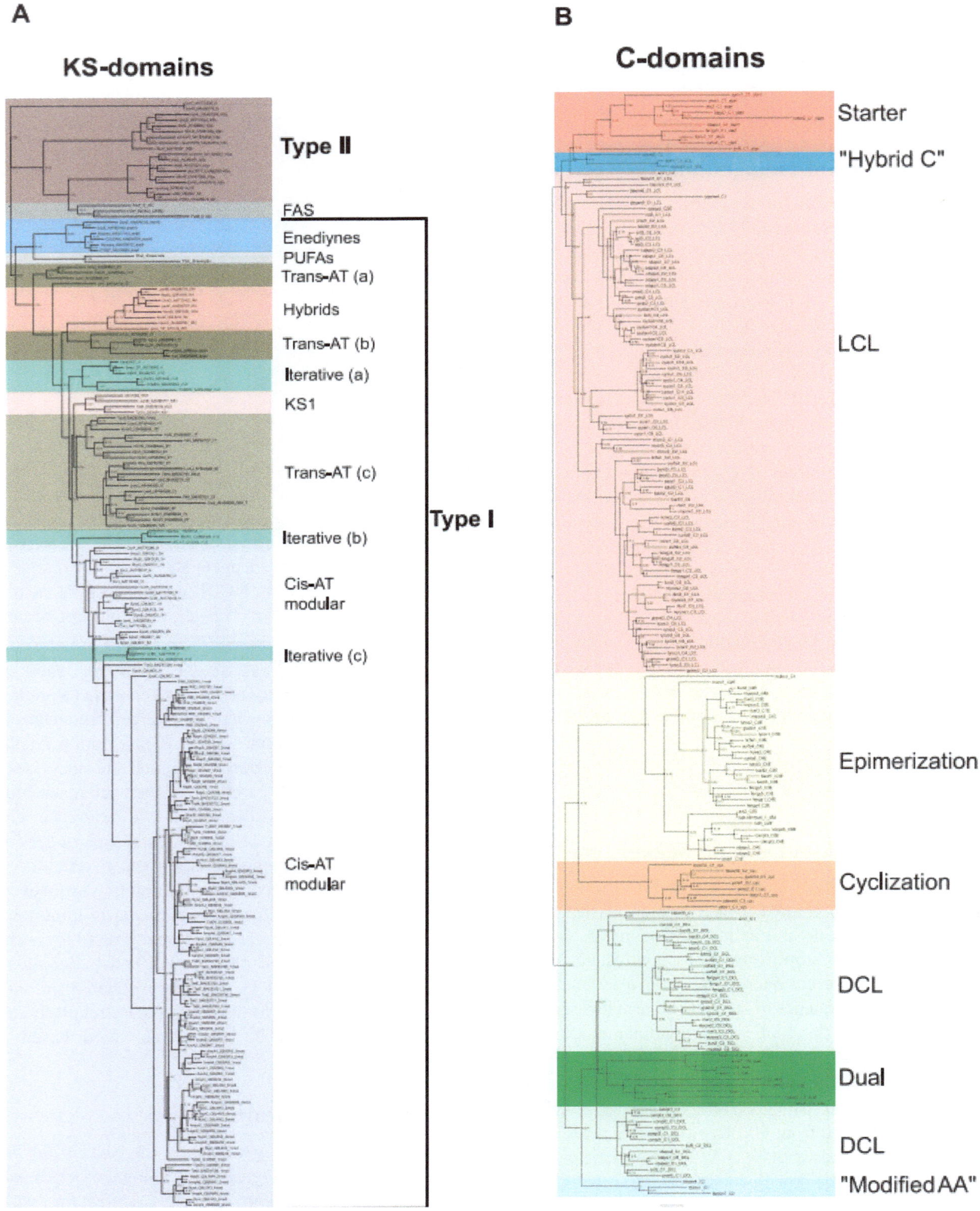

Figure 3: Phylogeny based domain classification. A) KS domain phylogeny. Polyphyletic groups are distinguished by letters. B) C domain phylogeny.

Table 1: KS domain classification

Type	Class	Description	Product (example)
I	Enediyne	Iteratively acting, builds typical 9 or 10 membered enedyines.	Enediyne (calicheamicin)
	Trans-AT	Module lacks cognate AT domain; this activity is provided instead by a discrete protein encoded in *trans*.	Polyketide/macrolide (leinamycin)
	Cis-AT	Multi-domain module that includes AT domain.	Polyketide/macrolide (erythromycin)
	Hybrid	Catalyzes a condensation reaction between an amino acid and an acyl extender unit in a NRPS/PKS pathway.	Peptide-polyketide (microcystin)
	Iterative	Domain is used multiple times in a cyclic fashion.	Polycyclic polyketide (aflatoxin)
	PUFA	Produces long chain fatty acids containing more than one double bond.	Polyunsaturated fatty acid (omega-3-fatty-acid)
	KS1	Occurs in the first module of multimodular genes, includes typical starter KSs (KSQ) as well as KS domains that incorporate unusual precursors.	Polyketide, peptide-polyketide (salinosporamide)
II	Type II	Each domain occurs on a discrete protein.	Aromatic polyketide (actinorhodin)
	FAS	Involved in fatty acid biosynthesis (eg., FabB and FabF from bacteria).	Fatty acid (palmitic acid)

doi:10.1371/journal.pone.0034064.t001

In the case of C domains, the sequences generally clade based on substrate specificity, i.e. the stereochemistry of the amino acids incorporated and the subsequent tailoring reactions they perform (Figure 3B). Eight clades are identified in the tree of which six are functionally characterized. The characterized clades are comprised of LCL domains, which catalyze a peptide bond between two L-amino acids, DCL domains, which link an L-amino acid to a growing peptide ending with a D-amino acid, starter C domains, which acylate the first amino acid with a β-hydroxy-carboxylic acid, cyclization domains, which catalyze both peptide bond formation and the subsequent cyclization of cysteine, serine or threonine residues, epimerization (E) domains, which switch the chirality of the last amino acid in the growing peptide, and dual E/C domains, which catalyze both epimerization and condensation reactions. These six functional classes are well supported in the tree and largely monophyletic. Two experimentally uncharacterized clades are identified in the tree, one of which has been conditionally assigned the name "modified AA" (Figure 3B). This clade contains domains from the bleomycin and microcystin pathways. Although the biochemical function of these domains has not been experimentally defined, they appear to be involved in the modification of the incorporated amino acid, for example the dehydration of serine to dehydroalanine [57], [58]. C domains in the second functionally uncharacterized clade have been conditionally assigned the name "hybrid C". The three sequences in this clade (micro5, ituri1, and mycos1) are each located downstream of an aminotransferase domain and appear to be involved in the condensation of an amino acid to an aminated polyketide resulting in a hybrid PKS/NRPS secondary metabolite. The phylogenetic relationships of the KS and C domains in the reference dataset form the basis of the NaPDoS classification system and provide a framework within which new clades and biochemical functions can be discovered.

Genome Analyses

As a positive control, NaPDoS was used to analyze the genome sequence of *Streptomyces avermitilis* strain MA-4680. This analysis revealed 67 KS and 15 C domains (Table S1), which encompass all of the PKS, NRPS, and hybrid PKS/NRPS gene clusters that were reported to contain these domains [59]. NaPDoS also correctly identified all of the KS and C domains in the complete genome sequences of *S. tropica* (strain CNB-440) and *S. arenicola* (strain CNS-205) [50]. NaPDoS was then tested on four draft *Salinispora* genome sequences. These low coverage drafts were generated using pre-Titanium 454 technology (avreage read length 244 bp) and yielded poor assemblies and a large number of contigs (Table S2). There was no evidence that any biosynthetic gene clusters had been completely assembled based on the analysis of flanking regions and comparisons with pathways that appeared common with the CNB-440 and CNS-205 sequences [50]. None-the-less, NaPDoS successfully detected 18–36 KS domains and 5–14 C domains in the un-annotated FASTA files generated for each of the four draft genomes (Table S2). More than half (56%) of these sequences showed no significant BLAST matches to domains associated with biochemically characterized biosynthetic genes and thus could not be linked to specific secondary metabolic products. More significantly, 8 KS and 9 C domains detected in the four draft sequences were not observed in the two closed*Salinispora* genomes (Table S3). These sequences (KS7-14 and C5-13) cover a broad range of domain classes and indicate considerable new biosynthetic potential among a group of closely related strains. Two C domains fell into the "Modified AA" clade, which has yet to be experimentally characterized. Given that the upstream A domain specifies serine in both cases, it can be predicted that this domain results in the incorporation of dehydrated serine (ie., dehyroalanine) into the non-ribosomal peptide. This hypothesis has not yet been tested, but is supported by the reference sequences in this clade, which perform similar dehydration reactions.

Interestingly, two KS domains with close matches (89% and 94%) to those associated with the biosynthesis of salinosporamide A [60] were observed in "*S. pacifica*" strain CNT-133. This was unexpected given that compounds in this series had previously been reported exclusively from *S. tropica* [61]. This observation subsequently led to the discovery of a new compound in the salinosporamide series [5] and a rare window into pathway evolution in two closely related bacterial species [37]. Furthermore, a KS domain that shares close sequence identity with domains involved in the biosynthesis of tylactone in *Streptomyces fradiae* [62] was detected in strain CNH-643 (Table S3). Subsequent chemical studies revealed the production of several new tylactone derivatives by this strain (unpublished data). The same four draft genome sequences were also analyzed using antiSMASH [27], a sophisticated pipeline that can make structure predictions for a diverse range of secondary metabolic pathways. While antiSMASH worked well on the two complete *Salinispora* genomes, NaPDoS consistently detected more KS domains in the draft genomes (Table S4). While this is not surprising given that NaPDoS is specifically designed for this purpose, it nonetheless highlights the value of the sequence tag approach when working with draft genome sequences that contain many unassembled contigs.

Metagenomic analyses

NaPDoS was further tested on metagenomic data sets generated from a Minnesota farm soil and whale fall [51]. While the numbers of KS domains detected in both datasets are similar (Table S5), removing redundant sequences reveals a higher diversity of KS domains in the soil sample. The majority of the whale fall KS domains were classified as FASs suggesting they are associated with fatty acid biosynthesis. In contrast, nearly half of the KS domains detected in the Minnesota farm soil appear to be involved in secondary metabolite biosynthesis. These results are in agreement with a previous study in which these datasets were manually screened for type I PKSs [38]. All of the sequences shared <70% identity to the reference database or NCBI BLAST matches associated with experimentally characterized pathways and thus no predictions could be made about the structures of the potential secondary metabolic products. None-the-less, the majority of the KS domains detected could be rapidly classified by NaPDoS. The incorporation of these domains into a phylogenetic tree containing the reference sequences led to the reclassification of some and the prediction that others are functionally distinct from KS domains (Tables S6 and S7). These sequences were likely detected due to the low stringency at which the NaPDoS BLAST analyses were performed on the meta-data and is a positive indication that the KS analysis was comprehensive. The reclassification of some sequences emphasizes the importance of incorporating phylogeny into the analyses.

DISCUSSION

Rapid advances in DNA sequencing technologies are providing unprecedented opportunities to incorporate DNA sequence data into the natural product discovery process. The effective use of this information requires bioinformatic tools that can rapidly analyze large datasets in the context of a wide array of complex biosynthetic paradigms. While a number of excellent bioinformatic tools targeting secondary metabolism have been developed [22], [25], [27], they are largely predicated on accurate gene and operon assembly, something that has often proven challenging to obtain given the modular and highly repetitive nature of many genes involved in secondary metabolism [8]. This challenge can become especially problematic in the case of metagenomic analyses of complex microbial communities.

The Natural Product Domain Seeker (NaPDoS) is a web-based bioinformatic tool that was developed to detect and classify KS and C domains from a wide variety of sequence data. The use of domain sequence tags as proxies for the biosynthetic genes in which they reside is based on the well established and highly informative phylogentic relationships they maintain. These relationships form the foundation of the NaPDoS classification system and provide a rapid mechanism to delineate secondary metabolite biosynthetic gene richness and diversity within a genome or environmental sample. Short sequence tags (e.g. 600 bp) can be effectively analyzed using NAPDoS and thus minimum coverage, next generation sequence assemblies are well suited for this tool. The resulting estimates of biosynthetic potential can be used to guide more extensive sequencing efforts or targeted operon assembly. In cases where query sequences closely match domains derived from experimentally characterized biosynthetic pathways (eg., >90% sequence identity), it has proven possible to make accurate predictions about the structural class of the secondary metabolite(s) produced [37], [63]. The low stringency of the HMM searches and the ability to adjust the internal BLAST parameters provides opportunities to detect more highly divergent KS and C domains associated with secondary metabolism as well as domains that are not associated with secondary metabolism (e.g. fatty acid biosynthesis) and thus all results should be carefully scrutinized. As the number of experimentally characterized biosynthetic pathways increases, this approach will provide an increasingly effective method to "de-replicate", i.e. to identify strains that have the greatest potential to produce known compounds.

There is ample evidence that the mechanistic diversity of polyketide and non-ribosomal peptide assembly is considerably greater than that currently recognized [14], [64], and thus it can be expected that the NaPDoS classification system will need to evolve as new phylogenetic lineages are linked to specific biochemical functions and enzyme architectures. There is considerable preliminary evidence that the classes defined here will be further delineated once more experimentally characterized sequence data is obtained. For example, the current KS1 clade includes traditional starter KSs (KSQ) as well as domains from the salinosporamide (stro1024) and jamaicamide (JamE) pathways, which are involved in the incorporation of unusual extender units [8], [65]. Likewise, the Type II clade includes deeply branching KS domains derived from CurC and JamG that are predicted to be involved in decarboxylation as opposed to condensation reactions [65], [66]. A third example is the Iterative (a) class, which include traditional iterative KSs as well as those involved in the biosynthesis of polycyclic tetramate macrolactams [67]. Finally, the *trans*-AT (b) clade is comprised of KS sequenced derived from what appears to be an evolutionarily independent lineage of *trans*-AT sequences as well as genes associated with *beta*-branching [28], [55]. Despite the potential oversimplification of the current classification system, it provides a useful method to estimate the numbers and functional types of biosynthetic genes present in complex data sets.

Despite poor assembly, a large number and diversity of KS and C domains were detected among the four draft *Salinispora* genome sequences. Seventeen of these domains were not observed in either of the two complete *Salinispora* genomes providing evidence of the considerable biosynthetic variability that may occur among closely related strains. In addition, two C domains fell into the "Modified AA" clade, a lineage whose biochemical function has yet to be experimentally characterized. While the metagenomic datasets revealed similar total numbers of KS domains, the classification of those domains revealed dramatic differences in functional types. Analyses such as these provide insight into the potential significance of secondary chemistry in mediating population and community dynamics while at the same time identifying environments that can be prioritized for secondary metabolite discovery efforts.

Traditional natural product discovery paradigms have become increasingly inefficient [68] and are rapidly moving towards approaches that capitalize on access to DNA sequence data [69]. NaPDoS is a publically available bioinformatic tool that capitalizes on the well-established phylogenetic relationships of KS and C domains. It provides a rapid method to make informed interpretations of secondary metabolism based on small sequence tags extracted from a variety of data types including poorly assembled, next generation datasets. A major application of NaPDoS is the exploration of sequence space and the identification of new domain lineages, which have a high probability of being associated with new mechanisms of secondary metabolite biosynthesis. Prioritizing these lineages for experimental characterization will facilitate the discovery of new biochemistry and represents a rationale approach to secondary metabolite discovery.

At present, NaPDoS is optimized for the identification and classification of bacterial PKS and NRPS genes. Nonetheless, it is possible for NaPDoS to identify eukaryotic KS and C domains given their shared evolutionary history with prokaryotic homologs. The results obtained for non-bacterial sequences should be interpreted with caution however, as the reference database has not been adequately populated with these sequences to provide a robust classification system. Future plans include the expansion of NaPDoS to include additional eukaryotic sequences and subgroups within the FAS and PUFA lineages, the later of which were recently shown to cluster phylogenetically based on functional type [70]. Additional goals are to include type III PKSs, which were originally found in plants but are now known to occur in a wide range of bacteria [71]. These PKSs are distantly related to types I and II and thus will require a separate alignment and analysis pipeline. The inclusion of additional secondary metabolite families, such as terpenoids, alkaloids, and ribosomal peptides, is also conceivable.

AUTHOR CONTRIBUTIONS

Conceived and designed the experiments: NZ SP KP EA PJ. Performed the experiments: NZ SP KP. Analyzed the data: NZ SP KP JB EA PJ. Contributed reagents/materials/analysis tools: NZ SP KP JB EA PJ. Wrote the paper: NZ PJ.

REFERENCES

1. Bentley SD, Chater KF, Cerdeno-Tarraga AM, Challis GL, Thomson NR, et al. (2002) Complete genome sequence of the model actinomycete Streptomyces coelicolor A3(2). Nature 417: 141–147.
2. Ikeda H, Ishikawa J, Hanamoto A, Shinose M, Kikuchi H, et al. (2003) Complete genome sequence and comparative analysis of the industrial microorganism Streptomyces avermitilis. Nat Biotech 21: 526–531.
3. Challis GL (2008) Genome mining for novel natural product discovery. J Med Chem 51: 2618–2628.
4. Lautru S, Deeth RJ, Bailey LM, Challis GL (2005) Discovery of a new peptide natural product by Streptomyces coelicolor genome mining. Nat Chem Biol 1: 265–269.
5. Eusta´quio AS, Nam S-J, Penn K, Lechner A, Wilson MC, et al. (2011) The discovery of salinosporamide K from the marine bacterium "Salinispora pacifica" by genome mining gives insight into pathway evolution. ChemBioChem 12: 61–64.
6. Hornung A, Bertazzo M, Dziarnowski A, Schneider K, Welzel K, et al. (2007) A genomic screening approach to the structure-guided identification of drug candidates from natural sources. ChemBioChem 8: 757–766.
7. Winter JM, Behnken S, Hertweck C (2011) Genomics-inspired discovery of natural products. Curr Opin Chem Biol 15: 22–31.
8. Udwary DW, Zeigler L, Asolkar RN, Singan V, Lapidus A, et al. (2007) Genome sequencing reveals complex secondary metabolome in the marine actinomycete Salinispora tropica. Proc Natl Acad Sci 104: 10376–10381.
9. Baker DD, Chu M, Oza U, Rajgarhia V (2007) The value of natural products to future pharmaceutical discovery. Nat Prod Rep 24: 1225–1244.
10. Newman DJ, Cragg GM (2007) Natural products as sources of new drugs over the last 25 years. Journal of Natural Products 70: 461–477.
11. McPherson JD (2009) Next-generation gap. Nat Methods 6: S2–5.
12. Hertweck C (2009) The biosynthetic logic of polyketide diversity. Angew Chem Int Ed Engl 48: 4688–4716.
13. Finking R, Marahiel MA (2004) Biosynthesis of non-ribosomal peptides. Annual Review of Microbiology 58: 453–488.
14. Shen B (2003) Polyketide biosynthesis beyond the type I, II and III polyketide synthase paradigms. Curr Opin Chem Biol 7: 285–295.
15. Weissman KJ (2004) Polyketide biosynthesis: understanding and exploiting modularity. Philosophical Transactions of the Royal Society of London Series A, Mathematical, Physical and Engineering Sciences 362: 2671–2690.
16. Lautru S, Challis GL (2004) Substrate recognition by nonribosomal peptide synthetase multi-enzymes. Microbiology 150: 1629–1636.
17. Sieber SA, Marahiel MA (2005) Molecular mechanisms underlying nonribosomal peptide synthesis: Approaches to new antibiotics. Chemical Reviews 105: 715–738.
18. Fischbach MA, Walsh CT (2006) Assembly-line enzymology for polyketide and nonribosomal peptide antibiotics: logic, machinery, and mechanisms. Chem Rev 106: 3468–3496.

19. Ansari MZ, Yadav G, Gokhale RS, Mohanty D (2004) NRPS-PKS: a knowledge-based resource for analysis of NRPS/PKS megasynthases. Nucleic Acids Res 32: W405–413.

20. Rausch C, Weber T, Kohlbacher O, Wohlleben W, Huson DH (2005) Specificity prediction of adenylation domains in nonribosomal peptide synthetases (NRPS) using transductive support vector machines (TSVMs). Nucleic Acids Res 33: 5799–5808.

21. Yadav G, Gokhale RS, Mohanty D (2009) Towards prediction of metabolic products of polyketide synthases: an in silico analysis. PLoS Comput Biol 5: e1000351.

22. Starcevic A, Zucko J, Simunkovic J, Long PF, Cullum J, et al. (2008) ClustScan: an integrated program package for the semi-automatic annotation of modular biosynthetic gene clusters and in silico prediction of novel chemical structures. Nucleic Acids Research 36: 6882–6892.

23. McAlpine JB, Bachmann BO, Piraee M, Tremblay S, Alarco AM, et al. (2005) Microbial genomics as a guide to drug discovery and structural elucidation: ECO-02301, a novel antifungal agent, as an example. Journal of Natural Products 68: 493–496.

24. Moss SJ, Martin CJ, Wilkinson B (2004) Loss of co-linearity by modular polyketide synthases: a mechanism for the evolution of chemical diversity. Natural Product Reports 21: 575–593.

25. Bachmann BO, Ravel J (2009) Methods for in silico prediction of microbial polyketide and nonribosomal peptide biosynthetic pathways from DNA sequence data. Methods Enzymol 458: 181–217.

26. Jenke-Kodama H, Dittmann E (2009) Bioinformatic perspectives on NRPS/ PKS megasynthases: advances and challenges. Nat Prod Rep 26: 874–883.

27. Medema MH, Blin K, Cimermancic P, de Jager V, Zakrzewski P, et al. (2011) antiSMASH: rapid identification, annotation and analysis of secondary metabolite biosynthesis gene clusters in bacterial and fungal genome sequences. Nucleic Acids Research 39: W339–W346.

28. Nguyen T, Ishida K, Jenke-Kodama H, Dittmann E, Gurgui C, et al. (2008) Exploiting the mosaic structure of trans-acyltransferase polyketide synthases for natural product discovery and pathway dissection. Nat Biotechnol 26: 225–233.

29. Rausch C, Hoof I, Weber T, Wohlleben W, Huson DH (2007) Phylogenetic analysis of condensation domains in NRPS sheds light on their functional evolution. BMC Evol Biol 7: 78.

30. Eisen JA (1998) A phylogenomic study of the MutS family of proteins. Nucleic Acids Res 26: 4291–4300.

31. Eisen JA, Fraser CM (2003) Phylogenomics: intersection of evolution and genomics. Science 300: 1706–1707.

32. Metsa-Ketela M, Salo V, Halo L, Hautala A, Hakala J, et al. (1999) An efficient approach for screening minimal PKS genes from Streptomyces. FEMS Microbiol Lett 180: 1–6.

33. Moffitt MC, Neilan BA (2003) Evolutionary affiliations within the superfamily of ketosynthases reflect complex pathway associations. J Mol Evol 56: 446–457.

34. Jenke-Kodama H, Sandmann A, Mu¨ller R, Dittmann E (2005) Evolutionary implications of bacterial polyketide synthases. Mol Biol Evol 22: 2027–2039.

35. Ginolhac A, Jarrin C, Robe P, Perriere G, Vogel T, et al. (2005) Type I polyketide synthases may have evolved through horizontal gene transfer. J Mol Evol 60: 716–725.

36. Gontang EA, Fenical W, Jensen PR (2007) Phylogenetic diversity of grampositive bacteria cultured from marine sediments. Appl Environ Microbiol 73: 3272–3282.

37. Freel KC, Nam S-J, Fenical W, Jensen PR (2011) Evolution of secondary metabolite genes in three closely related marine actinomycete species. Appl Environ Microbiol 77: 7261–7270.

38. Foerstner KU, Doerks T, Creevey CJ, Doerks A, Bork P (2008) A computational screen for type I polyketide synthases in metagenomics shotgun data. PLoS ONE 3: e3515.

39. Roongsawang N, Lim SP, Washio K, Takano K, Kanaya S, et al. (2005) Phylogenetic analysis of condensation domains in the nonribosomal peptide synthetases. FEMS Microbiol Lett 252: 143–151.

40. Edgar RC (2004) MUSCLE: multiple sequence alignment with high accuracy and high throughput. Nucleic Acids Res 32: 1792–1797.

41. Thompson JD, Gibson TJ, Plewniak F, Jeanmougin F, Higgins DG (1997) The CLUSTAL_X windows interface: flexible strategies for multiple sequence alignment aided by quality analysis tools. Nucleic Acids Res 25: 4876–4882.

42. Maddison WP, Maddison DR (2009) Mesquite: a modular system for evolutionary analysis. Version 2.71 ed.

43. Dereeper A, Guignon V, Blanc G, Audic S, Buffet S, et al. (2008) Phylogeny.fr: robust phylogenetic analysis for the non-specialist. Nucleic Acids Res 36: W465–469.

44. Guindon S, Gascuel O (2003) A simple, fast, and accurate algorithm to estimate large phylogenies by maximum likelihood. Systematic Biol 52: 696–704.

45. Altschul SF, Gish W, Miller W, Myers EW, Lipman DJ (1990) Basic local alignment search tool. Journal of Molecular Biology 215: 403–410.

46. Finn RD, Mistry J, Tate J, Coggill P, Heger A, et al. (2009) The Pfam protein families database. Nucleic Acids Res 38: D211–222.

47. Junier T, Zdobnov EM. The Newick utilities: high-throughput phylogenetic tree processing in the UNIX shell. Bioinformatics 26: 1669–1670.

48. Goldberg SMD, Johnson J, Busam D, Feldblyum T, Ferriera S, et al. (2006) A Sanger/pyrosequencing hybrid approach for the generation of high-quality draft assemblies of marine microbial genomes. Proceedings of the National Academy of Sciences 103: 11240–11245.

49. Margulies M, Egholm M, Altman WE, Attiya S, Bader JS, et al. (2005) Genome sequencing in microfabricated

high-density picolitre reactors. Nature 437: 376–380.

50. Penn K, Jenkins C, Nett M, Udwary DW, Gontang EA, et al. (2009) Genomic islands link secondary metabolism to functional adaptation in marine Actinobacteria. ISME J 3: 1193–1203.

51. Tringe SG, von Mering C, Kobayashi A, Salamov AA, Chen K, et al. (2005) Comparative metagenomics of microbial communities. Science 308: 554–557.

52. Jenke-Kodama H, Dittmann E (2009) Evolution of metabolic diversity: insights from microbial polyketide synthases. Phytochemistry 70: 1858–1866.

53. Ridley CP, Lee HY, Khosla C (2008) Evolution of polyketide synthases in bacteria. Proc Natl Acad Sci U S A 105: 4595–4600.

54. Jenke-Kodama H, Sandmann A, Muller R, Dittmann E (2005) Evolutionary implications of bacterial polyketide synthases. Mol Biol Evol 22: 2027–2039.

55. Piel J (2010) Biosynthesis of polyketides by trans-AT polyketide synthases. Natural Product Reports 27: 996–1047.

56. Jenke-Kodama H, Dittmann E (2005) Combinatorial polyketide biosynthesis at higher stage. Mol Syst Biol 1: 2005–0025.

57. Tillett D, Dittmann E, Erhard M, von Dohren H, Borner T, et al. (2000) Structural organization of microcystin biosynthesis in Microcystis aeruginosa PCC7806: an integrated peptide-polyketide synthetase system. Chem Biol 7: 753–764.

58. Du L, Sanchez C, Chen M, Edwards DJ, Shen B (2000) The biosynthetic gene cluster for the antitumor drug bleomycin from Streptomyces verticillus ATCC15003 supporting functional interactions between nonribosomal peptide synthetases and a polyketide synthase. Chem Biol 7: 623–642.

59. Nett M, Gulder TA, Kale AJ, Hughes CC, Moore BS (2009) Function-oriented biosynthesis of beta-lactone proteasome inhibitors in Salinispora tropica. J Med Chem 52: 6163–6167.

60. Eustaquio AS, McGlinchey RP, Liu Y, Hazzard C, Beer LL, et al. (2009) Biosynthesis of the salinosporamide A polyketide synthase substrate chloroethylmalonyl-coenzyme A from S-adenosyl-L-methionine. Proc Nat Acad Sci 106: 12295–12300.

61. Jensen PR, Williams PG, Oh DC, Zeigler L, Fenical W (2007) Species-specific secondary metabolite production in marine actinomycetes of the genus Salinispora. Appl Environ Microbiol 73: 1146–1152.

62. Cundliffe E, Bate N, Butler A, Fish S, Gandecha A, et al. (2001) The tylosinbiosynthetic genes of Streptomyces fradiae. Antonie Van Leeuwenhoek 79: 229–234.

63. Gontang E, GaudI´ncio S, Fenical W, Jensen P (2010) Sequence-based analysis of secondary-metabolite biosynthesis in marine actinobacteria. Appl Environ Microbiol 76: 2487–2499.

64. Wenzel SC, Muller R (2005) Formation of novel secondary metabolites by bacterial multimodular assembly lines: deviations from textbook biosynthetic logic. Curr Opin Chem Biol 9: 447–458.

65. Edwards DJ, Marques BL, Nogle LM, McPhail K, Goeger DE, et al. (2004) Structure and biosynthesis of the Jamaicamides, new mixed polyketidepolypeptide neurotoxins from the marine cyanobacterium Lyngbya majuscula. Chemistry & Biology 11: 817–833.

66. Chang Z, Sitachitta N, Rossi JV, Roberts MA, Flatt PM, et al. (2004) Biosynthetic pathway and gene cluster analysis of Curacin A, an antitubulin natural product from the tropical marine cyanobacterium Lyngbya majuscula. Journal of Natural Products 67: 1356–1367.

67. Blodgett JAV, Oh D-C, Cao S, Currie CR, Kolter R, et al. (2010) Common biosynthetic origins for polycyclic tetramate macrolactams from phylogenetically diverse bacteria. Proceedings of the National Academy of Sciences 107: 11692–11697.

68. Li JW, Vederas JC (2009) Drug discovery and natural products: end of an era or an endless frontier? Science 325: 161–165.

69. Davies J (2011) How to discover new antibiotics: harvesting the parvome. Curr Opin Chem Biol 15: 5–10.

70. Shulse CN, Allen EE (2011) Widespread occurrence of secondary lipid biosynthesis potential in microbial lineages. PLoS One 6: e20146.

71. Moore BS, Hopke JN (2001) Discovery of a new bacterial polyketide biosynthetic pathway. Chembiochem 2: 35–38.

CHAPTER 16

ANALYSIS AND COMPARISON OF VERY LARGE METAGENOMES WITH FAST CLUSTERING AND FUNCTIONAL ANNOTATION

Weizhong Li

BACKGROUND

The emerging field of metagenomics enables a more comprehensive understanding of environmental microbial communities [1, 2, 3, 4, 5, 6, 7, 8, 9]. However, metagenomic data consists of enormous numbers of fragmented sequences that challenge data analysis methodologically and computationally. To address these challenges, new methods and resources have been developed, such as simulated datasets[10], IMG/M[11], CAMERA[12], MG-RAST[13], taxonomy tools[14, 15], statistical comparison[16], functional diversity analysis[17], binning [18, 19, 20] and so on.

The Rapid Analysis of Multiple Metagenomes with a Clustering and Annotation Pipeline (RAMMCAP) presented herein aims to address the particular computational challenges imposed by the huge size and great diversity of metagenomic data. The primary goal is to significantly reduce the computational effort in sequence comparison, as large-scale comparison of metagenomic sequences has become extremely time-consuming. For example, the protein analysis of the Global Ocean Sampling (GOS) study[2] took more than one million CPU hours.

Metagenomic datasets may contain many novel genes that don't show any homology to existing genes. For example, only ~10% of the sequences in the "Metagenomic Profiling of Nine Biomes" (BIOME) study [9] match known functional genes. Novel genes in metagenomic datasets have not been used in many studies with homology-based gene prediction and analysis, so the second goal of RAMMCAP is to explore whole datasets and make use of the novel sequences. Because the *ab initio* gene finding approaches developed for complete genomes work poorly with fragmented DNA sequences, recently, several new gene prediction methods were developed for short DNA sequences with high sensitivity and specificity, such as Metagene[21], MetageneAnnotator[22], and Neural Networks[23]. In RAMMCAP, ORFs are called with either Metagene or simple six reading frame translation; both methods can identify novel genes.

Since more and more metagenomes will be available in the future, the third goal of RAMMCAP is to provide a new way to compare metagenomes from various environmental conditions and to identity and visualize the statistically significant differences between metagenomes.

In this paper, RAMMCAP was implemented and applied to the two largest metagenomic collections. The first set, GOS [1, 2], features 7.7 million ~800 base Sanger reads from 44 samples. A second, the Biomes [9] set, has 14.6 million ~100 base 454 reads from 45 microbiomes and 42 viromes samples. With moderate computational effort, RAMMCAP can quickly analyzed these huge datasets and obtained many novel results that could not be achieved by other existing methods.

RESULTS AND DISCUSSION

Implementation

RAMMCAP is illustrated in Figure 1. Cluster analysis is a key approach in this pipeline. Our previous ultra-fast sequence clustering algorithm CD-HIT [24, 25, 26] was modified to handle large metagenomic datasets. Using the DNA version of CD-HIT, the metagenomic reads from one or more metagenomes are clustered together at 95% sequence identity over 80% of length (clustering parameters can be adjusted by users) to identify clusters of unique genomic sequences, referred to as read clusters. It takes ~1 hour to cluster a million 200 base reads.

Citation: Weizhong Li, Analysis and comparison of very large metagenomes with fast clustering and functional annotation, doi: 10.1186/1471-2105-10-359.

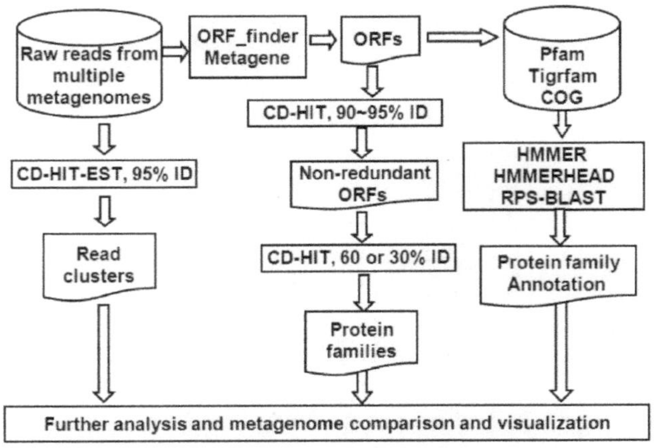

Figure 1: Metagenomic data analysis pipeline RAMMCAP.

ORFs are collected from sequence reads with ORF_finder, a ORF calling program implemented here by six reading frame translation in a similar way as the GOS study[2]. Within each reading frame, an ORF starts at the beginning of a read or the first ATG after a previous stop codon; it ends at the first stop codon or the end of that read. The minimal length of ORFs can be specified by users. ORFs can also be called from sequence reads with program Metagene[21]. Since these sequence reads are short, a predicted ORF maybe a portion of a complete ORF. An ORF may also be a translation from a non-coding frame: such an ORF is called a spurious ORF, as defined in the original GOS study [2]. The GOS study also introduced a spurious ORF detection method using nonsynonymous to synonymous substitution test, which is available along with a recent GOS clustering study [27]. This method is not integrated within RAMMCAP, but it can be used independently to identify the spurious ORFs predicted here.

ORFs are first clustered at 90-95% identity to identify the non-redundant sequences, which are further clustered to families (ORF clusters) at a conservative threshold, so that each cluster contains sequences of the same or similar function. A 30% sequence identity indicates significant similarities for full-length proteins. Since ORFs from Sanger reads are long enough, so they are clustered at 30% identity over 80% of ORF length. ORFs from 454 reads are much shorter; they are clustered at 60% identity over 80% of ORF length. ORF clusters are used for functional studies. The size of an ORF cluster is the number of its non-redundant sequences. For one million ORFs, it takes a few CPU hours to cluster at 60% identity and ~100 CPU hours at 30% identity.

The clustering method in RAMMCAP is quite different from the clustering method in the GOS study [2] and its incremental update [27], which generated core clusters by all-against-all BLAST search and then merged core clusters into final clusters using sequence profile methods. The final clusters in the GOS study are large and contain sequences of very remote similarities, whereas the clustering method employed here only tries to generate very conservative clusters.

ORFs are annotated from Pfam and Tigrfam with HMMER[28] (accelerated with Hammerhead[29]), and from COG with RPS-BLAST[30]. Hits must be with e-values ≤ 0.001, and meet the significant scores in case of HMMER searches. This annotation process only takes ~100 CPU hours for one million ORFs.

Optionally, ORF annotation may be performed quickly by running only the representative sequence of each ORF cluster and then transferring the annotation to other members in that cluster. But transferred annotation may be wrong in some cases, for example, where the target ORF has fewer domains than the source ORF (the representative). Since the annotation process is very fast, it is preferable to run all the ORFs for more accurate annotation.

Statistical comparison of metagenome

In many metagenomic projects, multiple samples from different environmental conditions were studied. This manuscript describes a novel way to compare metagenomes and visualize their differences. Since sequences from multiple metagenomes are clustered into families or classified into reference families (Pfam, Tigrfam, or COG), metagenomes can be compared by their occurrence profiles across all the families or selected families of interests.

Here, an occurrence profile coefficient, $r_{AB} = N_{A \cap B}/N_{A \cup B}$, is defined as the similarity measure between two metagenomes A and B. $N_{A \cap B}$ is the number of families that are found in both A and B above a minimal occurrence cutoff (defined later) without significant difference. $N_{A \cup B}$ is $N_{A \cap B}$ plus the number of families that occur in one metagenome significantly higher than in another metagenome. The value of r_{AB} is between 0-1, with 0 representing no overlap and 1 indicating a perfect match between A and B.

Let N_A and N_B be the number of sequences in A and B, let H_A and H_B be the number of sequences that occur in family H. One question is whether the difference between H_A and H_B is statistically significant. Rodriguez-Brito *et al* introduced a method to test such statistical significance of differences between two metagenomes [16]. Rodriguez-Brito's method used a large amount (in order of 10^5) of simulations by randomly picking a certain number of sequences from A, B, and A+B to generate distributions and analyze it, so it is very time-consuming.

In this paper, the *z* test for two independent proportions[31] is adopted to test the statistical significance of differences between two metagenomes. Given N_A, N_B, H_A, and H_B, there are three occurrence rates $P_A = H_A/N_A$, $P_B = H_B/N_B$, and $P = (H_A + H_B)/(N_A + N_B)$. The statistical significance between A and B can be described by:

$$z = \frac{P_A - P_B}{\sqrt{P(1-P)\left(\frac{1}{N_A} + \frac{1}{N_B}\right)}}$$

This method just needs calculation of a single equation, but it produces near identical results as the Rodriguez-Brito's method. Comparisons between the Rodriguez-Brito's method and the *z* test method under several different combinations of N_A, N_B, H_A, and H_B is shown in Figure 2.

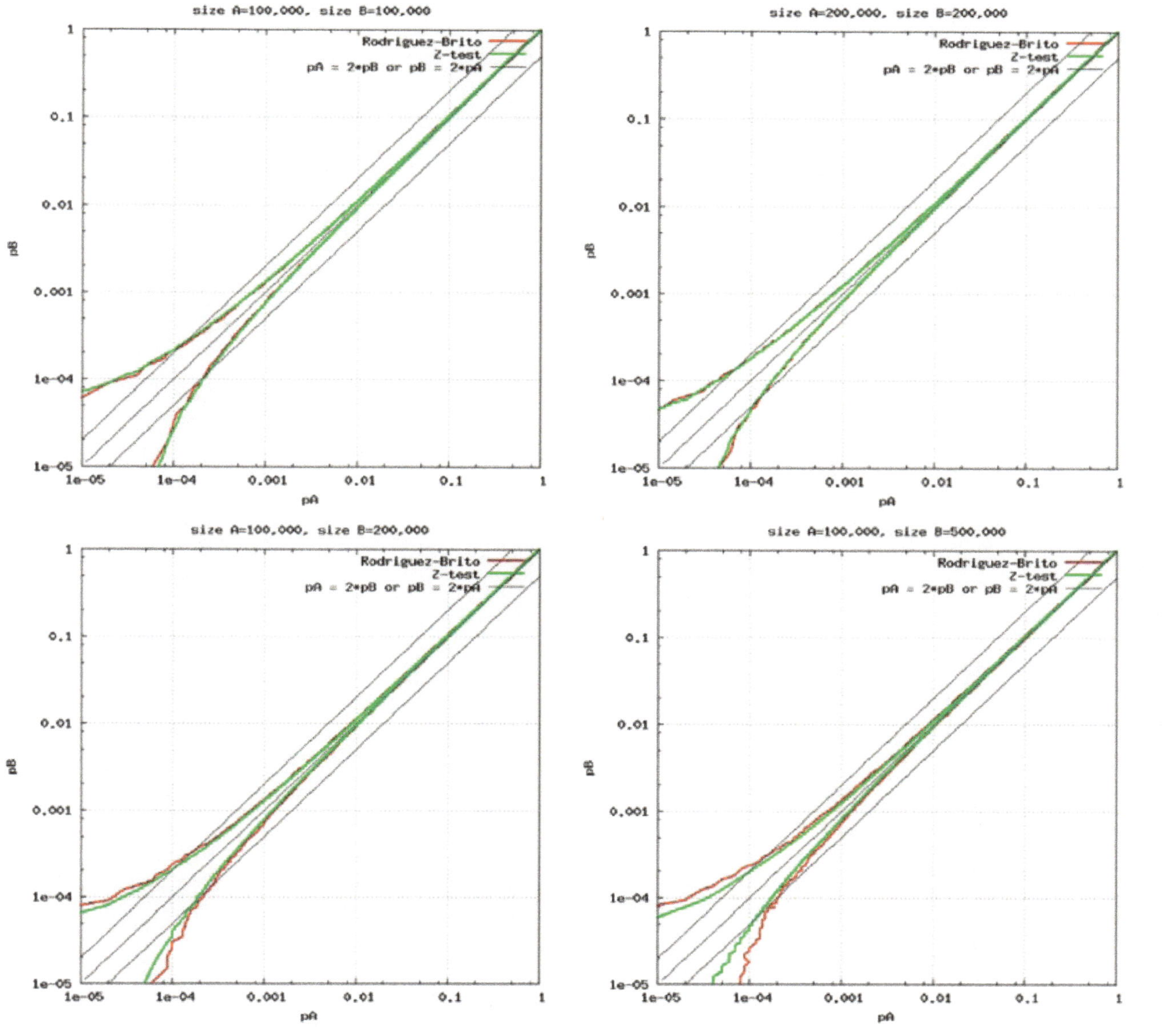

Figure 2: Comparison between Rodriguez-Brito's method and z test method. The *x*-axis and *y*-axis are occurrence rate P_A and P_B of two samples A and B. The 4 plots are made with different combination of sample size N_A and N_B as indicated in each plot. Red lines and green lines are calculated with Rodriguez-Brito's method and *z* test method respectively. Difference of A and B outside the area enclosed by a pair of red (or green) lines is statistically significant at 0.95 confidence level. This figure shows that when P_A and P_B become big enough (such as >0.001), a very small difference between them will be counted as significant.

In this manuscript, H_A is considered significantly higher than H_B if both (1) the *z* score satisfied a user defined confidence level such as 0.95, and (2) $P_A \geq f \times P_B$, where f (f>1) is also a user defined parameter, called significant factor.

The statistical significance cannot be established at very low occurrences, so the low occurrence families are excluded from $N_{A \cap B}$ in calculation of r_{AB}. The minimal occurrence cutoff of a family H within metagenome A is defined as the minimal number of H_A to produce a significant *z* score when A is compared to another metagenome B where $N_A = N_B$ and $H_B = 0$. It can be obtained that:

$$H_A \geq \frac{z^2}{1+z^2/2N_A}$$

Since $2N_A >> z^2$, the minimal cutoff of H_A is 4 at 0.95 confidence level (z = 1.96), and 7 at 0.99 confidence level (z = 2.58). The occurrence profile coefficients are calculated for all the metagenome pairs, and the output matrix is used to hierarchically cluster the metagenomes.

Performance of ORF prediction

ORF_finder and Metagene [21] were evaluated with simulated metagenomic reads generated with MetaSim software[32] from the completed microbial genomes released between January and May 2009 from NCBI. Four datasets (Sim100, Sim200, Sim400 and Sim800) of 1 million reads each with average length of 100, 200, 400 and 800 bases were generated to simulate the current sequencing techniques. The error rates for both 454 (>3%) and Sanger (1~2%) defined by MetaSim are much higher than the reported error rates [33, 34], so the exact model with default parameters were used in MetaSim for all simulated datasets. If a simulated read overlaps with an annotated ORF by NCBI at least 30 amino acid, the overlapped part is used as a true ORF. A predicted ORF by either ORF_finder or Metagene is a true-positive if at least 50% of it overlaps with a true ORF within the same reading frame. The ratios of true-positives relative to all true ORFs (sensitivity) and to predicted ORFs (specificity) were used as a performance measure (Table 1). Four ORF cutoff lengths (30, 40, 50, and 60 amino acids) were applied for ORF_finder. This analyses show that Metagene has high sensitivity and specificity (~94% and ~92%) for reads of at least 200 bases. But its sensitivity drops to 59% for 100 base sequences. ORF_finder has very high sensitivity (>99%) and very low specificity (<37%) at 30 amino acid cut off length; a longer cutoff length produces a higher specificity but lowers the sensitivity.

Table 1: Sensitivity and specificity of Metagene and ORF_finder on simulated metagenomic datasets

	Sensitivity/specificity (%)			
Datasets **Method**	**Sim100**	**Sim200**	**Sim400**	**Sim800**
Metagene	59.07/92.58	94.31/92.32	93.84/91.91	94.02/93.25
ORF_finder, cut off = 30aa	99.93/36.56	99.74/36.73	99.78/34.49	99.92/32.83
ORF_finder, cut off = 40aa		96.74/46.65	97.89/43.81	99.13/41.54
ORF_finder, cut off = 50aa		91.65/54.82	94.04/52.51	96.88/50.21
ORF_finder, cut off = 60aa		73.66/61.27	88.94/59.84	93.34/57.98

Testing on clustering

RAMMCAP was applied to GOS[2] and BIOME[9], the two largest metagenomic collections. Prior to the full development of RAMMCAP, we clustered the 17.4 million GOS ORFs released by the original GOS study [2] at 30% identity and published the detailed clustering results [35]. Here the BIOME reads and BIOME ORFs, which have 14.6 and 24.6 million sequences respectively, were clustered at 95% and 60% identity (Figure 3). The previous GOS clusters were added in Figure 3 to show the difference between GOS and Biome (Figure 3a, b).

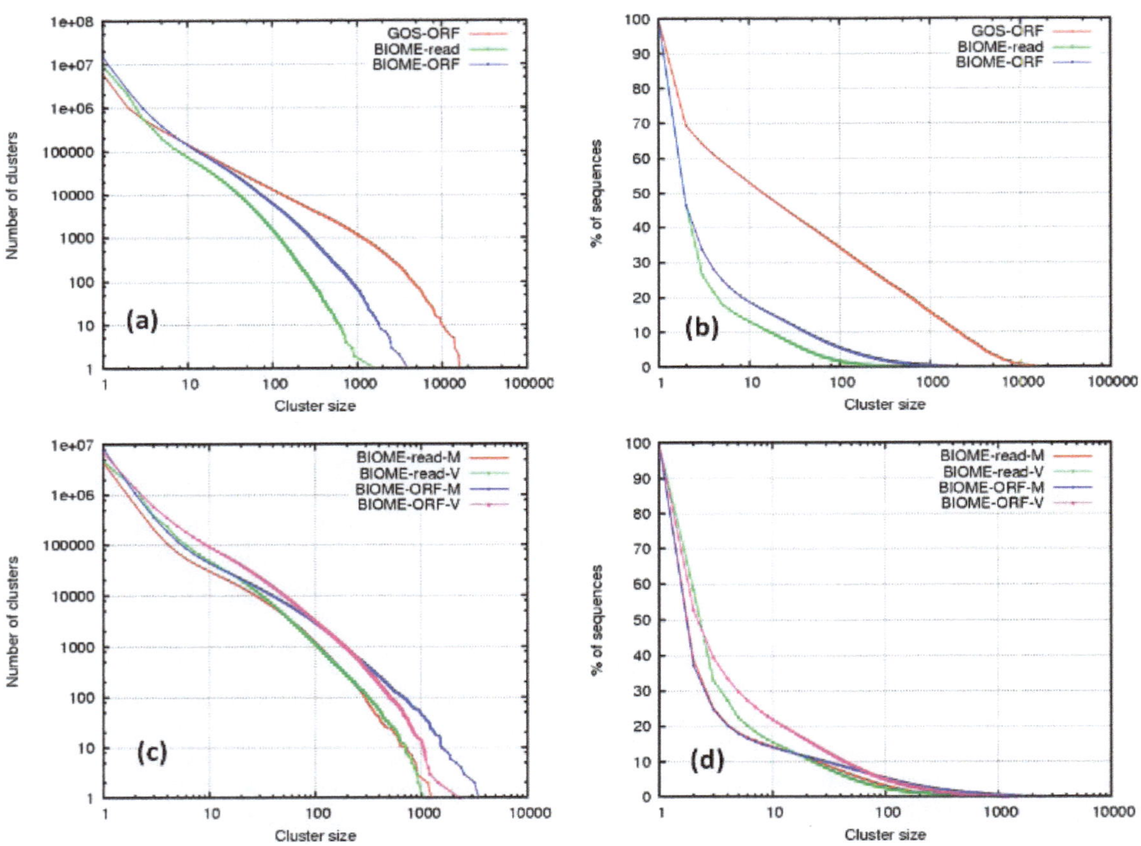

Figure 3: Distribution of clusters and sequences by cluster size. The *x*-axis is the cluster size *X*. The *y*-axis in left figures (a and c) is the number of clusters of size at least *X*; the *y*-axis in right figures (b and d) is the percentage of total sequences included in the clusters of size at least *X*. Clustering analyses were also made separately for the microbiomes and the viromes. So, together there are seven clustering experiments: GOS-ORF, BIOME-read, BIOME-ORF, BIOME-read-M, BIOME-read-V, BIOME-ORF-M, and BIOME-ORF-V (where M and V stand for microbiomes and viromes).

GOS ORFs have more than 1000 clusters that contain ≥ 1000 non-redundant sequences; BIOME ORFs have less than 100 such clusters. About 70% GOS ORFs, 46% BIOME reads, and a similar percent BIOME ORFs are found in non-singleton clusters. Within the BIOME datasets (Figure 3c, d), the microbiomes have more large clusters compared to viromes; suggesting that microbial sequences are more conserved than viral sequences.

Clustering analysis is a powerful tool to recover protein families and to discover the novel ones; it helps to recognize spurious ORFs. Clustering tends to put real ORFs into large clusters and leaves spurious ones in small or singleton clusters because spurious ORFs have more random features. If the ORFs with Pfam, Tigrfam, or COG matches are considered true ORFs, then 93% of these true GOS ORFs are found in clusters of size ≥ 10, which is only 1.3% total GOS clusters; here cluster size is the number of non-redundant sequences in a cluster. Further, 28% of the true BIOME ORFs are in 1.0% of top clusters of size ≥ 5. Many large clusters without any homology to known proteins are found, which may shed light on novel families of environment specific functions.

Validation of clustering quality

ORFs called from metagenomic reads are short and fragmented. In addition, errors such as frame shifting and wrong gene boundary may occur due to sequencing errors. Therefore a conservative threshold is used in producing ORF clusters to ensure that a cluster contains sequences of the same or similar function. The quality of clustering was evaluated with Pfam, the manually curated classification of protein families. The domain sequences in Pfam models (release 22.0) were extracted from the alignments and were clustered at 30% identity using the clustering protocol in RAMMCAP.

Not all the Pfam sequences were used. Very short Pfam families <30 amino acids were excluded because most of these families were built by sequence patterns rather than similarities. Some Pfam families are overlapping, for those families, short ones were excluded. Since Pfam families were built with very sensitive HMM models, the sequences within same families can be very diverse, even at sequence identities very much below 30%. Therefore it is anticipated that divergent sequences from a large Pfam family may be placed in separate clusters. The goals of clustering are: (1) to generate homogeneous clusters whose sequences are from the same Pfam families, (2) to cover most sequences in a small number of large clusters. The distributions of clusters of Pfam domain sequences are shown in Figure 4. A cluster is considered good if >95% of its members are from the same Pfam family. It is observed that most sequences (>97%) are in good clusters (Figure 4a), which are ~30 times more than bad clusters (Figure 4b). Although there are singleton clusters, but non-singleton clusters still cover 94% of sequences, and clusters of size ≥ 5 cover 85% of sequences.

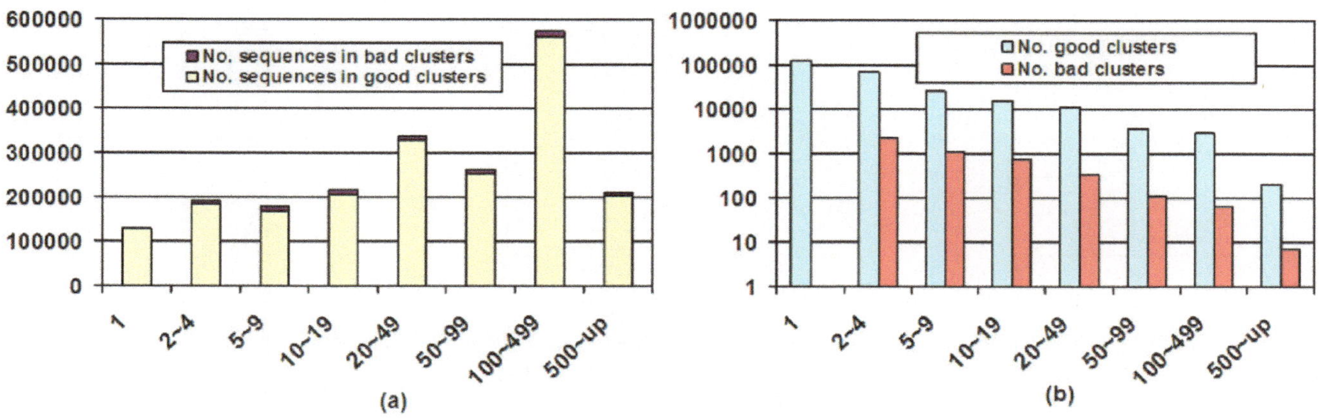

Figure 4: Distribution of clusters of Pfam sequences. The *x*-axis is cluster size. The *y*-axis in (a) is the number of sequences, and the *y*-axis in (b) is the number of clusters.

Testing on clustering-based comparison of metagenomes

Statistical comparisons of GOS and BIOME metagenomes based on the occurrence profile coefficient calculations using results of clusters are shown in Figure 5. The GOS samples show great overlaps, but all unique samples are classified as outliers, such as GS033 (hypersaline), GS020 (fresh water), and GS025 (reef, different filter size), from other marine samples. The sample-specific clusters may shed light on functional aspects related to the environment for further studies. The BIOME samples intersect much less, but notable overlaps are found between pairing samples, such as Fish-M vs Fish-V, and Coral-M vs Coral-V. The differences between GOS and BIOME samples reflect that the BIOME samples are more diverse.

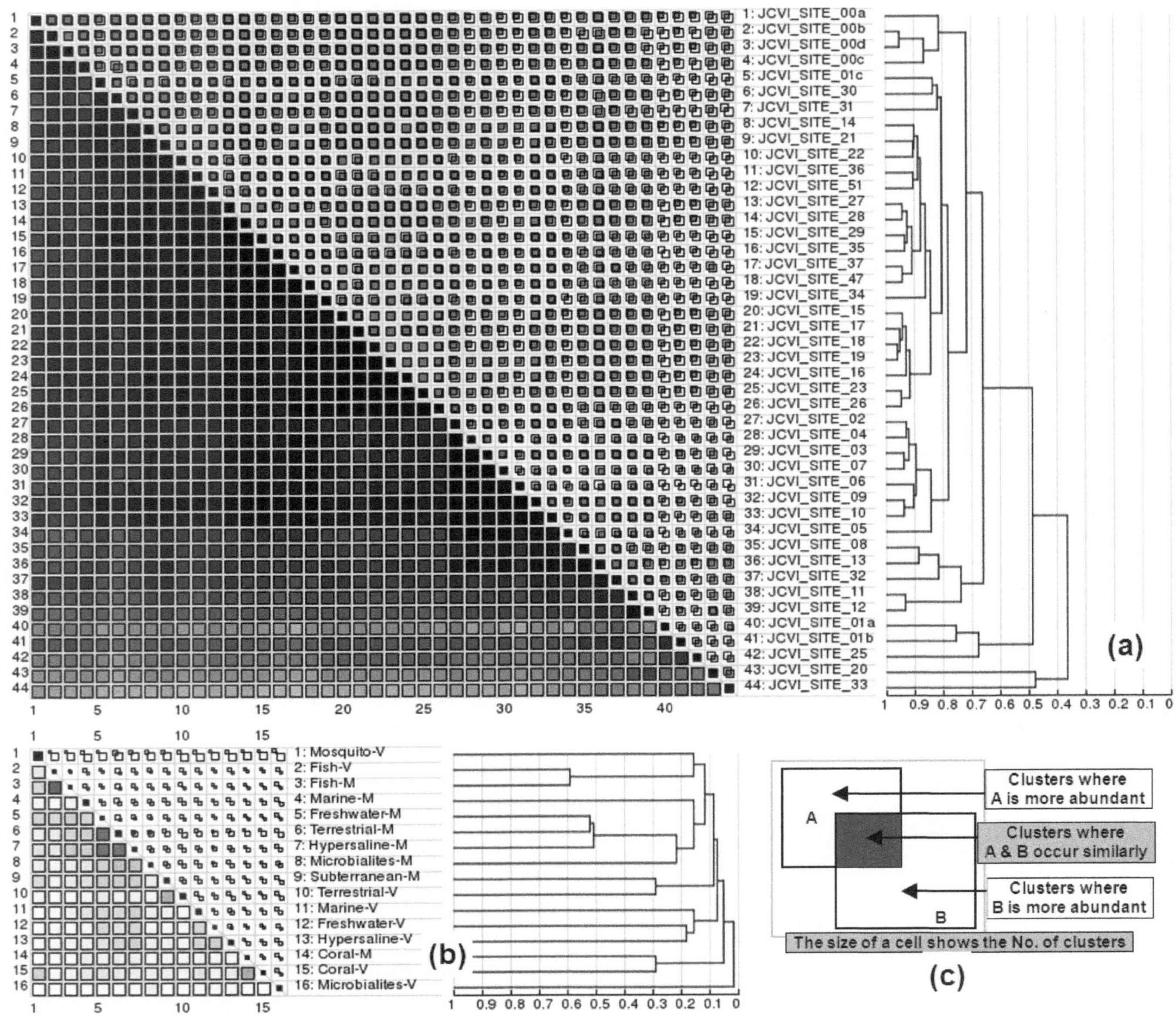

Figure 5: Similarity matrices of metagenomes. Squares along the diagonal represent the number of clusters where a sample occurs. Grayscale squares below the diagonal represent the occurrence profile coefficients r_{AB} between two samples with a darker color indicating a greater similarity. Cells above the diagonal show the unique and overlapping clusters, explained in **(c)**. Hierarchical clustering of samples based on the matrix is shown with vertical gridlines indicating the value of the coefficient where two nodes are merged. Matrices are made for GOS ORF clusters **(a)** and BIOME ORF clusters **(b)** with significant a factor $f = 2$ at 0.95 confidence level. The BIOME samples are grouped by biome type, such as Coral-M, which stands for coral microbiomes sample.

Testing on protein family-based comparison of metagenomes

Annotations from Pfam, Tigrfam, and COG are used for metagenome comparison in a similar way that is applied in clusters. These analyses show over- and under-represented families between samples. The protein families of these three databases are all organized into super families: clans in Pfam, role categories in Tigrfam, and functional classes in COG. Therefore, metagenome comparison can be made under a specific super family of interest, which is a unique feature of this study. Samples were compared systematically under all super families and many significant differences were found. Here we show a few interesting examples (Figure 6). GOS samples share an extremely similar occurrence profile across the 86 families of COG class F (nucleotide transport and metabolism), which suggests this class is highly conserved across the world's ocean. The least conserved class, aside from functional unknown classes and a few tiny classes, is T (Signal transduction mechanisms). Similar observations were obtained for Biome samples, but the intrinsic diversity of BIOME sequences introduces more non-overlapping families.

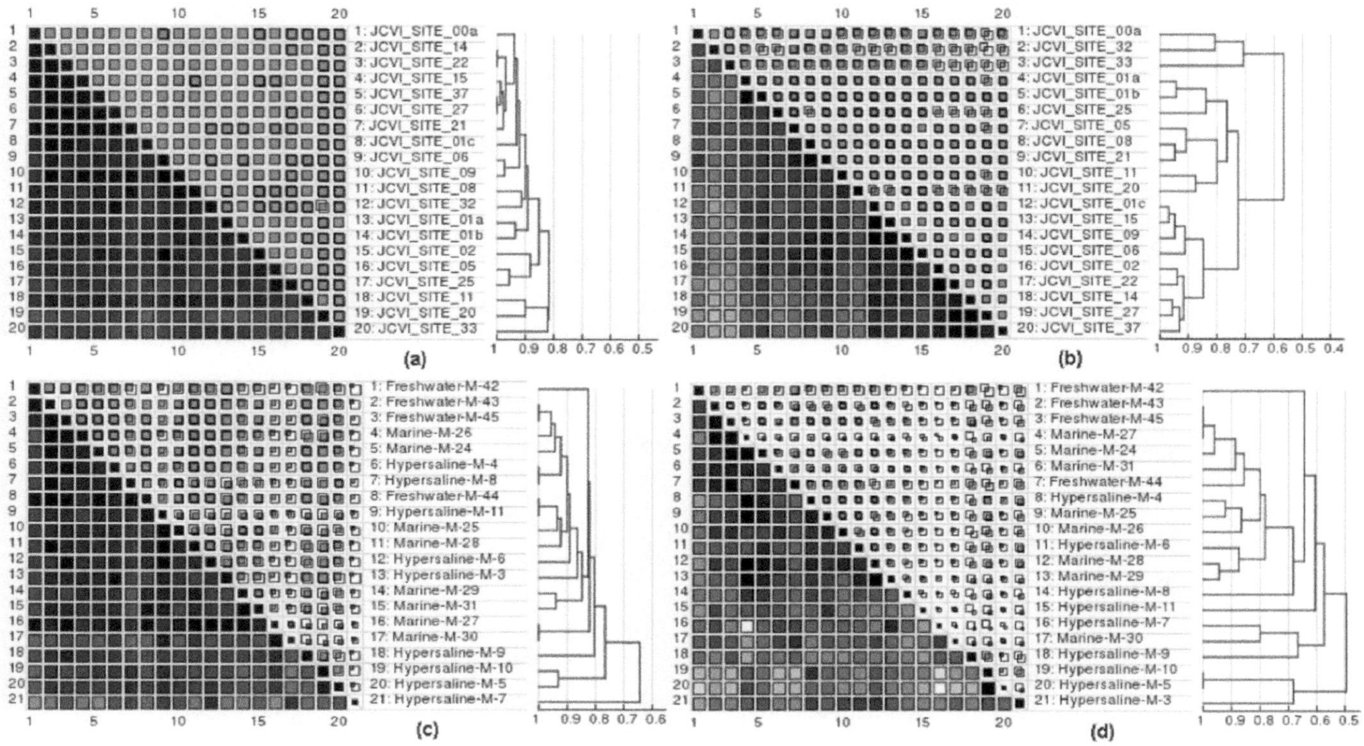

Figure 6: Similarity matrices of metagenomes based on families of two COG classes. Matrices are for GOS on COG class F**(a)**, GOS on class T **(b)**, BIOME on F **(c)**, and BIOME on T **(d)** respectively with significant a factor $f = 2$ at 0.95 confidence level. Because GOS samples are microbial marine samples, only the microbial (non-viral) water samples from the BIOME data was used. Further, a representative subset from GOS samples was selected so that the figures of GOS and BIOME are similar in size.

CONCLUSION

The CPU time for clustering GOS ORFs, BIOME reads, and BIOME ORFs were 9600, 125, and 513 hours, respectively. GOS ORFs cost relatively more, but still two orders of magnitude less than the original GOS study[2]. The annotation for GOS ORFs and BIOME ORFs took 3800 and 1560 hours. Through clustering analysis, many novel families can be identified and can be used in metagenome comparison. The RAMMCAP software and pre-calculated results are available at http://tools.camera.calit2.net/camera/rammcap/.

ACKNOWLEDGEMENTS

This work was supported by Gordon and Betty Moore Foundation through the CAMERA projecthttp://camera.calit2.net and NIH grant 1R01RR025030 from National Center for Research Resources. The CAMERA team provided much technical supports, especially Dr. Ying Huang who gave advices on statistical method. Forest Rohwer from San Diego State University provided the early access to the Nine Biomes data.

REFERENCES

1. Rusch DB, Halpern AL, Sutton G, Heidelberg KB, Williamson S, Yooseph S, Wu D, Eisen JA, Hoffman JM, Remington K, *et al.*: The Sorcerer II Global Ocean Sampling Expedition: Northwest Atlantic through Eastern Tropical Pacific. *PLoS Biol*2007, 5(3):e77. 10.1371/journal.pbio.0050077
2. Yooseph S, Sutton G, Rusch DB, Halpern AL, Williamson SJ, Remington K, Eisen JA, Heidelberg KB, Manning G, Li W, *et al.*: The Sorcerer II Global Ocean Sampling Expedition: Expanding the Universe of Protein Families. *PLoS Biol* 2007, 5(3):e16. 10.1371/journal.pbio.0050016
3. Gill SR, Pop M, Deboy RT, Eckburg PB, Turnbaugh PJ, Samuel BS, Gordon JI, Relman DA, Fraser-Liggett CM, Nelson KE: Metagenomic analysis of the human distal gut microbiome. *Science* 2006, 312(5778):1355–1359. 10.1126/science.1124234
4. DeLong EF, Preston CM, Mincer T, Rich V, Hallam SJ, Frigaard NU, Martinez A, Sullivan MB, Edwards R, Brito BR, *et al.*: Community genomics among stratified microbial assemblages in the ocean's interior. *Science* 2006, 311(5760):496–503. 10.1126/science.1120250
5. Angly FE, Felts B, Breitbart M, Salamon P, Edwards RA, Carlson C, Chan AM, Haynes M, Kelley S, Liu H, *et al.*: The marine viromes of four oceanic regions. *PLoS Biol* 2006, 4(11):e368. 10.1371/journal.pbio.0040368
6. Tringe SG, von Mering C, Kobayashi A, Salamov AA, Chen K, Chang HW, Podar M, Short JM, Mathur EJ, Detter JC, *et al.*: Comparative metagenomics of microbial communities. *Science* 2005, 308(5721):554–557. 10.1126/science.1107851
7. Venter JC, Remington K, Heidelberg JF, Halpern AL, Rusch D, Eisen JA, Wu D, Paulsen I, Nelson KE, Nelson W, *et al.*: Environmental genome shotgun sequencing of the Sargasso Sea. *Science* 2004, 304(5667):66–74. 10.1126/science.1093857

8. Tyson GW, Chapman J, Hugenholtz P, Allen EE, Ram RJ, Richardson PM, Solovyev VV, Rubin EM, Rokhsar DS, Banfield JF: Community structure and metabolism through reconstruction of microbial genomes from the environment. *Nature*2004, 428(6978):37–43. 10.1038/ nature02340Dinsdale EA, Edwards RA, Hall D, Angly F, Breitbart M, Brulc JM, Furlan M, Desnues C, Haynes M, Li L, *et al.*: Functional metagenomic profiling of nine biomes. *Nature* 2008, 452(7187):629–632. 10.1038/ nature06810
9. Mavromatis K, Ivanova N, Barry K, Shapiro H, Goltsman E, McHardy AC, Rigoutsos I, Salamov A, Korzeniewski F, Land M,*et al.*: Use of simulated data sets to evaluate the fidelity of metagenomic processing methods. *Nat Methods* 2007, 4(6):495–500. 10.1038/ nmeth1043Markowitz VM, Ivanova NN, Szeto E, Palaniappan K, Chu K, Dalevi D, Chen IM, Grechkin Y, Dubchak I, Anderson I, *et al.*: IMG/M: a data management and analysis system for metagenomes. *Nucleic Acids Res* 2008, (36 Database):D534–538.
10. Seshadri R, Kravitz SA, Smarr L, Gilna P, Frazier M: CAMERA: A Community Resource for Metagenomics. *PLoS Biol* 2007, 5(3):e75. 10.1371/journal.pbio.0050075
11. Aziz RK, Bartels D, Best AA, DeJongh M, Disz T, Edwards RA, Formsma K, Gerdes S, Glass EM, Kubal M, *et al.*: The RAST Server: rapid annotations using subsystems technology. *BMC Genomics* 2008, 9: 75. 10.1186/1471-2164-9-75
12. Krause L, Diaz NN, Goesmann A, Kelley S, Nattkemper TW, Rohwer F, Edwards RA, Stoye J: Phylogenetic classification of short environmental DNA fragments. *Nucleic Acids Res* 2008, 36(7):2230–2239. 10.1093/nar/ gkn038
13. Huson DH, Auch AF, Qi J, Schuster SC: MEGAN analysis of metagenomic data. *Genome Res* 2007, 17(3):377–386. 10.1101/gr.5969107
14. Rodriguez-Brito B, Rohwer F, Edwards RA: An application of statistics to comparative metagenomics. *BMC Bioinformatics*2006, 7: 162. 10.1186/1471-2105-7-162
15. Schloss PD, Handelsman J: A statistical toolbox for metagenomics: assessing functional diversity in microbial communities. *BMC Bioinformatics* 2008, 9: 34. 10.1186/1471-2105-9-34
16. Teeling H, Meyerdierks A, Bauer M, Amann R, Glockner FO: Application of tetranucleotide frequencies for the assignment of genomic fragments. *Environ Microbiol* 2004, 6(9):938–947. 10.1111/j.1462-2920.2004.00624.x
17. Teeling H, Waldmann J, Lombardot T, Bauer M, Glockner FO: TETRA: a web-service and a stand-alone program for the analysis and comparison of tetranucleotide usage patterns in DNA sequences. *BMC Bioinformatics* 2004, 5: 163. 10.1186/1471-2105-5-163
18. McHardy AC, Martin HG, Tsirigos A, Hugenholtz P, Rigoutsos I: Accurate phylogenetic classification of variable-length DNA fragments. *Nat Methods* 2007, 4(1):63–72. 10.1038/nmeth976
19. Noguchi H, Park J, Takagi T: MetaGene: prokaryotic gene finding from environmental genome shotgun sequences.*Nucleic Acids Res* 2006, 34(19):5623–5630. 10.1093/nar/gkl723
20. Noguchi H, Taniguchi T, Itoh T: MetaGeneAnnotator: detecting species-specific patterns of ribosomal binding site for precise gene prediction in anonymous prokaryotic and phage genomes. *DNA Res* 2008, 15(6):387–396. 10.1093/dnares/dsn027
21. Hoff KJ, Tech M, Lingner T, Daniel R, Morgenstern B, Meinicke P: Gene prediction in metagenomic fragments: a large scale machine learning approach. *BMC Bioinformatics* 2008, 9: 217. 10.1186/1471-2105-9-217
22. Li W, Jaroszewski L, Godzik A: Clustering of highly homologous sequences to reduce the size of large protein databases.*Bioinformatics* 2001, 17(3):282–283. 10.1093/bioinformatics/17.3.282
23. Li W, Jaroszewski L, Godzik A: Tolerating some redundancy significantly speeds up clustering of large protein databases.*Bioinformatics* 2002, 18(1):77–82. 10.1093/bioinformatics/18.1.77
24. Li W, Godzik A: Cd-hit: a fast program for clustering and comparing large sets of protein or nucleotide sequences. *Bioinformatics* 2006, 22(13):1658–1659. 10.1093/ bioinformatics/btl158
25. Yooseph S, Li W, Sutton G: Gene identification and protein classification in microbial metagenomic sequence data via incremental clustering. *BMC Bioinformatics* 2008, 9: 182. 10.1186/1471-2105-9-182
26. Eddy SR: Profile hidden Markov models. *Bioinformatics* 1998, 14(9):755–763. 10.1093/bioinformatics/14.9.755
27. Portugaly E, Johnson S, Ninio M, Eddy S: Improved HMMERHEAD for Better Sensitivity. *RECOMB 07 Poster, San Fransico*2007.
28. Altschul SF, Madden TL, Schaffer AA, Zhang J, Zhang Z, Miller W, Lipman DJ: Gapped BLAST and PSI-BLAST: a new generation of protein database search programs. *Nucleic Acids Res* 1997, 25(17):3389–3402. 10.1093/nar/25.17.3389
29. Sheskin D: *Handbook of parametric and nonparametric statistical procedures.* 3rd edition. Boca Raton, Fla.: Chapman & Hall/CRC; 2004.
30. Richter DC, Ott F, Auch AF, Schmid R, Huson DH: MetaSim: a sequencing simulator for genomics and metagenomics.*PLoS ONE* 2008, 3(10):e3373. 10.1371/ journal.pone.0003373
31. Quinlan AR, Stewart DA, Stromberg MP, Marth GT: Pyrobayes: an improved base caller for SNP discovery in pyrosequences. *Nat Methods* 2008, 5(2):179–181. 10.1038/nmeth.1172
32. Shendure J, Ji H: Next-generation DNA sequencing. *Nat Biotechnol* 2008, 26(10):1135–1145. 10.1038/nbt1486
33. Li W, Wooley JC, Godzik A: Probing metagenomics by rapid cluster analysis of very large datasets. *PLoS ONE* 2008, 3(10):e3375. 10.1371/journal.pone.0003375.

CHAPTER 17

PHYDESIGN: AN ONLINE APPLICATION FOR PROFILING PHYLOGENETIC INFORMATIVENESS

Francesc López-Giráldez and Jeffrey P Townsend

BACKGROUND

Due to advances in sequencing technologies and decreasing costs, the number of genomes sequenced for species across the tree of life is increasing dramatically. Tools and databases for selecting single-copy orthologous loci [1, 2] and designing successful primers for them [3, 4] are available. However, orthology assessments for multiple genomes can provide thousands of candidate loci to sequence, and yet only a few of those have been commonly used as markers for phylogenetic studies [5]. To address the challenge of locus selection for sequencing in designing a phylogenetic study, Townsend [6] proposed a metric that provides a quantitative prediction of phylogenetic signal across historical times. Based on estimates of rates across sites, the phylogenetic informativeness metric facilitates prioritization of loci even when the taxa of interest have never been sequenced for a given locus. To estimate phylogenetic informativeness, prior data on the molecular evolutionary pattern of a locus is required. This prior information may be derived from three potential sources: 1) preliminary data on the candidate loci from a well-studied subset of the taxa of interest; 2) data on the candidate loci from a well-studied sister clade; or 3) comparative genomic data from sequenced genomes within and/or outside the clade of interest. Using this information for a number of classic and recent data sets for different time scales, the method successfully recapitulates the qualitative utility of loci [6, 7, 8, 9,10]. However, no software or automated methodology to apply the phylogenetic informativeness metric has been available.

Citation: Francesc López-Giráldez and Jeffrey P Townsend, PhyDesign: an online application for profiling phylogenetic informativeness, DOI: 10.1186/1471-2148-11-152.

To facilitate wider application of phylogenetic informativeness analysis, we have developed the online application PhyDesign. PhyDesign features:

- Online accessibility, platform independence, and immediate access to software updates.
- User-friendly graphical interface with rich process feedback.
- Multiple locus analyses of amino acid and DNA alignments and multiple epoch integration.
- High quality graphical outputs in SVG format easily edited for final publication.
- Spreadsheet outputs including site rate estimates and phylogenetic informativeness values in rank-order of priority for the epochs of interest, as well as tables to reproduce the profiles for further processing.

PhyDesign is freely accessible online at http://phydesign.townsend.yale.edu. A manual and sample files to be used with the application, as well as a FAQ section, can be also found at the site. In addition, the source code and a Perl module to calculate the profiles are available for download.

IMPLEMENTATION

To estimate phylogenetic informativeness profiles [6], the PhyDesign application consists of 3 components:

- A form to upload information and choose an application to calculate the evolutionary rates for each alignment site.
- A table listing the evolutionary rate results.
- A graphical interface to plot the phylogenetic profiles and calculate integration values.

On the client-side, these components of PhyDesign are coded in JavaScript/AJAX using jQuery libraries, facilitating dynamic communication between the server and the user without page reloading. Online input undergoes validation to detect possible errors before engaging the next step in the process. To communicate validation success or failure and its causes, unobtrusive temporary message boxes are featured. In addition, the web site uses modal windows to provide optional advanced settings. On the server-side, PhyDesign is implemented as a collection of Perl scripts and modules, integrated with third-party software used to calculate the evolutionary rates for each site in the alignment.

To run, PhyDesign requires at minimum an internet browser equipped with JavaScript, and SVG for the graphical output. The latest versions of major modern browsers have native support for JavaScript and for rendering SVG markup directly. Internet Explorer 8 and older versions require a plug-in to render SVG content. SVG format, as vector graphics format, is resolution-independent and easy to edit for final publication with common vector graphic editors such as Adobe Illustrator (commercially available) and Inkscape (open source). The online application has been tested as rendered by Internet Explorer, Firefox, Chrome and Safari. Although graphical representations of phylogenetic informativeness profiles are typically best kept to lower numbers of loci for purposes of display, PhyDesign was able to profile and produce quantitative results for 2000 loci of a length of 1000 bp each. For a higher volume of data, a Perl module to calculate the profiles is available for download.

Input form

To obtain the site rate distributions for each locus, two data entries are needed: (1) an alignment of loci of interest, pruned to contain a set of taxa for which the tree topology is fairly well known, and (2) an ultrametric tree for those taxa. The ultrametric tree can be a chronogram (an ultrametric tree with branch lengths proportional to time). Alternatively, branch lengths can be in unspecified molecular evolutionary units. PhyDesign reads amino acid and DNA alignments in NEXUS, FASTA, and Phylip formats, and trees in NEXUS and Newick formats. Formats are carefully checked via custom and BioPerl modules [11]. Data partitioning is permitted via NEXUS format. Once files are uploaded and checked, PhyDesign provides an interface for selecting and parameterizing any one of several third-party programs that estimate the site rates - HyPhy [12], DNArates [Olsen, unpublished] and Rate4site [13]. DNArates and Rate4site are called by the application "as is"; a *de novo* HyPhy script was implemented to estimate rates under the full diversity of time-reversible models. Due to the ability to specify of the evolutionary model and its parameters, HyPhy is the recommended and default option for DNA sequences. For amino acid sequences, Rate4site is available. To facilitate extensive analyses of large datasets, a link to the rates results is sent by email. A site rate file with the proper formatting will be generated after analysis of the alignments and the corresponding ultrametric tree. Supplied alignments are deleted from the server immediately after the site rates are calculated, and the site rate results are stored in the server for 24 hours. During this time, results can only be accessed by the link submitted. It is convenient, then, to save the site rate file for future use. Alternatively, if the rate distribution for each locus is known, PhyDesign also accepts direct supply of these rates to obtain the phylogenetic informativeness profiles.

RESULTS AND DISCUSSION

Output of rate vectors

The first result provided features a table with basic information about each locus analyzed, including the program used to analyze the rates, locus length, the number of sites for which a substitution rate could be calculated, and the number of faulty sites for which this calculation was not possible. Two files are offered as downloads: (1) a compressed file containing individual rate files for each locus, and (2) a single file containing rate site vectors for all loci. The latter can be downloaded for future uploading in the rate vectors form, eliminating the need to repeat rate calculations. Partitions to be profiled and the colors for representing them may then be specified interactively.

Profiles and area calculations

The second result provided is a graphical representation of phylogenetic informativeness and calculations of quantitative measures of informativeness. Two graphs are shown in the main section: the ultrametric tree and, aligned to it, the phylogenetic informativeness profiles (Figure 1).

These graphs can be readily downloaded in manipulable SVG format as displayed. In addition, a downloadable spreadsheet file with the profile data points is provided in the left panel, so that profiles can be replotted or reanalyzed with diverse software. The left panel facilitates further customization of the plots, including selection of loci to display, colors of profiles, adjustment of the range of time values, integration over different epochs, and downloading of the results. Integrating phylogenetic informativeness over specific epochs provides a metric for ranking loci. Integration values will be largest for the loci that have the highest probability of substitution in the given epoch that will not be obscured by subsequent evolution. Note, however, that phylogenetic informativeness plots display predicted signal and do not account for phylogenetic noise (homoplasy) caused by convergence or parallelism in divergent lineages. Thus, quantitative results should be thoughtfully considered in light of homoplasy that is likely to arise, significantly diminishing utility during epochs deeper than the peak of informativeness for a given profile. All informativeness values can be calculated on a net or per site basis. While the net phylogenetic informativeness quantifies signal as a whole, it is more subject to phylogenetic noise [14, 15] than is phylogenetic informativeness per site, which maximizes both informativeness and cost-effectiveness, and more effectively minimizes noise. It also of conceptual interest to characterize the phylogenetic informativeness per site to compare relative power of genes without the confounding influence of gene length. A combination of shorter genes with a sequencing effort equal to that of a longer gene can lead to better results.

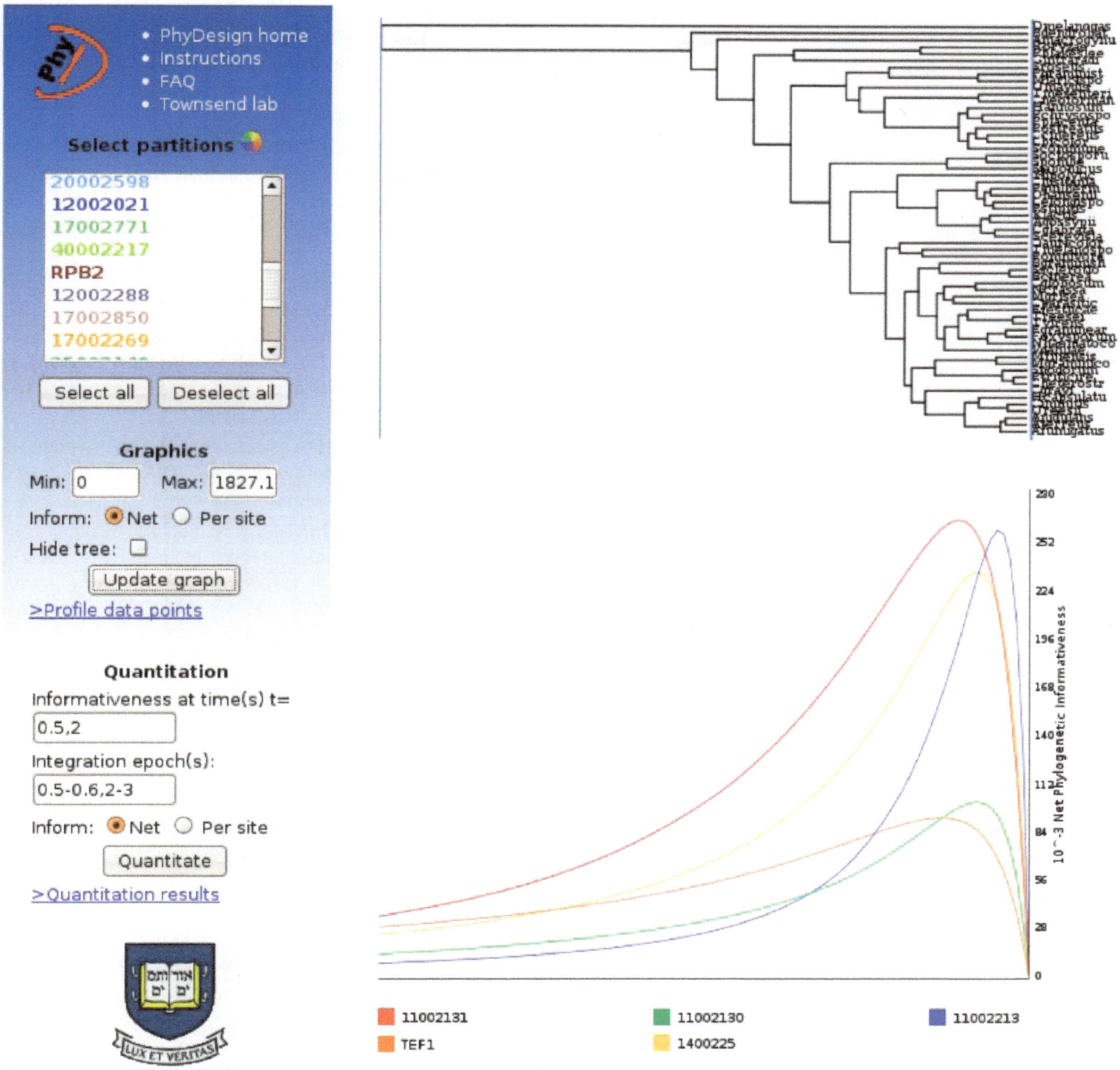

Figure 1: Screenshot of PhyDesign. Phylogenetic informativeness profiles are shown for five loci, aligned to a corresponding chronogram. The side bar is the interface through which the user can select loci and colors, adjust the range of time values, integrate over different epochs, and download the results.

CONCLUSIONS

By providing these profiles, PhyDesign facilitates locus prioritization, increasing the efficiency of sequencing for phylogenetic purposes compared to traditional studies with more laborious and low capacity screening methods, as well as increasing the accuracy of phylogenetic studies. Future website implementations will include the latest theoretical advances developed in our research group, expanding the current phylogenetic informativeness methodology to quantify the effects of parallelism and convergence, as well as quantifying the utility of taxon addition [16].

ACKNOWLEDGEMENTS

We are grateful to S.L.K. Pond for helping with the HyPhy batch file design and E. Roberts for helping with the JavaScript programming. We also thank two anonymous reviewers for their thoughtful and useful comments.

REFERENCES

1. Ranwez V, Delsuc F, Ranwez S, Belkhir K, Tilak MK, Douzery EJ: OrthoMaM: a database of orthologous genomic markers for placental mammal phylogenetics.

BMC Evol Biol. 2007, 7: 241-10.1186/1471-2148-7-241.

2. Marthey S, Aguileta G, Rodolphe F, Gendrault A, Giraud T, Fournier E, Lopez-Villavicencio M, Gautier A, Lebrun MH, Chiapello H: FUNYBASE: a FUNgal phYlogenomic dataBASE. BMC Bioinformatics. 2008, 9: 456-10.1186/1471-2105-9-456.
3. Boutros R, Stokes N, Bekaert M, Teeling EC: UniPrime2: a web service providing easier Universal Primer design. Nucleic Acids Res. 2009, W209-13. 37 Web Server
4. Contreras-Moreira B, Sachman-Ruiz B, Figueroa-Palacios I, Vinuesa P: primers4clades: a web server that uses phylogenetic trees to design lineage-specific PCR primers for metagenomic and diversity studies. Nucleic Acids Res. 2009, W95-W100. 37 Web Server
5. Cummings MP, Meyer A: Magic bullets and golden rules: data sampling in molecular phylogenetics. Zoology. 2005, 108 (4): 329-336. 10.1016/j.zool.2005.09.006.
6. Townsend JP: Profiling phylogenetic informativeness. Syst Biol. 2007, 56 (2): 222-231. 10.1080/10635150701311362.
7. Mahon BC, Neigel JE: Utility of arginine kinase for resolution of phylogenetic relationships among Brachyuran genera and families. Mol Phylogenet Evol. 2008, 48 (2): 718-727. 10.1016/j.ympev.2008.05.005.
8. Townsend JP, López-Giráldez F, Friedman R: The phylogenetic informativeness of nucleotide and amino acid sequences for reconstructing the vertebrate tree. J Mol Evol. 2008, 67 (5): 437-447. 10.1007/s00239-008-9142-0.
9. Schoch CL, Sung GH, López-Giráldez F, Townsend JP, Miadlikowska J, Hofstetter V, Robbertse B, Matheny PB, Kauff F, Wang Z, Gueidan C, Andrie RM, Trippe K, Ciufetti LM, Wynns A, Fraker E, Hodkinson BP, Bonito G, Groenewald JZ, Arzanlou M, de Hoog G, Crous PW, Hewitt D, Pfister DH, Peterson K, Gryzenhout M, Wingfield MJ, Aptroot A, Suh SO, et al: The Ascomycota Tree of Life: A Phylum-wide Phylogeny Clarifies the Origin and Evolution of Fundamental Reproductive and Ecological Traits. Syst Biol. 2009, 58 (2): 224-239. 10.1093/sysbio/syp020.
10. Tekle YI, Grant JR, Kovner AM, Townsend JP, Katz LA: Identification of new molecular markers for assembling the eukaryotic tree of life. Mol Phylogenet Evol. 2010, 55 (3): 1177-1182. 10.1016/j.ympev.2010.03.010.
11. Stajich JE, Block D, Boulez K, Brenner SE, Chervitz SA, Dagdigian C, Fuellen G, Gilbert JGR, Korf I, Lapp H, Lehväslaiho H, Matsalla C, Mungall CJ, Osborne BI, Pocock MR, Schattner P, Senger M, Stein LD, Stupka E, Wilkinson MD, Birney E: The Bioperl toolkit: Perl modules for the life sciences. Genome Res. 2002, 12 (10): 1611-1618. 10.1101/gr.361602.
12. Pond SLK, Frost SDW, Muse SV: HyPhy: hypothesis testing using phylogenies. Bioinformatics. 2005, 21 (5): 676-679. 10.1093/bioinformatics/bti079.
13. Mayrose I, Graur D, Ben-Tal N, Pupko T: Comparison of site-specific rate-inference methods for protein sequences: empirical Bayesian methods are superior. Mol Biol Evol. 2004, 21 (9): 1781-1791. 10.1093/molbev/msh194.
14. Klopfstein S, Kropf C, Quicke DLJ: An evaluation of phylogenetic informativeness profiles and the molecular phylogeny of diplazontinae (Hymenoptera, Ichneumonidae). Syst Biol. 2010, 59 (2): 226-241. 10.1093/sysbio/syp105.
15. Townsend JP, Leuenberger C: Point of View: Taxon sampling and the optimal rates of evolution for phylogenetic inference. Syst Biol. 2011, 60 (3): 358-365. 10.1093/sysbio/syq097.
16. Townsend JP, Lopez-Giraldez F: Optimal selection of gene and ingroup taxon sampling for resolving phylogenetic relationships. Syst Biol. 2010, 59 (4): 446-457. 10.1093/sysbio/syq025.

CHAPTER 18 PRIMERVIEW: HIGH-THROUGHPUT PRIMER DESIGN AND VISUALIZATION

Damien M. O'Halloran

Keywords: PCR, Primer design, Genotyping, Perl, NGS, Sequencing.

BACKGROUND

With the advent of next generation sequencing (NGS) technologies, there has been an explosion in the volume of genomic data available to researchers. NGS provides a platform to rapidly sequence genomes, and offers new ways to unlock the genomes of species that are difficult to maintain. Creative federal incentives in the US (National Human Genome Research Institute - http://www.genome.gov/10000368) have contributed to an unprecedented drop in the costs involved in sequencing a genome from ~ $100,000 in 2002 to ~ $5000 in 2013 [1], which in effect has converted a field that was previously dominated by consortiums into an open playing field where small individual labs can participate. However, to prevent individual researchers from becoming caught in the maelstrom of this new genomic era, it is imperative to develop open source and user-friendly tools to help investigators study this volume of data. Designing primers to validate candidate genes from RNA-seq projects as well as developing diagnostic tools for the genomes of recently sequenced species, are examples of routine tasks faced by researchers in tackling NGS related data. Primer design, and in particular primer design *en masse*, also becomes essential for researchers working with multi-gene families, or metagenomic samples [2], as well as many other PCR based applications including primer walking, assembly PCR, digital PCR, ligation PCR, nested PCR, and quantitative PCR. Therefore, as the volume of genomic data continues to increase, so does the scale of experiments related to its analysis, and this is particularly true for primer design.

Here I describe a Perl module called *PrimerView* that is straightforward to implement or plug into larger pipelines, and enables the user to automate the process of primer design for DNA datasets of any size. Often, a visual readout of primer position on the target sequence is the fastest and most helpful way to validate the distribution and position of primers, and to this end PrimerView includes graphical outputs for each primer mapped to its target sequence. Each primer/target sequence pair is aligned and converted into a JPEG formatted file for easy visualization (other formats are also available). A PNG format file is also generated by PrimerView to depict the distribution of all designed primers across each input sequence. PrimerView uses the popular Bioperl [3] modules to align primers to the target sequence using the alignment software MUSCLE [4], and also to convert the alignment from CLUSTAL [5] format into graphical files. PrimerView may be particularly helpful for researchers working with large datasets where primers must be efficiently designed for many genes, as well as for various PCR applications including primer walking and assembly PCR reactions, where a graphical output can quickly help users determine primer coverage and distribution.

IMPLEMENTATION

PrimerView is written using Perl and has been tested successfully on both Windows command prompt as well as UNIX. PrimerView uses the Bioperl [3] dependencies Bio::Align::Graphics, Bio::Graphics, and Bio::SeqFeature::Generic to generate graphical output, which are all freely available from CPAN (http://www.cpan.org/). The alignment software MUSCLE [4] is used for a single iteration to map each designed primer to the inputted sequence, and this alignment is then converted into JPEG and PNG images depicting the position and distribution of all primers across each inputted sequence. PrimerView is a package with a constructor subroutine called "new" that allows the user to run the module by instantiating a PRIMERVIEW object. Separate subroutines for primer design, alignment, and conversion to graphical

Citation: Damien M. O'Halloran, "PrimerView: high-throughput primer design and visualization," DOI: 10.1186/s13029-015-0038-2.

output, are called from a script called 'primerview_driver.pl'. The input for PrimerView is any number of sequences in FASTA format [6]. A sample sequence file called 'test_seqs.fasta' is included in the download. The main primer design subroutine of PrimerView requires various parameters, which can be collected from the command-line. Default settings for each parameter will be invoked in the absence of command-line arguments, with the exception of the input filename which must be provided. These options (a through to k) are as follows: [−a filename *e.g.* test_seqs.fasta] [−b 5′ search area, integer] [−c 3′ search area, integer] [−d primer length max, integer] [−e primer length min, integer] [−f GC clamp Y or N] [−g upper GC%, integer] [−h lower GC%, integer] [−i upper Tm, integer] [−j lower Tm, integer] [−k specificity to the entire input file (Y) or just the specific sequence (N), Y or N]. The '-b' and '-c' flags refer to the five prime or three prime search areas across which PrimerView will scan for appropriate primers within each sequence; if the user wants to scan the entire length of the sequence, these flags can be set to the total sequence length in nucleotides. Features of the basic algorithm for PrimerView have been described previously and use nearest neighbor thermodynamic calculations to determine primer T_m values [7, 8, 9]. Example settings to execute PrimerView are: ">perl primerview_driver.pl -a test_seqs.fasta".

RESULTS AND DISCUSSION

PrimerView is an easily implemented Perl package that automates the design of forward and reverse primers from datasets of any size, while generating graphical output in JPEG and PNG formats for each designed primer mapped to the target sequence. For the JPEG generated file, other output formats are available for the graphical output by simply changing the output file extension in the 'graphic' subroutine within 'PRIMERVIEW.pm' to 'png' or 'gif' (see CPAN page for Bio::Align::Graphics for more details -http://search.cpan.org/~cjfields/BioPerl/Bio/Align/Graphics.pm). The function performed by PrimerView provides automation of a routine task, with user-defined features to provide more custom usage, especially in larger pipelines. By generating plots that map each designed primer against the target sequence provides easy and fast validation controls for researchers to examine the distribution and position of each primer. Figure 1a shows a representative image from the conversion of a MUSCLE program derived alignment in CLUSTAL format into a JPEG file to generate a primer map for a designed primer to the target sequence. The sequence name is provided in the user input file as the FASTA format header line, and a primer name that refers to its start position in base pairs on the target sequence is also highlighted, as well as the position of the primer. Figure 1b depicts an example output from PrimerView of the PNG file generated, illustrating the distribution of each primer mapped to the scaled input sequence. The primers are denoted as arrowed glyphs pointing in the direction of synthesis with each primer named by its starting position.

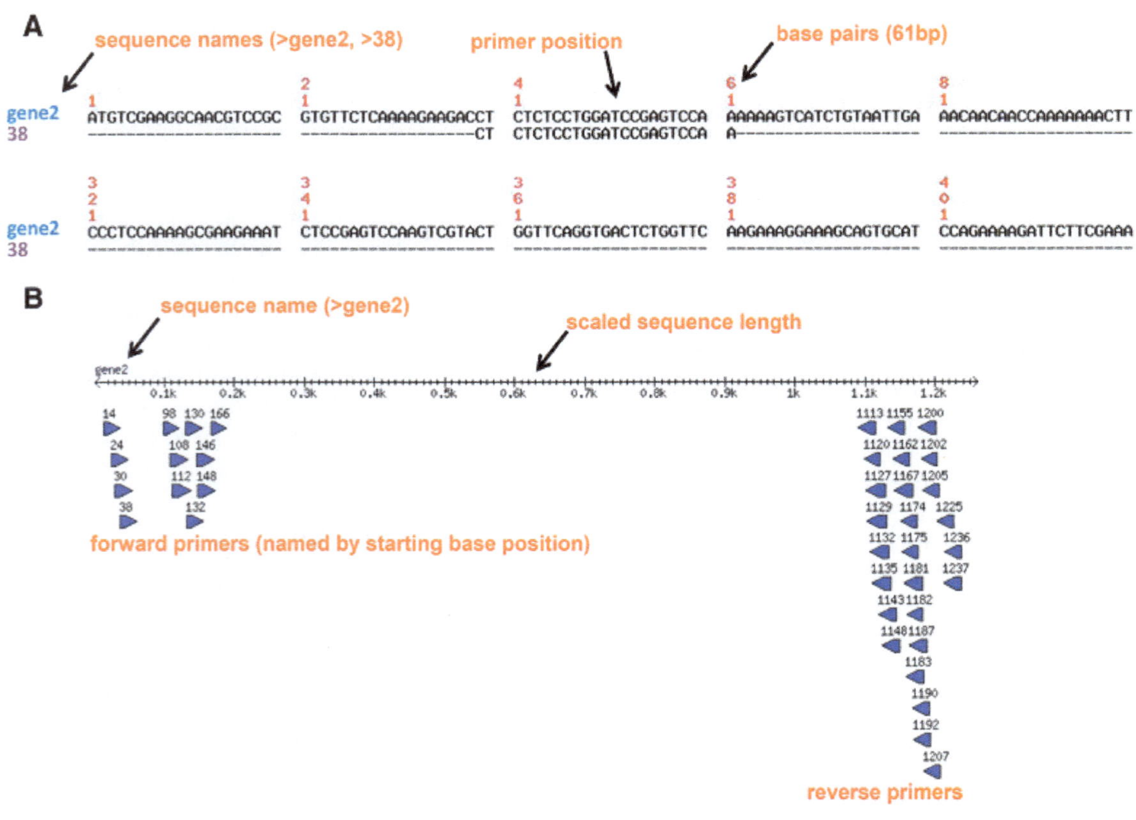

Figure 1: Screen-shot of the graphical outputs generated by PrimerView. **a** PrimerView designs primers for datasets of any size and generates a JPEG format file of the alignment for each primer illustrating its position on the target sequence. The user inputted sequence name (derived from the FASTA description line) is shown as well as the base pair positions and location of the designed primer. **b** A PNG file is also generated by PrimerView that maps the distribution of all designed primers onto the scaled target sequence. The target sequence is represented by a number line that equates to nucleotide position. The gene name is indicated above the number line and the name of each primer is represented by its starting position in nucleotides

During testing and validation of PrimerView, files containing varying numbers of sequences (10, 20, 50, 100, 200, 400, 1000, and 20,284) were provided as input to PrimerView, and ran using default settings to generate PNG file primer distribution output. The output from these performance tests is shown in Fig. 2a by plotting the program run-time for files containing increasing numbers of sequences (y-axis) against execution time in seconds (x-axis). A linear fit (m*x + b) reveals an R^2 value of 0.9994, however, fitting the relationship with a quadratic equation (a*x^2 + b*x + c) yields an R^2 value of 1.0. Testing also included a comparison between the primer T_m values returned from PrimerView to that of Primer3 [10] (Fig. 2b). To perform this test only thecalcTm subroutine within PrimerView was called for 400 primers designed using Primer3 [10]. A robust correlation (R^2 value of 0.97) was observed between each method, both of which employ nearest-neighbour parameter sets. Finally, validations of numerous primer pairs returned using PrimerView were tested using the MFE-Primer 2.0 in silico PCR tool (http://biocompute.bmi.ac.cn/CZlab/MFEprimer-2.0/) [11, 12]. An example output from one primer pair validation test is shown in Fig. 2c, and in all test cases each primer returned from PrimerView exhibited correct orientation and specificity.

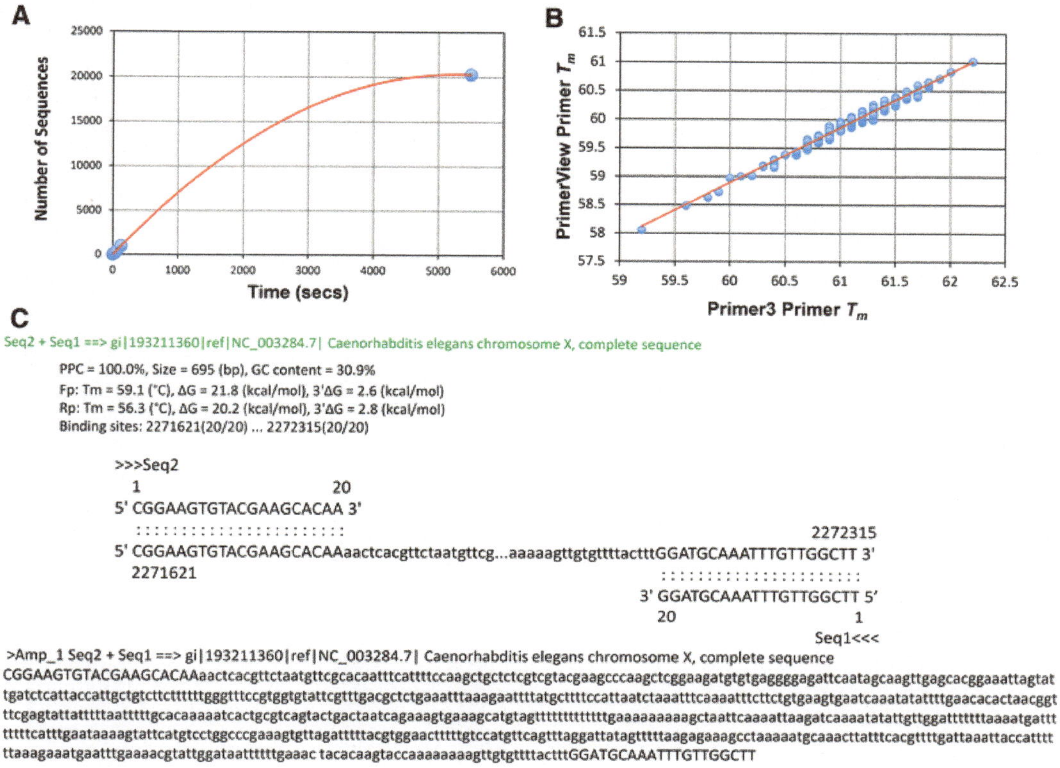

Figure 2: Testing and validation of PrimerView. **a** Performance testing of PrimerView. Files containing different numbers of sequences (10, 20, 50, 100, 200, 400, 1000, 20,284 – obtained from WormBase ftp site:ftp://ftp.wormbase.org/pub/wormbase/) were provided as input to PrimerView and ran using default settings to generate PNG file primer distribution outputs. Fitting the relationship between sequence number and run-time with a quadratic equation yields an R^2 value of 1.00. **b** Comparison of primer T_m values between PrimerView with Primer3 [10]. To perform these tests only the calcTm subroutine within PrimerView was called for 400 primers designed using Primer3 [10]. A robust correlation (R^2 value of 0.97) was observed between each method, both of which employ nearest-neighbour parameter sets. **c** Specificity and orientation of primers returned using PrimerView was tested using the MFE-Primer 2.0 in silico PCR tool (http://biocompute.bmi.ac.cn/CZlab/MFEprimer-2.0/) [11, 12]. Sample output from one test showing the sequence file (NC_003284.7) in FASTA format, and displaying all sequence matches in the database that lie between and include the primer pair (Seq2 and Seq1). The FASTA header details the region in the database as well as the primers; the body of the FASTA sequence is capitalized in areas where the primer sequence matches the database sequence and in lower-case elsewhere.

CONCLUSIONS

By handling both single sequence and multi-sequence input, PrimerView facilitates automated primer design for specific targets as well as large gene datasets. Although many other primer design tools exist such as Primer3, BatchPrimer3, and PerlPrimer [10, 13, 14], the utility of PrimerView are the graphical outputs that can quickly and easily depict the distribution of all primers across a target sequence from multi-sequence input. Generating graphical outputs that map each designed primer to the target sequence is an efficient means of quickly validating the spread of primers across a target.

ACKNOWLEDGMENTS

I would like to thank The George Washington University Columbian College of Arts and Sciences, GW Office of the Vice-President for Research, and the Department of Biological Sciences for Funding.

AUTHORS' CONTRIBUTIONS

DO'H conceived the idea for PrimerView, wrote the code, and wrote the manuscript.

REFERENCES

1. *Hayden EC. The $1,000 genome. Nature. 2014;507(7492):294–5.*
2. *Contreras-Moreira B, Sachman-Ruiz B, Figueroa-Palacios I, Vinuesa P. primers4clades: a web server that uses phylogenetic trees to design lineage-specific PCR primers for metagenomic and diversity studies. Nucleic Acids Res. 2009;37(Web Server issue):W95–100.*
3. *Stajich JE, Block D, Boulez K, Brenner SE, Chervitz SA, Dagdigian C, et al. The Bioperl toolkit: Perl modules for the life sciences. Genome Res. 2002;12(10):1611–8.*
4. *Edgar RC. MUSCLE: multiple sequence alignment with high accuracy and high throughput. Nucleic Acids Res. 2004;32(5):1792–7.*
5. *Thompson JD, Higgins DG, Gibson TJ. CLUSTAL W: improving the sensitivity of progressive multiple sequence alignment through sequence weighting, position-specific gap penalties and weight matrix choice. Nucleic Acids Res. 1994;22(22):4673–80.*
6. *Lipman DJ, Pearson WR. Rapid and sensitive protein similarity searches. Science. 1985;227(4693):1435–41.*
7. *Li K, Brownley A, Stockwell TB, Beeson K, McIntosh TC, Busam D, et al. Novel computational methods for increasing PCR primer design effectiveness in directed sequencing. BMC Bioinformatics. 2008;9:191-2105-9-191.*
8. *Rychlik W, Spencer WJ, Rhoads RE. Optimization of the annealing temperature for DNA amplification* in vitro. *Nucleic Acids Res. 1990;18(21):6409–12.*
9. *O'Halloran DM. STITCHER: a web resource for high-throughput design of primers for overlapping PCR applications. BioTechniques. 2015;58(6):325–8.*
10. *Untergasser A, Cutcutache I, Koressaar T, Ye J, Faircloth BC, Remm M, et al. Primer3–new capabilities and interfaces. Nucleic Acids Res. 2012;40(15):e115.*
11. *Qu W, Shen Z, Zhao D, Yang Y, Zhang C. MFEprimer: multiple factor evaluation of the specificity of PCR primers. Bioinformatics. 2009;25(2):276–8.*
12. *Qu W, Zhou Y, Zhang Y, Lu Y, Wang X, Zhao D, et al. MFEprimer-2.0: a fast thermodynamics-based program for checking PCR primer specificity. Nucleic Acids Res. 2012;40(Web Server issue):W205–8.*
13. *You FM, Huo N, Gu YQ, Luo MC, Ma Y, Hane D, et al. BatchPrimer3: a high throughput web application for PCR and sequencing primer design. BMC Bioinformatics. 2008;9:253-2105-9-253.*
14. *Marshall O. Graphical design of primers with PerlPrimer. Methods Mol Biol. 2007;402:403–14.*

CHAPTER 19

A PARSIMONY APPROACH TO BIOLOGICAL PATHWAY RECONSTRUCTION/INFERENCE FOR GENOMES AND METAGENOMES

Yuzhen Ye, Thomas G. Doak

INTRODUCTION

Microbial whole genome sequencing has become a routine practice in recent years, because of the rapid advances of DNA sequencing technologies [1]. One of the first analyses that biologists attempt, once they obtain a complete genome sequence, is to *reconstruct* the biological pathways encoded by the organism, which is usually accomplished *in silico* by mapping the protein coding genes onto reference pathway collections, such as KEGG [2] or SEED [3], based on their homology to reference genes with previously characterized functions. For example, KAAS, the pathway annotation system based on the KEGG database [4], first annotates K numbers (each K number represents an ortholog group of genes, and is directly linked to an object (a biochemical step) in the KEGG pathway map), and then reconstructs pathways based on the assigned K numbers. Similarly, the RAST server (and MG-RAST) first annotates FIG families and then maps the identified FIG families onto the SEED subsystems[5],[6]. These automatic methods are promising for the analysis of most genomes, although they may leave "holes" in the reconstructed pathways, due to either *missing* genes (i.e. the genes are non-homologous to reference genes of the same specific functions, and thus cannot be identified by a homology-based method, or were simply not annotated as ORFs by annotation pipelines) [7], or alternative and novel pathways (i.e. the target organism adopts variant pathways, which are different from the reference pathway, to accommodate a specific niche or lifestyle) [8]. After all, many bacterial genomes have fewer than 60% of their genes assigned to a proposed function [9],[10].

Citation: Ye Y, Doak TG (2009) A Parsimony Approach to Biological Pathway Reconstruction/Inference for Genomes and Metagenomes. PLoS Comput Biol 5(8): e1000465. doi:10.1371/journal.pcbi.1000465.

We note that *pathway reconstruction* is essential for understanding the biological functions that a newly sequenced genome encodes. For instance, in a recently published report, the coupling of N_2 fixation to cellulolysis was revealed within protist cells in the termite gut, based solely on the *in silico* pathway reconstruction of the complete genome sequence of a bacterial endosymbiont [11].

Moreover, pathway reconstruction based on some new high throughput techniques must provide conclusions from explicitly incomplete information, which poses fresh challenges. For example, in a typical *proteomics* experiment, the proteins represent a particular biological sample collected under a specific physiological condition or from a specific tissue (e.g. from yeast cells after the heat shock), which are in high enough abundance to be identified by tandem mass spectrometry [12],[13]. Based on these data, one may ask, what biological pathways were activated (or suppressed) under the physiological condition? A similar, but more complicated case is pathway analysis of *metagenomic* data, to characterize the aggregate metabolic processes of microbial communities in a given environment [14]. Metagenomic profiling data can be viewed as a sampling of the genomic sequences from many kinds of microbes living in a specific environment. Again, the incompleteness of the data makes it difficult to reconstruct the entire pathways encoded by a metagenome. Nevertheless, it is becoming routine to "reconstruct" pathways for proteomic [15] and metagenomic data [16],[17], by best similarity matches (often derived from BLAST searches): a pathway is inferred to be absent or present in a dataset if highly confident homolog protein hits identify one or more of the protein functions associated with the pathway in other organisms.

In addition to the problems that arise from incomplete data, existing methods of pathway reconstruction or inference may over-estimate the number of pathways because of redundancy in the protein-pathway, at four levels. First, different pathways may share the same biological functions. The partition of pathways (as the entire cellular network is partitioned into several hundreds of biological pathway entities in KEGG database) is extremely important for understanding of biological processes, even though there is only a single large biological network within any cell and all pathways are to some extent connected [18]. It is not surprising that many pathways defined in the pathway databases are overlapping. Second, some proteins carry out multiple biological functions [19], *e.g.* through different protein domains, active sites, or substrate specificities. Third, neither organisms nor communities are closed boxes, and the products or intermediates of pathways may be exogenously supplied. Finally, homology-based protein searching may map one protein to multiple homologous proteins with different biological functions (i.e. paralogous proteins). In summary, it cannot be safely concluded that a pathway is present, even if one or more proteins are mapped to it.

Even for single complete genomes, pathway reconstruction does not always give a clear picture of the biological functions in an organism, and human curation and experimental verification is often needed [20],[21]. We illustrate this by a rather extreme example found in the pathway analysis of the human genome. The KEGG pathway annotation of the human genome includes the reductive carboxylate cycle, with proteins annotated to 6 steps in this pathway (http://www.genome.jp/kegg-bin/show_organism?menu_type=pathway_maps&org=hsa) (as of July 2nd, 2009). The Calvin cycle is the most common method of carbon fixation, while the reductive carboxylate cycle is an alternative carbon fixation pathway, currently found only in certain autotrophic microorganisms. In fact, the reductive carboxylate cycle is essentially the reverse of the Krebs cycle (citric acid or tricarboxylic acid cycle), the final common pathway in aerobic metabolism for the oxidation of carbohydrates, fatty acids and amino acids, so they share reactions and functional roles. For this reason, the proteins responsible for the normal function of the Krebs cycle can be mistakenly taken as evidence for the existence of a reductive carboxylate cycle in the human genome.

Here we propose a pathway reconstruction/inference method in which we do not attempt to reconstruct entire pathways from a given set of protein sequences (e.g. identified in a proteomics experiment, or encoded by the sequences sampled in a metagenomic project), but to determine the minimal set of biological pathways that must exist in the biological system to explain the input protein sequences sampled from it. In this context, we note *pathway inference* might be a more suitable terminology than *pathway reconstruction*. However, considering that pathway inference has been used in a different context to infer networks or pathways from gene express data [22], and pathway reconstruction is commonly used in the field, we use both pathway inference and pathway reconstruction in this paper. To address the issues of both incomplete data, and pathway redundancy, we formulate a parsimony version of the pathway reconstruction/inference problem, called MinPath (*Min*imal set of *Path*ways), which can be roughly described as the following: given a set of reference pathways and a set of proteins (and their predicted functions) that can be mapped to one or more pathways, we attempt to find the minimum number of pathways that can explain all proteins (functions) (seeFig. 1). Although this problem is NP-hard in general, we provide an integer programming (IP) framework to solve it.

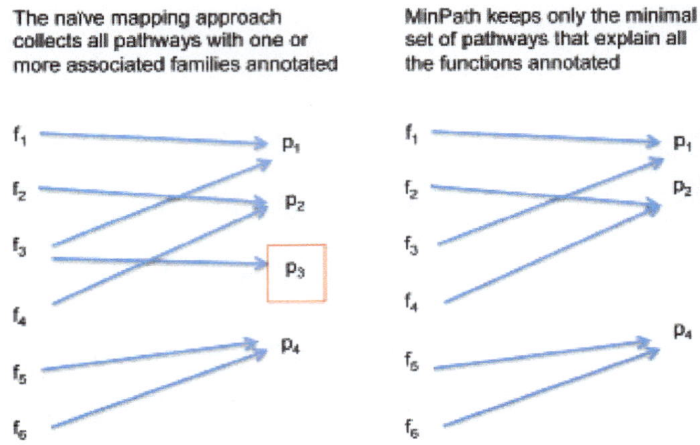

Figure 1: Schematic illustration of the MinPath method. Assume 6 families (or orthologous groups, f_1, ..., f_6) are identified from a given sample of genes (e.g., the genes could be from a genome, or sampled from a metagenome). The naïve mapping approach (shown on the left) will lead to a reconstruction with 4 pathways annotated (p_1, p_2, p_3, and p_4). Due to the overlapping nature of the biological pathways (see text for more details), pathway p_3 shares function f_3 with pathway p_2. We claim that only three pathways, p_1, p_2, and p_3 are sufficient to explain the existence of the 6 families annotated in the dataset, and a conservative reconstruction of pathways should have only 3 pathways (shown on the right). As we show in the paper, such a conservative estimation of pathways provides a more reliable estimation of the functional diversity of a sample.

We focus on analyzing complete genomes in this study because there is a relatively good understanding of the pathways that actually exist in organisms with completely sequenced genomes (as compared to the emerging metagenomes), making this analysis a good test of our method. Besides, the pathway annotations of these genomes are still far from perfection, as in the example of a carbon fixation pathway in the human genome (as well as chickens, mosquitoes, etc). We also applied MinPath to the analyses of several metagenomic datasets, to demonstrate the potential applications of MinPath in metagenome annotation.

RESULTS

We first revisited the pathway reconstruction of individual genomes using MinPath. The results show that MinPath gave a conservative, but reliable estimation of the pathways of a genome, and therefore the functional ability/diversity encoded by a genome. In addition, MinPath found suspicious pathways in the KEGG database. We then applied MinPath to a set of metagenomic datasets, and the results indicate that the current estimation of functional diversity/ability of studied microbial communities might be overestimated.

Pathway Reconstruction for Genomes

Overview of the performance of MinPath.

The function annotations of the genes from individual genomes were extracted from the KEGG database, and used as input for MinPath to reconstruct the pathways encoded by each genome. A total of 854 genomes were studied, and the overall performance of MinPath is shown in Fig. 2, compared with the curated KEGG pathways and the pathway reconstructions produced by the naïve mapping approach (see METHODS for details). The comparison shows that MinPath gives an estimation of functional diversity (measured by the number of pathways constructed) that is closer to the curated KEGG database, as compared to simple pathway construction based on the appearance of families. MinPath gives a more conservative estimation of the pathways than even KEGG in most genomes (with fewer annotated biological pathways), but we would like to argue that even some of the pathways collected in KEGG should be removed (such as the ascorbate and aldarate metabolism pathway in human, as we discuss below).

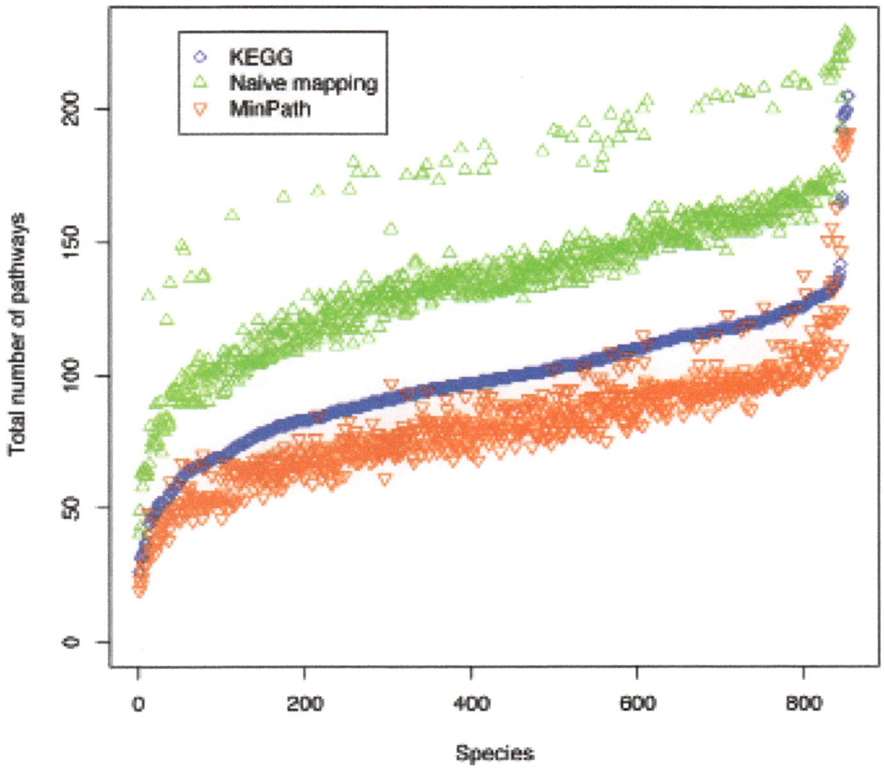

Figure 2: Comparison of the number of pathways reconstructed for various genomes by different methods. The coloring schema is as following: MinPath (red triangles), naïve mapping approach (green), and the pathway annotation maintained in KEGG database after human evaluation (blue).

The human genome.

For the human genome, there are 205 predicted KEGG pathways (as of December 2008), while the naïve mapping approach identifies 227 pathways. MinPath identified only 191 pathways—these pathways are necessary and sufficient to explain all the annotated human proteins in the KEGG database.

Many of the pathways that are identified by the naïve mapping approach are spurious and are not curated in the KEGG database (e.g. the penicillin and cephalosporin biosynthesis pathway, and the two-component general and the type II secretion systems), indicating that MinPath can be applied to remove pathways that are otherwise mistakenly annotated using the naïve mapping approach. More examples are listed in the supplementary website.

Some of the pathways that are curated in the KEGG database are marked by MinPath as spurious (see Table 1). For example, the ascorbate and aldarate metabolism pathway (Fig. 3) is annotated in KEGG as a biological pathway in human, but not by MinPath. In humans there are only three functions (out of 24) annotated for this pathway and these three functions are not unique to the pathway: EC 1.2.1.3 (aldehyde dehydrogenase 2 family) is involved in 15 other pathways, EC 1.1.1.22 (UDP-glucose dehydrogenase) is involved in three other pathways, and myo-inositol oxygenase (EC:1.13.99.1) is involved in both this pathway and the inositol phosphate metabolism pathway. Based on the sparseness of the genes assigned to this pathway and their ubiquitous nature, and the fact that humans require vitamin C in the diet, we believe that the ascorbate and aldarate metabolism pathway should be removed from the pathways reconstructed for the human genome.

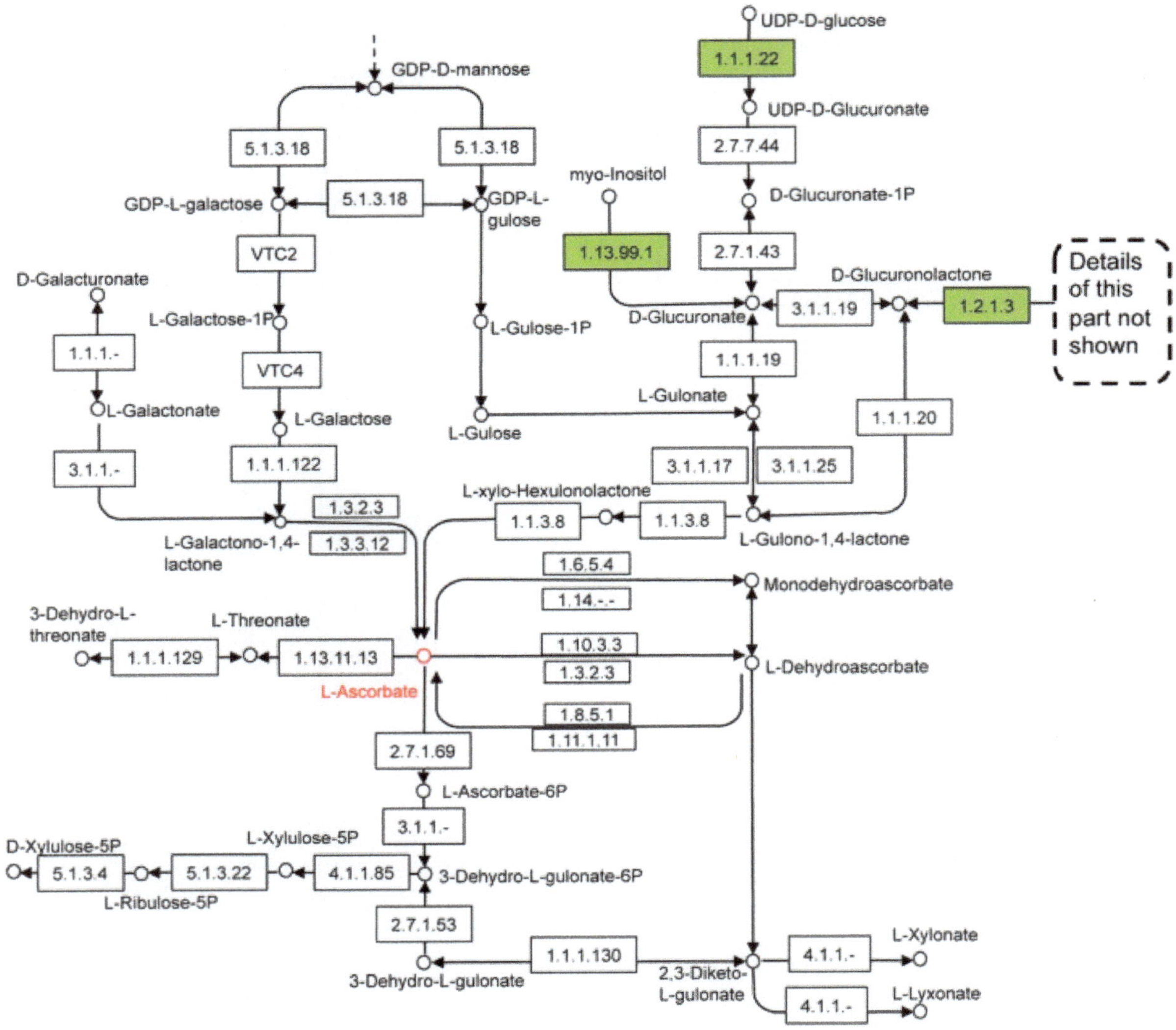

Figure 3: The ascorbate and aldarate metabolism pathway, eliminated by MinPath. The diagram was prepared based on the corresponding KEGG pathway (ID=00053), and only part of the pathway is shown for clarity. The three enzymes that are annotated in the human genome are highlighted in green, even though none of these enzymes are unique to this pathway.

Table 1: Selected spurious pathways of the human genome that are incorrectly identified by the naïve mapping approach

KEGG ID	Pathway description	Possible reason for being falsely identified by the naïve mapping approach	Removed by MinPath?	Additional notes
00053	ascorbate and aldarate metabolism	pathway redundancy (same function involves in multiple pathways)	yes	humans can not synthesize ascorbic acid (vitamin C)
00290	valine, leucine and isoleucine biosynthesis	pathway redundancy	yes	all three are essential amino acids in humans
00521	streptomycin biosynthesis[a]	pathway redundancy	yes	see table note[a]
00720	reductive carboxylate cycle	pathway redundancy	yes	it is a CO_2 fixation pathway found in photosynthetic bacteria

[a]Steptomycin biosynthesis is not listed for the human genome (http://www.genome.jp/kegg-bin/show_organism?menu_type=pathway_maps&org=hsa) in KEGG; but there are 5 functional roles from this pathway annotated in the human genome based on the KEGG annotation, including K00844, K01092, K01710, K01835, and K01858.

doi:10.1371/journal.pcbi.1000465.t001

The Escherichia coli genome.

MinPath identified 98 biological pathways that are sufficient to explain all the identified functions encoded by the *E.coli* genome. It is a conservative estimation, as compared to the 125 pathways for *Escherichia coli K-12 MG1655* collected in the KEGG database, and 158 pathways that have at least one or more associated functions identified in the genome. Refer to the supplementary webpage for the details of the pathway reconstructions and their comparison for *E.coli*. It is obvious that the naïve mapping approach leads to an inflated estimate of biological pathways in *E.coli*—the list even includes several biological pathways involved with human cancer, including renal cell carcinoma (pathway ID 05211), prostate cancer (pathway ID 05215), and bladder cancer (pathway ID 05219). These pathways were wrongly annotated because one or more predicted functions in these pathways are also involved in other pathways. For example, fumarate hydratase (ko:K01679) is involved in the renal cell carcinoma pathway, as well as in the citrate cycle, a fundamental pathway present in most bacteria. Based on the identification of this enzyme alone, the naïve mapping approach predicted the presence of the renal cell carcinoma pathway in *E. coli*, which obviously cannot be true. MinPath removed these spurious pathways from the list of constructed pathways, without human curation.

We argue that KEGG predictions also overestimate the biological pathway encoded by *E. coli*genome, e.g. the mitochondrial fatty acid elongation pathway and the bile acid biosynthesis pathway (see Table 2).

Table 2: Selected spurious pathways of the *E. coli* genome (collected in KEGG) eliminated by MinPath

KEGG ID	Pathway description	Functions involved	Removed by MinPath?	Justification	Additional notes
00062	fatty acid elongation in mitochondria	K00022	yes	K00022 is shared by this pathway and 6 other pathways	*E.coli* has no mitochondria
00521	bile acid biosynthesis	K00001 K00632	yes	K00001 is shared by several other pathways, including the glycolysis pathway; K00632 is shared by the fatty acid metabolism pathway and others.	bile acids are steroid acids found predominantly in the bile of mammals

doi:10.1371/journal.pcbi.1000465.t002

Pathway Reconstruction for Metagenomes

We used MinPath to re-analyze the biological pathways of several metagenomes [17], which were previously analyzed by a naïve mapping approach. The results are summarized in Table 3. We used both the KEGG and SEED databases in this experiment. For KEGG pathways, we did local BLAST searches, using the criteria as shown in [16] for KO family identification. For SEED subsystems, the FIG annotations were downloaded from the MG-RAST server (http://metagenomics.theseed.org/).

Table 3: Comparison of biological pathway reconstruction based on MinPath and the naïve mapping approach for selected metagenomes[a]

Environmental samples	Naïve mapping (KEGG)[c]	MinPath (KEGG)	Naïve mapping (SEED)[d]	MinPath (SEED)
Coral-Mic (7)[b]	188/232[e]	109/171	497/629	186/392
Coral-Vir (6)	174/211	105/140	594/667	285/441
Marine-Mic (8)	221/236	146/174	695/730	488/577
Marine-Vir (10)	213/236	154/175	680/733	507/599
Freshwater-Mic (4)	196/220	137/165	678/739	460/601
Freshwater-Vir (4)	113/154	57/90	392/559	112/283
Hyper-saline-Mic (9)	196/221	146/170	724/763	558/650
Hyper-saline-Vir (12)	164/181	105/137	613/697	347/510

[a]metagenomes sampled from different environments [17] (-Mic, and -Vir are for microbial and viral metagenomes, respectively, as shown in the table).
[b]microbial metagenomes sampled from coral, with the total number of sequencing datasets shown in the brackets.
[c]based on the KEGG pathways (the KEGG database used in this study was downloaded in Dec, 2008, which has 345 pathways).
[d]based on the SEED subsystems (we used FIGfams release 6, which has more subsystems than reported in [17], and the total number of subsystems included is 898).
[e]the two numbers present the total number of pathways (or subsystems) found in at least two of the datasets (e.g., two out of 7 for Coral-Mic), and in at least one of the datasets for each environmental location, respectively.
doi:10.1371/journal.pcbi.1000465.t003

For all the datasets we tested, MinPath reduced the total number of annotated pathways (or subsystems) significantly (as shown in Table 3). For example, for the metagenome sampled from a coral microbial community (Coral-Mic), there are in total 232 KEGG biological pathways annotated in at least one of the 7 sequencing datasets. Based on MinPath, however, only 160 KEGG biological pathways are sufficient to explain all the functions predicted for these datasets. These results indicate that the naïve mapping of the biological pathways from predicted functions may overestimate the biological pathways (so the functional diversity) of those microbial communities, and we need to be cautious when interpreting the results from such an analysis [16],[17].

We also show the details of pathway reconstruction for a single sequence dataset from the coral biome (4440319.3.dna.fa). The naïve mapping approach identified 224 KEGG pathways, whereas MinPath identified only 143 KEGG pathways. The pathways eliminated by MinPath include the inositol metabolism pathway, the androgen and estrogen metabolism pathway, the caffeine metabolism pathway, etc (see more examples at the supplementary website). Obviously, comparisons of microbial communities or other biomes will be more telling if spurious pathways are eliminated, and our results suggest that as many as 40% of the 224 pathways could be wrong.

DISCUSSION

We have developed the MinPath approach to provide more conservative—but more reliable—estimations of biological pathways from a sequence dataset, and applied this approach to revisit the biological pathway reconstruction problem for genomes as well as metagenomes. Our results show that without further post-processing of the reconstructed pathways, the naïve mapping strategy may overestimate the biological pathways that are encoded by a genome or metagenome, which could jeopardize any conclusions drawn from the constructed biological pathways (such as the metabolic diversity/ capacity of an environmental microbial or viral community, as measured by the Shannon Index) [16],[17], or other downstream analysis based on constructed pathways [23]. It was noted in [16] that most of the microbial communities in that study were approaching saturation for known pathways: more conservative estimates of pathways for each environment may allow real functional differences between the samples to be detected.

Note that MinPath is not designed to directly improve the still imperfect definition of pathways and/or functions in databases such as KEGG or SEED. For example, as a result of how some pathways are grouped in the KEGG database, peptidoglycan biosynthesis is listed for the human genome by KEGG annotation and MinPath does not eliminate this pathway from the list of annotated pathways from human genome. In this sense, efforts are still needed to improve the elucidation and annotation of extent biochemical pathways. But given a database of reference pathways, we feel that MinPath provides a sensible method for inferring the pathways represented in biological sequence samples.

MATERIALS AND METHODS

First we will briefly describe the naïve mapping approach that is commonly used in current automatic biological pathway reconstruction services (e.g., the KAAS and RAST servers), as well as for pathway reconstruction for metagenomic sequences. Then we present a novel minimal pathway reconstruction approach based on a simple yet efficient algorithm for solving this problem.

The Naïve Mapping Approach to Pathway Reconstruction

Pathway reconstruction has become routine in functional annotation of genomes and metagenomes, in which KEGG pathways (or other biological pathways such as SEED subsystems) are reconstructed based on homology. KEGG and SEED databases collect pathways (or subsystems) curated by experts, each pathway/subsystem consisting of a series of functional roles (enzymes, transporters, etc). Pathway reconstruction consists of two key steps: (1) predicting the functions (represented by protein families) of proteins encoded by the DNA sequences, which is often achieved by similarity searching of the predicted proteins against reference proteins from previously characterized genomes; and (2) predicting the presence or absence of pathways in the query dataset, based on the identified functions associated to the pathways. Conventional pathway reconstruction usually adopts simple criterion in this second step (herein referred to as the naïve mapping approach), i.e., a pathway is considered to be present if one or more functions in the pathway are identified in the first step. We have shown in this paper that this approach may lead to the identification of spurious pathways and an overestimation of functional ability, which motivated us to develop a novel approach to pathway reconstruction based on the parsimony principle presented below.

Minimal Pathway Reconstruction Problem

We define the *minimal pathway reconstruction problem* as the following: given a list of functions annotated for a set of genes (which can be an incomplete set, as we encounter in metagenomic analysis, or a nearly complete set, as in complete genome analysis), find the *minimal* set of pathways that include all given functions (see Fig 1). Note that this formulation is different from the conventional formulation of the pathway reconstruction problem, which attempts either to reconstruct the complete pathways encoded by a given genomic dataset (in a sense, the pathway holes should to be minimized), or to identify the set of pathways that have at least one associated function annotated (i.e., the naïve mapping approach).

Integer Programming Algorithm

We use integer programming to solve the minimal pathway reconstruction problem. Linear programming (LP) is an algorithm for finding the maximum or minimum of a linear function of variables (objective function) that are subject to linear constraints [24]. Simplex and interior point methods are widely used for solving LP problems. The related problem of integer programming (IP) requires some or all of the variables to take integer (whole number) values. Some of the most powerful algorithms for finding exact solutions of combinatorial optimization problems [25] are based on IP. LP and IP have been applied to many fields in the biological sciences, such as the maximum contact map overlap problem for protein structure comparison[26], optimal protein threading [27], probe design for microarray experiments [28], and the pathway variant problem [8].

Here we transform the minimal pathway reconstruction problem to an integer programming problem: Denote the number of functions (protein families) that are annotated in a dataset as n. Let the total number of putative pathways which have at least one component function annotated be p. Denote the mapping of protein functions to the pathways as M, where $M_{ij}=1$ if function i is involved in pathway j, otherwise 0 (note one function may map to multiple pathways or subsystems). Denote if a pathway j is selected in the final list or not as P_j, with $P_j=1$ if selected, $P_j=0$ otherwise. The set of pathways with $P_i=1$ composes the minimal set of pathways that can explain all the functions that are annotated for a dataset.

The objective function for integer programming is,

$$\min \sum_{j=1}^{p} P_j$$

$$s.t. \quad \sum_{j=1}^{p} M_{ij} P_j \geq 1 \quad \forall i \in [1,n]$$

i.e., our goal is to find the minimum number of pathways that can explain all the functions carried by at least one protein from a dataset.

Protein Function and Function Annotation

We use the KO and FIG protein families defined in the KEGG database and the SEED subsystems, respectively, for this study. Many of the mappings of KO families to KEGG pathways were done manually in the KEGG database. These families are the basic units for pathway reconstruction (or subsystem reconstruction in SEED), in which a pathway (or a subsystem) is composed of a list of functional roles.

Implementation Details

We use the GLPK package (GNU Linear Programming Kit;http://www.gnu.org/software/glpk/glpk.html) for solving the integer-programming problem; all the other functions are implemented in Python.

The input for MinPath is a list of protein families (*e.g.*, KO and FIG families) annotated in a given dataset of genes (from a genome, or a metagenome), and the output is the list of pathways reconstructed/inferred for the dataset.

Note that in some cases two pathways may share most of their functional roles (for example, the biosynthesis and degradation pathway of the same biological molecule, such as the lysine biosynthesis and degradation pathways). MinPath will keep one of these pathways, because that is sufficient to explain the functional roles identified. We added a post-processing step here to add those pathways that have more than 50% of their functional roles identified back to the pathway pool, even when these functional roles appear in another pathway that is already predicted by MinPath.

Benchmarking Experiments

We revisited the pathway reconstruction for the 854 genomes in the KEGG database (as of December, 2008) that have at least 20 KEGG pathways annotated for each of these genomes. For these genomes, the function (or protein families) annotations were downloaded from the KEGG database (ftp://ftp.genome.jp/pub/kegg/release/current/).

We also applied MinPath to reanalyze the pathways for nine biome metagenomic datasets[17]. The FIG family annotations for the metagenomic sequences were downloaded from the MG-RAST server (http://metagenomics.theseed.org/). We conducted the KO family annotations of the sequences based on the best blast hits with E-value cutoff of 1e-5, a typical E-value cutoff used for KEGG pathway reconstruction in metagenomes [16].

ACKNOWLEDGMENTS

We thank Dr. Haixu Tang for inspiring discussions, and reading the manuscript. We thank Drs. Alex Rodriguez and Ross Overbeek for their help with using the Figfam database.

REFERENCES

1. Morozova O, Marra MA (2008) Applications of next-generation sequencing technologies in functional genomics. Genomics 92: 255–264.
2. Kanehisa M, Goto S (2000) KEGG: kyoto encyclopedia of genes and genomes. Nucleic Acids Res 28: 27–30.
3. Overbeek R, Begley T, Butler RM, Choudhuri JV, Chuang HY, et al. (2005) The subsystems approach to genome annotation and its use in the project to annotate 1000 genomes. Nucleic Acids Res 33: 5691–5702.
4. Moriya Y, Itoh M, Okuda S, Yoshizawa AC, Kanehisa M (2007) KAAS: an automatic genome annotation and pathway reconstruction server. Nucleic Acids Res 35: W182–185.
5. Aziz RK, Bartels D, Best AA, DeJongh M, Disz T, et al. (2008) The RAST Server: rapid annotations using subsystems technology. BMC Genomics 9: 75.
6. Meyer F, Paarmann D, D'Souza M, Olson R, Glass EM, et al. (2008) The metagenomics RAST server - a public resource for the automatic phylogenetic and functional analysis of metagenomes. BMC Bioinformatics 9: 386.
7. Osterman A, Overbeek R (2003) Missing genes in metabolic pathways: a comparative genomics approach. Curr Opin Chem Biol 7: 238–251.
8. Ye Y, Osterman A, Overbeek R, Godzik A (2005) Automatic detection of subsystem/pathway variants in genome analysis. Bioinformatics 21: Suppl 1i478–486.
9. Sivashankari S, Shanmughavel P (2006) Functional annotation of hypothetical proteins - A review. Bioinformation 1: 335–338.
10. Friedberg I, Jambon M, Godzik A (2006) New avenues in protein function prediction. Protein Sci 15: 1527–1529.
11. Hongoh Y, Sharma VK, Prakash T, Noda S, Toh H, et al. (2008) Genome of an endosymbiont coupling N2 fixation to cellulolysis within protist cells in termite gut. Science 322: 1108–1109.
12. Gilchrist A, Au CE, Hiding J, Bell AW, Fernandez-Rodriguez J, et al. (2006) Quantitative proteomics analysis of the secretory pathway. Cell 127: 1265–1281.
13. Koller A, Washburn MP, Lange BM, Andon NL, Deciu C, et al. (2002) Proteomic survey of metabolic pathways in rice. Proc Natl Acad Sci U S A 99: 11969–11974.
14. Galperin M (2004) Metagenomics: from acid mine to shining sea. Environ Microbiol 6: 543–545.
15. Yates J, Ruse CI, Nakorchevsky A (2009) Proteomics by Mass Spectrometry: Approaches, Advances, and Applications. Annu Rev Biomed Eng Epub ahead of print.
16. Turnbaugh PJ, Hamady M, Yatsunenko T, Cantarel BL, Duncan A, et al. (2009) A core gut microbiome in obese and lean twins. Nature 457: 480–484.
17. Dinsdale EA, Edwards RA, Hall D, Angly F, Breitbart M, et al. (2008) Functional metagenomic profiling of nine biomes. Nature 452: 629–632.
18. Okuda S, Yamada T, Hamajima M, Itoh M, Katayama T, et al. (2008) KEGG Atlas mapping for global analysis of metabolic pathways. Nucleic Acids Res 36: W423–426.
19. Rosin FM, Watanabe N, Lam E (2005) Moonlighting vacuolar protease: multiple jobs for a busy protein. Trends Plant Sci 10: 516–518.
20. Francke C, Siezen RJ, Teusink B (2005) Reconstructing the metabolic network of a bacterium from its genome. Trends Microbiol 13: 550–558.
21. Oberhardt MA, Puchalka J, Fryer KE, Martins dos Santos VA, Papin JA (2008) Genome-scale metabolic network analysis of the opportunistic pathogen Pseudomonas aeruginosa PAO1. J Bacteriol 190: 2790–2803.
22. Ourfali O, Shlomi T, Ideker T, Ruppin E, Sharan R (2007) SPINE: a framework for signaling-regulatory pathway inference from cause-effect experiments. Bioinformatics 23: i359–366.

23. Gianoulis TA, Raes J, Patel PV, Bjornson R, Korbel JO, et al. (2009) Quantifying environmental adaptation of metabolic pathways in metagenomics. Proc Natl Acad Sci U S A 106: 1374–1379.

24. Bertsimas D, Tsitsiklis JN (1997) Introduction to Linear Optimization. Nashua: Athena Scientific.

25. Cook WJ, Cunningham WH, Pulleyblank WR, Schrijver A (1998) Combinatorial Optimization. New York: John Wiley and Sons.

26. Caprara A, Carr R, Istrail S, Lancia G, Walenz B (2004) 1001 optimal PDB structure alignments: integer programming methods for finding the maximum contact map overlap. J Comput Biol 11: 27–52.

27. Xu J, Li M, Kim D, Xu Y (2004) RAPTOR: optimal protein threading by linear programming. J Bioinform Comput Biol 1: 95–117.

28. Klau GW, Rahmann S, Schliep A, Vingron M, Reinert K (2004) Optimal robust non-unique probe selection using Integer Linear Programming. Bioinformatics 20: i186–i193.

CHAPTER 20 ESPRIT: ESTIMATING SPECIES RICHNESS USING LARGE COLLECTIONS OF 16S RRNA PYROSEQUENCES

Yijun Sun, Yunpeng Cai, Li Liu, Fahong Yu, Michael L. Farrell, William McKendree and William Farmerie

INTRODUCTION

The latest development of massively parallel pyrosequencing technology enables ultra-deep sequencing of complex microbial populations rapidly and inexpensively (1,2). For example, 454 Life Sciences GS FLX systems (Branford, CT, USA) (3) can finish a full pyrosequencing run with more than 400K sequences within one day of operation (http://www.454.com/products-solutions/system-features.asp). It allows researchers to study genetic materials recovered directly from environmental samples, bypassing the needs for isolation and lab cultivation of individual species, and thus opens a new window to probe the hidden world of microbial communities. This technique has been successfully used in several 16S rRNA-based metagenomics analyses of various environments. For example, Sogin et al. (4) provided one of the first global indepth descriptions of microbial diversities and their relative abundance in the ocean, and Keijser et al. (5) were among the first to study oral microbial populations. It has been shown that the microbial diversities are at least one order of magnitude larger than previously reported. These estimation results, however, were computed through extrapolation. In order to obtain more accurate estimates, surveys that are several orders of magnitude larger than those reported in the literature may be required to uncover sequences from minor components (4,5). However, analyzing large collections of 16S ribosomal sequences poses a serious computational challenge for existing algorithms.

In this article, we focus on taxonomy independent analysis where sequences are classified into operational taxonomic units (OTUs) of specified sequence variations, based on which various ecological metrics are estimated. Typically, sequences with <3% dissimilarity are assigned to the same species, while those with <5% dissimilarity are assigned to the same genus, although these distinctions are controversial (6,7,8). One outstanding challenge of taxonomy independent analysis is the alignment of sequences of a sample for the calculation of sequence variations. One commonly used method in the literature is multiple sequence alignment (MSA) [see, for example, (4,5,9)]. While significant improvement has been made in the last decade to reduce the computational complexity of MSA [e.g. MUSCLE (10) and MAFFT (11)], it is still computationally intractable to align hundreds of thousands of sequences. Moreover, the use of MSA for aligning hypervariable regions of 16S rRNA gene has not been well justified in the literature. MSA is used to infer homologous segments of input sequences. An underlying assumption is that input sequences should share some similarities, which may not be valid for 16S rRNA-based studies that target on hypervariable regions of rRNA genes (e.g. V6 and V3 regions). We experimentally found through a benchmark study that the use of MSA leads to an inflated estimate of genetic distances and microbial diversities. Another major issue of taxonomy-independent analysis is the high computational complexity and memory requirement associated with assigning sequences into OTUs. DOTUR (8) is a commonly used algorithm for this purpose. It conveniently integrates the tasks of performing clustering and statistical interference of species richness into one program, which frees researchers from manually manipulating a large distance matrix. However, DOTUR does not scale well to handle extremely large data sets. The recently released MOTHUR significantly improved the computational performance of DOTUR. As with DOTUR, MOTHUR loads a distance matrix into memory before proceeding to perform clustering. Hence, it does not fundamentally address the computational issue associated with processing massive pyrosequencing data. Given a full run of 454 data, a full distance matrix can be as

---

Citation: Yijun Sun, Yunpeng Cai , Li Liu , Fahong Yu , Michael L. Farrell , William McKendree and William Farmerie, ESPRIT: estimating species richness using large collections of 16S rRNA pyrosequences, doi:10.1093/nar/gkp285.

large as 1000 GB. Even if we remove duplicated sequences and sequence pairs that have a pairwise distance larger than a cutoff value (say 0.1), the resulting distance matrix in a sparse format can be 20 GB in size, which is too big to be directly loaded into the memory in most computers. There exist a few algorithms that used various strategies to overcome the above mentioned computational issues. Two typical methods are FastGroupII (12) and RDP-Pyro (13). By submitting data to their web applications and through personal communications, we found that these methods can work efficiently only for small- or medium-scale data sets.

In this article, we proposed a new algorithm, referred to as ESPRIT, which addresses several computational limitations with prior work by using parallel computing. ESPRIT uses the Needleman–Wunsch algorithm (14) to optimally align each pair of sequences. Through a benchmark study, we demonstrated that global pairwise alignment can provide a much more accurate estimate of microbial richness than multiple alignment. A more important reason, however, is that pairwise alignment allows for parallel computing, while most MSA algorithms can only be used in a single computer. We also developed a new cluster algorithm, referred to as Hcluster within the ESPRIT framework, to handle large-scale clustering problems. Unlike a brute-force method, Hcluster groups sequences into OTUs on-the-fly, while keeping track of linkage information, to overcome memory limitations. By assigning a computational task to hundreds of nodes, ESPRIT is not computationally constrained by the number of sequences to be analyzed, but by the capacity of a computer cluster. Hence, ESPRIT is very suitable for applications such as global ocean survey of microbial populations that may require the analysis of data collected from thousands of locations.

We conducted large-scale experiments to demonstrate the effectiveness of the proposed algorithm. We first performed a simulation study by using a data set generated by sequencing PCR amplicon libraries of known 16S rRNA genes. To our knowledge, this is probably the first benchmark study reported in the literature to assess the estimation accuracy of an algorithm for taxonomy-independent analysis. Our experimental results justified the use of global pairwise alignment for the purpose of estimating microbial community compositions, and showed that the commonly used MUSCLE+DOTUR pipeline may overestimate biodiversity. We then applied ESPRIT to eight seawater samples. The results are very consistent with that of the benchmark study. We finally used ESPRIT to analyze a recently collected air sample consisting of about 350K short reads. It is stated in (15) that microbial communities may be too diversified to be practically tested by amplification and sequencing of 16S rRNA genes. Our results suggest that at least for air samples, it is computationally practical to use the current sequencing technology to conduct 16S rRNA-based biodiversity surveys.

ESPRIT ALGORITHM

The algorithm consists of four modules: (i) removes low-quality reads using various criteria, (ii) computes pairwise distances of reads, (iii) groups reads into OTUs at different dissimilarity levels and (iv) performs statistical inference to estimate species richness. We developed two versions of ESPRIT, one for personal computers (PC) and one for computer clusters (CC). The PC version can process several tens of thousands sequences within a few minutes, while the CC version is able to analyze several hundreds of thousands of reads within one day. The source code and user guide are freely available at http://www.biotech.ufl.edu/people/sun/esprit.html. The readers who have no access to a CC to analyze their data may contact the corresponding author.

Removing low quality reads

The error rate of 454 Life Sciences GS FLX systems is estimated to be five errors per kilobase (http://www.454.com/products-solutions/system-features.asp). However, only a small number of reads account for most of the errors (16), and these outlier reads may be classified at certain dissimilarity levels as rarely occurrent OTUs (i.e. clusters containing only a few members). This will lead to an overestimate of microbial richness in subsequent analyses, since many ecological metrics are computed based on the numbers of rarely occurrent OTUs. To minimize the effects of random sequencing errors, we remove the reads that contain ambiguous nucleotides (N), and those with more than one mismatch with the PCR primer at the beginning of a read. Also, we eliminate the sequences with atypical lengths. The default setting of the algorithm is to retain the reads with a length within 1 SD from the mean length. In order to reduce computational complexity, by following the strategy used in (17), if two sequences are identical or one sequence is a subset of the other, only the longer sequence is retained, and the number of occurrence of each retained sequence is recorded and used in the statistical inference. A similar trimming procedure was also used in (16). ESPRIT allows users to bypass this module to use a sample filtered by a customized trimming procedure.

Computing pairwise distances

We use the Needleman–Wunsch algorithm to optimally align each pair of sequences in a sample, and the quickdist algorithm (4) to compute pairwise distances. More specifically, each pairwise distance equals mismatches, including indels, divided by a sequence length. To avoid overestimating distances between sequences from rapidly diverging variable regions, end gaps are ignored and gaps of any length are treated as a single evolutionary event or mismatch.

The naive application of the Needleman–Wunsch algorithm to an environmental sample, however, is computationally too expensive. For example, given one full run of 454 data with >400K reads, it is estimated that it would take about 3 years to align all 80 billion pairs of sequences, and need about 1000 GB to store the resulting distance matrix in the PHYLIP format. However, we notice that for the purpose of estimating biodiversity, only the sequence pairs with distances <0.10 are of most interest, which only account for a small fraction of all possible pairs (about 1–5%, Figure 1S). This means possibly 20–100 folds of speedup. Moreover, by removing unwanted pairs, distance information can be stored in a sparse matrix that requires much less memory. In order to rapidly identify the sequence pairs of interest, we compute the *k*-mer distance of each pair of sequences. We give below a brief description of how it works. A *k*-mer, also known as *k*-tuple, is a sub-sequence consisting of *k* possible nucleotides. By specifying the value of k, a complete alphabet Ω of k-mer is constructed. The number of occurrence of each k-mer in alphabet Ω, also called genomics profile (18,19), is computed for each sequence. Then, the k-mer distance of a pair of sequences is derived as

$$d = 1 - \sum_{i=1}^{|\Omega|} \min(f_1(i), f_2(i))/(\min(L_1, L_2) - k + 1), \quad 1$$

where $f_1(i)$ and $f_2(i)$ are the numbers of the occurrence of the i-th k-mer of two sequences, respectively, and L_1 and L_2 are the lengths of two sequences, respectively. It has been shown that k-mer distances are highly correlated with genetic distances (20). The concept of k-mer counting has been used for various applications, including sequence alignment (10), phylogenetic analysis (21,22) and detection of horizontal gene transfer (23). We used this technique to remove unwanted sequence pairs, and found that by using a k-mer distance of 0.4, most of the sequence pairs of interest can be rapidly identified (Figure 1a). The default setting in ESPRIT is 0.5. Figure 1b shows that removing sequence pairs with k-mer distances larger than the default parameter has a negligible impact on the estimation accuracy for the distance levels <0.1. It should be noted that though the default k-mer threshold works well for all of the 10 data sets tested in this article, it is possible that the parameter may vary for different applications. Hence, we provide an auxiliary program in the software package that allows users to determine the parameter by comparing k-mer distances against genetic distances using a small subset of their samples.

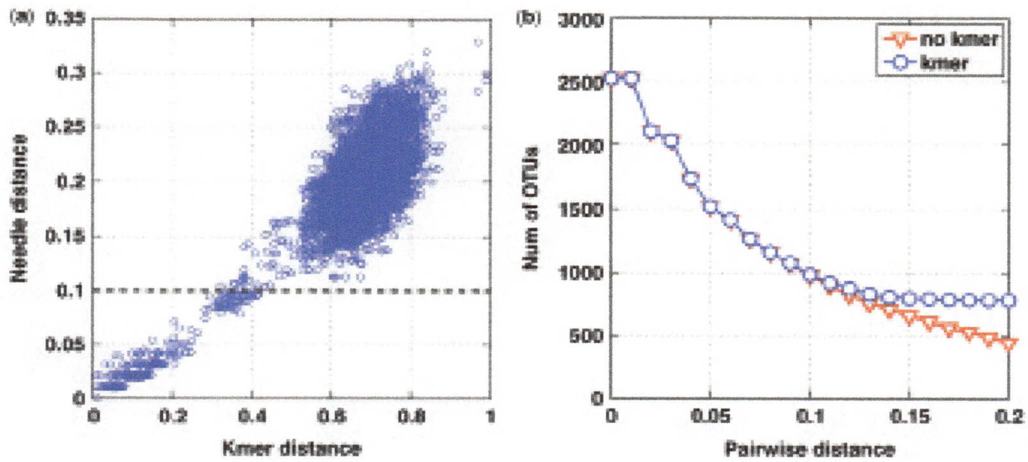

Figure 1: (**a**) k-mer distances are highly correlated with genetic distances. (**b**) Removing sequence pairs with k-mer distances larger than the default threshold has a negligible impact on the estimation accuracy for the distance levels of interest. The experiment was performed on the 53R seawater sample (see Results Section).

In addition to a PC version, we also developed a CC version for calculating both k-mer and genetic distances based on globally aligned sequences. Given one full run of 454 sequences, by using 100 computer nodes, it takes only 3–5 h to finish this module, depending on the availability of computer nodes.

Assigning sequences into OTUs

After a distance matrix is computed, complete-link hierarchical clustering (24) is performed to assign sequences into OTUs of defined sequence variations. Although ESPRIT uses a sparse distance matrix, given a few runs of 454 reads, the size of the matrix can still be quite large. We spent considerable efforts on algorithm development in order to overcome the memory issue, and devised a new clustering algorithm, referred to as Hcluster. The basic idea of the algorithm is to first sort pairwise distances in an ascending order, and then process the distance information on-the-fly. At each step, we classify the clusters being analyzed into 'active' or 'inactive' clusters. Active clusters are those with known distance information to other clusters, but the information is not enough to decide whether to merge them with other clusters. Inactive clusters, on the other hand, are those with no information at all, or those already merged with other clusters. We only need to maintain the linkage information for active clusters, which is updated at the time when new distance information is processed. We have conducted a large-scale experiment that demonstrated that Hcluster performed very well in the presence of several hundreds of thousands of reads. The accuracy of Hcluster has been benchmarked against DOTUR and MOTHUR. The three methods yielded the exactly same results (Figure 6S).

Clustering is a common problem in bioinformatics. We provide Hcluster as a standalone algorithm in the software package. This algorithm may be useful for other applications where large-scale clustering is needed [e.g. taxonomy-dependent analysis (13)].

Statistical inference of species richness

ESPRIT supports three richness estimators: rarefaction analysis (25), Chao1 (26) and ACE (27,28). Rarefaction allows the calculation of the species richness for a given number of sampled individuals and constructs so-called rarefaction curves. The curve is the number of observed OTUs as a function of the number of sampled sequences. In case of a steep slope, it means that a large fraction of the species diversity is not sampled yet, and more exhaustive sampling will yield a significant number of additional species. Chao1 and ACE are two abundance-based coverage estimators that predict the species richness based on the number of rarely occurrent OTUs. The cluster information generated by ESPRIT allows users to compute other ecological metrics, to derive a consensus sequence of each cluster, and to align the sequences of rarely occurrent OTUs against a database, which may lead to the identification of new organisms.

RESULTS

We conducted large-scale experiments to demonstrate the effectiveness of the newly proposed ESPRIT algorithm. When the PC version of ESPRIT was used, the experiment was performed on a server with eight E5345 Xeon CPUs and 16 GB memory operated on Linux 5.2 system. When the CC version was used, the experiment was performed on a CC administrated by the High-Performance Computing Center at the University of Florida. The detailed computer configuration can be found athttp://wiki.hpc.ufl.edu/index.php/Operating_Environment.

Benchmark study

The purpose of the simulation study is 2-fold: (i) to investigate which strategy, pairwise or MSA, is more suitable for taxonomy-independent analysis, and (ii) to assess the estimation accuracy of the ESPRIT algorithm. Although MUSCLE+DOTUR has been used in several metagenomics studies, its performance has never been benchmarked in the literature. This is in part due to the fact that under current technologies, one may never know precisely the ground truth information of the compositions of a microbial community. The benchmark information, however, is critical to evaluate the performance of an algorithm performed on real-world data and to make a meaningful comparison of taxonomic distributions of different environments.

The simulation data consisted of about 340K short sequences, generated by pyrosequencing two PCR amplicon libraries of 43 known 16S rRNA gene fragments using the Roche GS20 system. It was originally used in (16) to study the prebased error rate of the system. We applied both ESPRIT and MUSCLE+DOTUR to the 43 reference gene sequences to estimate the numbers of OTUs defined at various distance levels, also known as lineage-through-time curve in the literature (8). Parameters -maxiters 1 -diags 1 and -sv were used in MUSCLE. The two so-obtained curves served as the ground truth to benchmark the performance of the two algorithms. In order to study how the two algorithms perform in the presence of sequencing errors, we generated two data sets by mapping each read to the 43 reference sequences and retaining the reads that have less than 3% or 5% mismatches with the closest reference sequence. To eliminate statistical variations, each algorithm was run 10 times for each data set. In each run, 10K reads were randomly sampled from each data set. The lineage-through-time curves, averaged over 10 runs, are plotted in Figure 2. Since the ground truth curves generated by ESPRIT and MUSCLE+DOTUR are very similar, for ease of presentation, only the curve of ESPRIT is plotted. From the figure, we observe that while both algorithms always lead to an overestimate of the number of OTUs due to the presence of sequencing errors, pairwise alignment can provide a more accurate estimate of microbial richness than multiple alignment. For example, for the first data set, at the 0.05 distance level, the numbers of OTUs estimated by ESPRIT and MUSCLE+DOTUR are 44 (95% CI: 43–45) and 92 (95% CI: 82–102), respectively, while the ground truth is 42 (Figure 2a).

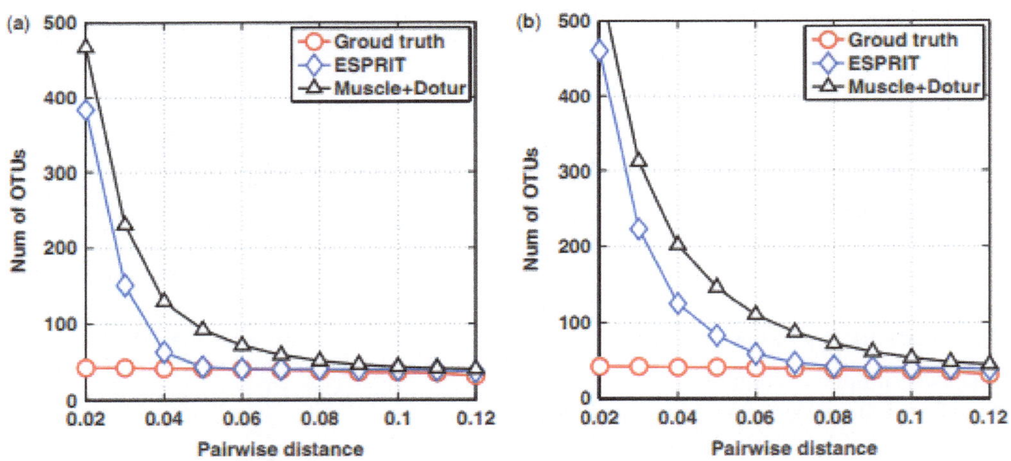

Figure 2: Lineage-through-time curves generated by using ESPRIT and MUSCLE+DOTUR algorithms performed on simulation data with each read containing up to (**a**) 3% and (**b**) 5% sequencing errors.

One may wonder whether the disparity between the results of the two methods is due to the compromise on the parameter settings of MUSCLE. To answer this question, we repeated the experiment by using the default parameter of MUSCLE (-maxiters 16) (29). The results are reported in Figure 3. Though computationally expensive, using the default parameter results in a much better alignment in terms of the sum of pairwise alignment scores. Interestingly, we observe that the estimate obtained by using the default parameter is much worse than that obtained by using the simple parameter. This can be explained by the fact that MSA aims to minimize the sum of pairwise alignment scores, ignoring the fact that a large proportion of sequence pairs originate from distantly related OTUs. Since many existing methods used MSA for taxonomy-independent analysis, this issue merits further investigation, which is presented. We also conducted a benchmark study comparing the prediction performance of ESPRIT with those of NAST (30) and RDP-Pyro (13) in Section 2S.

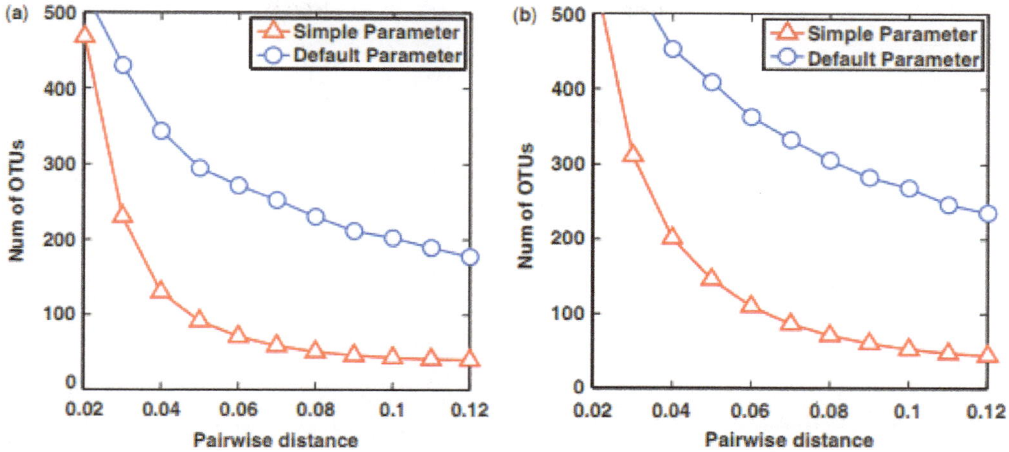

Figure 3: Lineage-through-time curves generated by MUSCLE+DOTUR using simple or default parameters performed on simulation data with each read containing up to (**a**) 3% and (**b**) 5% sequencing errors.

Experiments on eight seawater samples

We applied ESPRIT to reanalyze eight seawater samples downloaded from (4). The DNA materials within environmental samples were collected from eight different locations in the Atlantic and Pacific Oceans, respectively, as a part of efforts to develop a global description of microbial diversities in the ocean. A total of 118 000 PCR amplicons were sequenced that covered the V6 hyperviable region of rRNAs. The 454 reads have undergone a systematic trimming process. Thus, we bypassed the trimming procedure. The number of sequences for each sample ranges from 5000–17 666 (see Table 1). The interested reader may refer to (4) for a detailed description of the data and sample preparations.

Table 1: Running time of the PC version of ESPRIT performed on eight seawater samples

	Data sets							
	112R	115R	137	138	53R	55R	FS312	FS396
Number of reads	9282	11 005	13 097	14 374	5000	13 902	4835	17 666
CPU time	2 m 54 s	4 m 13 s	4 m 28 s	6 m 19 s	59 s	7 m 31 s	49 s	6 m 34 s

We reran MUSCLE+DOTUR by using the parameters provided by (4). The default parameters of ESPRIT were used. Due to space limitations, only the results of sample FS396 are presented in the main text. FS396 contains 17 666 reads after trimming, and is the largest data set in size among the eight samples. The lineage-through-time curves, ACE and Chao1 estimates at three different distance levels are depicted in Figure 4. We observe that ESPRIT gives a much lower estimate of species richness than the competing method, which is consistent with the result of the simulation study described above. For example, at the distance of 0.05, the number of observed OTUs and the ACE and Chao1 estimates obtained by ESPRIT are only about 50–70% of those obtained by MUSCLE+DOTUR. This becomes more evident in Figure 5, where the results of the rarefaction analysis of FS396 are presented. Though the existing pipeline overestimates the diversity of microbes, the conclusion of (4) still holds that the microbial diversity of seawater samples is much larger than previously reported, and the steep slopes of the rarefaction curves, even at relatively large genetic distances (e.g. 0.1), suggest that a large fraction of species have not been sampled yet. We notice that the disparity between the results of the two methods is much larger than that in the simulation study (Figures 2, 4, S8 and S9). One possible reason is that the sequences in the seawater samples are more diverse than those in the simulation data (see Section 1S for a detailed discussion).

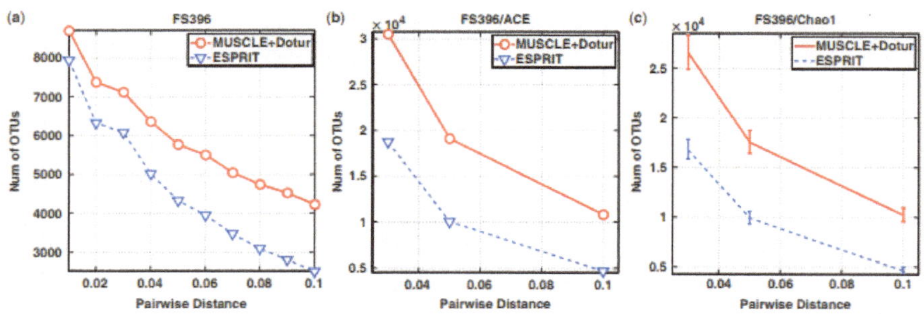

Figure 4: (**a**) Lineage-through-time curves, (**b**) ACE and (**c**) Chao1 estimates generated by using ESPRIT and MUSCLE+DOTUR algorithms performed on the FS396 data. Error bars of Chao1 estimates represent the 95% confidence interval.

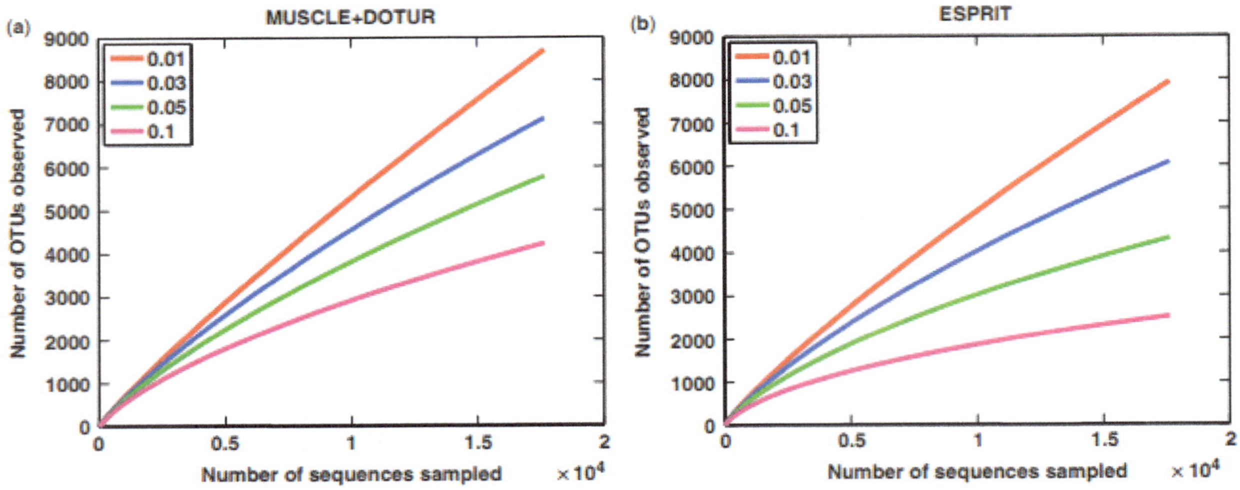

Figure 5: Rarefaction curves generated by using (**a**) MUSCLE+DOTUR and (**b**) ESPRIT performed on the FS396 data.

ESPRIT is computationally very efficient. As can be seen from Table 1where the CPU times of the PC version of ESPRIT performed on the eight seawater samples are recorded, ESPRIT can process several tens of thousands of sequences within a few minutes. As both the results of (4) and our reanalysis of their samples suggest deeper sequencing is required for accurate estimation of microbial richness of environmental samples, the CC version of ESPRIT is expected to be computationally more efficient than the PC version, as we will see in the next section.

Experiments on an air sample

We applied ESPRIT to an air sample recently collected from Iraq. The small subunit rRNA gene fragments that cover the V6 hyperviable region of rRNAs were amplified, cloned and pyrosequenced using 454 Life Sciences systems. A total of 348 952 sequences of an average of 275 nt in length were obtained. This is one of the largest 16s rRNA-based biodiversity surveys conducted using air samples. The number of sequences is one order of magnitude larger than those of the seawater samples we considered in the previous section.

The CC version of ESPRIT with the default parameters was used. After the trimming procedure, 40 928 (about 13%) reads were removed as low-quality reads. Figure 6 presents the results of the rarefaction analysis of the OTUs defined at four different sequence variations, ranging from 0.01 to 0.1. We observe that, in contrast with the results of the seawater samples, the rarefaction curves saturate even at a relatively small genetic distance of 0.03, indicating that additional sampling may not lead to significantly increased estimates of total species diversity. The results of ACE and Chao1 estimators are reported in Table 2. At the distance levels of 0.05 and 0.1, the numbers of observed OTUs are already very close to the ACE and Chao1 estimates. These experimental results suggest that, at least for air samples, the current sequencing technology is sufficient to conduct 16S rRNA-based biodiversity surveys. The experiment was performed on a small CC consisting of 100 nodes, and it took about 10 h to finish the entire analysis.

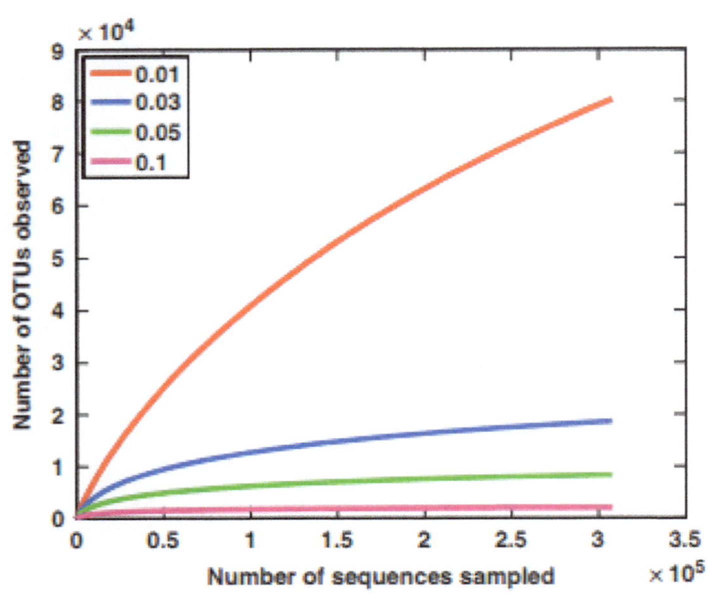

Figure 6: Rarefaction analysis of an air sample. Rarefaction curves are shown for OTUs with sequence variations that do not exceed 1, 3, 5 or 10%.

Table 2: The number of observed OTUs, the ACE and Chao1 estimates of an air sample at four different distance levels

	Pairwise distance			
	0.01	0.03	0.05	0.1
OTUs	80 238	18 686	8344	2109
ACE	147 266	23 894	9664	2262
Chao1	138 376	23 921	9748	2293
Upper	139 911	24 302	9932	2362
Lower	136 881	23 566	9585	2242

The 95% CIs of the Chao1 estimates are also provided.

DISCUSSIONS

Equipped with next-generation sequencing technology, researchers now start to sequence many millions of sequences for applications such as global ocean surveys and epidemiological studies with many patients. Parallel computing that distributes computation to hundreds or even thousands of nodes seems at this time to be the only viable approach. We have demonstrated that the newly proposed ESPRIT algorithm can process several hundreds of thousands of sequences within 10 h by using a relatively small CC. To our knowledge, no existing methods can efficiently handle such large data. The two key components of ESPRIT are the use of pairwise instead of multiple sequence alignment, which allows for parallel computing, and the development of Hcluster that performs hierarchical clustering using online learning to address the memory issue. We have demonstrated that MSA is not only computationally expensive, but also tends to overestimate microbial diversity (see Section 1S). In order to process many millions of reads, researchers may need to collaborate with a supercomputing center (e.g. IBM Roadrunner which has about 20 000 nodes). Our ESPRIT algorithm can easily be modified to work on any cluster. The computational complexity of ESPRIT is quadratic with respect to the number of reads. At Roadrunner's full capacity, we estimate that it will take 1 s, 10 h and 40 h to process 1, 50 and 100 million reads, respectively. Of course, this is an overoptimistic estimation, since it focuses only on computational complexity and assumes that the resources of the entire supercomputing center are available for this singular task. ESPRIT, however, does provide a promising direction for analyzing large collection of 16S rRNA data. We will continue to optimize the performance of ESPRIT and release the software to the community. We believe that ESPRIT will evolve to be a powerful tool for metagenomics study.

ACKNOWLEDGEMENTS

We thank the editor and three reviewers for their valuable comments to improve the quality of the article. The simulation study presented in Section 1S was inspired by one of the reviewers' comment.

Conflict of interest statement. None declared.

REFERENCES

1. Eisen,J.A. (2007) Environmental shotgun sequencing: its potential and challenges for studying the hidden world of microbes. PLoS Biol., 5, e82.
2. Rothberg,J.M. and Leamon,J.H. (2008) The development and impact of 454 sequencing. Nat. Biotechnol., 26, 1117–1124.
3. Margulies,M., Egholm,M., Altman,W.E., Attiya,S., Bader,J.S., Bemben,L.A., Berka,J., Braver man,M.S., Chen,Y.J., Chen,Z. et al. (2005) Genome sequencing in microfabricated high-density picolitre reactors. Nature, 437, 376–380.
4. Sogin,M.L., Morrison,H.G., Huber,J.A., Welch,D.M., Huse,S.M., Neal,P.R., Arrieta,J.M. and Herndl,G.J. (2006) Microbial diversity in the deep sea and the underexplored "rare biosphere". Proc. Natl Acad. Sci. USA, 103, 12115–12120.
5. Keijser,B., Zaura,E., Huse,S.M., van der Vossen,J., Schuren,F., Montijn,R.C., ten Cate,J.M. and Crielaard,W. (2008) Pyrosequencing analysis of the oral microflora of healthy adults. J. Dent. Res., 87, 1016–1020.
6. Borneman,J. and Triplett,E.W. (1997) Molecular microbial diversity in soils from eastern Amazonia: evidence for unusual microorganisms and microbial population shifts associated with deforestation. Appl. Environ. Microbiol., 63, 2647–2653.
7. Sait,M., Hugenholtz,P. and Janssen,P.H. (2002) Cultivation of globally distributed soil bacteria from phylogenetic lineages previously only detected in cultivation-independent surveys. Environ. Microbiol., 4, 654–666.
8. Schloss,P.D. and Handelsman,J. (2005) Introducing DOTUR, a computer program for defining operational taxonomic units and estimating species richness. Appl. Environ. Microbiol., 71, 1501–1506.
9. Roesch,L.F.W., Fulthorpe,R.R., Riva,A., Casella,G., Hadwin,A.K.M., Kent,A.D., Daroub,S.H., Camargo,F.A.O., Farmerie,W.G. and Triplett,E.W. (2007) Pyrosequencing enumerates and contrasts soil microbial diversity. ISME J., 1, 283–290.
10. Edgar,R.C. (2004) MUSCLE: multiple sequence alignment with high accuracy and high throughput. Nucleic Acids Res., 32, 1792–1797.
11. Katoh,K, Kuma,K., Toh,H. and Miyata,T. (2005) MAFFT version 5: improvement in accuracy of multiple sequence alignment. Nucleic Acids Res., 33, 511–518.
12. Yu,Y., Breitbart,M., McNairnie,P. and Rohwer,F. (2006) FastGroupII: a web-based bioinformatics platform for analyses of large 16S rDNA libraries. BMC Bioinformatics, 7, 57.
13. Cole,J.R., Wang,Q., Cardenas,E., Fish,J., Chai,B., Farris,R.J., Kulam-Syed-Mohideen,A.S., McGarrell,D.M., Marsh,T., Garrity,G.M. et al. (2009) The Ribosomal Database Project: improved alignments and new tools for rRNA analysis. Nucleic Acids Res., 37, D141–D145.
14. Needleman,S.B. and Wunsch,C.D. (1970) A general method applicable to the search for similarities in the amino acid sequence of two proteins. J. Mol. Biol., 48, 443–453.
15. Gans,J., Woilinsky,M. and Dunbar,J. (2005) Computational improvements reveal great bacterial diversity and high metal toxicity in soil. Science, 309, 1387–1390.

16. Huse,S.M., Huber,J.A., Morrison,H.G., Sogin,M.L. and Welch,D.M. (2007) Accuracy and quality of massively parallel DNA pyrosequencing. Genome Biol., 8, R143.

17. Huber,J.A., Welch,D.B.M., Morrison,H.G., Huse,S.M., Neal,P.R., Butterfield,D.A. and Sogin,M.L. (2007) Microbial population structures in the deep marine biosphere. Science, 318, 97–100.

18. Karlin,S. and Burge,C. (1995) Dinucleotide relative abundance extremes: a genomic signature. Trends Genet., 11, 283–290.

19. Karlin,S., Mrazek,J. and Campbell,A. (1997) Compositional biases of bacterial genomes and evolutionary implications. J. Bacteriol., 179, 3899–3913.

20. Edgar,R.C. (2004) Local homology recognition and distance measures in linear time using compressed amino acid alphabets. Nucleic Acids Res., 32, 380–385.

21. Mavromatis,K., Ivanova,N., Barry,K., Shapiro,H., Goltsman,E., McHardy,A.C., Rigoutsos,I., Salamov,A., Korzeniewski,F., Land,M. et al. (2007) Use of simulated data sets to evaluate the fidelity of metagenomic processing methods. Nat. Methods, 4, 495–500.

22. Sun,Y., Yu,F., Liu,L. and Farmerie,W. (2007) Estimating microbial population densities based on genomic signatures. Proc. Intl. Conf. Bioinform. Comput. Biol., 1, 163–168.

23. Dalevi,D., Dubhashi,D. and Hermansson,M. (2006) Bayesian classifiers for detecting HGT using xed and variable order markov models of genomic signatures. Bioinformatics, 22, 517–522.

24. Duda,R.O., Hart,P.E. and Stork,D.G. (2001) Pattern Classification, 2nd edn. Wiley, New York.

25. Hurlbert,S.H. (1971) The non-concept of species diversity: a critique and alternative parameters. Ecology, 52, 577–586.

26. Chao,A. (1984) Non-parametric estimation of the number of classes in a population. Scand. J. Stat., 11, 265–270.

27. Chao,A. and Lee,S.M. (1992) Estimating the number of classes via sample coverage. J. Am. Stat. Assoc., 87, 210–217.

28. Chao,A., Ma,M.C. and Yang,M.C.K. (1993) Stopping rules and estimation for recapture debugging with unequal failure rates. Biometrika, 80, 193–201.

29. Edgar,R.C. (2004) MUSCLE user guide. Technical Report. Available at http://wwwdrive5.com/muscle/docs.htm.

30. DeSantis,T.Z., Hugenholtz,P., Keller,K., Brodie,E.L., Larsen,N., Piceno,Y.M., Phan,R. and Andersen,G.L. (2006) NAST: a multiple sequence alignment server for comparative analysis of 16S Rrna genes. Nucleic Acids Res., 34, W394–W399.

SECTION VI: METATRANSCRIPTOMICS

CHAPTER 21

SIMULTANEOUS ASSESSMENT OF SOIL MICROBIAL COMMUNITY STRUCTURE AND FUNCTION THROUGH ANALYSIS OF THE META-TRANSCRIPTOME

Tim Urich, Anders Lanzen, Ji Qi, Daniel H. Huson, Christa Schleper, Stephan C. Schuster

INTRODUCTION

Soils cover almost all of the terrestrial area on Earth and have an indispensable ecological function in the global cycles of carbon, nitrogen and sulfur. Due to their physico-chemical complexity with many micro-niches, they teem with bio-diversity, both phylogenetically and functionally. A single gram of soil has been estimated to contain thousands to millions of different bacterial, archaeal and eukaryotic species [1], [2] interwoven in extremely complex food webs. Communities of soil microbes carry out a multitude of small-scale processes that underlie many environmentally important functions. However, the explicit functional and ecological roles of individual taxa remain uncertain because most microbes withstand laboratory cultivation [3], [4]. Therefore the most basic questions in microbial ecology "who is out there?" and "what are they doing?" are still often unanswered for many environments and for many microbial taxa. Ideally, especially the second question requires simultaneously information about the identity of taxa within a community and about functional processes performed. While soils seem to harbor the most complex microbial communities, these considerations apply to many other environments as well, like e.g. oceans and sediments.

With metagenomic technologies new dimensions in the characterization of complex microbial communities have been reached. Large scale shotgun sequencing approaches allow the discovery of many novel genes found in the environments independent of cultivation efforts[5]–[8]. Sequencing of large genomic inserts that contain phylogenetic "anchors" allows a direct link to the microbial taxon. However, in almost all of the metagenomic studies, a separate accompanying molecular typing method-usually based on PCR-amplified 16S rRNA genes-is needed to characterize the gene discoveries in the context of the microbial phylogenetic diversity of the sample [7]–[9].

These PCR-based typing methods-though very powerful, in particular when combined with the novel pyro-sequencing technology-[10], [11] have some well-known drawbacks: (1) bias is introduced by primers and/or exponential amplification; (2) simultaneous quantitative assessment of all three domains of life is impossible and (3) persistence of free DNA can bias the measurement of community responses to environmental changes.

Furthermore, DNA-based metagenomic and diversity studies do not allow us to draw conclusions on the expression state of the genes and therefore the functional role of genes or organisms in the investigated environment remains uncertain. In analogy to postgenomic studies of cultivated organisms, a logical next step in the metagenomic area therefore includes meta-transcriptomic technology.

First attempts to study the transcription of genes in environmental samples have been performed. They involved specific purification steps to selectively enrich for mRNA of bacteria or eukarya by depleting ribosomal RNA or enriching for polyA-tailed mRNA, respectively [12],[13]. A more large scale (pyro-)sequencing approach was recently adapted for use with bacterial and archaeal mRNA from an environmental marine sample [14]. It involved an *in vitro* amplification step to maintain small sample size and fast preparation.

In order to overcome some of the limitations of the approaches mentioned above, we explore here the possibility to analyse the total RNA pool of a community, as it is naturally enriched not only in functionally but also taxonomically relevant molecules, i.e. mRNA and rRNA, respectively.

Citation: Urich T, Lanzén A, Qi J, Huson DH, Schleper C, Schuster SC (2008) Simultaneous Assessment of Soil Microbial Community Structure and Function through Analysis of the Meta-Transcriptome. PLoS ONE 3(6): e2527. doi:10.1371/journal.pone.0002527.

This offers the opportunity to link community structure and function in a single experiment, and to reach beyond the community's genomic potential as assessed in DNA-based methods, towards its *in situ* activity. Furthermore, the use of total RNA avoids extensive cleaning or amplification steps of mRNA molecules, enabling fast preparation even from difficult samples, such as soils, which are notoriously enriched for humic acids and other substances inhibiting molecular biological applications.

We have published initial results of the approach earlier in the context of characterizing the transcriptional activity of particular organisms from soil [15]. Here we present and validate the "double-RNA" approach for in-depth characterization of a soil microbial community by studying mRNA and rRNA molecules simultaneously from the same sample. Our approach ensures (i) fast preparation of the RNA in the light of the short half life of mRNA molecules, it avoids (ii) biases introduced by PCR or cloning procedures and (iii) obtains qualitative and quantitative information simultaneously of genes from organisms of all three domains of life, the archaea, bacteria and eukarya. Since highly parallel sequencing was used for this approach, which produced many, but short reads (100 bp), we have set up an appropriate bioinformatic analysis pipeline that allows to reliably extract in-depth functional and taxonomic information from this dataset.

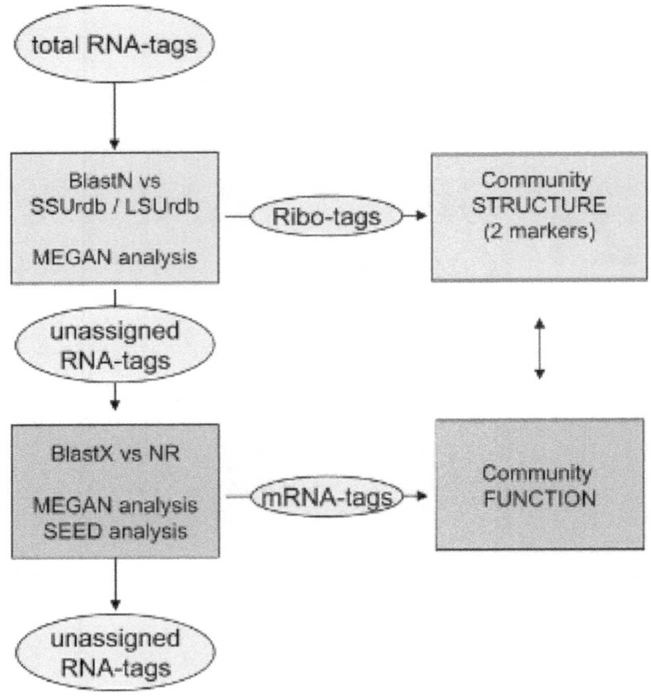

Figure 1: Overview about the double RNA analysis pipeline.

RESULTS AND DISCUSSION

Community profiling based on SSU and LSU rRNA-tags

The sandy soil ecosystem studied here stems from a conservation area and is comparably poor in nutrients and of neutral pH. It has been the subject of a variety of studies. Many molecular data are therefore available which facilitate result interpretation [15]–[20]. We generated cDNA from the transcriptome of this soil community by a non-targeted approach applying random-hexamer primed reverse transcription on total RNA. The cDNA was subjected directly to pyrosequencing without any prior PCR or cloning steps (see materials and methodsfor details). Several potentially biasing steps were therefore avoided. One experiment resulted in 258,411 RNA-derived sequence tags of ~98 bp. We set up a two-step analysis process to identify rRNA- and mRNA-derived tags ("ribo-tags" and "mRNA-tags", respectively) for efficiently and reliably obtaining taxonomic and functional information (Fig. 1). In the first step, all RNA-tags were compared against a small subunit and a large subunit rRNA reference database (SSUrdb and LSUrdb, respectively) that we compiled from publicly available sources (SILVA[21], RDP-II[22], Genbank nucleotide; see materials and methods for details). These two databases contain sequences from all three domains of life This approach enables the first simultaneous taxonomic analysis of communities based on the two most commonly used taxonomic marker molecules. The output file was analyzed with the program MEGAN[23]. In the second stage, all unassigned RNA-tags were compared against the Genbank non-redundant protein database to identify mRNA-tags.

The validity of our taxonomic analysis was tested with two simulated datasets composed of 43 small or large subunit rRNAs from bacterial (32), archaeal (5) and eukaryotic (6) representatives. Both databases, though very different in size and sequence composition, produced a similarly high taxonomic resolution power not only at the taxonomic level of domain and phylum, but also at the level of order for most of the species tested. These results also showed that neither reference database introduced a major artificial "community" shift; rather, both correctly reflected the test set's "community" structure. In addition, none of the ribo-tags were taxonomically wrongly assigned. Even with a ribo-tag length of 100 bp the power of resolution sufficed for our approach (this has also been demonstrated by other simulations, e.g. [24], [25]). Additionally, we simulated a more natural situation by removing all reference sequences similar to test rRNAs from the databases, according to similarity distributions of ribo-tags from our soil dataset against the reference databases. The identified thresholds reflected the median similarity of SSU and LSU ribo-tags to their best BLAST match in the reference databases. No significant decrease in taxonomic resolution was observed for the SSUrdb and no ribo-tag was incorrectly assigned The much smaller LSUrdb assigned 5% of the ribo-tags incorrectly. When removing all reference sequences ≥86% similar to test sequences from the SSUrdb (to simulate a situation reflecting to the lowest decile similarity for SSU ribo-tags observed in the soil dataset), 5% of the ribo-tags were incorrectly assigned. These results indicated a robust performance of the taxonomic binning approach. Encouragingly, the taxonomic resolution power will only improve further as (1) read lengths increase –250 bp are now possible, and 450 bp will soon be achievable; and (2) rRNA reference databases compile ever more sequences.

The soil dataset contained 193,219 RNA-tags which had significant BLAST hits against the rRNA reference databases (Table 1 and Fig 2). This dataset–which was generated without any PCR or cloning steps–was two to three orders of magnitude larger than traditional rDNA clone libraries, and was also far larger than recently published soil-derived pyro-sequencing datasets[26].

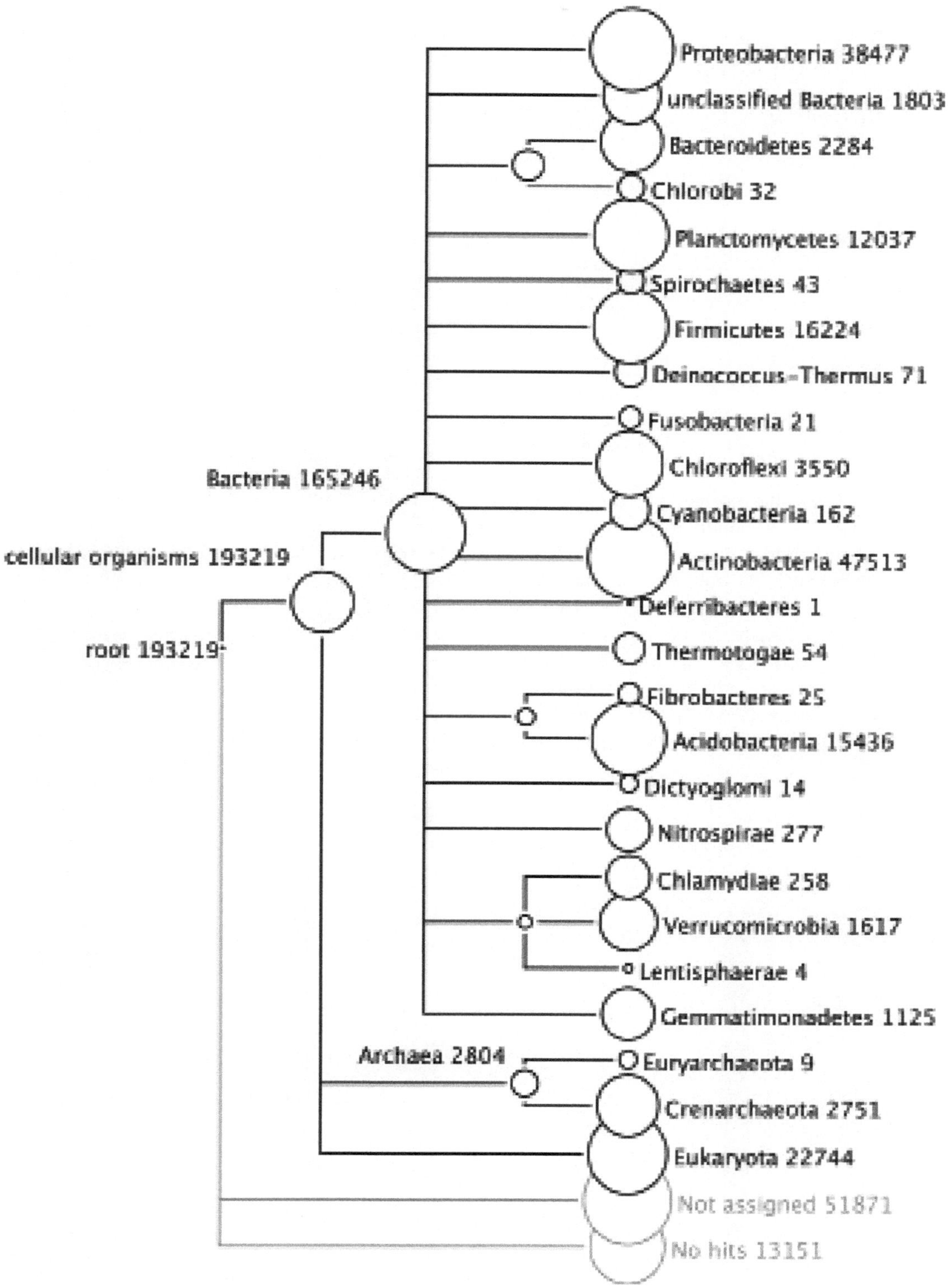

Figure 2: In-depth taxonomic community profiling based on SSU and LSU rRNA. MEGAN tree with the taxonomic affiliation of ribo-tags identified by BLASTN of all RNA-tags against our SSUrdb and LSUrdb according to the NCBI taxonomy. The numbers and sizes of the circles at the tree nodes refer to the ribo-tags affiliated with the respective taxon (absolute cutoff: BLASTN bit score 86, relative cutoff 10% of BLASTN top hit).

Table 1: Statistics of soil community RNA-tag analysis

	No. of RNA-tags	% of RNA-tags
Total RNA-tags	258,411	100
Total ribo-tags	193,219	74.8
SSU ribo-tags	99,061	38.3
LSU ribo-tags	94,165	36.4
mRNA-tags	21,133	8.2
unassigned RNA-tags	44,059	17.0

Note that the number of total ribo-tags differs from the sum of SSU and LSU ribo-tags. Seven ribo-tags contained regions with significant similarity against both databases (not shown). These putative chimeras were likely produced from SSU and LSU derived cDNA fragments during second-strand cDNA synthesis.

doi:10.1371/journal.pone.0002527.t001

The SSUrdb and LSUrdb reported very similar relative community proportions on the domain level, with 10.3% and 13.3% of ribo-tags stemming from Eukarya, 87.2% and 83.8% from Bacteria and 1.5% and 1.4% from Archaea, respectively (Fig. 3a), which shows the reliability of our analytical approach. We further confirmed the experimental reproducibility by performing two independent cDNA syntheses from the same RNA pool (Fig. 3b, see materials and methods for details). In addition, abundances measured previously for some bacterial and archaeal groups in the same environment using various quantitative real-time PCR or metagenomic methodologies were in agreement with this study [15]–[19]. One percent of the SSU and LSU ribo-tags were consistently not sorted into one of the three domains, instead being classified as "cellular organisms". A closer inspection revealed that these represented short ribo-tags with comparably low taxonomic resolution power (not shown), rather than sequence tags of deeply rooting lineages or even of a "fourth domain of life". In principle, though, this non-targeted approach has the potential to identify lineages, which are currently not seen by primer-based PCR methodology, like e.g. the Nanoarchaeota [27].

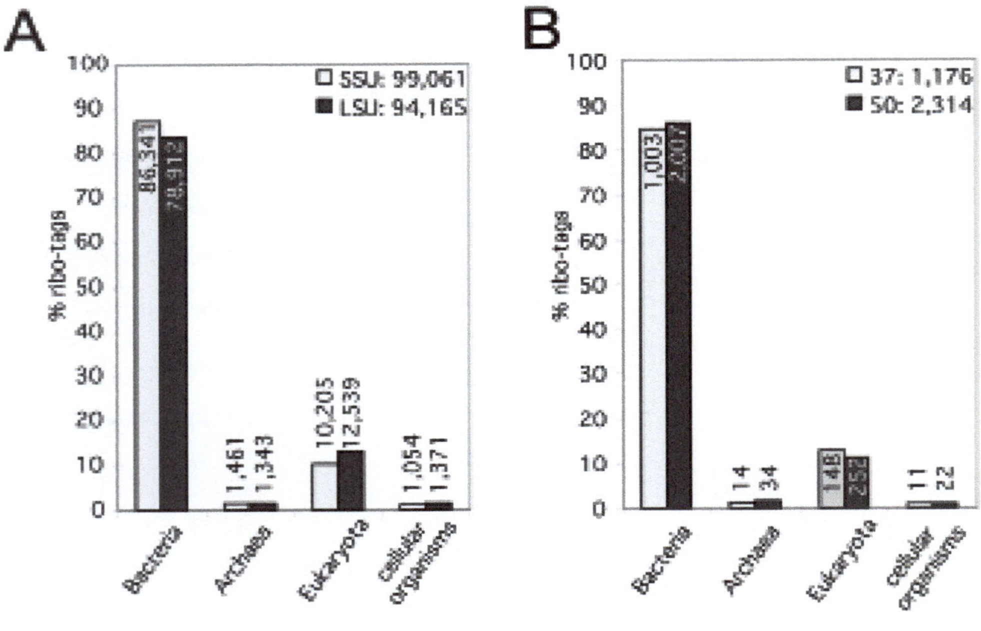

Figure 3: a, Relative ribo-tag distribution of the different cellular domains of life in the SSUrdb and LSUrdb. Absolute numbers are additionally given. Both reference databases report similar fractions, with bacteria-derived ribo-tags being most prominent. Note that approximately 1% of the ribo-tags are not affiliated to any domain by the reference databases. b, Comparison of relative ribo-tag distribution from two independent cDNA syntheses at 37°C (n=4141 RNA-tags) and 50°C (n=1985 RNA-tags) performed on the same RNA pool. Values represent the mean of SSU and LSU ribo-tags.

Diversity and relative abundances of Bacteria and Archaea

With 165,246 ribo-tags (85.5% of total ribo-tags), bacteria were by far the most abundant domain in the sample; they were also extraordinarily diverse. Nineteen out of 24 validly described bacterial phyla were identified (Fig. 2) in addition to 20 candidate divisions. The latter are recently discovered, deep-branching, poorly characterized groups. The dominant bacterial phyla in our sample, such as Actinobacteria (47,513 ribo-tags) and Proteobacteria (38,477 ribo-tags) are typically found in high numbers in soil microbiotas. While both databases showed similar results for the five most abundant phyla (Fig. 4a), incongruencies arose, where LSUrdb sequences were underrepresented (e.g. Chloroflexi, Verrucomicrobia, Bacteroidetes), or completely missing. Even rare phyla, such as Chlorobi and Dictyoglomi (ca. 0.02% of all bacterial tags) were reliably detected by both rRNA databases. It is also noteworthy that the analysis remains congruent at a more detailed taxonomic resolution (class and order, Figures 4b and 4c) for the taxonomic groups, which are comparably well covered in both databases (e.g. Proteobacteria, Firmicutes, Actinobacteria). Approximately 2% (1803) of the bacterial SSU ribo-tags were distributed over 20 candidate divisions. The numbers of ribo-tags ranged from one to several hundred (SPAM, OP8, OP10, NKB19, VC2), or in other words, from indications of the presence of a taxon, to a high confidence including a wealth of sequence information.

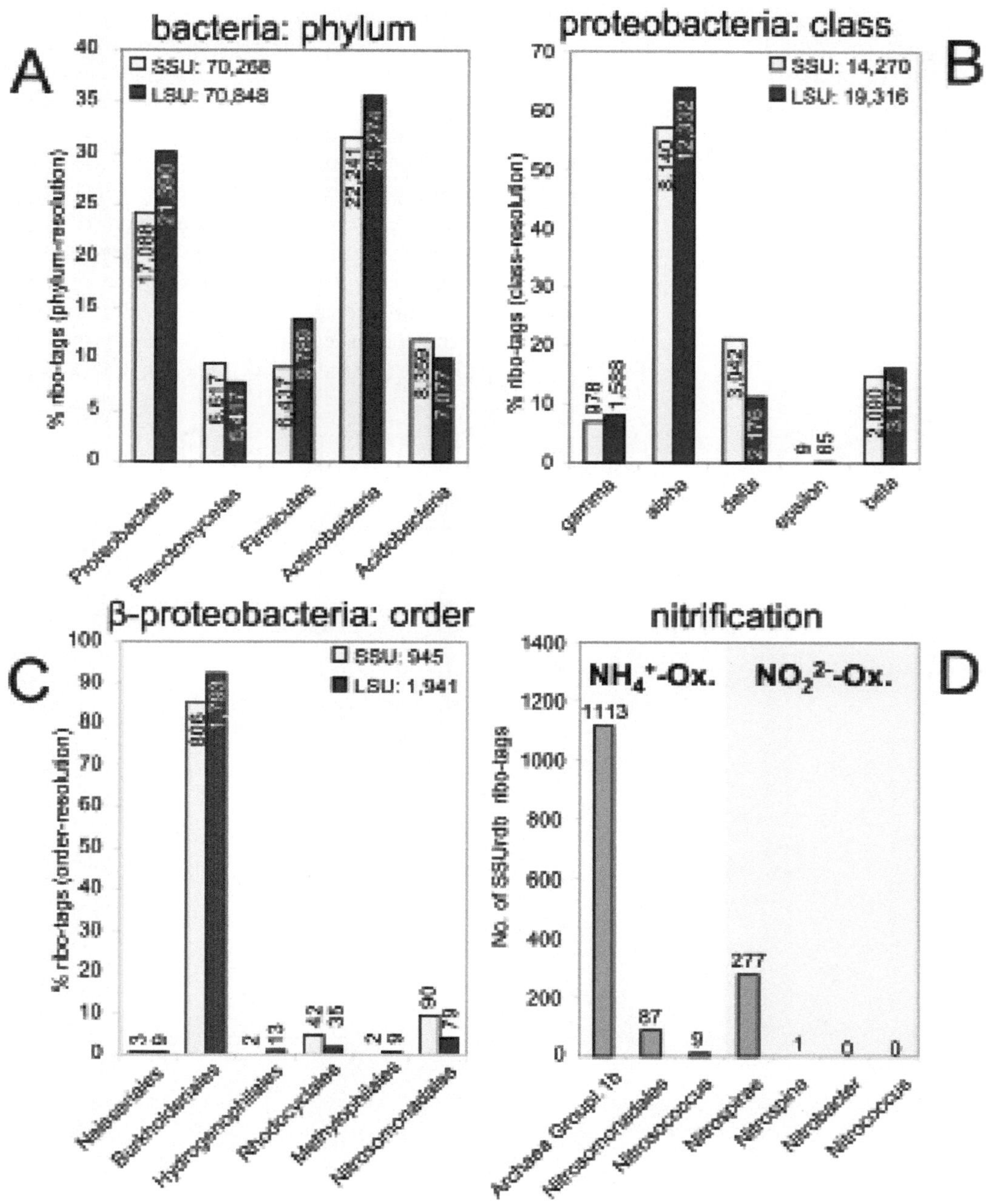

Figure 4: Relative distribution of bacterial SSU and LSU ribo-tags at different taxonomic resolutions and the abundance of prokaryotic groups involved in the nitrification process. a, fraction of phylum-resolution SSU and LSU ribo-tags affiliated to the five numerically dominant bacterial phyla (>5% of bacterial ribo-tags). Absolute numbers are additionally given. b, fraction of class-resolution SSU and LSU ribo-tags affiliated to the five proteobacterial classes. c, fraction of order-resolution SSU and LSU ribo-tags affiliated to the identified beta-proteobacterial orders. d, numbers of SSU ribo-tags from the seven archaeal and bacterial taxa involved in nitrification.

In-depth taxonomic profiling enables the analysis of microbial communities from various functional perspectives. We demonstrate this for the process of nitrification, the conversion of ammonia to nitrate via nitrite. Both SSU-based and LSU-based analyses consistently identified the Crenarchaeal candidate division GroupI.1b as the predominant archaeal taxon in the soil sample; this is consistent with earlier studies of the same habitat [16] [15]. Members of this group were recently identified as important players in ammonia oxidation [15], [20]. So it is perhaps not surprising that community-wide analysis of the different groups of bacteria and archaea harboring ammonia- and nitrite-oxidation capabilities (Fig. 4d) indicated that the groupI.1b Crenarchaeota were, at least from a numerical standpoint, the major ammonia-oxidizing group in our sample. The ribo-tag ratios between this group and ammonia-oxidizing bacteria were very similar to archaeal and bacterial amoA transcript ratios determined from the same cDNA preparation (12 vs. 16[15]). The subsequent nitrite-oxidation step appeared to be mainly performed by members of the Nitrospirae bacterial phylum. This result would have required seven independent quantitative real-time PCR assays to target the groups involved in nitrification when using traditional methods, whereas it represents only one of a variety of results obtained in this non-targeted approach. Additionally the respective functional groups are displayed in the context of the whole community.

As opposed to PCR-dependent approaches [10], [11], [26] that are confined to specific regions in the SSU rRNA molecule, our method allows the reassembly *in silico* of a full length "composite community" or consensus rRNA sequence for certain taxa. To illustrate this, the 1502 bp large SSU rRNA gene for groupI.1b of Crenarchaeota was assembled from 1105 ribo-tags. The resulting rRNA sequence differed by 3.1 % and 5.4 %, respectively, from the SSU rRNA sequences of the two fosmid clones, 29i4 and 54d9, isolated earlier from this habitat[17], [20].

Diversity and relative abundances of Eukaryotes

Although Eukaryotes are major players in soils and strongly influence the prokaryotic community structure, their diversity and abundance has received comparably little attention in molecular studies [28]. They account for ~11% of the ribo-tags (i.e. cellular biomass) in the soil community. Due to in-depth sampling, we obtained 22,740 eukaryotic ribo-tags, with more than 80% of the tags having a taxonomic resolution at least to the kingdom level. The numerically dominant kingdoms were the Fungi, known as the major destruents in soil, as well as plants (Viridiplantae) and Metazoa (Fig. 5a), with ~50%, ~20% and ~10% of the ribo-tags being consistently assigned by both reference databases. Within the Fungi, the phylum Ascomycota accounted for two thirds of the ribo-tags (Fig. 5b), followed by the phyla Glomeromycota and Basidiomycota.

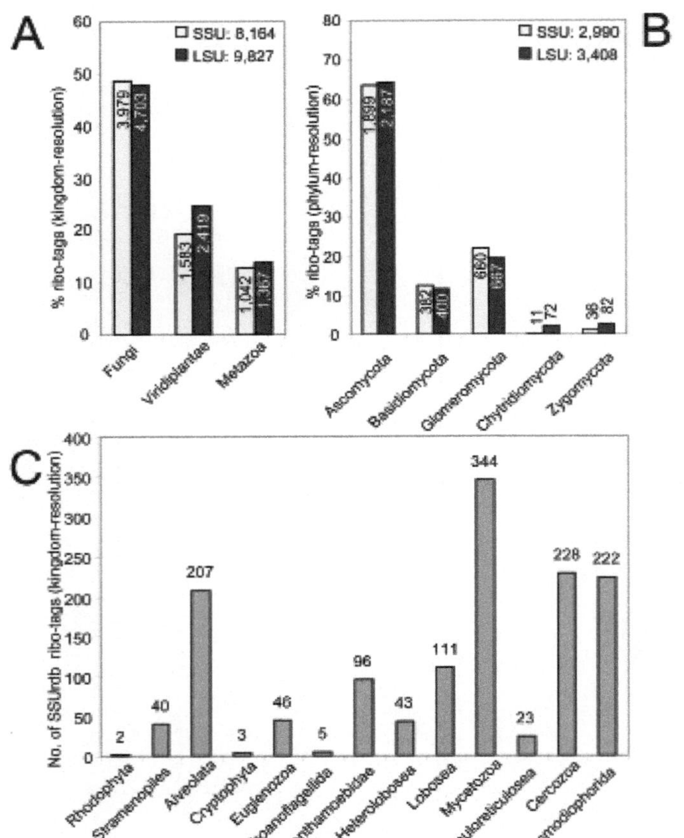

Figure 5: Relative and quantitative distribution of eukaryotic ribo-tags. a, fraction of kingdom-resolution SSU and LSU ribo-tags affiliated to the three numerically dominant eukaryotic kingdoms. b, fraction of phylum-resolution SSU and LSU ribo-tags affiliated to the five fungal phyla detected. The distribution was highly consistent in the SSUrdb and LSUrdb. c, numbers of SSU ribo-tags affiliated to the various protist kingdoms.

As grazers, Protists play important roles in the soil food web by regulating microbial populations. Few LSU rRNA sequences are currently available for protists. Therefore, we estimated the Protist community composition using only the dataset of 1321 SSU ribo-tags (Fig. 5c). This is, to our knowledge, still the biggest molecular dataset of a protist community generated so far. It displayed the presence of slime molds (Mycetozoa) as the most abundant group, followed by Cercozoa, Plasmodiophorida and Alveolata. In addition, Lobosea, Acanthamoebidae, Heterolobosea, Euglenozoa and Stramenopiles were present with at least 40 ribo-tags in the sample.

The non-targeted and in-depth approach resulted in a dataset, which enabled a broad and holistic view onto a community and covered many of the trophic levels present. We are aware that the community profiles presented here are derived from the counting of ribosomal molecules, but not their genes, as is the case in DNA-based approaches. The ribosome content can vary considerably in the cells of an organism reflecting its physiological state; it also differs between taxa. Currently, few data are available; *E. coli* cells have been estimated to contain between 6,800 and 72,000 ribosomes, depending on the growth phase [29], unicellular eukaryotes likely have more ribosomes than microorganisms, and still higher is the ribosomal content of multi-cellular organisms like fungi, metazoans or plants. We consider the amount of ribo-tags as determined in this study to be a measure of a taxon's cellular biomass within a community [30], [31].

Global functional analysis of putative mRNA-tags

The 65,192 RNA-tags that did not give a significant hit against the rRNA reference databases were aligned against the Genbank non-redundant protein database. Homologues to 21,133 RNA-tags were found, showing that a considerable amount of assignable mRNA had been reversely transcribed (Table 1). We subjected those presumable mRNA-tags (2.1 Mbp) to a global functional analysis using the SEED database[32] and compared the meta-transcriptomic data to metagenomic data from (1) the same soil habitat (4.3 Mbp[19]) and (2) a different farm soil (145 Mbp[7]). Overall, the DNA-based (metagenomic) functional repertoire in both soils was surprisingly similar, as judged from the relative distribution of the functional subsystems (Figure 6). This indicates that a generally similar "pool of functions" is present in both soil communities, which consequentially indicates that functional investigations of soil communities based on DNA might always give similar global patterns. In contrast to this, categories involved in RNA and protein metabolism (transcription, translation, protein folding and degradation) were significantly over-represented in the meta-transcriptome compared to both metagenomes, as one would expect to see for active organisms. Further differences between the transcriptome and the metagenomes were related to carbohydrate metabolism: while transcripts of proteins involved in the aerobic degradation of mono-, di- and oligo-saccharides and amino-sugars seemed to be less frequent than suggested by the metagenomes, transcripts for fermentation, degradation of sugar alcohols and CO_2-fixation were equally represented.

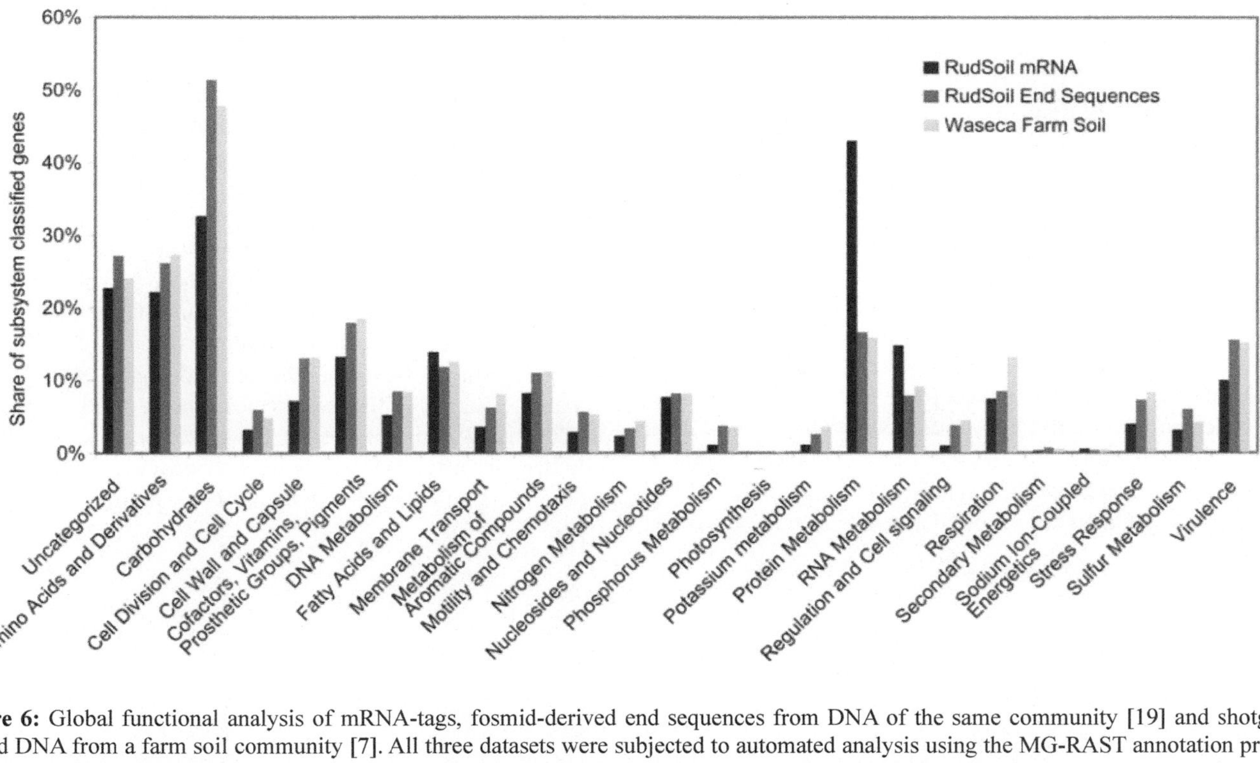

Figure 6: Global functional analysis of mRNA-tags, fosmid-derived end sequences from DNA of the same community [19] and shotgun-cloned DNA from a farm soil community [7]. All three datasets were subjected to automated analysis using the MG-RAST annotation procedure at the SEED (http://metagenomics.theseed.org). Percentages are expressed as the number of mRNA-tags assigned to a subsystem category, divided by the total number of mRNA-tags assigned to subsystems.

Metagenomic analyses frequently assign protein-encoding genes to taxonomic groups by comparing them against the content of sequenced genomes to derive a taxonomic community profile [5], [6]. The simultaneously obtained rRNA and mRNA data provided us with the unique opportunity to validate these procedures. When we compared the ribo-tag profile with the community profile derived from taxonomic binning of the mRNA-tags using MEGAN, we observed a considerable shift for the five dominant bacterial phyla. These differences correlated strongly with the number of sequenced genomes of the different phyla (Fig. 7). This suggests that taxonomic binning solely based on protein encoding genes currently generates an artificial bias against groups with few sequenced genomes, and correspondingly over-represents phyla with many sequenced genomes. This problem, which is inherent to all metagenomic studies, will likely be overcome as more genome sequences of less represented phyla become available.

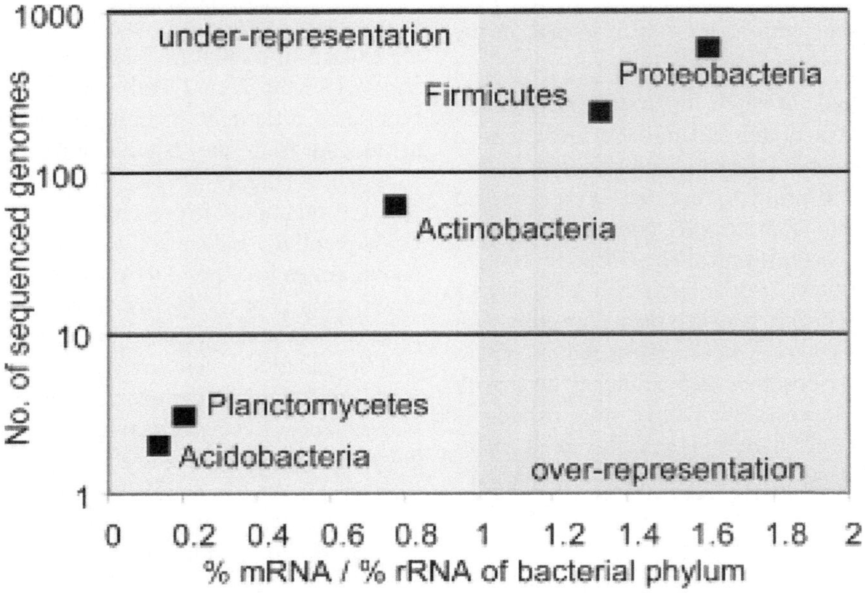

Figure 7: Logarithmic bi-variance plot of the number of publicly available genomes (as of September 2007) of the five numerically dominant bacterial phyla, as judged from the ribo-tags, versus the mRNA-tag based over- and under-representation, compared to the mean of SSU and LSU ribo-tag fraction. A ratio of one means that mRNA- and ribo-tags report the same fraction for the respective phylum.

Probing the metabolism of an uncultured low-abundant group

1.7% of the mRNA-tags (ca. 36 kb) were identified as being of archaeal origin, similar to 1.5% of the ribo-tags. These metatranscriptomic data allowed a first glimpse into the *in situ* activity of the yet uncultured soil Crenarchaeota from groupI.1b. Homologues to more than 80 mRNA-tags were functionally annotated in the databases. Besides transcripts of typical archaeal house-keeping gene products, those involved in ammonia oxdiation were predominant (Fig. 8); 13 mRNA-tags were derived from transcripts of the key metabolic enzyme ammonia monooxygenase (*amoA and amoC*). Furthermore, mRNA-tags of a putative copper-containing nitrite reductase (*nirk*) gene [20] indicated that this enzyme–as postulated for ammonia oxidizing bacteria–could be involved in the process of ammonia oxidation either under aerobic or anaerobic conditions[33]. These findings again hint for ammonia oxidation being the main energy metabolism in soil Crenarchaeota [34]. In addition, ten mRNA-tags could be related to the potential carbon metabolism. One mRNA-tag was derived from a homologue of methyl-malonyl-CoA mutase (MCM) and two from 4-hydroxybutyryl-CoA dehydratase (4-HBDH) homologues. These two gene products are, together with Acetyl-CoA/Propionyl-CoA carboxylase, diagnostic for a CO_2fixation pathway recently characterised in hyperthermophilic Crenarchaeota and suggested for marine crenarchaeota [35]. This indicates that a similar pathway of CO_2 fixation might act in the soil crenarchaeota. Taken together, the metatranscriptomic data provide evidence for a chemolithoautotrophic lifestyle of this yet poorly characterised group.

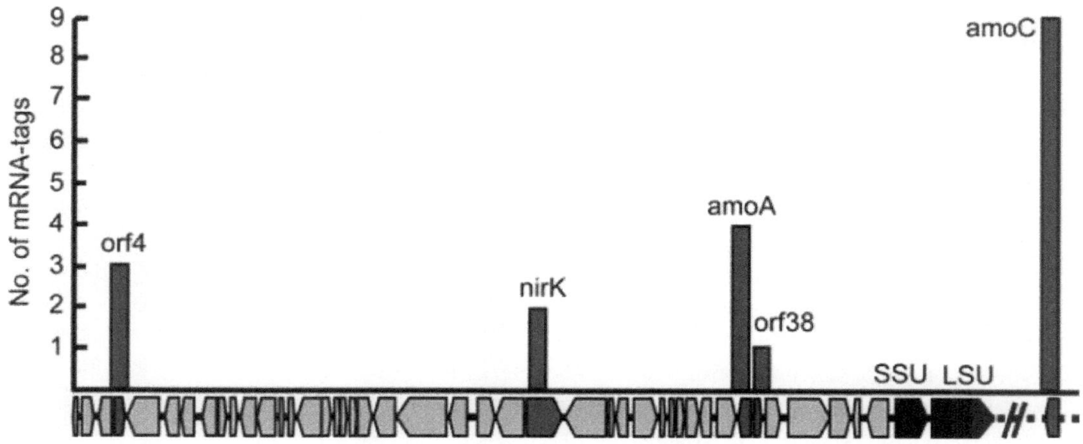

Figure 8: Abundance-dependent plot of presumably archaeal mRNA-tags onto the crenarchaeal fosmid clone 54d9 isolated from the same soil habitat (accession number: AJ627422). The x-axis represents the annotated open reading frames (orfs). Note that an *amoC*gene was not found on the fosmid and is therefore indicated as loosely affiliated.

With the non-targeted randomly primed reverse transcription approach, we can also assemble complete or nearly complete "composite" genes of specific lineages. A near full-length "composite" *amoC* transcript from the mRNA-tags of this analysis has been assembled. The deduced amino acid sequence covered 146 out of 189 positions (77%) of the archaeal sponge symbiont *Cenarchaeum symbiosum amoC* homolog [36] and had 88% identity to *C. symbiosum amoC*, similar to the 84% identity for *amoA* [15], [20].

CONCLUSIONS

In conclusion, we have presented a rapid experimental and analytical approach that uses rRNA and mRNA to characterize microbial community structure and *in situ* function in-depth and simultaneously. This methodology will help (1) to identify microbial groups in complex communities; (2) to relate taxonomic groups to their ecological function (as demonstrated for the soil crenarchaeota); and (3) to efficiently monitor structural and functional community shifts caused by environmental changes. Furthermore, this approach enables for the first time the simultaneous quantitative assessment of the abundance of members of all three domains of life. The analytical power of this approach will continuously improve as sequence read lengths increase and as rRNA reference and genome databases continuously grow. We believe that the analysis presented here could be used in parallel with mRNA enrichment procedures [13], [14] to ensure an efficient global analysis of the activity of naturally occurring assemblages, with one approach covering all trophic levels and domains as well as reasonable numbers of mRNA tags in one sequencing step (total RNA) and the other tool adding further in depth information from the enriched mRNA fraction.

MATERIALS AND METHODS

Site description and soil sample processing

The soil samples were obtained from a sandy lawn in the environmentally protected area "Am Rotböll" (Germany, Hessen, 49°55′34″N 8°37′21.6″E, [15]). The soil is nutrient-poor with a pH of 7.06, and with water extractible organic nitrogen (WEON) and carbon (WEOC) of 0.9 and 7.6 mg kg dry weight soil^{-1}, respectively. The top soil was sampled at a depth of 0–10 cm in January 2006 after a snow thawing period (soil temperature 5.5°C). Three replica within a distance of 1 m were withdrawn and kept in open plastic bags in the dark at 4°C for 40 hours. 100 g of the soil samples were sieved (2 mm mesh size) and equal amounts of the triplicates were subsequently mixed. Six grams of soil were then processed for RNA extraction.

cDNA synthesis

Nucleic acid extractions (RNA and DNA) were performed using a modification of the method of[37], and as briefly described in [15]. 0.5 ml of both CTAB buffer and phenol:c hloroform:isoamylalcohol (25:24:1, pH 6.8) were added to a lysing matrix E tube (Q-Biogene) containing 0.5 g of soil. The cells were lysed in a FastPrep machine (Q-Biogene) at speed 5.5 for 30 seconds, followed by nucleic acid precipitation with PEG 8000. Total nucleic acids were subjected to Dnase treatment and the remaining total RNA was used as template for random hexamer-primed reverse transcription performed independently at 37°C and 50°C, respectively. The successful and complete hydrolysis of DNA was probed by archaeal and bacterial *amoA* specific quantitative real time PCR, which detected specific products in the cDNA samples after reverse transcription but not in parallel treated RNA samples, where no cDNA synthesis had taken place ([15], data not shown). The subsequently generated double-stranded (ds) cDNA ranged from approximately 100 to 1500 bp in length (not shown). Approximately one microgram of ds cDNA was generated.

Pyrosequencing

Sequencing was performed as described previously [15]. The 37°C and 50°C samples were kept separate during pre-sequencing processing. Independent test-sequencing was performed on a Roche GS20 sequencer (Roche Applied Sciences/454 Life Sciences, Branford, CT) for both samples, producing small subsets of RNA-tags (37°C=4141 and 50°C:

n=1985). Both samples were subsequently pooled for one full run resulting in 314,041 sequences [15]. Re-analysis of the sequences resulted in 258,411 high-quality RNA-tags.

Reference Databases

Two rRNA reference sequence databases were constructed for community structure analysis using MEGAN [23]. The sequences therein were linked to taxa in the NCBI Taxonomy (as of June 2007), extended with 28 archaeal candidate divisions [38]. The Small- and Large Subunit rRNA Reference Databases (hereafter SSUrdb and LSUrdb, respectively) were constructed by combining sequences from several public databases. The SSUrdb includes sequences from the RDP-II release 9.39 [22] (bacteria), the SILVA SSURef database release 89 [21] (eukaryotes and archaea), and sequences from the dataset described in [38] (archaea). The LSUrdb includes sequences from the Silva LSURef database release 90[21]. Bacterial SSU rRNA sequences were retrieved from RDP-II [22]. More than 112,000 high quality sequences from isolates and uncultured strains were downloaded as FASTA files. The selection was made according to their taxonomic affiliation in the NCBI taxonomy. Sequences with low taxonomic resolution (e.g. "unclassified bacteria", "environmental samples") were mostly omitted until the "order" taxonomic level. Only in phyla with comparably few reference sequences and/or a relatively poorly refined taxonomy (e.g. the phyla Acidobacteria and Verrucomicrobia, and various candidate divisions), those sequences were included. All sequences were screened for vector contamination using the *cross_match* program from the Phred/Phrap package [39] against NCBI's UniVec database. All identified vector subsequences were removed from the database sequences. In some RDPII derived sequences, the entry contained more than the SSU rRNA gene, i.e. flanking regions and the LSU rRNA gene. In order to remove those, all sequences longer than 1,550 bp were aligned to the remainder of the sequences in the database (those within the expected length of an SSU rDNA gene), using BLASTN. Based on these alignments, any regions without significant similarity ($E \leq 1e-6$) were cropped, leaving only the SSU rRNA gene.

All eukaryotal and most archaeal SSU rRNA sequences were retrieved from the SILVA project[21]. The SSURef database version 89 was retrieved as an ARB file. 24,197 aligned eukaryotal sequences including 331 mitochondrial and 641 plastid sequences were analyzed and chosen for our database based on a correct taxonomic affiliation in the NCBI taxonomy and a comparatively high taxonomic resolution using the software package ARB [40]. Sequences were exported as FASTA files using a filter (including positions between 1000 and 43284 of the ARB alignment) for eliminating sequence information not belonging to the SSU rRNA gene.

The archaeal part of SSUrdb reference database contained 944 sequences from cultured strains, extracted from the SILVA SSURef database, again based on a correct and high resolution taxonomic affiliation in the NCBI taxonomy. Many archaeal lineages today consist exclusively of uncultured representatives. Those are not well resolved in the NCBI taxonomy, but deposited as "uncultured archaea" or "environmental samples", which prevents an informative taxonomic grouping of sequences belonging to those lineages. To overcome this, we have extended the NCBI taxonomy with 28 archaeal candidate divisions as described in[38] and exchanged the taxonomic affiliation of the effected sequences in the SSUrdb accordingly. Altogether 544 sequences are distributed over those groups (extracted from the ARB dataset used in [38];). The archaeal part therefore consists of 1,490 sequences and the whole SSUrdb of 135,160 sequences in total.

The large subunit reference database (LSUrdb) was generated from the SILVA database project. The LSURef database version 90 was retrieved as ARB file. The sequences therein were analyzed and chosen based on a correct taxonomic affiliation in the NCBI taxonomy reflected in the sequence alignment and a comparatively high taxonomic resolution using ARB. Sequences were exported as FASTA files using a filter (including positions between 66155 and 129011 of the ARB alignment) for eliminating sequence information not belonging to the LSU rRNA gene. Vector sequences were removed as described above. The LSUrdb consists of 6,247 sequences, of which 2,759 belong to Bacteria, 130 to Archaea (107 cultured and 23 uncultured) and 3,358 Eukaryotes (including 122 mitochondrial and 491 plastid sequences).

As MEGAN utilizes the FASTA header in the BLAST output to identify the corresponding taxon (see below), the headers of all files in the reference databases contain the Genbank, RefSeq or EMBL accession number as identifier, to enable taxon identification with MEGAN (see below).

Taxonomic assignment of RNA-tags using MEGAN

All RNA-tags in the sample were compared to both of the rRNA reference databases using the NCBI *blastall* implementation of BLASTN (default parameters, except setting the maximum number of hits to 100). Analysis of BLAST output files was performed using the MEGAN software version 1beta18 [23]. This software reads the results of a BLAST comparison as input and attempts to place each read on a node in the NCBI taxonomy. This is performed by the LCA algorithm that assigns each RNA-tag to the lowest common ancestor in the taxonomy from a subset of the best scoring matches in the BLAST result (absolute cutoff: BLAST bitscore 86, relative cutoff: 10% of the top hit). RNA-tags that have no BLAST matches are assigned to the special node "no hits" and those unassigned due to algorithmic reasons (e.g. below an applied threshold) are placed on the special node "unassigned". The result of the analysis is displayed as a tree representation of the NCBI taxonomy (as of June 2007). To enable identification of the species involved in the BLAST hits, we employed lookup-tables that map GI accessions to taxon IDs obtained from the NCBI website.

Database refinement, analysis pipeline adjustment and testing

In order to evaluate the performance and stability of our taxonomic binning method using MEGAN and the generated reference databases, we used LSU and SSU rRNA test sets containing simulated ribo-tags. Those consisted of 100 bp long sequence fragments and were obtained by randomly cropping out a 100 bp window from a given rRNA sequence. This procedure was repeated 200 times resulting in a statistical coverage of 13.3 for each position in SSU rRNA (for an assumed length of 1500 bp) and 6.7 for LSU rRNA sequences (for an assumed length of 3000 bp) within the test set. Datasets were iteratively compared by BLASTN against the reference database(s) and analyzed using MEGAN.

Simulated SSU ribo-tag datasets were aligned to the LSUrdb and vice versa, to estimate the potential of "cross-contamination", i.e. the assignment of SSU rRNA-derived ribo-tags as LSU rRNA ribo-tags and vice versa. No such assignments were observed above a BLASTN score of 71 bits. Manual inspections of the alignments from the simulated data lead us to choose 86 bits as the minimum BLASTN score for an RNA-tag to be considered as ribo-tag. These alignments are always longer than 40 bases and correspond to an e-value of 2e-16 for SSUrdb alignments. In addition, a relative threshold was introduced in the taxonomic binning procedure applying MEGAŃs LCA algorithm. Here, ten percent was chosen as cut-off, which means that the taxonomic information of all reference sequences on the BLAST hit list which are within 10% of the score of the BLAST top hit were included in the taxonomic binning of the respective ribo-tag. Increasing this cutoff to 20% or more, i.e. making the taxonomic assignment of a ribo-tag less strict, led to a very minor shift in taxonomic resolution (data not shown), showing that the chosen relative cut-off is already rather relaxed.

The rRNA reference databases were screened for taxonomically wrongly assigned reference sequences using simulated ribo-tag datasets. Since those consisted of ribo-tags of known origin, an assignment congruent with the taxonomic resolution power of the database for the respective taxon was expected. Where results deviated from these expectations, they were manually inspected. Especially the BLAST hit lists of ribo-tags with poor taxonomic resolution (mainly the taxonomic levels "cellular organism", "domain" and several prokaryotic "phyla") were analysed in detail. For example, a single acidobacterial reference sequence wrongly affiliated to another bacterial phylum would result in a drastic under-representation of acidobacteria, because the LCA algorithm would group the acidobacterial ribo-tags at the bacterial domain level. In the analysis of a natural community, this would introduce a strong artificial community shift, biasing against acidobacteria. Suspicious reference sequences were compared by BLAST against NCBI non-redundant nucleotide database and removed from the reference database. The test set was subsequently compared to the refined reference database, to verify an improved performance for the respective taxon and/or to identify additional wrongly affiliated reference sequences. Applying this iterative procedure, more than 600 bacterial sequences with wrong affiliations in the NCBI taxonomy were removed from the SSUrdb.

In order to extend the sensitivity and robustness analysis, we attempted to simulate the situation where ribo-tags would be as similar to the database as in median case, by removing entries from the database. Database filtering was carried out by aligning the full-length SSU and LSU rRNA sequences from seven selected test species to all entries in the respective reference database, using BLASTN with modified scoring parameters (mismatch penalty=1, gap open cost=2, gap extension cost=1). In all cases where the full-length test sequence showed higher similarity than the threshold to a database sequence, this sequence was removed from the database. The procedure was carried out independently for each test species with the respective sample median similarities used as thresholds (98% for the SSUrdb and 93% for the LSUrdb), thus producing 7 filtered versions of the LSUrdb and SSUrdb. For the SSUrdb, the lowest decile similarity (86%) was also removed in a separate filtering. The simulated SSU and LSU ribo-tags from the test species were then aligned to respective filtered database versions using BLASTN and taxonomically assigned using MEGAN, as described above.

Assembly of a "composite community" rRNA sequence

Using MEGAN, all 1,113 putative SSU ribo-tags assigned to the archaeal GroupI.1b were extracted from the dataset. The assembly program CAP3 [41] was then used to attempt to assemble the data (default parameters) which resulted in a single contiguous sequence contig with a length of 1,502 bp assembled from 1105 of the extracted ribo-tags.

Functional analysis of putative mRNA-tags

All RNA-tags without a significant similarity against the rRNA reference databases (BLAST score threshold below 86 bits), were translated in all six reading frames and aligned to the NCBI non-redundant protein database (release of June 25, 2007) using BLASTX [42] and analysed with MEGAN. The resulting tags with at least one BLAST alignment producing a bitscore of 30 or above were assigned as putative mRNA transcripts. Manual inspection revealed that sequences with BLAST hits close to a bitscore of 30 produce apparently meaningful alignments. In addition, the coverage of the ribosomal reference databases, compared to the true distribution of all existing ribosomal sequences, is expected to be much better than the coverage of Genbank nr compared to all existing protein coding sequences. This assumption, together with the short lengths of the translated reads (approximately 30 amino acids) means that possibly true mRNA transcripts would be missed, were we to use a higher bitscore limit.

Putative mRNA sequences were annotated using the MG-RAST (Meta Genome Rapid Annotation using Subsystem Technology; v1.2) server at the Argonne National Library (http://metagenomics.nmpdr.org), using subsystem-based annotation based on the SEED database [32]. Subsystems are groups of genes, or functional roles, acting together in a biological process, e.g. in a metabolic pathway. These are grouped into subsystem categories. The MG-RAST annotation pipeline assign some putative genes to more than one subsystem, i e predicts that these genes have multiple functions. In such cases, each assignment was counted in the statistic Total Subsystem Assignments (so that the gene was counted more than once). As a consequence of this, the sum of the relative subsystem counts add up to more than 100%. In addition to the putative mRNA-tags, subsystem annotation of two metagenome datasets was also carried out using the same methodology. The first of these contains genomic sequences from a sample collected previously from the same site[19] and the second from farm soil [7]. For each subsystem, the relative population was calculated, i.e. the putative genes assigned to a particular subsystem divided by the total number of genes assigned to subsystems. In addition to manual comparison of relative populations between the samples putative mRNA-tag dataset and genomic sequence from the same site, these datasets were compared statistically using the method described in[43]. In order to rule out that the difference in abundance was not an artifact caused by the shorter length of the putative mRNA-tags, compared to the longer genomic reads obtained using Sanger sequencing, the genomic reads from [19] were fragmented to a number of shorter reads and re-analysed using the MG-RAST server. From each genomic read, seven fragments were randomly generated with a length identical to the average length of the mRNA-tags (98bp), such that the total sequence length was close to that of the un-fragmented sample. The distribution of subsystem categories for the short fragments did not show any deviances larger that 10% compared to the original dataset, for the subsystems or subsystem categories abundant in the mRNA sample (data not shown). This showed that the observed abundance is not an artifact caused by shorter read length of the mRNA-tags. A study where 454 transcripts were functionally categorized and compared to Sanger sequenced ESTs in the plant *Medicago*, also showed that 100bp length is sufficient for meaningful functional classification comparable to traditional ESTs [44].

The putative mRNA-tags were taxonomically binned by analysing the output file of the BLASTX comparison against the NCBI nr protein database with MEGAN. The same parameters as described above for the ribo-tag analysis were applied (except a minimum score of 30 bits). The resulting taxonomic community profile was manually compared to that from the corresponding ribo-tag analyses. In addition, higher score cutoffs (40 and 50 bit) were applied on the mRNA-tags dataset, which gave essentially the same community profile (not shown).

Nine mRNA-tags which were found to have similarity to the ammonia monooxygenase subunit C gene (*amoC*) of archaea, were translated to protein sequence using the reading frames corresponding to the BLASTX hits, and aligned to the *Cenarchaeum symbiosum* AmoC protein sequence using CLUSTAL X [45].

Data deposition

The SSUrdb and LSUrdb are made available for download as compressed fasta files fromhttp://www.bioinfo.no/services/community-profiling. The sequences reported in this paper have been submitted to GenBank, under accession number SRA001014.

ACKNOWLEDGMENTS

We thank the Bergen Center for Computational Science for access to the computer cluster. We acknowledge the invaluable work of the different ribosomal database projects by providing curated rRNA sequences to the public. We wish to thank the anonymous reviewers of a previous reviewing process for their very instructive comments. TU acknowledges the participants of the EuroDiversity workshop on Microbial diversity and Ecosystem functioning (Lunz, Austria) for valuable discussions.

AUTHOR CONTRIBUTIONS

Conceived and designed the experiments: SS TU CS. Performed the experiments: TU. Analyzed the data: TU AL CS. Contributed reagents/materials/analysis tools: SS DH JQ TU AL. Wrote the paper: SS TU AL CS.

REFERENCES

1. Torsvik V, Ovreas L, Thingstad TF (2002) Prokaryotic diversity–magnitude, dynamics, and controlling factors. Science 296: 1064–1066.
2. Gans J, Wolinsky M, Dunbar J (2005) Computational improvements reveal great bacterial diversity and high metal toxicity in soil. Science 309: 1387–1390.
3. Prosser JI (2007) Microorganisms Cycling Soil Nutrients and Their Diversity. In: van Elsas JD, Jansson JD, Trevors JT, eds. Modern Soil Microbiology, Second Edition. Boca Raton: CRC Press. pp 237–262.
4. Fierer N, Bradford MA, Jackson RB (2007) Toward an ecological classification of soil bacteria. Ecology 88: 1354–1364.
5. Martin-Cuadrado AB, Lopez-Garcia P, Alba JC, Moreira D, Monticelli L, et al. (2007) Metagenomics of the deep mediterranean, a warm bathypelagic habitat. PLoS ONE 2: e914.
6. Rusch DB, Halpern AL, Sutton G, Heidelberg KB, Williamson S, et al. (2007) The Sorcerer II Global Ocean Sampling expedition: northwest Atlantic through eastern tropical Pacific. PLoS Biol 5: e77.
7. Tringe SG, von Mering C, Kobayashi A, Salamov AA, Chen K, et al. (2005) Comparative metagenomics of microbial communities. Science 308: 554–557.
8. Venter JC, Remington K, Heidelberg JF, Halpern AL, Rusch D, et al. (2004) Environmental genome shotgun sequencing of the Sargasso Sea. Science 304: 66–74.
9. Gill SR, Pop M, Deboy RT, Eckburg PB, Turnbaugh PJ, et al. (2006) Metagenomic analysis of the human distal gut microbiome. Science 312: 1355–1359.
10. Huber JA, Welch DB, Morrison HG, Huse SM, Neal PR, et al. (2007) Microbial population structures in the deep marine biosphere. Science 318: 97–100.
11. Sogin ML, Morrison HG, Huber JA, Welch DM, Huse SM, et al. (2006) Microbial diversity in the deep sea and the underexplored "rare biosphere". Proc Natl Acad Sci U S A 103: 12115–12120.
12. Bailly J, Fraissinet-Tachet L, Verner MC, Debaud JC, Lemaire M, et al. (2007) Soil eukaryotic functional diversity, a metatranscriptomic approach. ISME J 1: 632–642.
13. Poretsky RS, Bano N, Buchan A, LeCleir G, Kleikemper J, et al. (2005) Analysis of microbial gene transcripts in environmental samples. Appl Environ Microbiol 71: 4121–4126.
14. Frias-Lopez J, Shi Y, Tyson GW, Coleman ML, Schuster SC, et al. (2008) Microbial community gene expression in ocean surface waters. Proc Natl Acad Sci U S A 105: 3805–3810.
15. Leininger S, Urich T, Schloter M, Schwark L, Qi J, et al. (2006) Archaea predominate among ammonia-oxidizing prokaryotes in soils. Nature 442: 806–809.
16. Ochsenreiter T, Selezi D, Quaiser A, Bonch-

Osmolovskaya L, Schleper C (2003) Diversity and abundance of Crenarchaeota in terrestrial habitats studied by 16S RNA surveys and real time PCR. Environ Microbiol 5: 787–797.

17. Quaiser A, Ochsenreiter T, Klenk HP, Kletzin A, Treusch AH, et al. (2002) First insight into the genome of an uncultivated crenarchaeote from soil. Environ Microbiol 4: 603–611.

18. Quaiser A, Ochsenreiter T, Lanz C, Schuster SC, Treusch AH, et al. (2003) Acidobacteria form a coherent but highly diverse group within the bacterial domain: evidence from environmental genomics. Mol Microbiol 50: 563–575.

19. Treusch AH, Kletzin A, Raddatz G, Ochsenreiter T, Quaiser A, et al. (2004) Characterization of large-insert DNA libraries from soil for environmental genomic studies of Archaea. Environ Microbiol 6: 970–980.

20. Treusch AH, Leininger S, Kletzin A, Schuster SC, Klenk HP, et al. (2005) Novel genes for nitrite reductase and Amo-related proteins indicate a role of uncultivated mesophilic crenarchaeota in nitrogen cycling. Environ Microbiol 7: 1985–1995.

21. Pruesse E, Quast C, Knittel K, Fuchs BM, Ludwig W, et al. (2007) SILVA: a comprehensive online resource for quality checked and aligned ribosomal RNA sequence data compatible with ARB. Nucleic Acids Res.

22. Cole JR, Chai B, Farris RJ, Wang Q, Kulam-Syed-Mohideen AS, et al. (2007) The ribosomal database project (RDP-II): introducing myRDP space and quality controlled public data. Nucleic Acids Res 35: D169–172.

23. Huson DH, Auch AF, Qi J, Schuster SC (2007) MEGAN analysis of metagenomic data. Genome Res 17: 377–386.

24. Wang Q, Garrity GM, Tiedje JM, Cole JR (2007) Naive Bayesian classifier for rapid assignment of rRNA sequences into the new bacterial taxonomy. Appl Environ Microbiol 73: 5261–5267.

25. Liu Z, Lozupone C, Hamady M, Bushman FD, Knight R (2007) Short pyrosequencing reads suffice for accurate microbial community analysis. Nucleic Acids Res.

26. Roesch LFW, Fulthorpe RR, Riva A, Casella G, Hadwin AKM, et al. (2007) Pyrosequencing enumerates and contrasts soil microbial diversity. The ISME Journal 1: 283–290.

27. Huber H, Hohn MJ, Rachel R, Fuchs T, Wimmer VC, et al. (2002) A new phylum of Archaea represented by a nanosized hyperthermophilic symbiont. Nature 417: 63–67.

28. Fierer N, Breitbart M, Nulton J, Salamon P, Lozupone C, et al. (2007) Metagenomic and Small-Subunit rRNA Analyses Reveal the Genetic Diversity of Bacteria, Archaea, Fungi, and Viruses in Soil. Appl Environ Microbiol 73: 7059–7066.

29. Bremer H, Dennis PP (1996) Modulation of chemical composition and other parameters of the cell by growth rate. In: Neidhardt FC, Curtiss RI, Ingraham JL, Lin ECC, Low KB, et al., eds. Escherichia coli and Salmonella: cellular and molecular biology. Washington, D.C.: ASM Press. pp 1553–1569.

30. Schaefer M (1990) The soil fauna of a beech forest on limestone: trophic structure and energy budget. Oecologia 82: 128–136.

31. Weller R, Ward DM (1989) Selective Recovery of 16S rRNA Sequences from Natural Microbial Communities in the Form of cDNA. Appl Environ Microbiol 55: 1818–1822.

32. Overbeek R, Begley T, Butler RM, Choudhuri JV, Chuang HY, et al. (2005) The subsystems approach to genome annotation and its use in the project to annotate 1000 genomes. Nucleic Acids Res 33: 5691–5702.

33. Beaumont HJ, Lens SI, Reijnders WN, Westerhoff HV, van Spanning RJ (2004) Expression of nitrite reductase in Nitrosomonas europaea involves NsrR, a novel nitrite-sensitive transcription repressor. Mol Microbiol 54: 148–158.

34. Konneke M, Bernhard AE, de la Torre JR, Walker CB, Waterbury JB, et al. (2005) Isolation of an autotrophic ammonia-oxidizing marine archaeon. Nature 437: 543–546.

35. Berg IA, Kockelkorn D, Buckel W, Fuchs G (2007) A 3-hydroxypropionate/4- hydroxybutyrate autotrophic carbon dioxide assimilation pathway in Archaea. Science 318: 1782–1786.

36. Hallam SJ, Mincer TJ, Schleper C, Preston CM, Roberts K, et al. (2006) Pathways of carbon assimilation and ammonia oxidation suggested by environmental genomic analyses of marine Crenarchaeota. PLoS Biol 4: e95.

37. Griffiths RI, Whiteley AS, O'Donnell AG, Bailey MJ (2000) Rapid method for coextraction of DNA and RNA from natural environments for analysis of ribosomal DNA- and rRNA-based microbial community composition. Appl Environ Microbiol 66: 5488–5491.

38. Schleper C, Jurgens G, Jonuscheit M (2005) Genomic studies of uncultivated archaea. Nat Rev Microbiol 3: 479–488.

39. Ewing B, Hillier L, Wendl MC, Green P (1998) Base-calling of automated sequencer traces using phred. I. Accuracy assessment. Genome Res 8: 175–185.

40. Ludwig W, Strunk O, Westram R, Richter L, Meier H, et al. (2004) ARB: a software environment for sequence data. Nucleic Acids Res 32: 1363–1371.

41. Huang X, Madan A (1999) CAP3: A DNA sequence assembly program. Genome Res 9: 868–877.

42. Altschul SF, Madden TL, Schaffer AA, Zhang J, Zhang Z, et al. (1997) Gapped BLAST and PSI-BLAST: a new generation of protein database search programs. Nucleic Acids Res 25: 3389–3402.

43. Rodriguez-Brito B, Rohwer F, Edwards RA (2006) An application of statistics to comparative metagenomics. BMC Bioinformatics 7: 162.

44. Cheung F, Haas BJ, Goldberg SM, May GD, Xiao Y, et al. (2006) Sequencing Medicago truncatula expressed sequenced tags using 454 Life Sciences technology. BMC Genomics 7: 272.

45. Jeanmougin F, Thompson JD, Gouy M, Higgins DG, Gibson TJ (1998) Multiple sequence alignment with Clustal X. Trends Biochem Sci 23: 403–405.

CHAPTER 22

METATRANSCRIPTOMICS REVEALS THE DIVERSITY OF GENES EXPRESSED BY EUKARYOTES IN FOREST SOILS

Coralie Damon, Frederic Lehembre, Christine Oger-Desfeux, Patricia Luis, Jacques Ranger, Laurence Fraissinet-Tachet, Roland Marmeisse

INTRODUCTION

In terrestrial ecosystems, plant litter degradation is a key ecological feature which controls not only the equilibrium between soil carbon storage and CO_2 release in the atmosphere, but also the release of essential soil nutrients trapped in dead plant biomass such as organic forms of phosphorus and nitrogen.

In ecological studies, litter degradation is often estimated by measuring parameters such as soil respiration [1], litter mass loss [2] or the activities of specific microbial enzymes in soil extracts [3]. In microbiology, the degradation of plant-derived compounds such as lignocellulose has been studied using a few microbial model species and has recently led to the sequencing of the genomes of different saprotrophic fungal species which use different strategies to degrade plant material [4]–[7], thus revealing the full enzymatic machinery implicated in this process.

Under natural conditions, litter degradation is generally carried out by consortia of species that either act simultaneously or replace one another on a common piece of plant debris in a sometimes predictable manner and not by a single microbial species [8]. It can therefore be anticipated that the molecular machinery deployed to completely mineralize litter in the field is far more complex and diverse than the machinery observed in a single microbial genome. In addition, it is likely that the diversity of this machinery is partly controlled by litter chemistry and complexity and therefore by plant community composition. Indeed, by affecting litter quality and soil physicochemical properties, plant cover could influence microbial community composition and/or diversity [9], [10], and select different microbial taxa that may possess different degradation machineries.

By allowing access to the genome contents of the different microorganisms present in a common environment (metagenomics) or to the set of genes they express (metatranscriptomics), environmental genomics offers a novel opportunity to decipher at the molecular level, complex ecological processes such as plant organic matter degradation, thus bridging the gap between global field measurements and targeted genomic approaches. These approaches have been carried out on a variety of animal digestive systems [11]–[17] or model composts [18] and DNA and/or RNA sequencing of their associated microbial communities has lead to the characterization of numerous members of carbohydrate-active enzyme families whose frequencies differ depending on the origin of the sequenced metagenomic DNA/RNA[11], [15].

Since the primary degradation of lignocellulosic materials in soil is believed to be carried out essentially by fungi, we present a first metatranscriptomic analysis of forest soils focussing specifically on eukaryotic organisms whose polyadenylated mRNA can be separated from prokaryotic and non coding ones, thanks to their poly-A tails [13], [19], [20]. Poly-A mRNA were directly extracted from beech (*Fagus sylvatica*, a broadleaf deciduous angiosperm species) and spruce (*Picea abies*, an evergreen gymnosperm species) forest soils, converted into cDNA then cloned and 10,000 of them per library sequenced using the Sanger technology. The sequence datasets were specifically analysed with respect to the diversity of enzymes implicated in the breakdown of lignocellulosic material and downstream monosaccharide membrane transporters, as well as enzymes and transporters involved in the mobilization of the organic forms of other elements (phosphorus, nitrogen). In addition, since many microbial enzymes produced for soil nutrient acquisition, and/or toxic soil compounds inactivation (e.g. P450 monooxygenases) are

Citation: Damon C, Lehembre F, Oger-Desfeux C, Luis P, Ranger J, Fraissinet-Tachet L, et al. (2012) Metatranscriptomics Reveals the Diversity of Genes Expressed by Eukaryotes in Forest Soils. PLoS ONE 7(1): e28967. doi:10.1371/journal.pone.0028967.

also of biotechnological interest, we performed a global evaluation of the sequence dataset with respect to its biotechnological potential.

RESULTS

Sequence datasets (Table 1)

Table 1: Characteristics of the sequence datasets

	Spruce soil	Beech soil
No. of sequenced clones	10,000	10,000
No. of contaminating rRNA squences	684	765
No. of "good quality cDNAs"	8606	7905
Average length of cDNA sequences (nucleotides)	482	430
cDNA GenBank/EMBL/DDJB Accession Nos.	FR697056–706058	FR706059–714330
% cDNA in clusters (No. of clusters)	9% (248)	9% (260)
No. of singletons	7851	7222
% cDNA with BLASTX hits (GenBank nr)	48%	39.5%
% cDNA with GO terms	39%	31.5%
% cDNA with GO "Biological process" terms	25.5%	23.5%
% cDNA with E.C. No. (No. of E.C. Nos.)	13% (321)	12% (313)
% cDNA with taxonomic annotation (MEGAN)	46%[d]/50%[s]	37%[d]/42%[s]

[d]values obtained using the default settings of the MEGAN software ("Min.support" parameter set to five);
[s]values obtained using the MEGAN software with "Min.support" parameter set to one.
doi:10.1371/journal.pone.0028967.t001

cDNA libraries of ~9 10^5 and 5 10^5 plasmid clones were obtained using poly-A mRNA extracted from beech and spruce soils, respectively. A significant proportion of cDNA inserts were between 400 and 700 bp-long, as estimated by agarose gel electrophoresis (64% and 48% of cDNA for beech and spruce respectively). Following sequencing and cleaning of the sequences (removing bad quality sequences, poly-A tails, vector and adaptor sequences as well as contaminating rRNA sequences, and discarding sequences shorter than 100 nt), 7905 (beech) and 8606 (spruce) cDNAs were retained for further analyses. Contamination levels by eukaryotic and bacterial rRNA sequences represented at most 8.8% of the sequences (beech). These rRNA sequences could have been cloned because they processed A-rich sequence stretches, and therefore may not reflect the taxonomic diversity of the soil microbial communities. Finally the average read length was 430 bp for beech and 482 bp for spruce cDNAs.

For spruce and beech cDNA sequence datasets, 9% of the sequences grouped into clusters which, for more than 70% of them, contained only two sequences. As a result, the rarefaction curves plotting the numbers of clusters *versus* the numbers of sequences had the shapes of straight lines suggesting that the datasets represented a small proportion of the total metatranscriptome sequence diversities. Only two of the 12 clusters grouping at least 8 sequences corresponded to known genes. One, from the beech library, encoded a fungal hydrophobin, a class of proteins which coat the external surface of hyphae and spores, and the other, from the spruce library, a putative chitinase.

Taxonomic diversity of the soil eukaryotic communities

For each forest soil, the taxonomic composition of the eukaryotic community contributing to the soil RNA pool was independently estimated using two sequence datasets: (i) 18S rDNA fragments PCR amplified from reverse-transcribed soil RNA, and (ii) the sequenced cDNAs. Sequences were assigned to one of eight major eukaryotic phyla, i.e. the Fungi, Metazoas, Plantae, Amoebozoas, Alveolata, Heterokonta, Rhizarias and Excavata (as defined in [21]). The last 5 phyla are collectively referred as the "protists" in the manuscript.

Whatever the sequence dataset and the studied soil, the Opisthokont group (i.e. essentially Fungi and Metazoas) dominated, representing 71% (beech) and 77% (spruce) of the unambiguously annotated PCR amplified 18S rDNA and 60–61% of the annotated cDNAs. Furthermore, among the Opisthokonts, fungal sequences were the most abundant, representing between 75 and 87% of the cDNAs attributed to this group.

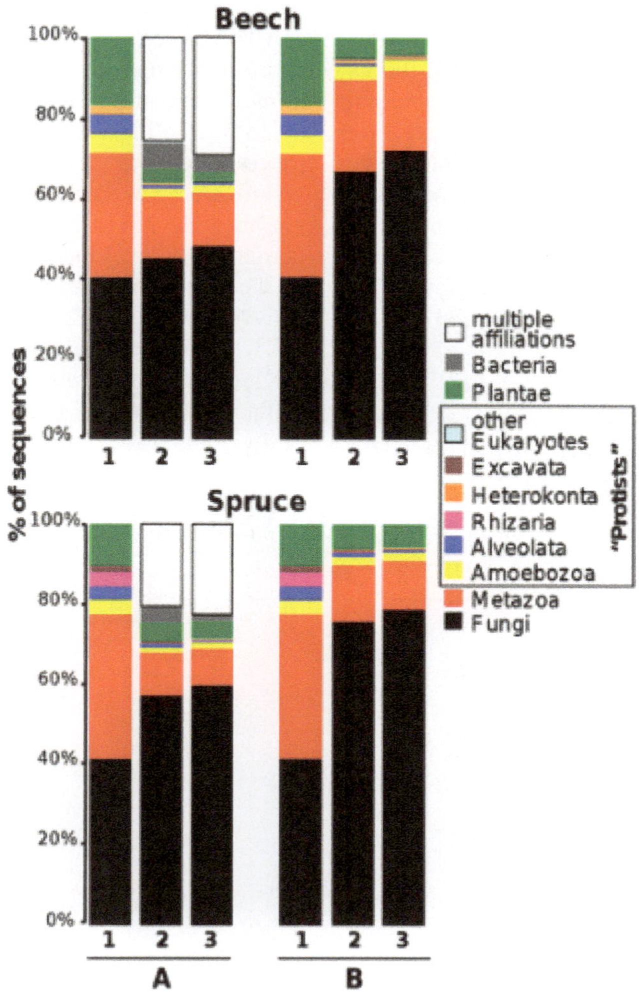

Figure 1: Taxonomic affiliation of the different environmental sequences. PCR amplified 18S rRNA (column 1) and cDNAs (columns 2 and 3) from spruce and beech soils were attributed to one of eight eukaryotic phyla or to the Bacteria. The "multiple affiliation" category contains sequences which could not be unambiguously placed in one of the other categories. (A), Results for all sequences which had a homolog in the GenBank/EMBL/DDJB database; (B), same analysis after removing cDNA sequences from bacteria and cDNA with uncertain "multiple affiliation" for a direct comparison between 18S rRNA and cDNA datasets. For the annotation of cDNA sequences using MEGAN, two different parameters were used; the "stringent analysis" (column 2) corresponded to the "Min Support" parameter set to default value of 5 which was set to 1 in the "less stringent" analysis (column 3).

Sequences from each "protist" group were identified at least once in each of the different beech or spruce sequence datasets and protist sequences contributed to the different datasets between 3% (cDNAs) and 12% (amplified 18S rDNA) of the unambiguously annotated sequences. It should however be stressed that the use of a single primer pair to amplify eukaryotic rDNA genes is known to capture only a fraction of the true diversity [22] and that amplification of 18S sequences from several groups of unicellular eukaryotes requires group-specific primers [23].

Using the default parameters and the BLASTX results against the NCBI-nr protein database as input file, the MEGAN software attributed a taxonomic annotation to 46% (spruce) and 37% (beech) of the cDNAs. Among these annotated cDNAs, only 2.5% were attributed to "protists" and 22–29% could not be unambiguously attributed to a single taxonomic group (multiple affiliation category, Figure 1). We asked if under-representation of specific taxonomic groups, such as the protists, and a high prevalence of sequences with multiple potential affiliations may have resulted from the settings of the MEGAN software. Indeed, using default parameters, MEGAN could take into consideration a too high number of BLAST hits (a maximum of five) to infer a taxonomic affiliation thus underestimating the abundance of taxonomic groups poorly represented in sequence databases. This hypothesis was tested by lowering the "Min Support" parameter from 5 to 1 (relaxed setting). Although this modification increased the number of sequences attributed to protists by 50% (which still represented at most 3.5% of the cDNAs); this did not change the overall taxonomic distribution of the cDNAs (Figure 1).

Global functional annotation of the cDNAs

Using an e-value threshold of 10^{-6}, 48% (spruce) and 39.5% (beech) cDNA sequences returned a positive hit in BLASTX searches against the GenBank nr protein database. Among the sequences with no homolog in this preliminary analysis, an additional 227 (spruce) and 238 (beech) contained a registered protein motif identified using InterProScan.

Using Blast2GO, a Gene Ontology (GO) term (associated to either a biological process, a molecular function or a cellular component GO category) could be assigned to 31.5% (beech) and 39% (spruce) of the cDNAs. The five most abundant "biological process" and "molecular function" GO categories were the same for the beech and spruce datasets and altogether accounted for respectively 44% and 43% of the sequences annotated with either a "biological process" and "molecular function" GO term. The two most abundant categories were related to protein synthesis (ribosome biogenesis and translation) and contained, for example, ribosomal proteins encoding sequences.

Finally, 12–13% of the sequences in each datasets encoded proteins with enzymatic activities to which an Enzyme Commission (E.C.) number could be associated (Table 1). Two hundred different E.C. numbers were placed into 117 (spruce) and 112 (beech) different KEGG pathways. Several of these pathways such as glycolysis, tricarboxylic acids (TCA) cycle, nitrogen and sulfur metabolisms, amino acid biosynthesis or degradation (Table 2) are essential to appreciate basic soil processes such as global microbial activities and soil nutrient assimilation. For several of these pathways we identified key enzymes that represent the obligate entry points of major soil nutrients in fungal metabolism. This is for example the case of the glutamine synthase (E.C. 6.3.1.2)/NADPH-glutamate dehydrogenase (E.C. 1.4.1.4)/glutamate synthase (E.C. 1.4.1.13) for ammonium; the adenylyl-sulfate kinase (E.C. 2.7.1.25) and the sulfate adenylyltransferase (E.C. 2.7.7.4) for sulfate and aldehyde reductase (E.C. 1.1.1.21) for pentoses (xylose, arabinose) which are end products of hemicelluloses hydrolysis .

Table 2: A list of KEGG pathways relevant to soil nutrient (C, N, S) utilisation and microbial metabolism for which different cDNA sequences from the spruce and beech soil could be affiliated

	No. of cDNAs	
KEGG pathway	**Beech**	**Spruce**
Nitrogen metabolism	13 (16.4)	17 (19.7)
Alanine, aspartate and glutamate metabolism	19 (24.0)	26 (30.2)
Arginine and proline metabolism	20 (25.3)	22 (25.6)
beta-Alanine metabolism	6 (7.6)	13 (15.1)
Cysteine and methionine metabolism	10 (12.7)	19 (22.1)
Glycine, serine and threonine metabolism	10 (12.7)	14 (16.3)
Histidine metabolism	4 (5.1)	4 (4.6)
Lysine biosynthesis	6 (7.6)	2 (2.3)
Lysine degradation	7 (8.9)	10 (11.6)
Phenylalanine metabolism	16 (20.2)	14 (116.3)
Phenylalanine, tyrosine and tryptophan biosynthesis	6 (7.6)	8 (9.3)
Tryptophan metabolism	8 (10.1)	12 (13.9)
Tyrosine metabolism	10 (12.7)	13 (15.1)
Valine, leucine and isoleucine biosynthesis	3 (3.8)	6 (7.0)
Valine, leucine and isoleucine degradation	18 (22.8)	15 (17.4)
Citrate cycle (TCA cycle)	19 (24.0)	29 (33.7)
Glycolysis/Gluconeogenesis	21 (26.6)	48 (55.8)
Pentose phosphate pathway	15 (19.0)	20 (23.2)
Pyruvate metabolism	20 (25.3)	21 (24.4)
Galactose metabolism	4 (5.1)	7 (8.1)
Fructose and mannose metabolism	12 (15.2)	18 (20.9)
Starch and sucrose metabolism	16 (20.2)	18 (20.9)
Sulfur metabolism	2 (2.5)	3 (3.5)

A list of some of the enzymes identified in several of these pathways is given in Table S3. Between parentheses are given the figures extrapolated to a common sample size of 10,000 cDNAs for both spruce and beech.
doi:10.1371/journal.pone.0028967.t002

Targeted annotation of the cDNAs

Our annotation effort concentrated on gene categories encoding proteins with biochemical activities directly linked to the breakdown of plant compounds and other complex nutrients (lignocellulose, chitin, proteins but also secondary metabolites) and polypeptides involved in the mobilisation of soil nutrients such as transmembrane transporters. This targeted annotation was essentially performed through sequence homology searches against specialized protein databases, which compile members of each of these different protein categories, or by using reference sequences when such specialized databases did not exist (in the case of phytases and dioxygenases). Results were further filtered manually and/or by using the MetaBioME database [24] to remove enzymes with no obvious links to the studied processes (such as for example intracellular proteases component of the proteasome).

This annotation strategy identified a large variety of enzymes involved in the above-mentioned processes including many different enzymes classes active on most plant cell wall polymers (cellulose, hemicellulose, pectin and lignin) but also on starch and chitin which, as a component of arthropods exoskeleton and fungal cell wall, represents an abundant nitrogen source in forest soils (Table 3). Altogether, Glycoside Hydrolases (GH CAZymes) represented between 0.5% (beech) and 0.8% (spruce) of all cDNAs. In addition we also identified a set of enzymes responsible for the breakdown of non-cell wall organic molecules such as different classes of proteases, a phytase, a cutinase and a putative carotenoid ester lipase These latter enzymes were all highly similar to known excreted fungal enzymes.

Table 3: An illustration of the diversity of potential plant cell wall and other polysaccharides (starch, chitin) degrading enzymes identified among the beech and spruce soil ESTs

Category	Family	Potential enzyme activity (/substrate)[a]	beech	spruce
CAZymes[**]			**65[b,c] (82)**	**113[b,c] (131)**
	CBM[ns]	**Carbohydrate Binding Activity**	**3 (3.8)**	**10 (11.6)**
	CE[ns]	**Carbohydrate Esterases**	**3[c] (3.8)**	**9[c] (10.5)**
	CE1	xylan, pectin	1	1
	GT[ns]	**Glycosyltransferases**	**21 (26.5)**	**23 (26.7)**
	PL[ns]	**Polysaccharide Lyases**	**0 (0)**	**1 (1.2)**
	PL1	pectate lyase/pectin	0	1
	GH[*]	**Glycoside Hydrolases**	**38[c] (48.1)**	**70[c] (81.3)**
	GH5	Cellulase/cellulose	2	5
	GH6	endoglucanase/cellulose	1	0
	GH7	endoglucanase/cellulose	3	4
	GH9	endoglucanase/cellulose	1	0
	GH10	endo-1,4-β-xylanase/hemicellulose	0	1
	GH11	xylanase/hemicellulose	2	1
	GH12	endoglucanase/xyloglucan	0	1
	GH28	galacturonase/pectin	0	2
	GH13	α-amylase/starch	0	3
	GH15	glucoamylase/starch	1	1
	GH18	chitinase/chitin	3	6
	GH31	α-glucosidase/xyloglucan	0	1
	GH35	β-galactosidase/pectin	0	2
	GH44	endoglucanase/cellulose	0	1
	GH45	endoglucanase/cellulose	0	2
	GH51	α-L-arabinofuranosidase/pectin	0	1
	GH54	α-L-arabinofuranosidase/pectin	2	0
	GH61	Cu-metalloenzyme/cellulose	1	6
	GH76	α-1,6-mannanase/hemicellulose	0	2
	GH78	α-L-rhamnosidase/pectin	0	1
	GH93	exo-α-L-1,5-arabinanase/pectin	0	1
FOLymes[*]			**6 (7.6)**	**24 (27.9)**
	LO[ns]	**Lignin Oxidases/lignin**	**3 (3.8)**	**5 (5.8)**
	LO1	Laccase/Catechol oxydase	2	1
	LO2	peroxydase	1	1
	LO3	cellobiose dehydrogenase	0	1
	LDA[**]	**Lignin Degrading Auxiliary enzymes**	**3 (3.8)**	**19 (22)**
	LDA1	aryl-alcohol oxydase	0	3
	LDA3	glyoxal oxydase	0	7
	LDA5	galactose oxydase	0	1
	LDA6	glucose oxydase	0	1
	LDA7	benzoquinone reductase	3	5
	LDA8	alcohol oxydase	0	2

For the main categories, between parentheses are given the figures extrapolated to a common sample size of 10,000 cDNAs. Differences between Beech and Spruce were tested using Pearson Chi-square test;
**, $P<0.01$;
*, $0.01<P<0.05$;
ns, not significant, $P>0.05$.
[a]according to data from the CAZyme database, [18], [29] and [67].
[b]Excluding CBMs which can be associated to other CAZYmes.
[c]These figures include all CAZYmes, CE or GH identified in the datasets; not just the families illustrated in this table.
doi:10.1371/journal.pone.0028967.t003

Given that organic matter input in soil also results in the input of toxic secondary metabolites, we searched for enzymes classes potentially involved in their detoxification. While we only identified a single putative fungal dioxygenase (cathechol dioxygenase) among sequences from the spruce soil, we identified a total of 26 P450 monooxygenases belonging to 12 different families (as defined in the CYPED [25] database,). Among the CYPED homologous sequences, several did indeed contribute to the breakdown of toxic compounds such as benzoate or the flavonoid phytoalexin pisatin while others were part of biosynthetic pathways such as an O-methylsterigmatocystin oxidoreductase involved in aflatoxin production.

Besides their involvement in organic matter processing, most of these different enzymes are targets for the bioindustry. We therefore performed a global annotation of the environmental EST datasets with respect to their content in genes of biotechnological interest by blasting all sequences against the MetaBioME database which compiles sequences of "Commercially Useful Enzymes" (CUE). For both spruce and beech soils, more than 3% of the sequences returned a positive hit in this analysis and the two most represented CUE classes were the oxidoreductases and the hydrolases. About half of these sequences were from fungi

Both spruce and beech soil metatranscriptomes included significant proportions (6.2 and 4.8% respectively; E-value threshold of 10^{-6}) of putative membrane transporters as identified by blasting all cDNA sequences against the Transporter Classification Database (TCDB; Table 4)[26]. The most abundant category of fungal transporters involved in the assimilation of soil nutrients was the sugar porter one (T.C. 2.A.1.1) implicated in both hexoses (eg glucose, galactose, mannose) and pentoses (eg arabinose, xylose) assimilation, followed by the different families participating to amino acid and phosphate uptake. Concerning inorganic nutrients, besides phosphate transporters, we also identified ammonium transporters (T.C.1.A.11.) but neither nitrate (T.C.2.A.1.8) nor sulfate (T.C.2.A.53.1) ones.

Table 4: Diversity of fungal plasma membrane transporters, potentially involved in soil nutrient uptake, identified among the beech and spruce soil cDNAs

TCDB[a] family	description	beech	spruce
Sugar transporters			
2.A.1.1.-	Sugar porter family	1	31
Amino-acid transporters			
2.A.18.-	Amino acid/auxin permease (AAAP)	2	3
2.A.3.10.-	Amino Acid-Polyamine-Organocation (APC)	4	6(7)
Peptide transporters			
2.A.1.14.-	Anion:Cation Symporter (ACS)	1	0
2.A.67.-	Oligopeptide Transporter (OPT)	2	6
Phosphate transporters			
2.A.1.9.-	Phosphate: H^+ Symporter (PHS)	2(3)	6
Ammonium transporters			
1.A.11.-	Ammonia Channel Transporter (Amt) Family	1	3
Total transporters**		**14**	**56**

Figures represent the No. of unique sequences after clustering; figures between brackets give the total No. of sequences before clustering. For the total number of transporters, differences between Beech and Spruce were tested using Pearson Chi-square test;
**, $P<0.01$.
[a]The Transporter Classification Database (http://www.tcdb.org/).
doi:10.1371/journal.pone.0028967.t004

For most of the explored gene categories we consistently identified a higher number of homologous sequences in the spruce dataset compared to the beech one (Tables 3, and4). The most striking difference concerned the sugar porter family with 31 sequences in the spruce forest metatranscriptome (0.34% of the sequences) against only one in the beech forest data set (0.01%).

Identification of full-length CAZymes

Twelve presumably full-length Carbohydrate Active Enzymes (CAZyme) encoding cDNAs, belonging to families CE1 (accession no. FR865947), GH5 (FR865941), GH7 (FR865943-4), GH11 (FR865942), GH45 (FR865945), GH61 (FR865936-40) and PL1 (FR865946), were identified and sequenced entirely. The longest cDNA clone was 1671 bp-long and encoded a putative GH7 cellobiohydrolase. Except for the CE1 family member, which is thought to be an intracellular enzyme with S-formyl glutathione hydrolase activity, all other 11 predicted protein sequences possessed a N-terminal signal peptide in agreement with the potential polysaccharide degrading activities of the corresponding enzymes.

The putative taxonomic/phylogenetic origin of these 12 cDNAs was assessed using Bayesian (MrBayes) and/or maximum likelihood (PhyML) phylogenetic analyses of the deduced amino acid sequences. Protein sequences used in the phylogenetic analyses included not only the CAZyme best BLASTX hits from GenBank nr database but also homologous sequences identified among GenBank non fungal ESTs in addition to the gene models predicted from 57 recently released fungal and non fungal (animals, choanoflagellate) genome sequences

Each of the 12 full-length environmental sequences were homologous to fungal sequences and to sequences from other taxonomic groups either closely related to the Fungi, such as the Metazoa (GH7 and GH45) or the choanoflagellates (CE1), or to more distantly related groups such as the Plantae (GH5, PL1) or the Bacteria (GH5, GH11 and PL1). With the exception of the Carbohydrate Esterase family 1 (CE1) genes, all other studied CAZyme genes were not present in all genomes analysed and, when present, frequently occurred as gene families from which between one and three members per species were used for the phylogenetic analyses.

In many cases, sequences from one specific taxonomic group (eg Fungi or Bacteria) did not group together to form a single homogeneous and statistically well-supported clade (Figure 2,). This could be due to an insufficient number of phylogenetically-informative characters in the sequences or to a complex evolutionary history of these gene families characterised by frequent gene duplications, gene losses and potential horizontal gene transfers between distantly related taxonomic groups [27]). *In fine*, only the phylogeny of the CE1 family fitted to the overall phylogeny of the Fungi [28] and the CE1 environmental sequence most likely originated from an Ascomycota Pezizomycotina fungal species (). Several other environmental sequences, including putative pectin lyase (PL1,), xylanase (GH11,), cellobiohydrolase (GH7, Figure 2B), endocellulase (GH45,Figure 2A) and members of the copper-dependent oxidase GH61 family [29], did not group with other referenced sequences to form statistically well supported phyla. This suggests that these sequences originated from fungal or non-fungal taxonomic groups distantly related to those from which homologous sequences have already been characterised. Percentages of identical amino acid positions between the environmental sequences and their closest relatives in the phylogenetic analyses (Figures 2) supported this hypothesis for several sequences. These percentages ranged from as low as 35–38% for two GH61 and the PL1 sequence to 67–70% for a GH7 and the CE1 sequences

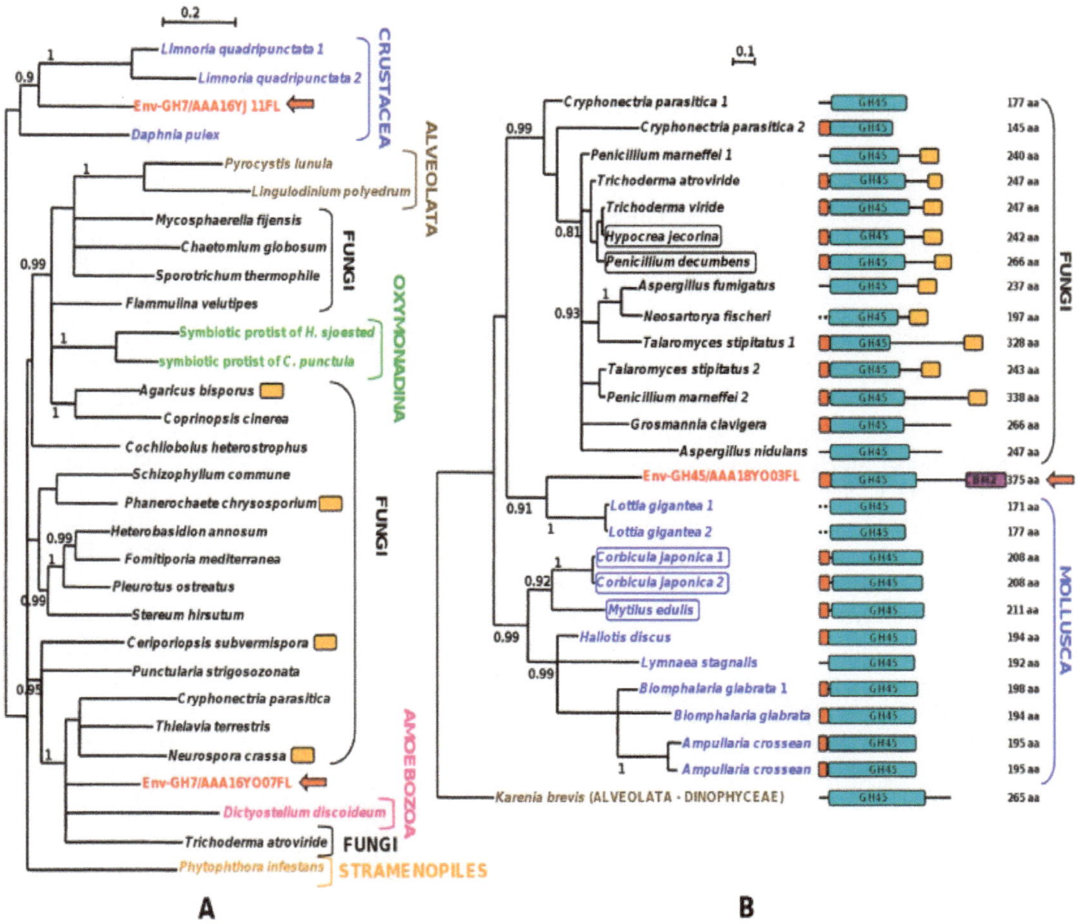

Figure 2: Putative phylogenetic origins of three environmental Glycoside Hydrolases belonging to families GH7 (A) and GH45 (B). Bayesian (MrBayes) phylogenetic trees include protein sequences from different taxonomic groups, each identified by a specific colour. Diagrams drawn to scale illustrate the modular structure of each of the different GH45 protein sequences. Red rectangles, potential signal peptides; blue rectangles, GH45 domains used for the phylogenetic analysis; orange rectangles, family one Carbohydrate Binding Modules (CBM1) characteristic of fungal GH45; purple rectangle, CBM2 module only found in the environmental sequence. Boxed species names indicate protein sequences for which an endoglucanase catalytic activity has been experimentally established [30]–[33]. Posterior branch probabilities above 0.8 are given; branches with less than 0.5 probability support were collapsed.

Two of the environmental CAZymes potentially originated from the soil fauna. This was the case for one of the two analyzed GH7 (putative cellobiohydrolase Env-GH7/16YJ11; Figure 2A) family members that grouped with the recently identified crustacean cellobiohydrolase polypeptides. This was also the case for the GH45 family member encoding a putative endocellulase (Figure 2B). This latter environmental sequence grouped with Molluscan GH45 sequences but differed from these sequences by the presence of a family two carbohydrate binding module (CBM2) at its C-terminal end (Figure 2B). Carbohydrate binding modules existed in several fungal GH45 proteins but they all belonged to the unrelated CBM1 family.

DISCUSSION

This report on the systematic sequencing of environmental eukaryotic cDNAs establishes the main features of this experimental approach. First, the specific analysis of polyadenylated mRNA allows access to the protein-coding gene pool of eukaryotes and efficiently counter-selects not only rRNA but also "non-eukaryotic" mRNA. This is in contrast to the studies published so far on the analysis of "global, essentially prokaryotic, metatranscriptomes"[34]–[38] which result in sequence datasets comprising as little as less than 1% [35]–[36] to often more than 50% of rRNA sequences despite the use of different protocols to eliminate these molecules [39], [40]. Furthermore, with less than 3% of cDNAs attributed to bacteria in both beech and spruce soil metatranscriptomes, polyadenylation of bacterial mRNA [41] does not seem to represent a challenge for the study of eukaryotic environmental mRNA. Moreover, the exact taxonomic origin of cDNAs attributed to bacteria should to be carefully evaluated as they may include genuine eukaryotic genes not yet identified in this taxonomic group such as bacterial genes recently acquired by eukaryotes by horizontal gene transfer; as for example recently suggested for nematodes carbohydrate active enzymes (CAZymes; [27]).

Secondly, among the cDNAs, between 52 and 60% have no homologs following BLASTX searches against the GenBank nr database. These figures, similar to the one reported for the Muskoxen rumen eukaryotic metatranscriptome [13], are higher than the percentages of orphan genes revealed in recently published genomic sequences from fungi and invertebrate phyla, that dominate the two studied soil eukaryotic communities. Figures vary widely from as low as roughly 5% for the entomopathogenic Ascomycota fungi *Metarhizum sp.* [42] to 20% for the symbiotic species *Tuber melanosporum* [43]. In the case of the insect *Pediculus humanus*[44] and the crustacean *Daphnia pulex* [45], two arthropods distantly related to other fully sequenced species from this phylum, the percentages of genes without homologs have been estimated to be of 10 and 36% respectively. Factors susceptible to accentuate the low percentage of cDNA taxonomic affiliation in eukaryotic metatranscriptomes include the presence of poorly studied protist groups in soils and also the large proportion of short sequences located at the 3′ end of the cDNAs which comprise untranslated 3′ UTRs. Future use of pyrosequencing or of other high-throughput sequencing technologies may increase the percentages of short reads, a factor that will be mitigated by a sharp increase in the number of sequences produced. All these figures must however be considered as transitory. Indeed, the current exponential increase in the number of available genome sequences in conjunction with a broader taxonomic sampling of the sequenced taxa, more representative of the soil biota, should have a direct and rapid positive impact on the taxonomic annotation of eukaryotic environmental sequences. More specifically it will allow us to carefully evaluate the contribution of protist taxa to the soil metatranscriptomes. Regarding functional annotation, better assignation will depend upon both (i) functional studies on individual unknown environmental sequences [46] and (ii) functional validation of unknown proteins from model organisms.

Despite the large proportion of housekeeping genes involved in basic cell maintenance mechanisms, both global (KEGG pathways) and targeted annotation identified a large variety of genes of interest in a context of soil ecosystem functioning and more specifically in the turnover of plant biomass and soil nutrient cycling and utilization. These nutrients encompass inorganic (ammonium, phosphates) and organic (aminoacids, peptides, phytate) forms of nitrogen and phosphorus, sulfates and simple sugars resulting essentially from either rhizodeposition or lignocellulose breakdown. We also noticed that genes representative of other key pathways such as the nitrate assimilation one (which comprises specific transporters, nitrate and nitrite reductases) were however missing in the datasets which could indicate that the corresponding nutrient represents a minor N source in the studied acidic forest soils.

Enzymes participating to plant cell wall deconstruction are of special interest not only from an ecological point of view to quantify and understand soil C turnover, but also because they are among the most widely used enzymes in the industry and are instrumental to the development of second generation biofuels. We identified among the beech and/or spruce soil cDNAs representatives of enzymes active on the main plant cell wall polymers: lignins, cellulose, hemicelluloses and pectins. Genes encoding carbohydrate active enzymes clearly outnumbered those encoding enzymes directly active on lignin although members of the three lignin oxidase families were identified in at least one of the two metatranscriptomes (Table 3). Glycoside Hydrolases (GH) encoding genes represented 0.46% (beech) and 0.78% (spruce) of the analyzed cDNAs. These figures are very similar to those reported for different "carbohydrate-adapted" metagenomes [18]. In a perspective of mining soil eukaryotic metatranscriptomes for such enzymes these values signify that the screening effort could be similar to the screening effort developed for metagenomes [14]. Additional sequencing effort as well as the study of replicate soil samples are nevertheless required before concluding on a potential effect of litter quality (beech *versus* spruce) on the relative abundance of transcripts related to plant cell wall degradation or nutrient assimilation (eg sugar transporters). It is indeed known that these two species differ with respect to the nature and proportions of polymers present in their cell walls [47], [48] and that spruce litter mineralizes more quickly compared to the beech one [49].

The use of Sanger sequencing of cloned cDNAs allowed us to easily identify several full-length genes among which several CAZyme ones. Phylogenetic analysis of the coded protein sequences suggests that many of them are distantly related to protein sequences present in databases. This further emphasizes the interest of eukaryotic metatranscriptomes as a source of novel enzymes with potentially interesting catalytic properties. This also indicates that organisms studied thus far at the molecular level in a context of plant cell wall deconstruction may not be representative of the organisms at work in the soil ecosystem. It is particularly interesting to consider that at least two out of the 12 full-length CAZymes (putative cellobiohydrolase and endocellulase) could originate not from fungi but from animals. A number of recent publications have indeed reported the existence of plant cell wall degrading enzyme genes in the genome of different groups of invertebrates which are represented among the soil fauna [27], [31], [33], [45], [50], [51]. This could indicate that in addition to their well recognized role in the fragmentation and mixing of plant litter, the soil fauna could also play a direct and potentially major role in the actual hydrolysis of plant carbohydrates in soils which adds up to hydrolysis by their gut microbiomes.

In conclusion, this first report of a eukaryote-specific analysis of soil metatranscriptomes highlights the potential of this experimental approach in two research fields. In the field of environmental biotechnology, eukaryotic metatranscriptomes represent diversified multigenome resources for many different gene categories used in the bioindustry. In this context, construction of environmental cDNA libraries whose potentially large inserts can be sequenced (this study) or directly expressed in heterologous hosts [20], [46], [52], [53]potentially represents the best experimental approach. In the field of environmental sciences, the study of eukaryotic metatranscriptomes will help us to reevaluate the contribution of the different soil eukaryotes to basic and essential soil processes such as organic matter degradation. It will also contribute to compare ecosystems which differ with respect to *i.e.*vegetation type, soil characteristics and climate and evaluate the relative contribution of these different variables on different soil processes. In this respect, high-throughput sequencing technologies should be implemented.

MATERIALS AND METHODS

Study site and soil sampling

Soil samples were taken from an environmental research observatory forest site located in central France (Breuil-Chenue forest, 47°18′10″N, 4°4′44″E, 638 m above sea level). This site, initially covered by a mixed broadleaved tree forest (*Fagus sylvatica*, *Quercus sessiliflora*,*Betula verrucosa*, *Corylus avelana*) was clear-cut and replanted in 1976 by separate, single forest tree species stands, among which spruce (*Picea abies*) and beech (*Fagus sylvatica*). On July the 10th 2007, 14 (spruce stand) and 16 (beech stand) soil samples (8 cm in diameter, 15 cm in depth) were collected along a systematic sampling grid. For each stand, the uppermost 3–7 cm thick organic matter-rich horizon of each core was sieved (2 mm mesh size) and mixed together (100 ml per core) to obtain a composite sample which was immediately frozen at −70 °C. Pedoclimatic parameters, soil characteristics and sampling strategy are detailed

RNA extraction, cDNA libraries construction and sequencing

Total RNA (~100 μg) was extracted from ~90 g of each of the two composite forest soil samples as described in [20] and [54]. Polyadenylated mRNA were separated from non-polyadenylated RNAs (rRNAs, mitochondrial and prokaryotic mRNAs, tRNAs) by affinity capture on paramagnetic beads coated with poly-dT (Dynabeads Oligo (dT) kit, Dynal). Unbound non-polyadenylated RNAs were recovered by ethanol precipitation. Reverse transcription was carried out following the SMART cDNA Library Construction Kit instructions (Clontech), using between 90 (beech) and 210 ng (spruce) of purified mRNA. cDNA were further amplified by long distance PCR (LD-PCR) as described in the SMART cDNA Library Construction Kit, by using 20 and 24 PCR cycles for spruce and beech respectively. After size-fractionation to remove cDNA smaller than 400 pb (CHROMA SPIN-400 Columns, Clontech), the cDNAs were digested with *Sfi*1 and ligated into the *Sfi*1-digested pDNR-LIB plasmid vector (Clontech).

Each cDNA library was introduced in electro-competent *E. coli* cells (DH10B, Invitrogen) and 10,000 randomly selected cDNA clones were sequenced from their 5′ end using the M13-20 primer and Sanger chemistry (Genoscope sequencing centre, Ivry, France).

18S rDNA gene libraries construction and sequencing

350 (beech) and 730 (spruce) ng of soil non-polyadenylated RNA were reverse transcribed using 0.2 μg of random hexamers and 200 U of M-MuLV Reverse Transcriptase according to the manufacturer instructions (MBI Fermentas). A *ca* 560 bp-long fragment located at the 5′-end of the eukaryotic 18S rDNA gene was amplified by PCR using primers Euk1A (CTGGTTGATCCTGCCAG) and Euk516R (ACCAGACTTGCCCTCC) described by [55]. PCR mixtures (25 μl) contained 200 nM of each primer, 200 μM of each dNTP, 1 mM $MgCl_2$, 0.25 mg.ml^{-1} of bovine serum albumin, 0.625 U of *Pfu* DNA polymerase, the appropriate buffer (Fermentas) and one tenth of the reverse-transcription mix. Amplification reaction started with an initial denaturation step of 3 min at 96°C, followed by 25 cycles comprising 45 sec at 96°C, 45 sec at 56°C and 2 min at 72°C, and finished with an elongation step of 5 min at 72°C. Amplification products of the expected size from 15 PCR tubes were pooled, isolated from an agarose gel (Nucleospin Extract kit, Macherey-Nagel) and ligated in the plasmid pCR-Blunt II-TOPO (Zero Blunt TOPO PCR Cloning kit, Invitrogen) that was used to transform electro-competent DH10B*E. coli* cells (Invitrogen).

For each rDNA library, the *ca* 560 bp rDNA inserts of 96 clones were entirely sequenced (Agowa Company, Berlin, Germany) using universal primer M13-20. Sequences were manually corrected and edited. BLASTn searches were performed against GenBank nr nucleotide database at NCBI (http://www.ncbi.nlm.nih.gov/) using default parameters except for word size which was set to 7. Sequences were analyzed with the rRNA Database Project CHECK_ CHIMERA program (http://rdp8.cme.msu.edu/). Potential chimeras were further analysed by blasting separately the two dissimilar segments of the sequences against GenBank. Confirmed chimeras and artefacts were eliminated from the sequence datasets. All remaining sequences were submitted to EMBL and are available under accession numbers FN393180–FN393221 (beech) and FN393323–FN393380 (spruce).

cDNA sequences cleaning and clustering

All cDNA sequences were trimmed using TIGR SEQCLEAN tool (http://www.tigr.org/tdb/tgi/software) with default parameters to eliminate poly-A sequences at the 3′end of cDNA, vector, adaptor and primer sequences and undetermined nucleotides. Sequences shorter than 100 nucleotides were eliminated. Contaminant ribosomal RNA sequences were identified by BLASTN analysis (cut-off threshold of E-value$\leq 10^{-10}$, word-size=11) of all sequences against LSUrdb and SSUrdb_SSURef100 libraries described in [38](available at http://services.cbu.uib.no/supplementary/community-profiling/), as well as against NCBI-nr nucleotide database (looking for "ribosomal RNA" or "rRNA" in the title). The resulting "cleaned" sequence dataset was submitted to EMBL and is available under accession numbers FR706059–714330 (beech) and FR697056–706058 (spruce).

Cleaned sequences were clustered using Cd-hit (http://cd-hit.org, [56]) using a 90% identity threshold. The resulting clusters were used to perform a rarefaction analysis using S. Holland's Analytical Rarefaction version 1.3 software (http://www.uga.edu/strata/software/).

Global cDNA sequences annotation

Cleaned cDNA sequences were queried (BLASTX) against NCBI nr protein database (http://www.ncbi.nlm.nih.gov/), using 2 different strategies, (i) Q-BLAST via BLAST2GO [57]with parameters, E-value$\leq 10^{-6}$ and overlapping length >33% to the corresponding best hit and, (ii) NetBlast 2.2.22 (E-value$\leq 10^{-6}$). As both analyses gave similar percentages of annotated cDNA sequences, we only considered the Q-BLAST results. In addition to raw BLASTX analyses, we also used BLAST2GO to perform Gene Ontology (GO, [58]), Enzyme code (E.C.) and InterPro (conserved patterns in sequences) annotations.

Targeted annotation of cDNAs

Cleaned sequence datasets were searched for sequences similar to genes coding for organic matter degradation enzymes and for enzymes of biotechnological interest, using BLASTX searches (default parameters) against various specialized databases. Searches were made against Carbohydrate Active Enzymes (CAZy, http://www.cazy.org/; [59]); and Fungal Oxidative Lignin Enzymes (FOLy; http://foly.esil.univ-mrs.fr/; [60]) databases to find plant cell wall degradation enzymes. cDNA similar to cytochrome P450 potentially involved in detoxication of plant secondary metabolites were searched against CYPED database (downloaded on October 2009, http://www.cyped.uni-stuttgart.de/, [61]). Lipases/esterases and proteases were searched against the Lipases Engineering database (LED, downloaded on December 2009; http://www.led.uni-stuttgart.de/, [62]) and MEROPS (downloaded on January 2010, http://merops.sanger.ac.uk/index.shtml, [63]) respectively. Membrane transporters (for sugar, amino acids, oligopeptides and phosphate) were searched against the Transporter Classification Database (downloaded on April 2010, http://www.tcdb.org/, [26]). For other enzymes (phytases and dioxygenases), for which no specific databases exist, reference protein sequence were extracted from GenBank/EMBL/DDJB database and used as queries to perform a tBLASTn analysis (default parameters) against our cDNA libraries.

For all these specific analyses, the functional annotations obtained were further confirmed using a BLASTX analysis (default parameters) of the corresponding cDNAs against GenBank nr and/or Swissprot protein databases. We only retained cDNA sequences similar to the fungal targeted enzyme using a E-value threshold of 0.001. For cytochrome P450, lipases, proteases, phytases and dioxygenases, identified cDNAs were further filtered by BLASTX (default parameter, E-value$\leq 10^{-6}$) against a curated database of commercially useful enzymes (MetaBioME, http://metasystems.riken.jp/metabiome/index.php, [24]) to retain only sequences similar to enzymes, which have known applications in industries. To identify full-length sequences, annotated cDNAs were translated and aligned with the closest protein sequences identified in Genbank/EMBL/DDJB to search for a putative start codon in the N-terminal region.

Taxonomic annotation of rRNA and cDNA sequences

For PCR-amplified 18S ribosomal sequences, BLASTN searches (default parameters, except word size set at 7) were performed against the GenBank nr nucleotide database at NCBI (http://www.ncbi.nlm.nih.gov/). 18S sequences were then attributed to major eukaryotic phyla (listed in Figure 1) by performing phylogenetic analyses (BioNJ and PhyM) using for the sequence alignments the best Blast hits and a set of reference sequences for each of the different phyla. Analyses (sequence alignments and phylogenetic analyses) were performed using Seaview [64].

Using BLASTX output files, cDNAs were taxonomically assigned using MEGAN V 3.5 (Metagenome analyser, www-ab.informatik.uni-tuebingen.de/software/megan, [65]). This program uses the result of a BLAST comparison and assigns each read to a taxon at a specific taxonomic level. All parameters of MEGAN were kept at default values. For the cDNA taxonomic assignation, an additional analysis was performed by setting the "min support" option (the minimum number of sequence reads that must be assigned to a taxon) to one instead of 5 used as default parameter.

Phylogenetic analyses

Protein sequences used for phylogenetic reconstructions were selected among BLAST results against different databases. Each of the environmental protein sequences were blasted against the GenBank/EMBL/DDJB nr protein database (BLASTX) and non human EST one (TBLASTN). Additional homology searches (BLASTX) were performed against the proteomes of fully sequenced organisms available through different websites. Depending on the number and taxonomic diversity of the best BLAST hits, between one/two (individual genome sequences) and five (GenBank/EMBL/DDJB) protein sequences were selected for the sequence alignments.

Alignments were performed on the phylogeny.fr platform (www.phylogeny.fr; [66]) using MUSCLE (default parameters) followed by the selection of phylogenetically informative regions using GBlocks (low stringency parameters). The resulting cured alignment were used for phylogenetic analyses using MrBayes (less than 30 sequences) and/or PhyML. Parameters for Bayesian analyses were set as followed: number of substitution type, 6 (GTR); substitution model, WAG; rates variation across sites, invariable plus Gamma; analyses were run for 100,000 generations, sampling tree every 10 generations, burning of 100. PhyML was run on the Seaview platform [64] using default parameters except for the WAG substitution model.

REFERENCES

1. Meier C, Bowman WD (2008) Links between plant litter chemistry, species diversity, and below-ground ecosystem function. Proc Natl Acad Sci U S A 105: 19780–19785.
2. Cornwell WK, Cornelissen JH, Amatangelo K, Dorrepaal E, Eviner VT, et al. (2008) Plant species traits are the predominant control on litter decomposition rates within biomes worldwide. Ecol Lett 11: 1065–1071.
3. Sinsabaugh RL, Lauber CL, Weintraub MN, Ahmed B,

Allison SD, et al. (2008) Stoichiometry of soil enzyme activity at global scale. Ecol Lett 11: 1252–1264.

4. Martinez D, Larrondo LF, Putnam N, Gelpke MD, Huang K (2004) Genome sequence of the lignocellulose degrading fungus Phanerochaete chrysosporium strain RP78. Nat Biotechnol 22: 695–700.
5. Martinez D, Berka RM, Henrissat B, Saloheimo M, Arvas M, et al. (2008) Genome sequencing and analysis of the biomass-degrading fungus Trichoderma reesei (syn. Hypocrea jecorina). Nat Biotechnol 26: 553–560.
6. Martinez D, Challacombe J, Morgenstern I, Hibbett D, Schmoll M, et al. (2009) Genome, transcriptome, and secretome analysis of wood decay fungus Postia placenta supports unique mechanisms of lignocellulose conversion. Proc Natl Acad Sci U S A 106: 1954–1959.
7. Jeffries TW, Grigoriev IV, Grimwood J, Laplaza JM, Aerts A, et al. (2007) Genome sequence of the lignocellulose-bioconverting and xylose-fermenting yeast Pichia stipitis. Nat Biotechnol 25: 319–26.
8. Frankland JC (1998) Fungal succession - unravelling the unpredictable. Mycol Res 102: 1–15.
9. Nielsen UN, Osler GHR, Campbell CD, Burslem DFRP, van der Wal R (2010) The influence of vegetation type, soil properties and precipitation on the composition of soil mite and microbial communities at the landscape scale. J Biogeogr 37: 1317–1328.
10. Zinger L, Lejon DPH, Baptist F, Bouasria A, Aubert S, et al. (2011) Contrasting diversity patterns of crenarchaeal, bacterial and fungal soil communities in an alpine landscape. PLoS ONE 6: e19950.
11. Brulc JM, Antonopoulos DA, Berg Miller ME, Wilson MK, Yannarell AC (2009) Gene-centric metagenomics of the fiber-adherent bovine rumen microbiome reveals forage specific glycoside hydrolases. Proc Natl Acad Sci U S A 106: 1948–1953.
12. Hess M, Sczyrba A, Egan R, Kim TW, Chokhawala H, et al. (2011) Metagenomic discovery of biomass-degrading genes and genomes from cow rumen. Science 331: 463–7.
13. Qi M, Wang P, O'Toole N, Barboza PS, Ungerfeld E, et al. (2011) Snapshot of the Eukaryotic gene expression in Muskoxen rumen – A metatranscriptomic approach. PLoS ONE 6: e20521.
14. Tasse L, Bercovici J, Pizzut-Serin S, Robe P, Tap J, et al. (2010) Functional metagenomics to mine human gut microbiome for dietary fiber catabolic enzymes. Genome Res 20: 1605–1612.
15. Pope PB, Denman SE, Jones M, Tringe SG, Barry K, et al. (2009) Adaptation to herbivory by the Tammar wallaby includes bacterial and glycoside hydrolase profiles different from other herbivores. Proc Natl Acad Sci U S A 107: 14793–798.
16. Tartar A, Wheeler M, Zhou X, Coy MR, Boucias DG, et al. (2009) Parallel metatranscriptome analyses of host and symbiont gene expression in the gut of the termite Reticulotermes flavipes. Biotechnol Biofuels 2: 25.
17. Warnecke F, Luginbu¨hl P, Ivanova N, Ghassemian M, Richardson TH, et al. (2007) Metagenomic and functional analysis of hindgut microbiota of a woodfeeding higher termite. Nature 450: 560–5.
18. Allgaier M, Reddy A, Park JI, Ivanova N, D'haeseleer P, et al. (2010) Targeted discovery of glycoside hydrolases from a switchgrass-adapted compost community. PLoS ONE 5: e8812.
19. Grant S, Grant WD, Cowan DA, Jones BE, Ma Y, et al. (2006) Identification of eukaryotic open reading frames in metagenomic cDNA libraries made from environmental samples. Appl Environ Microbiol 72: 135–143.
20. Bailly J, Fraissinet-Tachet L, Verner M-C, Debaud J-C, Lemaire M, et al. (2007) Soil eukaryotic functional diversity, a metatranscriptomic approach. ISME J 1: 632–642.
21. Lo´pez-Garcı´a P, Moreira D (2008) Tracking microbial biodiversity through molecular and genomic ecology. Res Microbiol 159: 67–73.
22. Jeon S, Bunge J, Leslin C, Stoeck T, Sunhee H, et al. (2008) Environmental rRNA inventories miss over half of protistan diversity. BMC Microbiol 8: 222.
23. Stephenson SL, Fiore-Donno AM, Schnittler M (2011) Myxomycetes in soil. Soil Biol Biochem 43: 2237–2242.
24. Sharma VK, Kumar N, Prakash T, Taylor TD (2010) MetaBioME: a database to explore commercially useful enzymes in metagenomics datasets. Nucl Ac Res 38: D468–D472.
25. Fischer M, Knoll M, Sirim D, Wagner F, Funke S, et al. (2007) The cytochrome P450 engineering database: a navigation and prediction tool for the cytochrome P450 protein family. Bioinformatics 23: 2015–2017.
26. Saier MH, Jr., Tran CV, Barabote RD (2006) TCDB: the transporter Classification Database for membrane transport protein analyses and information. Nucl Acids Res 34: D181–D186.
27. Danchin EG, Rosso MN, Vieira P, de Almeida-Engler J, Coutinho PM, et al. (2010) Multiple lateral gene transfers and duplications have promoted plant parasitism ability in nematodes. Proc Natl Acad Sci U S A 107: 17651–17656.
28. James TY, Kauff F, Schoch CL, Matheny PB, Hofstetter V, et al. (2006) Reconstructing the early evolution of Fungi using a six-gene phylogeny. Nature 443: 818–22.
29. Quinlan RJ, Sweeney MD, Leggio LL, Otten H, Poulsen J-CN, et al. (2011) Insight into oxidative degradation of cellulose by a copper metalloenzyme that exploits biomass components. Proc Natl Acad Sci U S A 108: 15079–15084.
30. Liu G, Wei X, Qin Y, Qu Y (2010) Characterization of the endoglucanase and glucomannanase activities of a glycoside hydrolase family 45 protein from Penicillium decumbens 114-2. J Gen Appl Microbiol 56: 223–229.
31. Sakamoto K, Toyohara H (2009) Molecular cloning of glycoside hydrolase family 45 cellulase genes from brackish water clam Corbicula japonica. Comp Biochem Physiol B Biochem Mol Biol 152: 390–396.
32. Saloheimo A, Henrissat B, Hoffre´n AM, Penttila¨ M (1994) A novel, small endoglucanase gene, egl5, from Trichoderma reesei isolated by expression in yeast. Mol

Microbiol 13: 219–228.

33. Xu B, Hellman U, Ersson B, Janson J-C (2000) Purification, characterization and amino-acid sequence analysis of a thermostable, low molecular mass endo-ß-1,4-glucanase from blue mussel, Mytilus edulis. Eur J Biochem 267: 4970–4977.

34. Frias-Lopez J, Shi Y, Tyson GW, Coleman ML, Schuster SC, et al. (2008) Microbial community gene expression in ocean surface waters. Proc Natl Acad Sci U S A 105: 3805–3810.

35. Gilbert J, Field D, Huang Y, Edwards R, Li W, et al. (2008) Detection of large numbers of novel sequences in the metatranscriptomes of complex marine microbial communities. PLoS ONE 3: e3042.

36. Gilbert JA, Field D, Swift P, Thomas S, Cummings D, et al. (2010) The taxonomic and functional diversity of microbes at a temperate coastal site: a 'multi-omic' study of seasonal and diel temporal variation. PLoS ONE 5: e15545.

37. Shrestha PM, Kube M, Reinhardt R, Liesack W (2009) Transcriptional activity of paddy soil bacterial communities. Environ Microbiol 11: 960–970.

38. Urich T, Lanzen A, Qi J, Huson DH, Schleper C, et al. (2008) Simultaneous assessment of soil microbial community structure and function through analysis of the metatranscriptome. PLoS ONE 3: e2527.

39. He S, Wurtzel O, Singh K, Froula JL, Yilmaz S, et al. (2010) Validation of two ribosomal RNA removal methods for microbial metatranscriptomics. Nature Met 10: 807–812.

40. Stewart FJ, Ottesen EA, DeLong EF (2010) Development and quantitative analyses of a universal rRNA-substraction protocol for microbial metatranscriptomics. ISME J 4: 896–907.

41. Sharkar N (1997) Polyadenylation of mRNA in prokaryotes. Ann Rev Biochem 66: 173–197.

42. Gao Q, Jin K, Ying S-H, Zhang Y, Xiao G, et al. (2011) Genome sequencing and comparative transcriptomics of the model entomopathogenic fungi Metarhizium anisopliae and M. acridum. PLoS Genet 7: e1001264.

43. Martin F, Kohler A, Murat C, Balestrini R, Coutinho PM, et al. (2010) Pe´rigord black truffle genome uncovers evolutionary origins and mechanisms of symbiosis. Nature 464: 1033–1038.

44. Kirkness EF, Haas BJ, Sun W, Braig HR, Perotti MA, et al. (2010) Genome sequences of the human body louse and its primary endosymbiont provide insights into the permanent parasitic lifestyle. Proc Natl Acad Sci U S A 107: 12168–12173.

45. Colbourne JK, Pfrender ME, Gilbert D, Thomas WK, Tucker A, et al. (2011) The ecoresponsive genome of Daphnia pulex. Science 331: 555–561.

46. Damon C, Vallon V, Zimmermann S, Haider MZ, Galeote V, et al. (2011) A novel fungal family of oligopeptide transporters identified by functional metatranscriptomics of soil eukaryotes. ISME J. doi: 10.1038/ismej.2011.67.

47. Sarkar P, Bosneaga E, Auer M (2009) Plant cell walls throughout evolution: towards a molecular understanding of their design principles. J Exp Bot 60: 3615–3635.

48. Scha¨del C, Blo¨chl A, Richter A, Hoch G (2010) Quantification of monosaccharide composition of hemicelluloses from different plant functional types. Plant Physiol Biochem 48: 1–8.

49. Moukoumi J, Munier-Lamy C, Berthelin J, Ranger J (2006) Effect of tree species substitution on organic matter biodegradability and mineral nutrient availability in a temperate topsoil. Ann For Sci 63: 763–771.

50. King AJ, Cragg SM, Li Y, Dymond J, Guille MJ, et al. (2010) Molecular insight into lignocellulose digestion by a marine isopod in the absence of gut microbes. Proc Natl Acad Sci U S A 107: 5345–5350.

51. Pauchet Y, Wilkinson P, Chauhan R, Ffrench-Constant RH (2010) Diversity of beetle genes encoding novel plant cell wall degrading enzymes. PLoS ONE 5: e15635.

52. Kellner H, Luis P, Portetelle D, Vandenbol M (2010) Screening of a soil metatranscriptomic library by functional complementation of Saccharomyces cerevisiae mutants. Microbiol Res 166: 360–368.

53. Findley SD, Mormile MR, Sommer-Hurley A, Zhang X-C, Tipton P, et al. (2011) Activity-based metagenomic screening and biochemical characterization of bovine ruminal protozoan glycoside hydrolases. Appl Environ Microbiol doi:10.1128/AEM.05925-11.

54. Damon C, Barroso G, Fe´randon C, Ranger J, Fraissinet-Tachet L, et al. (2010) Performance of the COX1 gene as a marker for the study of metabolically active Pezizomycotina and Agaricomycetes fungal communities from the analysis of soil RNA. FEMS Microbiol Ecol 74: 693–705.

55. Diez B, Pedros-Alio C, Massana R (2001) Study of genetic diversity of eukaryotic picoplankton in different oceanic regions by small-subunit rRNA gene cloning and sequencing. Appl Environ Microbiol 67: 2932–2941.

56. Li W, Godzik A (2006) Cd-hit: a fast program for clustering and comparing large sets of protein or nucleotide sequences. Bioinformatics 22: 1658–1659.

57. Conesa A, Go¨tz S, Garcı´a-Go´mez JM, Terol J, Talo´n M, et al. (2005) Blast2GO: A universal tool for annotation, visualization and analysis in functional genomics research. Bioinformatics 21: 3674–367.

58. The Gene Ontology Consortium (2000) Gene ontology: tool for the unification of biology. Nat Genet 25: 25–29.

59. Cantarel BL, Coutinho PM, Rancurel C, Bernard T, Lombard V, et al. (2009) The Carbohydrate-Active EnZyme database (CAZy): an expert resource for glycogenomics. Nucl Ac Res 37: D233–D238.

60. Levasseur A, Piumi F, Coutinho PM, Rancurel C, Asther M, et al. (2008) FOLy: an integrated database for the classification and functional annotation of fungal oxidoreductases potentially involved in the degradation of lignin and related aromatic compounds. Fung Genet Biol 45: 638–645.

61. Sirim D, Wagner F, Lisitsa A, Pleiss J (2009) The cytochrome P450 Engineering Database: integration of biochemical properties. BMC Biochem 10: 27.

62. Fischer M, Pleiss J (2003) The lipase engineering

database: a navigation and analysis tool for protein families. Nucl Ac Res 31: 319–321.

63. Rawlings ND, Barrett AJ, Bateman A (2009) MEROPS: the peptidase database. Nucl Ac Res 38: D227–D233.

64. Gouy M, Guindon S, Gascuel O (2010) SeaView version 4: a multiplatform graphical user interface for sequence alignment and phylogenetic tree building. Mol Biol Evol 27: 221–224.

65. Huson DH, Richter DC, Mitra S, Auch AF, Schuster SC (2009) Methods for comparative metagenomics. BMC Bioinformatics 10(Suppl 1): S12.

66. Dereeper A, Guignon V, Blanc G, Audic S, Buffet S, et al. (2008) Phylogeny.fr: a robust phylogenetic analysis for the non-specialist. Nucl Acids Res 36: W465–W469.

67. Coutinho PM, Andersen MR, Kolenova K, vanKuyk PA, Benoit I, et al. (2009) Post-genomic insight into the plant polysaccharide degradation potential of Aspergillus nidulans and comparison to Aspergillus niger and Aspergillus oryzae. Fung Genet Biol 46: S161–S169.

Section VII: Metaproteomics

CHAPTER 23

PROTEOMIC TOOLS TO DECIPHER MICROBIAL COMMUNITY STRUCTURE AND FUNCTIONING

Florence Arsène-Ploetze & Philippe N. Bertin & Christine Carapito

Keywords: Community proteomics, Metaproteomics, Mass spectrometry, Targeted proteomics, LC-SRM, Quantitative proteomics

INTRODUCTION

Microbial communities are complex biological assemblies, whose study has been difficult for a long time because a large fraction of the species is unknown (Bertin et al. 2008). Indeed, in any given environment only a small fraction of organisms present can actually be cultivated. Nowadays, these communities can be explored as a whole by using environmental genomics. Genomics aims to study the biology of microorganisms by analyzing the genetic information they contain. The concept of metagenomics, i.e., sequencing of the genomic DNA content of a community, has emerged over the past 15 years and provides a detailed inventory of genes from a community and thus of the potential capacities of microorganisms (Bertin et al. 2008). A significant reduction in sequencing costs combined to new broadband technologies has led to an explosion of such genomic programs. These approaches have resulted in an inventory of the genomic content of several communities but have also highlighted the high diversity found in nature (Schleper et al. 2005; Bertin et al. 2008; Lasken 2012; Parkhill 2013; McCann et al. 2014; Tseng and Tang 2014). To go further in the understanding of community functioning, these approaches are now often linked to functional genomic approaches allowing the characterization of genes expressed in an organism under specific conditions. Two approaches are commonly used in functional genomics of communities: metatranscriptomics to study messenger RNAs and metaproteomics or whole community proteomics to characterize proteins.

The metaproteomics approach has some advantages when compared to metatranscriptomics. Indeed, as proteins are more stable than RNAs (especially those originating from prokaryotes), the metaproteome content is supposed to be less affected by extraction procedures, and probably gives a better insight into the biological functions expressed in situ. Moreover, proteins are crucial effectors of the biological response of living organisms. The amount of these effectors in organisms varies at different levels and a modulation of their activity depends on a change of the corresponding genes expression, post-translational modifications, or proteolysis/protein turnover. Therefore, there has been a growing interest in studying expression of proteins of various microorganisms in their habitats and, in the field of microbial ecology, metaproteomics (Wilmes and Bond 2004) or community proteomics (Ram et al. 2005; Lacerda and Reardon 2009; Keller and Hettich 2009) has emerged to characterize in a global way the protein content of microbial communities (Hettich et al. 2012, 2013). Metaproteomics is used to obtain protein catalogues giving important information on the community activity or its structure, but also to compare protein contents in two different ecosystems by using quantitative metaproteomics or to complete or correct metagenomic data (community proteogenomics). In this review, global proteomic approaches used to decipher the physiology of a microorganism or the functioning of microbial communities will be presented with a particular focus on recent advances in global and targeted quantitative proteomics approaches.

Proteomic analysis workflow

The proteomic approach was first defined as a functional genomics approach allowing studying the protein expression pattern of one organism, i.e., to obtain a protein map of all proteins expressed by one organism grown in one particular condition. Such an approach was then developed to compare proteins expressed by an organism incubated in two different conditions. This approach, called differential proteomics,

Citation: Arsène-Ploetze F, Bertin PN, Carapito C, "Proteomic tools to decipher microbial community structure and functioning", doi 10.1007/s11356-014-3898-0.

requires quantifying protein amounts in each condition. Proteomics was thus complementary to genome sequencing, giving information on the non-model microorganism activities (Muller et al. 2007; Weiss et al. 2009). Finally, such an approach is now used at the community level (metaproteomics). The success of (meta)proteomic workflows depends essentially on three crucial steps: the efficiency of proteins extraction, the methods used to separate or fractionate proteins in a complex mixture, and the unambiguous identification of peptides/proteins from tandem mass spectrometry (MS/MS) data (Fig. 1). In addition, robust quantification methods are required in order to compare expression patterns in different conditions.

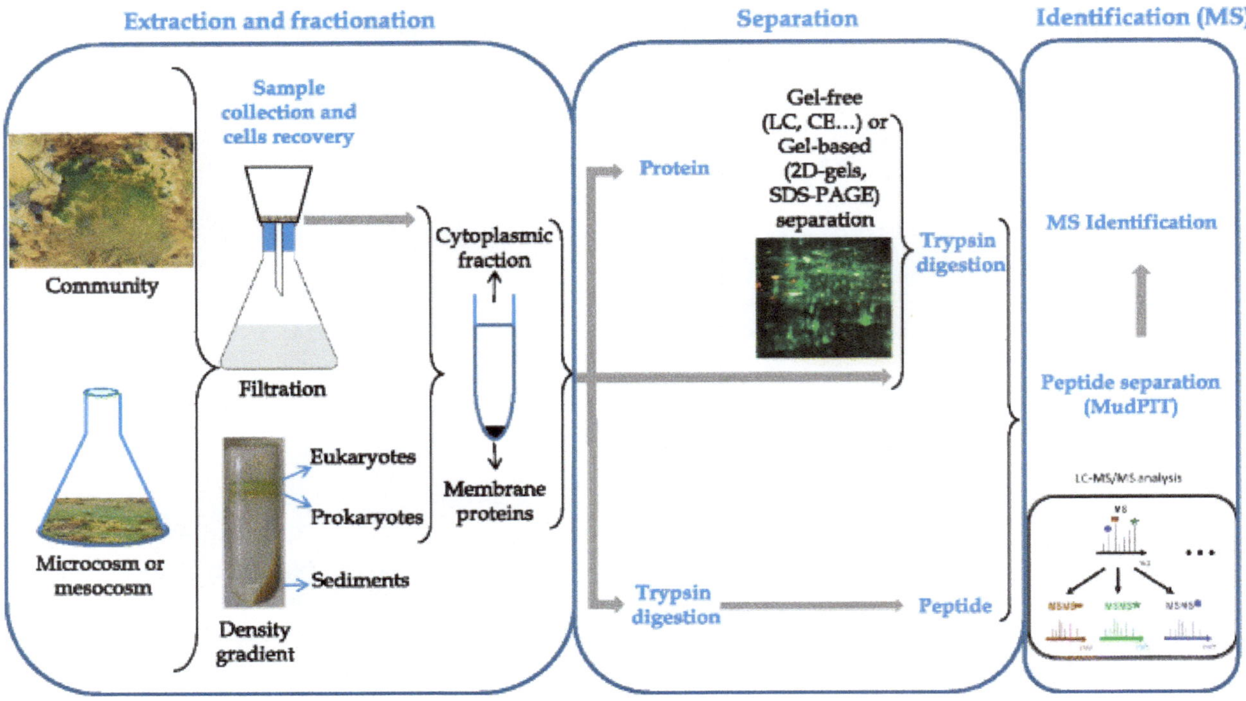

Figure 1: Proteomic workflow to study the structure, functioning, and interactions in complex communities

Protein extraction

To study the proteins expressed by microorganisms, a sufficient quantity of cells has to be collected in order to extract their protein content. Indeed, since there is no "PCR-analog" for proteins, enough material has to be extracted to allow the detection and identification of proteins without any "amplification" step. At least 20 to 50 µg of proteins are typically required to conduct a proteomics experiment. Working with lower amounts of material is delicate and often leads to very poor peptide recovery for further liquid chromatography-MS (LC-MS) analysis after enzymatic digestion and desalting. On most sensitive nanoLC-MS/MS systems, high attomole to low femtomole levels of peptides can be detected, lower limits obviously depending on the peptide mixture complexity and dynamic range. To reach a reasonable amount of starting protein material, the amount of samples to be treated directly depends on the characteristics of the studied site. As an example, in the sediments of the Carnoules AMD, 50 ml of sediments had to be treated to obtain around 10 µg of proteins (Arsène-Ploetze, unpublished data). Therefore, the cell collection step is crucial in metaproteomics and a low biomass collection often makes this approach unsuccessful. In the case of marine or freshwater context, cell collection is often performed by filtration (Fig. 1). The filtration often needs to be done on hundreds of liters of water to recover enough cells. When the community is within soil or sediments or associated to plants or animals, a preliminary step may be required in order to take off cells from biotic or abiotic surface or from surrounding particles. This step is particularly important in the case of microorganisms recovered from soil or sediments, which naturally contain interfering substances such as humic acids, inevitably extracted together with proteins (Bastida et al. 2009). Moreover, when the community is more complex, for example when it is composed of eukaryotes and prokaryotes, it may be necessary to split the different types of cells and to study only a fraction of the community (Fig. 1), or to preserve the cells for further analysis in laboratory. For example, the main oceanic microbial populations, namely *Synechococcus* cells, have been studied using proteomics after cell separation using microwave fixation and flow cytometry sorting (Mary et al. 2010). In several studies, it has been possible to separate microorganisms from sediments, but also bacteria from the eukaryotic population using a Nycodenz gradient (Fig. 1) and study the two populations separately (Bertin et al. 2011; Halter et al. 2012). Recently, successful protocols have been developed to extract proteins directly from soil without separating cells and particles (Chourey et al. 2010; Keiblinger et al.2012). Unbiased protein extraction requires optimization of lysis conditions (Fig. 1). Several protocols are usually tested, such as those described in (Cañas et al. 2007). Physical lysis methods are the most commonly used in the case of microorganisms, such as the use of glass beads grinding, sonication, alternating cycles of freezing and thawing, or high- and low-pressure cycles. The combination of these methods with enzymatic lysis or the use of detergents can improve the efficiency of cell lysis (Cañas et al. 2007).

Protein separation

Proteins can be separated prior to identification in order to fractionate very complex protein mixtures and analyze each protein fraction separately. As the complexity of environmental samples is very high with proteins present at widely spread concentration ranges, this strategy is often chosen for metaproteomic studies. Protein separation can be achieved by different methods among which the most commonly applied are electrophoresis on acrylamide gel (1D-polyacrylamide gel electrophoresis, SDS-PAGE, or 2D-gel-electrophoresis, 2D-GE), capillary electrophoresis (CE), or liquid chromatography (LC; Fig. 1). Historically, 2D-GE separation was most successfully applied in proteomics studies (Rabilloud et al. 2010). Proteins are separated in a first step according to their pI and in a second step as a function of their molecular weight. After separation, proteins are usually visualized with an organic dye (Coomassie blue), by reduction of a metal salt (silver nitrate) or a fluorescent labeling (Sypro, DeepPurple, ...). However, the 2D-GE approach has several limitations including the dynamic range of spots visible on the gels and the loss of membrane proteins due to their low solubility in 2D-GE solvents. For the preparation of membrane proteins requiring specific detergents, one-dimensional SDS-PAGE is often preferred (Laemmli 1970). Besides electrophoresis, gel-free strategies are now also commonly used to separate proteins such as multi-dimensional liquid or affinity chromatographies (Gundry et al. 2009). Many types and combinations of chromatographies have been explored in the field of proteomics to fractionate complex protein mixtures prior to identification. Each of these approaches has advantages and disadvantages and the choice of the best adapted separation method is highly dependent on the sample type to be analyzed. Early environmental proteomics researches used 2D-GE to separate proteins, but the achieved resolution was low (Wilmes and Bond 2004; Lacerda et al. 2007; Bruneel et al.2011). Recent protocols were optimized to obtain better 2D-GE performances (Kuhn et al.2011). SDS-PAGE or LC-MS/MS were also used (Benndorf et al. 2007; Wilmes et al. 2008a, b) allowing the authors to identify at least 10 times more proteins. In some cases, the high level of diversity of the communities and/or the wide dynamic range of abundances (both at the species as well as protein levels) make metaproteomics approaches rather difficult to apply. Therefore, for several years, metaproteomics studies were only successful when applied to communities with low levels of diversity. When a high level of diversity was observed, only the most abundant proteins could be identified. After protein separation and enzymatic digestion, the generated peptides are analyzed by MS and most commonly tandem MS (MS/MS) in order to precisely measure peptide masses and their associated fragments (in MS/MS mode).

Protein identification

The protein identification step of the general proteomics analysis workflow relies on the availability of high performance MS instrumentation generating sensitive and high resolution data but certainly also on sophisticated bioinformatic tools to extract most useful information from the generated data. The most widely used approach for protein identification is based on a prior enzymatic digestion of proteins into peptides and is thus called the "Bottom-up" strategy. "Shotgun" proteomics approaches can be used (Fig. 1) consisting in the straight enzymatic digestion of the complex protein mixtures without any prior fractionation; trypsin is the most commonly used enzyme which specifically cleaves the polypeptide chain after lysine and arginine residues. The peptides obtained are chromatographically separated by reverse phase liquid chromatography (LC) and analyzed by mass spectrometry (MS; Nesvizhskii 2010). Alternatively, a MudPIT (multidimensional protein identification technology) strategy can be applied, involving the use of multiple and various chromatographic separations of the peptides prior to their injection in the MS (Fränzel and Wolters 2011). The generated peptides are then analyzed by high-resolution mass spectrometers (MS), capable of ionizing and measuring peptide masses with a sub-ppm mass accuracy.

The general protein identification principle is based on the comparison of experimental masses to theoretical masses calculated from proteins present in protein sequence databases. Historically, the Peptide Mass Fingerprint (PMF) approach, based on tryptic peptides' masses measurements and their comparison with peptides' masses predicted from protein databases, was used to identify proteins. Though, this approach rapidly appeared to be not specific enough to unambiguously identify proteins because of the exponentially growing size of protein sequence databases. Nowadays, routine identification of proteins requires additional tandem MS (MS/MS) data, which allows accurately determining peptides' masses and their associated fragments (MS/MS). Numerous algorithms have been developed over the last 10 years to extract most reliable identifications from large-scale and high-throughput LC-MS/MS data by matching experimental mass lists and MS/MS spectra with in silico calculated mass lists (Nesvizhskii 2010). Additionally, strategies to evaluate false discovery rates (FDR) using target-decoy databases have been implemented in order to constantly attest for the overall quality of protein identifications (Elias and Gygi 2007). With the constant increase of the MS/MS sequencing pace of mass spectrometers and of the size of protein sequence databases (as a direct consequence of next generation DNA sequencing and automatic annotation of genomes), the MS/MS data interpretation step becomes highly informatic-resources consuming. In order to circumvent this bottleneck that MS/MS data interpretation commonly represents, software using distributed computing resources such as cloud or grid solutions have been developed (Carapito et al. 2014).

When analyzing a bacterial community as a whole, the genomes of the studied organisms are often unknown. In this case, identifications can be performed by using de novo sequencing that consists in the interpretation of MS/MS spectra to derive an amino acid sequence tag from each individual MS/MS spectrum. The derived sequence tags are then subjected to MS-BLAST (http://dove.embl-heidelberg.de/Blast2/msblast.html) to search for sequence homologies with orthologous proteins present in the databases (Carapito et al. 2006). This identification process is more time consuming, less high throughput than classical protein database searches and often requires manual verification of sequence tags and MS/MS spectra quality. Additionally, potential post-translational modifications (e.g., glycosylation or phosphorylation) may also complicate de novo sequencing. However, latest generation MS instruments resolutions, accuracies, and fragmentation quality, together with enhanced algorithms allow increasing the throughput and successful identification rates and reducing the FDR of de novo sequencing approaches (Carapito et al. 2014). Once peptide sequences have been successfully identified by either one of the previous approaches, they have to be grouped to protein functions and in this step resides the big challenge of protein inference that is especially complex when rich bacterial communities, including many close organisms, are studied (Seifert et al. 2013).

In order to improve protein identification, metaproteomics and metagenomics are nowadays often combined (Seifert et al. 2013). It is crucial however that the metagenomics data are of good quality allowing a robust assembling of sequences in order to obtain a large quantity of gene sequences and almost complete sequences for each gene. The recent transition in DNA sequencing from Sanger to next generation sequencing (NGS) approaches may reduce the efficiency of metaproteomics identification rates, since the assembly of genomic sequences via NGS is limited due to the short reads obtained. In some cases, although genome information is available, the identification of proteins can be compromised due to errors in sequence databases (frameshift, false start codon prediction), which is often the consequence of the automatic annotations of genomes. To avoid these problems resulting from genome data that are not expertly annotated, identification strategies to interpret MS/MS data directly in the complete, unannotated genome sequence have been developed. These approaches are at the origin of the proteogenomics field, defined as the use of proteomics results to improve genome annotations (Delalande et al. 2005; Gallien et al. 2009; Armengaud et al. 2014). Such an approach may be combined with labeling methods such as specific labeling of protein N-termini, allowing the identification of peptides corresponding to protein starts. For instance, such an N-terminomics strategy has been used to check or correct errors of translation initiation codon prediction or to validate signal peptides for secreted proteins in the case of microorganisms belonging to poorly documented genera (Bertaccini et al. 2013).

Protein quantification

Quantitative methods can be either relative or absolute (Fig. 2). Relative quantitative studies aim at determining the relative amounts of given proteins across different protein extracts, for instance across different growth conditions. A wide range of methods for relative quantification that differ in their accuracy, applicability, and sensitivity have been developed. Relative quantification methods do not allow intra-sample comparisons of abundances of peptide A to peptide B for example, as LC-MS response factors are highly variable from one peptide to another and depend on the peptides' sequences. On the contrary, absolute quantification implies that concentrations of peptides/proteins are determined thanks to the comparison of signal intensities with highly purified/quantified isotopically labeled peptides/proteins, ideally spiked into the samples in known amounts. Absolute quantification methods are most precise and robust when applied with limited sample fractionation, thus preferentially using gel-free approaches as described below. The most widely used approach today for absolute quantification is targeted proteomics using selected or multiple reaction monitoring mass spectrometry (SRM or MRM) as described at the end of this review (Fig. 2).

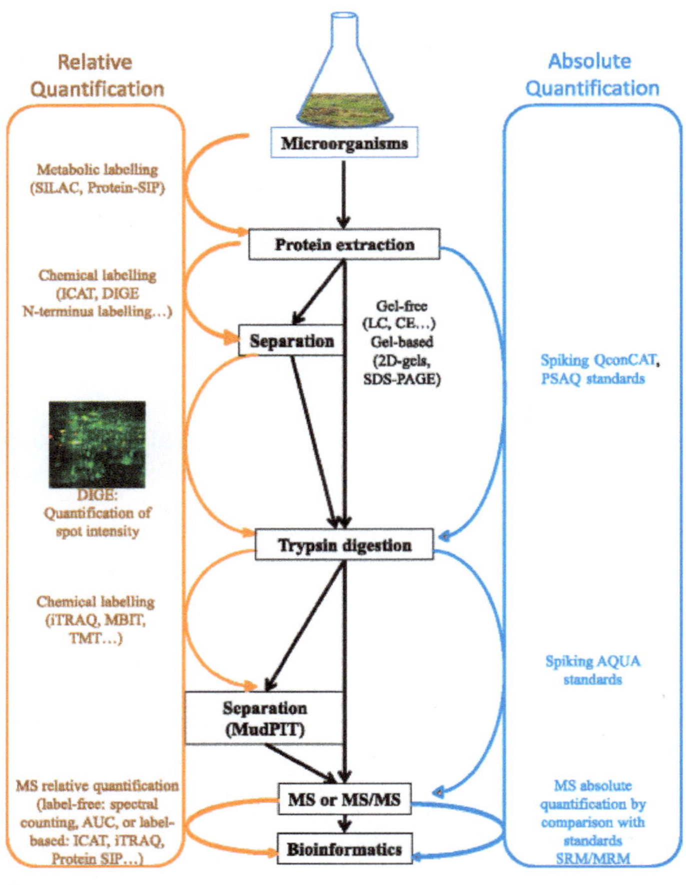

Figure 2: Protein quantification strategies. Quantitative methods can be either relative or absolute

Relative label-free quantification methods

Label-free quantitative approaches compare peptide spectral counts or peptide intensities between LC-MS/MS runs of samples (Nahnsen et al. 2013). The first approach, called spectral counting, relies on counting the number of MS/MS spectra acquired for a given protein. The second approach, called feature-based quantification, consists in calculating the relative intensities of extracted ion chromatograms (area under the curve, AUC) obtained using computational tools. These two approaches assume that the number of peptides identified for each protein, as well as their chromatographic intensity reflects the amount of protein present in each sample. Such approaches are useful when large numbers of samples are studied. However, they require high-quality and highly reproducible LC-MS. Because the samples are analyzed separately, this quantification may not be appropriate when small variations of protein amounts have to be observed between samples. Label-free methods have been used to quantify proteins from metaproteomic samples, using either signal intensities measured by MS for each peptide, spectral counts, or normalized spectral abundance factors (NSAF). For example, the amounts of proteins were compared using AUC in communities in aquifer during biostimulation (Callister et al. 2010) or by spectral counting on the same type of community (Wilkins et al. 2013). Such quantifications have also been performed to study the community found in oceans, in response to different nutrient concentrations (Morris et al. 2010), in AMD biofilm samples (Mueller et al. 2010), during litter decomposition (Schneider et al. 2012; Table 1). Recently, normalized intensities for each protein were calculated to quantify key markers in the field of human microbiome studies (Ferrer et al. 2013). Label-free quantification has also been used to estimate the efficiency of protein extraction in soil (Keiblinger et al. 2012).

Table 1: Examples of recent metaproteomics studies

Objective	Type of ecosystems	Type of proteomics approach	Highlights	Reference
Community structure	Contaminated environments	1D-GE or 2D-GE + LC-MS/MS	Analysis of the active community	Bruneel et al. (2011), Halter et al. (2011)
	AMD biofilm	Shotgun LC-MS/MS	Deciphering the differences between each ecotypes	Denef et al. (2010), Denef and Banfield (2012)
Functioning of ecosystems	Marine ecosystems (Ocean)	Shotgun LC-MS/MS	Effect of nutrient gradient on the community and expressed proteins	Morris et al. (2010), Sowell et al. (2011), Williams et al. (2013)
	Freshwater (Lake)	1D-GE + LC-MS/MS	Role of bacteria in carbon sulphur and nitrogen cycles	Ng et al. (2010), Lauro et al. (2011)
	Sediment impacted by AMDs	1D-GE + LC-MS/MS	Role of microorganisms in remediation, metaproteogenomics	Bertin et al. (2011), Halter et al. (2012)
	Forest soil	1D-GE + LC-MS/MS	Role of Fungi on leaf litter decomposition	Schneider et al. (2012)
	Crop Rhizospheric soil	2D-GE + LC-MS/MS	Interaction between plants and microorganisms	Wang et al. (2011)
	Semiarid solid	1D-GE + LC-MS/MS	Importance of the carbon and nitrogen fixation, nitrification, and photosynthesis	Bastida et al. (2014)
	Human microbiome	Shotgun LC-MS/MS	Markers of healthy and diseased state	Erickson et al. (2012), Ferrer et al. (2013)
Quantitative proteomics	Biofilms	Protein-SIP	Spatial and temporal variation in biofilm	Pan et al. (2011), Mueller et al. (2011), Belnap et al. (2011), Justice et al. (2014)
	Methanogenic consortium	Protein-SIP	Carbon flow	Morris et al. (2012)

Relative label-based quantification approaches

The 2D-GE approach is a standard proteomic approach to study relative differences in protein expression patterns when one organism is incubated in two different conditions, and therefore its adaptation capacities (Fig. 1). Because Coomassie-based staining and fluorescent labeling is correlated to the amount of proteins, it is possible to compare the protein amounts in extracts obtained in two different culture conditions by image analyses. To optimize such a comparison, the 2D Fluorescence Difference Gel Electrophoresis (DIGE) technology was developed 15 years ago (Unlü et al. 1997; Minden et al. 2009). This approach is based on the labeling of different protein extracts with different cyanines. Up to three protein samples are labeled with fluorescent dyes (for example Cy3, Cy5, Cy2) and mixed together prior to two-dimensional electrophoresis. After 2D-GE, proteins are detected by scanning the gel at excitation wavelength of each specific dye allowing the detection of each sample separately. This approach limits inter-gel variation of migration and is therefore more robust than Coomassie-stained 2D-GE. Such an approach was recently applied to community proteomics to analyze proteins specifically expressed in the presence of phenanthrene (Cébron et al. 2014).

Another way to estimate the amount of one protein in two different conditions consists in using metabolic labeling. The SILAC (stable isotopic labeling by amino acids in cell culture) approach is an example of a metabolic labeling approach in which proteins are labeled before protein extraction (Geiger et al. 2011). Incorporation of a stable isotope from any substrate labeled with^{13}C, ^{15}N, or ^{36}S into proteins is used in proteomic analysis of populations or communities and called protein-based stable isotope probing (protein-SIP; Seifert et al. 2012; von Bergen et al.2013). One substrate is replaced in the cell culture by a similar substrate but with substituted stable isotopic nuclei (e.g. deuterium, ^{13}C, ^{15}N). Two cell populations can then be compared as they are grown in culture media that are identical except that one of them contains a "light" and the other a "heavy" form of this particular substrate. After incorporation of the substituted stable isotopic nuclei into the proteins, and protein extraction, proteins from both cell populations are combined and analyzed together by MS. Pairs of chemically identical peptides of different stable-isotope composition can be differentiated in a mass spectrometer owing to their mass difference. The quantification is based on the intensity of labeled versus non-labeled peptide signals (relative isotope abundance, RIA, labeling ratio, LR, or shape of isotope pattern, (von Bergen et al. 2013)). Metabolic labeling from an isotopically labeled substrate (Protein-SIP) was used to study differences in microcosm community structure or composition (Table 1) and allowed studying the metabolic activity of the members within an ecosystem by tracking nutrient flow (von Bergen et al. 2013). Indeed, active microorganisms present in a community incorporate labeled substrates into proteins and the identification of labeled proteins gives information on the structure of the active community and on the change in its structure. Such an approach has been used for example to study spatial and temporal variation in biofilm structures (Pan et al. 2011; Mueller et al. 2011; Belnap et al. 2011; Justice et al. 2014) or to test the carbon flow in a methanogenic consortium (Morris et al. 2012).

Other methods consist in labeling samples with tags or stable isotopes (Christoforou and Lilley2012) and in analyzing these labeled samples using a gel-free strategy. For example, the principle of isobaric tagging approaches is to compare peptides obtained from two samples that differ only in their isotopic composition but behave identically during sample preparation, separation, and MS analysis. The technique called "isobaric tags for relative and absolute quantification" (iTRAQ) is such a multiplexed protein quantification technique (Evans et al.2012). Peptides from protein digestions are covalently labeled on N-termini and side chain amines with tags of varying fragment masses. These labeled samples are analyzed by nanoLC-MS/MS to identify labeled peptides. The fragmentation of the attached tag generates a low molecular mass reporter ion that can be used to relatively quantify the peptides and the proteins from which they originate. Although the term "absolute" is part of its name, this approach is a relative quantification method (Evans et al. 2012). Recently, such an approach has been combined with a DIGE approach to study the protein expression in the salivary glands of *Ixodes ricinus* ticks infected by various strains of *Borrelia burgdorferi* (Cotté et al. 2014). Similar approaches are the mass-balanced, $^{1}H/^{2}H$ isotope-coded dipeptide tag (MBIT) or the quantitative proteomics with tandem mass tags approach (TMT®; Thompson et al. 2003; Yoon et al. 2014). In these techniques, several tags can be used that contain four regions, namely a mass reporter region, a cleavable linker region, a mass normalization region, and a protein reactive group. The chemical structures of all the tags are identical but each contains isotopes substituted at various positions in such a way that the mass reporter and mass normalization regions have different molecular masses in each tag. The tags can therefore be distinguished after MS/MS fragmentation. The isotope-coded affinity tag (ICAT) approach also consists in tagging proteins with isotope-coded tags (Patton et al. 2002). ICAT reagents consist of a protein-reactive group, a linker region, and a biotin tag. Two different protein samples could therefore be labeled with different tags, combined for digestion, and the labeled peptides are enriched by affinity chromatography against the tag which is part of the labeling. These enriched peptides are then analyzed by LC-MS/MS together and differential mass-tagged peptide pairs are quantified to determine the relative levels of proteins from two samples.

Examples of proteomics approaches used in environmental microbiology

Several studies have been recently performed to characterize and quantify proteins expressed by microorganisms present within various ecosystems, such as soil, marine, and freshwater environments but also human or animal microbiome and plant-associated microorganisms (Siggins et al. 2012b). Such metaproteome datasets give important insight into microbial community structure, dynamics, and functioning (Table 1).

Proteomics is complementary to metagenomics to study community structures

The composition of a community is traditionally determined by the sequencing of the 16S rRNA gene. However, it is not always possible to affiliate bacteria using only 16S rRNA gene sequences (Schleifer 2009). Some microorganisms showing very similar 16S rRNA gene sequences turned out to belong to different taxa when other phylogenetic markers were used. Interestingly, the metaproteomics approach may give taxonomic information complementary to the 16S rRNA gene-based approach. Indeed, some identified peptides may be unique and specific to one species or subspecies, or semi-unique, i.e., specific to one genus. If a unique or semi-unique peptide is identified in a complex mixture, this information reveals that the species or members of this genus are present and active and thus provides valuable information on the community structure. The identification of signature peptides in orthologs has enabled the use of some proteins involved in conserved biological processes as taxonomic signatures. Such an approach was used to describe the active community at the genus level in the Carnoulès AMD (Bruneel et al. 2011), in a mildly arsenic contaminated creek (Halter et al. 2011) or in microcosms obtained from a community of phenanthrene-contaminated soil (Cébron et al. 2014).

Metaproteomic analysis coupled with deeply sampled community genomics has been a powerful tool to differentiate between close organisms and give a better view of the diversity found in a biofilm colonizing acid mine drainage (Lo et al. 2007; Simmons et al. 2008; Denef et al. 2009,2010). Metabolomics and metaproteomic analysis of such biofilm communities were combined and led to differentiate two bacterial species from the same genus and to highlight their specific function (Wilmes et al. 2010). Altogether strain-resolved expression patterns highlight that phylogenetically close microorganisms coexist in ecosystems, sometimes belonging to the same species but with less than 1 % divergence in their nucleotide sequences of genes encoding 16S rRNA (ecotypes). At a functional level, this microdiversity leads to functional diversity, since these strains play distinct roles (Wilmes et al. 2010; Denef et al. 2010). Moreover, several studies revealed that genome recombination occurred and is crucial for the adaptation of each ecotype and that such subtle genetic variations can lead to distinct ecological functions (Lo et al.2007; Denef et al. 2009, 2010; Denef and Banfield 2012).

Proteomic tools to study the functioning of communities

To study factors that may influence the community functioning, metaproteomics approaches were first performed on microcosms in laboratory-controlled conditions. Indeed, such an approach has been successfully used to analyze a laboratory-scale activated sludge system optimized for enhanced biological phosphorus removal (EBPR; Wilmes and Bond 2004). In those communities, proteins were identified belonging to an uncultured organism of the*Rhodocyclus* lineage known to accumulate polyphosphates (Wilmes and Bond 2004), but also originating from human or marine bacteria (Kuhn et al. 2011). Metaproteomics was recently applied to batch cultures in order to analyze an anaerobic microbial community degrading toluene or to study the effect of arsenic and phenanthrene on a bacterial community originating from an aged PAH and heavy-metal-contaminated soil (Jehmlich et al. 2010; Cébron et al.2014). Similarly, the effects of temperature and exposure to trichloroethylene (TCE) on proteins expressed by the community in laboratory-scale anaerobic conditions were analyzed (Siggins et al. 2012a). These studies illustrate that metaproteomics can be used not only to describe a community in laboratory-controlled microcosms but also to study its response to perturbations. Therefore, they highlight important functions optimizing bioengineering systems. Although such studies on laboratory-scale ecosystems are crucial to understand the functioning of these communities, studying microorganisms in laboratory conditions, even in microcosm conditions, may not reflect their particular adaptation capacities in their environmental niches. Therefore, several laboratories have optimized environmental proteomics approaches to study the functioning of microorganisms within their ecosystems. In some cases, protein abundances may not necessarily correlate with protein activities, since these crucial effectors may be present but inactive. Therefore, in order to have an integrated view of an ecosystem, metaproteomics is nowadays sometimes combined with other global analyses such as metabolomics.

Marine and freshwater ecosystems

Metagenomics datasets have revealed the high microbial diversity of marine and freshwater communities. For example, the spatial dynamics of bacterioplankton has been evaluated along the Chesapeake Bay, the largest estuary in the United States, and the proteins identified were shown to correlate with major microbial lineages, i.e., Bacteroides and Alphaproteobacteria, present in this ecosystem (Kan et al. 2005). More recently, thousands of proteins were identified originating from bacteria, archeae, and virus in several Seas or Oceans and were compared to proteins expressed in coastal water. Shift in nutrient transport, utilization, and energy transduction along a natural nutrient concentration gradient were observed, revealing different types of organisms and expressed functions, in particular, different transporters, between both ecosystems (Sowell et al. 2009, 2011; Morris et al. 2010; Williams et al. 2013). In freshwater environments, several studies have been recently performed. For example, the effects of carbon-stimulation and fermentation-based metabolism on biogeochemical cycling or bioremediation efficiency in aquifer were investigated (Wilkins et al. 2009, 2013; Callister et al. 2010; Chourey et al. 2013; Wrighton et al. 2014). Similarly, the functioning of microbial communities was analyzed in lakes such as the Ace Lake Antartica, pointing out the role of green sulphur bacteria, actinobacteria, and cyanobacteria in sulphur, carbon, and nitrogen cycles, in such ecosystems (Ng et al. 2010; Lauro et al. 2011).

Terrestrial ecosystems (soil and sediments)

Terrestrial ecosystems have been the subject of several studies to better understand carbon or nitrogen cycles or effects of toxic compounds on microbial community. In the Carnoulès arsenic-rich sediments, bacterial community analyses revealed that proteins involved in the biomineralization of iron and arsenic were expressed by *Acidothiobacillus ferrooxidans* and*Thiomonas*, respectively, which supports a major role of these microorganisms in the natural attenuation of this highly contaminated environment (Bertin et al. 2011). This approach also revealed that most proteins were expressed by uncultured microorganisms belonging to a novel phylum, i.e., "*Candidatus* Fodinabacter communificans". These bacteria may play an indirect but important role in the functioning of the ecosystem by recycling organic matter or providing other members with cofactors such as vitamins (Bertin et al. 2011). An additional study revealed that *Euglena mutabilis*, an abundant protist found in this AMD as well as in other AMDs, produces organic compounds that could serve as nutrients for bacteria (Halter et al.2012). Finally, in the Carnoulès ecosystem, the identification of a key protein involved in the natural remediation process, the rusticyanin enzyme, was crucial to complete a reconstructed genome from an *A. ferrooxidans*-like bacterium. Thus, the use of proteomic data to refine the annotation of metagenomic data is a recent application of metaproteomics and may be defined as "metaproteogenomics" (Bertin et al. 2011). Similarly, proteogenomics was useful to highlight the key role of cytochromes variants as well as their posttranslational modifications in AMD Biofilm (Singer et al. 2010).

Protein extraction has been the limiting step to develop metaproteomics approaches on soil, due to the presence of interfering substances, the high microbial diversity present in such ecosystems and the low amount of (meta)genomic information available on microorganisms present in soils (Bastida et al. 2009; Becher et al. 2013). Recently, new protocols have been developed to identify several hundreds of proteins in crop rhizospheric soil, forest soil, or semiarid soil

(Chourey et al. 2010; Wang et al. 2011; Keiblinger et al. 2012; Nicora et al. 2013; Bastida et al. 2014). Metaproteomics was also successfully used to identify the major fungi involved in leaf litter decomposition. In this study, proteins were analyzed by 1D-SDS-PAGE followed by liquid-chromatography and tandem mass-spectrometry (Schneider et al. 2012). A recent study was performed on three semiarid soils with different characteristics, testing different extraction methods (Bastida et al. 2014). This study revealed that depending on the protocols used, a taxonomic bias may be observed. Nevertheless, different community composition was observed among the three soils, and proteins involved in biogeochemical cycles of different elements were identified, revealing that the microbial communities from semiarid soils where organic carbon is limiting microbial growth, expressed proteins involved in photosynthesis, carbon and nitrogen fixation and in nitrification.

Eukaryotic host microbiomes (pathogens, symbionts, and commensals)

More recently, it has appeared that environmental proteomics may be crucial to identify not only proteins expressed in response to abiotic changes but also in response to biotic factors, such as those expressed by microbial hosts. The major difficulty in such studies is to distinguish microbial and host proteins, a difficulty that is reduced when the genome of both organisms is known. The second problem is to extract a sufficient amount of microbial proteins to be able to detect them. Successful studies allowed the identification of key proteins involved in virulence in several pathogens such as *Echinococcus granulosus* metacestode (Monteiro et al. 2010),*Clostridium perfringens* (Sengupta and Alam 2011), or *Anaplasma* or *Borrellia* when present in the tick vector (Ramabu et al. 2010; Cotté et al. 2014). Similarly, proteomics has been used to address the complex processes governing the interactions between symbiotic microorganisms and their host and vice versa, e.g., the adaptive response of plants interacting with mycorrhizae (Bona et al. 2011) and more recently proteins expressed by microorganisms existing in a crop rhizospheric soil (Wang et al. 2011).

Recent advances in metaproteomics also focused on animal or human microbiome. Metaproteomics was performed to study the higher termite hindgut microbiota (Burnum et al.2011), revealing that proteins involved in carbohydrate transport and metabolism, nitrogen fixation and assimilation, energy production, and amino-acid synthesis may play a role at least as important as those involved in cellulose degradation. The number of metaproteomics studies in the field of the human intestinal tract and oral cavity recently increased due to the progresses made to generate metagenomic data (Human Microbiome Project Consortium 2012). For example, catalogues comprising several thousands of proteins were obtained by shotgun proteomics from human fecal samples or oral cavity (Verberkmoes et al. 2009; Rooijers et al.2011; Kolmeder et al. 2012; Jagtap et al. 2012). In this field, metaproteomics will probably lead to the identification of protein markers of healthy or diseased states (Erickson et al. 2012; Juste et al. 2014), or to understand how microorganisms interact together or with their host in such ecosystems (Ferrer et al. 2013).

New developments in community proteomics: targeted proteomics approaches applied to community studies and environmental microbiology

For the past 20 years, global proteomic approaches have enabled the identification and relative quantification of ever-increasing lists of proteins, in all kinds of complex protein mixtures from plasma to cell lysates to bacterial communities. More recently, targeted proteomics approaches have been developed as an alternative to global approaches with the aim of trying to find ways to more precisely quantify a subset of proteins, even if this implies limiting the focus on a restricted number of proteins of interest. Selected or multiple reaction monitoring mass spectrometry (SRM or MRM) appeared to be the most promising approach to achieve this goal (Picotti and Aebersold 2012) and many applications emerged, mainly in the field of biomarker verification (Gillette and Carr 2013; Percy et al. 2014). To set up a targeted LC-SRM assay, the prior definition of a short list of proteins of interest is necessary. This is the main difference between global (in which the goal is to identify the largest number of peptides/proteins) and targeted approaches in which the protein targets must be defined prior to the experiment itself. Once the targets are defined, a few peptides whose sequences are specific for the target protein and visible in MS, called proteotypic peptides, will be selected to be used as tracers to quantify the corresponding protein.

Briefly, LC-SRM experiments are mainly conducted on triple quadrupole-type instruments, using the first quadrupole to select a precursor ion, the second quadrupole to fragment this precursor ion and the third quadrupole to select a specific fragment ion. A predefined pair of precursor/fragment ions is called a transition and multiple transitions are measured for each target peptide. The appropriate selection of proteotypic peptides and their most sensitive and specific (interference-free) transitions is a crucial step to reach sensitivity and specificity required for a robust and high-performance quantitative LC-SRM assay. To facilitate this selection, public libraries (atlases) built on synthetic peptides for reference proteomes, namely yeast, *Mycobacterium tuberculosis*, and human are available (http://www.srmatlas.org; Picotti et al. 2008; Schubert et al. 2013). Besides proteotypic peptide sequences for all proteins of these reference proteomes, the atlases also include important instrument parameters enabling a faster assay development, namely relative peptide retention times, optimal collision energies, or interface voltages. These instrument parameters need to be optimized in any case and this optimization is usually done on crude synthetic, ideally heavy labeled, peptides that need to be synthetized when information in the atlases is missing or incomplete. In the case of bacterial communities, no atlas is available and the possibility to get access to low-cost crude synthetic peptides is therefore very important.

A key factor for reliable quantification is the use of appropriate standards. Indeed, attempting to reach absolute quantification of peptides is possible only through simultaneous analysis by LC-SRM of endogenous peptides and labeled peptides added in known amounts. Highly purified and precisely quantified heavy labeled synthetic standards exist in several alternative forms: purified and quantified synthetic peptides (AQUA peptides, Gerber et al. 2003), concatemers of peptides (QconCAT, Beynon et al. 2005) and standard proteins biochemically identical to natural proteins (PSAQ, Protein Standard Absolute Quantification, Brun et al.2007). For precise absolute quantification, endogenous peptides of interest are quantified by calculating heavy/light peptide ratios thanks to spiked heavy-labeled synthetic peptides into the samples (Lange et al. 2008; Gallien et al. 2011; Fig. 2). Isotope dilution is used for the quantification of small molecules such as metabolites, xenobiotics, hormones, or pesticides with high precision (CV <5 %) for more than 30 years. Today, the main obstacles for the application of SRM to quantify peptides and proteins start to be overcome: (1) high sensitivity has been reached thanks to significant instrument improvements, (2) large dynamic ranges are accessible thanks to the implementation of reproducible fractionation steps prior to LC-SRM analysis, (3) multiplexing capacity has been increased thanks to significant advances in electronics, data acquisition, and software developed on triple quadrupole and high performance liquid chromatography instruments but also thanks to the implementation of retention time reference peptides workflows (Escher et al. 2012). Alternatively to LC-SRM methods, the high resolution of different instrument geometries (Quadrupole-Orbitrap or Quadrupole-Time Of Flight instruments) is explored to develop targeted methods for precise quantification. Rather than following isolated fragments from the precursors of interest as SRM does, these methods rely on the acquisition of full scan high resolution MS/MS fragmentation spectra of the peptides of interest (Gallien et al. 2012).

So far, most targeted proteomics applications dealt with biomarker studies and clinical proteomics (Hüttenhain et al. 2012; Gillette and Carr 2013; Percy et al. 2014). A very recent study demonstrated the application of LC-SRM assays to verify markers of Crohn's disease, discovered by 2D-DIGE experiments, in unfractionated gut microbiota (Juste et al. 2014). Besides, a study has demonstrated the ability to absolutely quantify proteins in complex environmental samples and mixed microbial communities (Werner et al. 2009). Such an approach will be probably extensively used in the next future to follow the activity of specific targeted bacteria within communities.

CONCLUSION

Nowadays metaproteomics is usually combined with other high-throughput "omics" methods, such as metatransciptomics, metametabolomics as well as more traditional methods of genetics, molecular biology and/or biochemistry. All generated data are combined in order to decipher ecosystem functioning and give an integrated view of biological objects in any environment, their roles, and relationships. Such global studies of microbial communities will be of great interest to investigate complex consortia and address the role of uncultivated microorganisms in microbial ecosystems. In the future, these approaches will be further improved in order to access proteins expressed by individual cells within a community, which will give important insight into the community structure and its functioning. The recent developments in quantitative and targeted community proteomics will open further possibilities to decipher which factors modify their dynamics. Therefore, proteomics should lead not only to a better understanding of ecosystems themselves, but also to the identification of new functions that can be exploited in biotechnological applications. This could lead to an optimal use of the properties of microorganisms and to a better understanding of how microorganisms colonize new ecological niches.

ACKNOWLEDGMENTS

Financial support came from the Université de Strasbourg (UdS), the Centre National de la Recherche Scientifique (CNRS), the EC_2CO program (INEE, CNRS), and the THIOFILM (ANR-12-ADAP-0013) project. Florence Arsène-Ploetze was supported by the CNRS ("mise en délégation pour activité de recherche au CNRS").

REFERENCES

1. Armengaud J, Trapp J, Pible O et al (2014) Non-model organisms, a species endangered by proteogenomics. J Proteomics 105C:5–18. doi:10.1016/j.jprot.2014.01.007
2. Bastida F, Hernández T, García C (2014) Metaproteomics of soils from semiarid environment: functional and phylogenetic information obtained with different protein extraction methods. J Proteomics 101:31–42. doi:10.1016/j.jprot.2014.02.006
3. Bastida F, Moreno JL, Nicolás C et al (2009) Soil metaproteomics: a review of an emerging environmental science. Significance, methodology and perspectives. Eur J Soil Sci 60:845–859. doi:10.1111/j.1365-2389.2009.01184.x
4. Becher D, Bernhardt J, Fuchs S, Riedel K (2013) Metaproteomics to unravel major microbial players in leaf litter and soil environments: challenges and perspectives. Proteomics 13:2895–2909. doi:10.1002/pmic.201300095
5. Belnap CP, Pan C, Denef VJ et al (2011) Quantitative proteomic analyses of the response of acidophilic microbial communities to different pH conditions. ISME J 5:1152–1161. doi:10.1038/ismej.2010.200
6. Benndorf D, Balcke GU, Harms H, von Bergen M (2007) Functional metaproteome analysis of protein extracts from contaminated soil and groundwater. ISME J 1:224–234. doi:10.1038/ismej.2007.39
7. Bertaccini D, Vaca S, Carapito C et al (2013) An improved stable isotope N-terminal labeling approach with light/heavy TMPP to automate proteogenomics data validation: dN-TOP. J Proteome Res 12:3063–3070. doi:10.1021/pr4002993
8. Bertin PN, Heinrich-Salmeron A, Pelletier E et al (2011) Metabolic diversity among main microorganisms inside an arsenic-rich ecosystem revealed by meta- and proteo-genomics. ISME J. doi:10.1038/ismej.2011.51
9. Bertin PN, Medigue C, Normand P (2008) Advances in environmental genomics: towards an integrated view of micro-organisms and ecosystems. Microbiology 154:347–359
10. Beynon RJ, Doherty MK, Pratt JM, Gaskell SJ (2005) Multiplexed absolute quantification in proteomics using artificial QCAT proteins of concatenated signature peptides. Nat Methods 2:587–589. doi:10.1038/nmeth774

11. Bona E, Marsano F, Massa N et al (2011) Proteomic analysis as a tool for investigating arsenic stress in *Pteris vittata* roots colonized or not by arbuscular mycorrhizal symbiosis. J Proteomics 74:1338–1350. doi:10.1016/j.jprot.2011.03.027
12. Bruneel O, Volant A, Gallien S et al (2011) Characterization of the active bacterial community involved in natural attenuation processes in arsenic-rich creek sediments. Microb Ecol 61:793–810. doi:10.1007/s00248-011-9808-9
13. Brun V, Dupuis A, Adrait A et al (2007) Isotope-labeled protein standards: toward absolute quantitative proteomics. Mol Cell Proteomics 6:2139–2149. doi:10.1074/mcp. M700163-MCP200
14. Burnum KE, Callister SJ, Nicora CD et al (2011) Proteome insights into the symbiotic relationship between a captive colony of *Nasutitermes corniger* and its hindgut microbiome. ISME J 5:161–164. doi:10.1038/ismej.2010.97
15. Callister SJ, Wilkins MJ, Nicora CD et al (2010) Analysis of biostimulated microbial communities from two field experiments reveals temporal and spatial differences in proteome profiles. Environ Sci Technol 44:8897–8903. doi:10.1021/es101029f
16. Cañas B, Piñeiro C, Calvo E et al (2007) Trends in sample preparation for classical and second generation proteomics. J Chromatogr A 1153:235–258. doi:10.1016/j.chroma.2007.01.045
17. Carapito C, Burel A, Guterl P et al (2014) MSDA, a proteomics software suite for in-depth Mass Spectrometry Data Analysis using grid computing. Proteomics 14:1014–1019. doi:10.1002/pmic.201300415
18. Carapito C, Muller D, Turlin E et al (2006) Identification of genes and proteins involved in the pleiotropic response to arsenic stress in *Caenibacter arsenoxydans*, a metalloresistant beta-proteobacterium with an unsequenced genome. Biochimie 88:595–606. doi:10.1016/j.biochi.2005.11.004
19. Cébron A, Arsène-Ploetze F, Bauda P et al (2014) Rapid impact of phenanthrene and arsenic on bacterial community structure and activities in sand batches. Microb Ecol 67:129–144. doi:10.1007/s00248-013-0313-1
20. Chourey K, Jansson J, VerBerkmoes N et al (2010) Direct cellular lysis/protein extraction protocol for soil metaproteomics. J Proteome Res 9:6615–6622. doi:10.1021/pr100787q
21. Chourey K, Nissen S, Vishnivetskaya T et al (2013) Environmental proteomics reveals early microbial community responses to biostimulation at a uranium- and nitrate-contaminated site. Proteomics 13:2921–2930. doi:10.1002/pmic.201300155
22. Christoforou AL, Lilley KS (2012) Isobaric tagging approaches in quantitative proteomics: the ups and downs. Anal Bioanal Chem 404:1029–1037. doi:10.1007/s00216-012-6012-9
23. Cotté V, Sabatier L, Schnell G et al (2014) Differential expression of *Ixodes ricinus* salivary gland proteins in the presence of the *Borrelia burgdorferi sensu lato* complex. J Proteomics 96:29–43. doi:10.1016/j.jprot.2013.10.033
24. Delalande F, Carapito C, Brizard J-P et al (2005) Multigenic families and proteomics: extended protein characterization as a tool for paralog gene identification. Proteomics 5:450–460. doi:10.1002/pmic.200400954
25. Denef VJ, Banfield JF (2012) In situ evolutionary rate measurements show ecological success of recently emerged bacterial hybrids. Science 336:462–466. doi:10.1126/science.1218389
26. Denef VJ, Kalnejais LH, Mueller RS et al (2010) Proteogenomic basis for ecological divergence of closely related bacteria in natural acidophilic microbial communities. Proc Natl Acad Sci U S A 107:2383–2390. doi:10.1073/pnas.0907041107
27. Denef VJ, VerBerkmoes NC, Shah MB et al (2009) Proteomics-inferred genome typing (PIGT) demonstrates inter-population recombination as a strategy for environmental adaptation. Environ Microbiol 11:313–325. doi:10.1111/j.1462-2920.2008.01769.x
28. Elias JE, Gygi SP (2007) Target-decoy search strategy for increased confidence in large-scale protein identifications by mass spectrometry. Nat Methods 4:207–214. doi:10.1038/nmeth1019
29. Erickson AR, Cantarel BL, Lamendella R et al (2012) Integrated metagenomics/metaproteomics reveals human host-microbiota signatures of Crohn's disease. PLoS ONE 7:e49138. doi:10.1371/journal.pone.0049138
30. Escher C, Reiter L, MacLean B et al (2012) Using iRT, a normalized retention time for more targeted measurement of peptides. Proteomics 12:1111–1121. doi:10.1002/pmic.201100463
31. Evans C, Noirel J, Ow SY et al (2012) An insight into iTRAQ: where do we stand now? Anal Bioanal Chem 404:1011–1027. doi:10.1007/s00216-012-5918-6
32. Ferrer M, Ruiz A, Lanza F et al (2013) Microbiota from the distal guts of lean and obese adolescents exhibit partial functional redundancy besides clear differences in community structure. Environ Microbiol 15:211–226. doi:10.1111/j.1462-2920.2012.02845.x
33. Fränzel B, Wolters DA (2011) Advanced MudPIT as a next step towards high proteome coverage. Proteomics. doi:10.1002/pmic.201100056
34. Gallien S, Duriez E, Crone C et al (2012) Targeted proteomic quantification on quadrupole-orbitrap mass spectrometer. Mol Cell Proteomics 11:1709–1723. doi:10.1074/mcp.O112.019802
35. Gallien S, Duriez E, Domon B (2011) Selected reaction monitoring applied to proteomics. J Mass Spectrom 46:298–312. doi:10.1002/jms.1895
36. Gallien S, Perrodou E, Carapito C et al (2009) Ortho-proteogenomics: multiple proteomes investigation through orthology and a new MS-based protocol. Genome Res 19:128–135. doi:10.1101/gr.081901.108
37. Geiger T, Wisniewski JR, Cox J et al (2011) Use of stable isotope labeling by amino acids in cell culture as a spike-in standard in quantitative proteomics. Nat Protoc

6:147–157. doi:10.1038/nprot.2010.192

38. Gerber SA, Rush J, Stemman O et al (2003) Absolute quantification of proteins and phosphoproteins from cell lysates by tandem MS. Proc Natl Acad Sci U S A 100:6940–6945. doi:10.1073/pnas.0832254100
39. Gillette MA, Carr SA (2013) Quantitative analysis of peptides and proteins in biomedicine by targeted mass spectrometry. Nat Methods 10:28–34. doi:10.1038/nmeth.2309
40. Gundry RL, White MY, Murray CI et al (2009) Preparation of proteins and peptides for mass spectrometry analysis in a bottom-up proteomics workflow. Curr Protoc Mol Biol Chapter 10:Unit10.25. doi:10.1002/0471142727.mb1025s88
41. Halter D, Cordi A, Gribaldo S et al (2011) Taxonomic and functional prokaryote diversity in mildly arsenic-contaminated sediments. Res Microbiol. doi:10.1016/j.resmic.2011.06.001
42. Halter D, Goulhen-Chollet F, Gallien S et al (2012) In situ proteo-metabolomics reveals metabolite secretion by the acid mine drainage bio-indicator, *Euglena mutabilis*. ISME J 6:1391–1402. doi:10.1038/ismej.2011.198
43. Hettich RL, Pan C, Chourey K, Giannone RJ (2013) Metaproteomics: harnessing the power of high performance mass spectrometry to identify the suite of proteins that control metabolic activities in microbial communities. Anal Chem 85:4203–4214. doi:10.1021/ac303053e
44. Hettich RL, Sharma R, Chourey K, Giannone RJ (2012) Microbial metaproteomics: identifying the repertoire of proteins that microorganisms use to compete and cooperate in complex environmental communities. Curr Opin Microbiol 15:373–380. doi:10.1016/j.mib.2012.04.008
45. Human Microbiome Project Consortium (2012) Structure, function and diversity of the healthy human microbiome. Nature 486:207–214. doi:10.1038/nature11234
46. Hüttenhain R, Soste M, Selevsek N et al (2012) Reproducible quantification of cancer-associated proteins in body fluids using targeted proteomics. Sci Transl Med 4:142ra94. doi:10.1126/scitranslmed.3003989
47. Jagtap P, McGowan T, Bandhakavi S et al (2012) Deep metaproteomic analysis of human salivary supernatant. Proteomics 12:992–1001. doi:10.1002/pmic.201100503
48. Jehmlich N, Kleinsteuber S, Vogt C et al (2010) Phylogenetic and proteomic analysis of an anaerobic toluene-degrading community. J Appl Microbiol 109:1937–1945. doi:10.1111/j.1365-2672.2010.04823.x
49. Juste C, Kreil DP, Beauvallet C et al (2014) Bacterial protein signals are associated with Crohn's disease. Gut. doi:10.1136/gutjnl-2012-303786
50. Justice NB, Li Z, Wang Y et al (2014) (15) N- and (2) H proteomic stable isotope probing links nitrogen flow to archaeal heterotrophic activity. Environ Microbiol. doi:10.1111/1462-2920.12488
51. Kan J, Hanson TE, Ginter JM et al (2005) Metaproteomic analysis of Chesapeake Bay microbial communities. Saline Syst 1:7. doi:10.1186/1746-1448-1-7
52. Keiblinger KM, Wilhartitz IC, Schneider T et al (2012) Soil metaproteomics—comparative evaluation of protein extraction protocols. Soil Biol Biochem 54:14–24. doi:10.1016/j.soilbio.2012.05.014
53. Keller M, Hettich R (2009) Environmental proteomics: a paradigm shift in characterizing microbial activities at the molecular level. Microbiol Mol Biol Rev 73:62–70. doi:10.1128/MMBR. 00028-08
54. Kolmeder CA, de Been M, Nikkilä J et al (2012) Comparative metaproteomics and diversity analysis of human intestinal microbiota testifies for its temporal stability and expression of core functions. PLoS One 7:e29913. doi:10.1371/journal.pone.0029913
55. Kuhn R, Benndorf D, Rapp E et al (2011) Metaproteome analysis of sewage sludge from membrane bioreactors. Proteomics 11:2738–2744. doi:10.1002/pmic.201000590
56. Lacerda CMR, Choe LH, Reardon KF (2007) Metaproteomic analysis of a bacterial community response to cadmium exposure. J Proteome Res 6:1145–1152. doi:10.1021/pr060477v
57. Lacerda CMR, Reardon KF (2009) Environmental proteomics: applications of proteome profiling in environmental microbiology and biotechnology. Brief Funct Genom Proteomics 8:75–87. doi:10.1093/bfgp/elp005
58. Laemmli UK (1970) Cleavage of structural proteins during the assembly of the head of bacteriophage T4. Nature 227:680–685
59. Lange V, Picotti P, Domon B, Aebersold R (2008) Selected reaction monitoring for quantitative proteomics: a tutorial. Mol Syst Biol 4:222. doi:10.1038/msb.2008.61
60. Lasken RS (2012) Genomic sequencing of uncultured microorganisms from single cells. Nat Rev Microbiol 10:631–640. doi:10.1038/nrmicro2857
61. Lauro FM, DeMaere MZ, Yau S et al (2011) An integrative study of a meromictic lake ecosystem in Antarctica. ISME J 5:879–895. doi:10.1038/ismej.2010.185
62. Lo I, Denef VJ, Verberkmoes NC et al (2007) Strain-resolved community proteomics reveals recombining genomes of acidophilic bacteria. Nature 446:537–541. doi:10.1038/nature05624
63. Mary I, Oliver A, Skipp P et al (2010) Metaproteomic and metagenomic analyses of defined oceanic microbial populations using microwave cell fixation and flow cytometric sorting. FEMS Microbiol Ecol 74:10–18. doi:10.1111/j.1574-6941.2010.00927.x
64. McCann JC, Wickersham TA, Loor JJ (2014) High-throughput methods redefine the rumen microbiome and its relationship with nutrition and metabolism. Bioinforma Biol Insights 8:109–125. doi:10.4137/BBI.S15389
65. Minden JS, Dowd SR, Meyer HE, Stühler K (2009) Difference gel electrophoresis. Electrophoresis 30:S156–S161. doi:10.1002/elps.200900098
66. Monteiro KM, de Carvalho MO, Zaha A, Ferreira HB (2010) Proteomic analysis of the*Echinococcus granulosus* metacestode during infection of its

intermediate host. Proteomics 10:1985–1999. doi:10.1002/pmic.200900506

67. Morris BEL, Herbst F-A, Bastida F et al (2012) Microbial interactions during residual oil and n-fatty acid metabolism by a methanogenic consortium. Environ Microbiol Rep 4:297–306. doi:10.1111/j.1758-2229.2012.00333.x

68. Morris RM, Nunn BL, Frazar C et al (2010) Comparative metaproteomics reveals ocean-scale shifts in microbial nutrient utilization and energy transduction. ISME J 4:673–685. doi:10.1038/ismej.2010.4

69. Mueller RS, Denef VJ, Kalnejais LH et al (2010) Ecological distribution and population physiology defined by proteomics in a natural microbial community. Mol Syst Biol 6:374. doi:10.1038/msb.2010.30

70. Mueller RS, Dill BD, Pan C et al (2011) Proteome changes in the initial bacterial colonist during ecological succession in an acid mine drainage biofilm community. Environ Microbiol 13:2279–2292. doi:10.1111/j.1462-2920.2011.02486.x

71. Muller D, Médigue C, Koechler S et al (2007) A tale of two oxidation states: bacterial colonization of arsenic-rich environments. PLoS Genet 3:e53. doi:10.1371/journal.pgen.0030053

72. Nahnsen S, Bielow C, Reinert K, Kohlbacher O (2013) Tools for label-free peptide quantification. Mol Cell Proteomics 12:549–556. doi:10.1074/mcp.R112.025163

73. Nesvizhskii AI (2010) A survey of computational methods and error rate estimation procedures for peptide and protein identification in shotgun proteomics. J Proteomics 73:2092–2123. doi:10.1016/j.jprot.2010.08.009

74. Ng C, DeMaere MZ, Williams TJ et al (2010) Metaproteogenomic analysis of a dominant green sulfur bacterium from Ace Lake, Antarctica. ISME J 4:1002–1019. doi:10.1038/ismej.2010.28

75. Nicora CD, Anderson BJ, Callister SJ et al (2013) Amino acid treatment enhances protein recovery from sediment and soils for metaproteomic studies. Proteomics 13:2776–2785. doi:10.1002/pmic.201300003

76. Pan C, Fischer CR, Hyatt D et al (2011) Quantitative tracking of isotope flows in proteomes of microbial communities. Mol Cell Proteomics 10:M110.006049. doi:10.1074/mcp.M110.006049

77. Parkhill J (2013) What has high-throughput sequencing ever done for us? Nat Rev Microbiol 11:664–665. doi:10.1038/nrmicro3112

78. Patton WF, Schulenberg B, Steinberg TH (2002) Two-dimensional gel electrophoresis; better than a poke in the ICAT? Curr Opin Biotechnol 13:321–328. doi:10.1016/S0958-1669(02)00333-6

79. Percy AJ, Chambers AG, Yang J et al (2014) Advances in multiplexed MRM-based protein biomarker quantitation toward clinical utility. Biochim Biophys Acta 1844:917–926. doi:10.1016/j.bbapap.2013.06.008

80. Picotti P, Aebersold R (2012) Selected reaction monitoring-based proteomics: workflows, potential, pitfalls and future directions. Nat Methods 9:555–566. doi:10.1038/nmeth.2015

81. Picotti P, Lam H, Campbell D et al (2008) A database of mass spectrometric assays for the yeast proteome. Nat Methods 5:913–914. doi:10.1038/nmeth1108-913

82. Rabilloud T, Chevallet M, Luche S, Lelong C (2010) Two-dimensional gel electrophoresis in proteomics: past, present and future. J Proteome 73:2064–2077. doi:10.1016/j.jprot.2010.05.016

83. Ramabu SS, Ueti MW, Brayton KA et al (2010) Identification of *Anaplasma marginale* proteins specifically upregulated during colonization of the tick vector. Infect Immun 78:3047–3052. doi:10.1128/IAI.00300-10

84. Ram RJ, Verberkmoes NC, Thelen MP et al (2005) Community proteomics of a natural microbial biofilm. Science 308:1915–1920. doi:10.1126/science. 1109070

85. Rooijers K, Kolmeder C, Juste C et al (2011) An iterative workflow for mining the human intestinal metaproteome. BMC Genomics 12:6. doi:10.1186/1471-2164-12-6

86. Schleifer KH (2009) Classification of Bacteria and Archaea: past, present and future. Syst Appl Microbiol 32:533–542. doi:10.1016/j.syapm.2009.09.002

87. Schleper C, Jurgens G, Jonuscheit M (2005) Genomic studies of uncultivated archaea. Nat Rev Microbiol 3:479–488. doi:10.1038/nrmicro1159

88. Schneider T, Keiblinger KM, Schmid E et al (2012) Who is who in litter decomposition? Metaproteomics reveals major microbial players and their biogeochemical functions. ISME J 6:1749–1762. doi:10.1038/ismej.2012.11

89. Schubert OT, Mouritsen J, Ludwig C et al (2013) The Mtb proteome library: a resource of assays to quantify the complete proteome of *Mycobacterium tuberculosis*. Cell Host Microbe 13:602–612. doi:10.1016/j.chom.2013.04.008

90. Seifert J, Herbst F-A, Halkjaer Nielsen P et al (2013) Bioinformatic progress and applications in metaproteogenomics for bridging the gap between genomic sequences and metabolic functions in microbial communities. Proteomics 13:2786–2804. doi:10.1002/pmic.201200566

91. Seifert J, Taubert M, Jehmlich N et al (2012) Protein-based stable isotope probing (protein-SIP) in functional metaproteomics. Mass Spectrom Rev 31:683–697. doi:10.1002/mas.21346

92. Sengupta N, Alam SI (2011) In vivo studies of *Clostridium perfringens* in mouse gas gangrene model. Curr Microbiol 62:999–1008. doi:10.1007/s00284-010-9821-0

93. Siggins A, Enright A-M, Abram F et al (2012a) Impact of trichloroethylene exposure on the microbial diversity and protein expression in anaerobic granular biomass at 37 °C and 15 °C. Archaea 2012:940159. doi:10.1155/2012/940159

94. Siggins A, Gunnigle E, Abram F (2012b) Exploring mixed microbial community functioning: recent advances in metaproteomics. FEMS Microbiol Ecol 80:265–280. doi:10.1111/j.1574-6941.2011.01284.x

95. Simmons SL, Dibartolo G, Denef VJ et al (2008) Population genomic analysis of strain variation in *Leptospirillum* group II bacteria involved in acid mine drainage formation. PLoS Biol 6:e177. doi:10.1371/journal.pbio.0060177

96. Singer SW, Erickson BK, VerBerkmoes NC et al (2010) Posttranslational modification and sequence variation of redox-active proteins correlate with biofilm life cycle in natural microbial communities. ISME J 4:1398–1409. doi:10.1038/ismej.2010.64

97. Sowell SM, Abraham PE, Shah M et al (2011) Environmental proteomics of microbial plankton in a highly productive coastal upwelling system. ISME J 5:856–865. doi:10.1038/ismej.2010.168

98. Sowell SM, Wilhelm LJ, Norbeck AD et al (2009) Transport functions dominate the SAR11 metaproteome at low-nutrient extremes in the Sargasso Sea. ISME J 3:93–105. doi:10.1038/ismej.2008.83

99. Thompson A, Schäfer J, Kuhn K et al (2003) Tandem mass tags: a novel quantification strategy for comparative analysis of complex protein mixtures by MS/MS. Anal Chem 75:1895–1904. doi:10.1021/ac0262560

100. Tseng C-H, Tang S-L (2014) Marine microbial metagenomics: from individual to the environment. Int J Mol Sci 15:8878–8892. doi:10.3390/ijms15058878

101. Unlü M, Morgan ME, Minden JS (1997) Difference gel electrophoresis: a single gel method for detecting changes in protein extracts. Electrophoresis 18:2071–2077. doi:10.1002/elps.1150181133

102. Verberkmoes NC, Russell AL, Shah M et al (2009) Shotgun metaproteomics of the human distal gut microbiota. ISME J 3:179–189. doi:10.1038/ismej.2008.108

103. Von Bergen M, Jehmlich N, Taubert M et al (2013) Insights from quantitative metaproteomics and protein-stable isotope probing into microbial ecology. ISME J 7:1877–1885. doi:10.1038/ismej.2013.78

104. Wang H-B, Zhang Z-X, Li H et al (2011) Characterization of metaproteomics in crop rhizospheric soil. J Proteome Res 10:932–940. doi:10.1021/pr100981r

105. Weiss S, Carapito C, Cleiss J et al (2009) Enhanced structural and functional genome elucidation of the arsenite-oxidizing strain *Herminiimonas arsenicoxydans* by proteomics data. Biochimie 91:192–203. doi:10.1016/j.biochi.2008.07.013

106. Werner JJ, Ptak AC, Rahm BG et al (2009) Absolute quantification of *Dehalococcoides*proteins: enzyme bioindicators of chlorinated ethene dehalorespiration. Environ Microbiol 11:2687–2697. doi:10.1111/j.1462-2920.2009.01996.x

107. Wilkins MJ, Verberkmoes NC, Williams KH et al (2009) Proteogenomic monitoring of*Geobacter* physiology during stimulated uranium bioremediation. Appl Environ Microbiol 75:6591–6599. doi:10.1128/AEM. 01064-09

108. Wilkins MJ, Wrighton KC, Nicora CD et al (2013) Fluctuations in species-level protein expression occur during element and nutrient cycling in the subsurface. PLoS One 8:e57819. doi:10.1371/journal.pone.0057819

109. Williams TJ, Wilkins D, Long E et al (2013) The role of planktonic *Flavobacteria* in processing algal organic matter in coastal East Antarctica revealed using metagenomics and metaproteomics. Environ Microbiol 15:1302–1317. doi:10.1111/1462-2920.12017

110. Wilmes P, Andersson AF, Lefsrud MG et al (2008a) Community proteogenomics highlights microbial strain-variant protein expression within activated sludge performing enhanced biological phosphorus removal. ISME J 2:853–864. doi:10.1038/ismej.2008.38

111. Wilmes P, Bond PL (2004) The application of two-dimensional polyacrylamide gel electrophoresis and downstream analyses to a mixed community of prokaryotic microorganisms. Environ Microbiol 6:911–920. doi:10.1111/j.1462-2920.2004.00687.x

112. Wilmes P, Bowen BP, Thomas BC et al (2010) Metabolome-proteome differentiation coupled to microbial divergence. MBio 1:e00246-10. doi:10.1128/mBio. 00246-10

113. Wilmes P, Wexler M, Bond PL (2008b) Metaproteomics provides functional insight into activated sludge wastewater treatment. PLoS One 3:e1778. doi:10.1371/journal.pone.0001778

114. Wrighton KC, Castelle CJ, Wilkins MJ et al (2014) Metabolic interdependencies between phylogenetically novel fermenters and respiratory organisms in an unconfined aquifer. ISME J 8:1452–1463. doi:10.1038/ismej.2013.249

115. Yoon H-J, Seo J, Shin SK (2014) Multi-functional MBIT for peptide tandem mass spectrometry. Mass Spectrom Rev. doi:10.1002/mas.21435

Section VIII: Metabolomics

CHAPTER 24

METABOLOME SEARCHER: A HIGH THROUGHPUT TOOL FOR METABOLITE IDENTIFICATION AND METABOLIC PATHWAY MAPPING DIRECTLY FROM MASS SPECTROMETRY AND USING GENOME RESTRICTION

A Ranjitha Dhanasekaran, Jon L Pearson, Balasubramanian Ganesan and Bart C Weimer

BACKGROUND

Bacterial metabolism impacts almost every aspect of our life. Microbial metabolism was exploited by early human civilization to create fermented foods and beverages [1,2]. The oldest known metabolically derived products from microbes include bread, cured meats, cheese, and beer [2-4]. Currently, metabolic engineering for the production of pharmaceuticals and bioactive compounds is giving way to discovery of novel metabolic pathways for production of alternative fuels [5-7]. Burgeoning needs to produce novel antibiotics for disease treatment and health supplements, such as amino-sugars and vitamins, also represent the metabolic end products that are genome encoded of an organism [8-11].

The virulence of bacterial pathogens is closely linked to their metabolism during infection, which is leading to metabolomic disease biomarkers that is pushing the boundaries of robust methods to quickly identify high throughput metabolomic data [12,13]. Cumulatively, the unusual metabolic networks of organisms in ecological niches are renewing interests in metabolites that highlight the lack of high throughput analysis tools for rapid compound identification when the compound is not included in a database. Unfortunately, rapid identification of multiple metabolites simultaneously is also lacking. However, if one considers an organism's genome to be a database of possible metabolic pathways and metabolite production, it enables customization of MS output analysis based on a specific organism. Approaching the genome as a metabolite database is being done using metabolic reconstruction methods in KEGG and Pathway Tools.

The metabolism of an organism changes during growth, survival, and persistence via complex gene expression changes. In many cases, metabolism begins with the transport of chemically diverse molecules for integration into biologically functional blocks. An organism's metabolic capability can be envisaged as a highly interconnected network of enzymatic reactions that provide energy, intermediates for macromolecular biosynthesis, cellular signaling, regulation of stress, and control of oxidation/reduction to ensure growth or survival [14]. Highly tuned regulatory mechanisms to modulate the metabolic network via gene expression and enzyme attenuation are needed to quickly adapt to local environmental changes. Evolution of genetic control and gene acquisition are critical to ensure the organism's survival in the near- and long-term [15]. Adaptation and genetic evolution results in new metabolic nodes in the interconnected network that modifies the intermediate and end product metabolism [14,16]. Of recent interests, metabolic engineering is largely dependent on understanding the metabolic network to regulate production of specific low molecular weight end products that often accumulate.

Low molecular weight metabolites, usually <1,000 Da (small molecules; Table 1), including sugars, lipids, fatty acids, amino acids, nucleotides, vitamins, and co-factors are usually the targets of metabolomics, which have bioactivity and lead to biomarker profiles (www.metacyc.org; [17]). An organism's metabolic demands are met by catabolism of complex macromolecules to the constituent small molecules (e.g. polysaccharides to sugars) or digestion of the molecules themselves (e.g. vitamins and amino acids) to end products.

Citation: Dhanasekaran AR, Pearson JL, Ganesan B1, Weimer BC, Metabolome searcher: a high throughput tool for metabolite identification and metabolic pathway mapping directly from mass spectrometry and using genome restriction, doi: 10.1186/s12859-015-0462-y.

The products of catabolism are reassembled through anabolic pathways into macromolecules of the organism to derive energy, oxidation/reduction regulation, pH control, and to maintain membrane potential that fuels transport functions. During growth catabolic and anabolic processes are regulated both genetically and biochemically to maintain a balance between growth and survival [18,19]. All of these activities are encoded in the genome, which provides an inherent genetic database of the possible metabolic compounds that an organism can produce during changing growth conditions.

Table 1: Metabolite distribution by molecular mass across metabolic encyclopaedias

Molecular weight range (Da)	Number of compounds in Metacyc	Number of compounds in KEGG
50-99	254	267
100-199	1,731	2,402
200-299	1,050	2,614
300-399	842	2,575
400-499	461	1,150
500-599	311	719
600-699	169	337
700-799	199	201
800-899	198	232
900-999	133	160
>1,000	119	224
Total	5,467	10,881

Metabolomics aspires to identify all the metabolites produced by an organism [20,21]. However, large data sets, limited identification databases, and limited MS parameters to differentiate small molecules are stumbling blocks for metabolomic analysis, which in turn limits the subsequent bioinformatic analysis and construction of biologically informative models [16,21]. Currently, NMR is of limited use for high throughput small molecule identification due to the lack of sensitivity and limited throughput, but is useful to elucidate the structures of unique metabolites [22,23]. However, NMR is very useful to track the metabolic fate of a small molecule with isotope labels, which provides information for a handful of metabolites once the entire compound list is narrowed to a specific set of metabolic intermediates [18]. Other post-separation detection techniques like photometric, electrochemical, and fluorescent detection are actively used to identify specific metabolites at a substantially reduced analytical scale, but the need to identify the set of compounds produced is overwhelmingly changing the goals of metabolite analysis [24-26]. Conversely, MS analysis, in addition to metabolic tracking estimates the masses of hundreds to thousands of small molecules within minutes and provides information on their relative levels in the sample [27-29], making it very useful for high throughput metabolome analysis. However, it lacks specific information as to the identity of the small molecules, which highlights the need to have curated databases for compound identification [21].

One approach to overcome the need to identify important molecules uses principal component analysis (PCA) to find changes with a specific treatment. From MS data acquisition this produces a reduced list of small molecules that are tagged as biomarkers [30,31]. Often the diagnostic peak is an unknown compound that is difficult to identify. Subsequently, more complex chemical analysis is used to determine the elemental composition of these biomarkers, which requires additional time, expertise, and often multiple instrumentation capabilities [32,33]. Biomarker identities are subsequently validated by standard compound injection to produce a compound library [22]. While this statistics-driven analytical approach favors method development for MS it ignores the underlying biochemistry and the importance of relatively minor changes of small molecules that can sometimes lead to misinterpretation of the biological impact with new small molecule production. This is especially prevalent for key metabolite classes like hormones, vitamins, and enzyme co-factors where small changes regulate large scale proteomic and metabolic fluctuations [23]. One way to overcome this limitation is to use tools that include all possible putative compounds generated directly from matched compound identities prior to statistical analysis. Subsequently, a significant list of putative compounds can be used for metabolic mapping to facilitate biological identity by linking compound identities to metabolic pathways and routes. Feist et al. [24] review the reconstruction approach with specific attention to metabolite identification.

Unfortunately, metabolite identification from hundreds to thousands of masses by searching a large compound database is a slow process that is ill defined relative to the specific search criteria that provides confident compounds assignments. GC-MS analysis often identifies compounds by comparison of MS spectra with large, well-established compound libraries (www.nist.gov). Such compound libraries for LC-MS analysis are available for only a small set of masses and are tightly linked to the LC conditions. Large compound databases such as Pubchem (http://pubchem.ncbi.nlm.nih.gov) and Chemspider (http://www.chemspider.com) allow searches of single masses and other query types, but they do not allow queries from large lists of masses or connect putative compounds to metabolic pathways. However, as the query list expands, as it does in metabolome data sets, data analysis using single queries becomes unrealistic for a timely and accurate analysis.

Multiple software suites are available for compound identification of mass spectrometry-based metabolite data that use mass spectral deconvolution and matching to reference databases. Some examples of full-fledged independent platforms are MetSign [25], MZmine 2 [26], MAVEN [27], and XCMS^2 [28], whereas MS Excel templates such as IDEOM [29], R packages like AStream [30] and MAIT [31], and web-applications like METLIN [32], XCMS Online [33], and MZedDB [34] are also available as web services. These tools offer either statistical or structural analyses of small molecule MS data and extract information from metabolic databases to create a list of compounds for their own localized database. For example, MetSign's compound database is formed from the cumulative compound collection of KEGG, HMDB, and LIPIDMAPS databases, MZmine 2's collection is from KEGG [35], HMP, and Pubchem compound, MAVEN uses KEGG, whereas MAIT, IDEOM and ASTREAM use unspecified databases. However, downstream of compound identification, they ignore the underlying biology and do not offer a mechanism to map the data back to the metabolic pathways. Further, they lack the flexibility of implementing user-defined parameters for database searches, as for example, electrospray ionization (ESI) parameters that are predefined in METLIN and MZedDB [34].

Querying large compound databases that contain millions of non-biological molecules can impede a researcher's ability to overlay a metabolic context onto metabolomic data [36]. Biologists are producing data at rates that outstrip the ability of analysts to examine the data set to uncover the biological importance. To keep pace with metabolome analysis, high throughput bioinformatic tools that bring compound identity and pathway relevance together to the biologist are crucial. This can be accomplished with: a) automated searches of metabolic databases to retrieve putative compound identification, b) large scale queries be performed seamlessly with MS output, c) provide users the flexibility of using multiple query types, and d) map query results to metabolic pathways, hence allowing data to be analyzed in a biological context.

The availability of over 1,000 annotated microbial genome sequences enables bioinformatic reconstruction (biocyc.org) of an organism's metabolic capability via the genome, which provides a broad network of metabolism that can be used to predict small molecule production [27,28]. Consequently, recent efforts have focused on uncovering the metabolic networks in many different biological systems [19,37]. Genome reconstructions of the metabolic pathways coupled to analytical methods, such as liquid chromatography (LC), gas chromatography (GC) and capillary electrophoresis with nuclear magnetic resonance spectroscopy (NMR) and mass spectrometry (MS) produces a new method to leverage genomic sequence to provide putative compound identification quickly [27,38].

In this study, a user-friendly web-based application called Metabolome Searcher to retrieve a list of small molecules identifications based on chemical formula, SMILES structure, and the monoisotopic mass was created using an organism's genome as a putative compound database. While single queries can be directly entered multiple queries with one or more query types can also be done using a text file containing the query list. One or more reference databases can be selected from the list against which the queries are performed. The output connects small molecules in a sample to metabolic databases via embedded links to specific metabolic pathways. The Metabolome Searcher's output allows researchers using metabolome data from different technologies to group the compound identifications into metabolic information so as to uncover the relevant biological function with multiple chemical criteria.

METHODS

The ProCyc webserver

We currently house a metabolic database webserver called ProCyc (www.usu.edu/westcent/procyc), which is an implementation of the Pathway Tools webserver (SRI Bioinformatics, Menlo Park, CA) with our own manually and automatically curated metabolic databases of interest. ProCyc houses over 47 metabolic database reconstructions of different classes of bacteria including probiotics, lactic acid bacteria, pathogens, and environmental bacteria that were reconstructed locally. The MetaCyc database and Human metabolism database are part of the basic installation of Pathway Tools software. Some of the reconstructed databases and the tier I/II databases of the basic software were used to exemplify the Metabolome Searcher implementation. This particular platform was chosen for its flexibility to immediately incorporate user-discovered pathways into the right metabolic databases.

Metabolic reference database creation

A Metabolic Reference Database (MRDB) of an organism is a flat file (tab-delimited, plain text file) that initially contains only the compound name, molecular formula, molecular weight, SMILES structure, and the respective pathways for all compounds extracted from Pathway Tools Pathway/Genome Database (PGDB) of that organism. The script to create the MRDB communicated with Pathway Tools [17] via the PerlCyc module (v1.1;www.arabidopsis.org/biocyc/perlcyc). The same approach was used to create an additional non-redundant database using Metacyc [17] and KEGG [35] (Table 2). The reference monoisotopic masses of individual elements were obtained from a publicly available compilation (Scientific Instrument Services, Inc., Ringoes, NJ;www.sisweb.com). Using the monoisotopic masses of individual elements, the monoisotopic masses of all compounds in the MRDB in their charged and neutral states were calculated based on their formulae. The MRDB was then modified to include the calculated monoisotopic masses, which is queried for compound identification and pathway mapping via the Metabolome Searcher's web interface.

Table 2: Organism-specific and general metabolic reference databases available for the Metabolome Searcher

Organism	ProCyc database	Metabolic reference database
Escherichia coli K12	EcoCyc	E. coli K12
Escherichia coli O157:H7	Ecoo157Cyc	E. coli O157:H7
Homosapiens	HumanCyc	Homo sapiens
KEGG Compounds		KEGG Compounds
Lactococcus lactis ssp. lactis IL1403	LlactisCyc[1]	L. lactis ssp. lactis IL1403
Lactococcus lactis ssp. cremoris SK11	LaccremoCyc[1]	L. lactis ssp. cremoris SK11
Lactobacillus acidophilus NCFM	LbacidCyc[1]	Lb. acidophilus NCFM
Lactobacillus johnsonii NCC 533	LbjohnCyc[1]	Lb. johnsonii NCC 533
Lactobacillus plantarum WCFS1	LbplanCyc[1]	Lb. plantarum WCFS1
Listeria monocytogenes EGDe	LmonoCyc[1]	Listeria monocytogenesEGDe
Mycobacterium bovis AF2122/97	MbovisCyc[1]	M. bovis AF2122/97
MetaCyc	MetaCyc	MetaCyc
Staphylococcus aureus Mu50	SaureusCyc[1]	S. aureus ssp. aureus Mu50
Saccharomyces cerevisiae S288C	YeastCyc[2]	S. cerevisiae S288C
Salmonella enterica ssp. enterica serovar Typhimurium LT2	Styp99287Cyc[3]	Salmonella typhimurium LT2

[1]PGDBs reconstructed, curated and hosted in ProCyc.

[2]Obtained from the Yeast genome database.

[3]Downloaded from the Pathway Tools registry of PGDBs.

Query input

The Metabolome Searcher allows the user to enter a single query by typing the name, formula, molecular/monoisotopic mass or SMILES structure, or multiple queries by uploading a query list within a file (Figure 1). This file contains masses and intensities of compounds as a tab-delimited text file. For mass searches, whether from a single entry or a file, the user selects the type (molecular weight or monoisotopic mass). Most MS systems contain software that enables data export to a text or an Excel file [25]. We used the QTof system (QTof Premier, Waters, MA) with MarkerLynx software for marker identification and analysis to test this approach. A MarkerLynx-derived text file was used without modification for the Metabolome Searcher query by submitting the file under the "MarkerLynx file" input on the interface (Figure 1). Alternately, analysis of output from other MS systems can be done using the "text file" option (Figure 1). While using the text file option, query values of any type, whether masses or specific compound names or a mixture of query types, were listed in the first column of the query file. Any headers, empty lines, and non-query values in the first column were removed prior to submission of data as a text file for matching. For both the file options, other information like statistics, marker quality, peak areas, peak heights, and concentrations across experiments and replicates were still retained in the file.

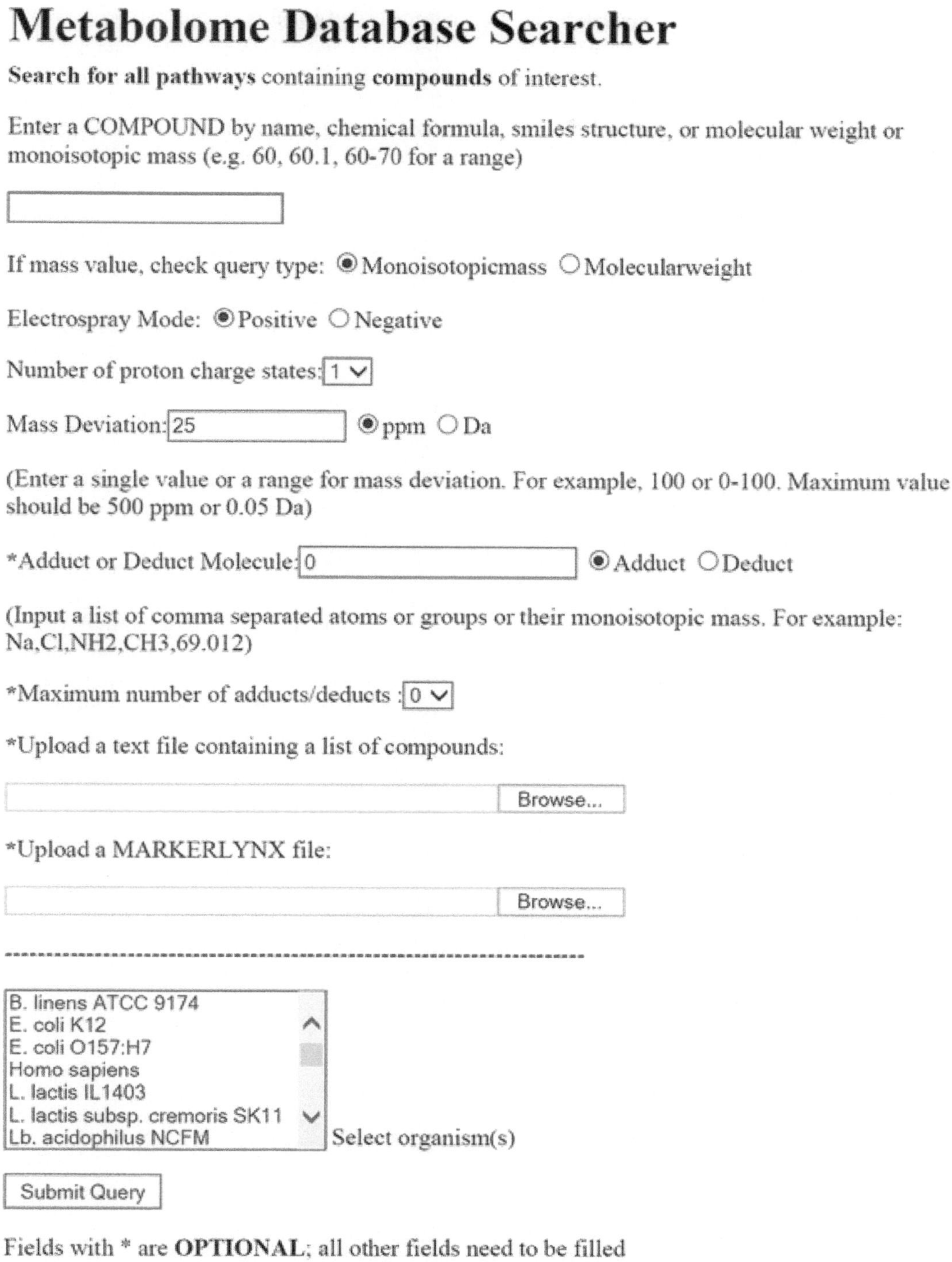

Figure 1: Metabolome searcher user interface screenshot.

Compound identification for MS analysis

For compound identification from monoisotopic masses, the user specifies the acceptable deviation from the theoretical masses (ppm or Da, under "Mass deviation"; Figure 1), the ionization mode (positive or negative, under "Electrospray mode"; Figure 1), the maximum number of charges (0-5; under the "Number of proton charge states"; Figure 1), and adducts (mass or formula; optional; under "Adduct or Deduct molecule" and "Maximum number of adducts/deducts"; Figure 1). The deviation value allows the software to obtain matches for queried masses within an acceptable range to narrow or expand the putative identification list. Acceptable mass deviation values may be experimentally determined or obtained from the literature based on a particular instrument and operating conditions [37].

Typically during MS analysis the molecules are detected by prior ionization with or by removal of protons (positive and negative mode, respectively) [23]. The MS settings are optimized to mainly produce singly charged ions. However, a molecule may still carry multiple charges depending on the MS settings [38]. The user can verify the charge state of compounds contained in the input list to recalibrate the MS settings by selecting different charge states during multiple search sessions.

Positively charged ionic species, such as sodium (Na+) and potassium (K+), or negative species, such as chloride (Cl−)

and formate (HCOO−), are also used during ionization due to their abundance in a sample. The addition of ionic species or adducts during ionization shifts the observed monoisotopic mass from that of the intact molecule plus/minus a proton [38]. These adducts can be specified either as individual elements or as partial functional groups in the "Adduct or Deduct molecule" textbox (Figure 1). Similar to adducts, if the user wishes to specify fragments lost during ionization or fragmentation the "Deduct" option can be selected. The user can also provide more than one adduct or deduct in the textbox simultaneously and specify the number of maximum possible adducts or fragments ("Maximum number of adducts/deducts" option).

Database Selection

MRDBs that contain metabolites from different PGDBs or the KEGG database along with calculated monoisotopic masses are used for the queries. MRDBs are included for user selection from the ones listed on the interface (Table 2) wherein the user can select single or multiple MRDBs for searching (Figure 1). If the user intends to query known metabolic pathways in an organism, the organism-specific MRDBs are provided for more specific and narrow options of possible compounds due to the known annotated pathways. However, if the intent is to discover new pathways unknown in a particular system, but identified in other organisms, or if an organism without a pre-constructed MRDB is being studied, the user can select a genotypically related organism's MRDB or the MetaCyc MRDB for matching. A user-generated PGDB can also be incorporated as an MRDB using the scripts defined above prior to the user defined query. The MRDBs were created in a flat file format to reduce complexity in processing and data handling such that newer MRDBs for other organisms can always be created in a consistent format and readily incorporated as per the user's need. Pathway Tools was selected as the main metabolic database platform to create MRDBs and link back to PGDBs due to its interactive features and user-level flexibility for metabolic database development and curation of whole genome PGDBs [17], while queries of an MRDB for the KEGG database [39] are also supported.

Database Searching

Once a text query has been submitted, the Metabolome Searcher determines whether a text input is the name of a compound, its chemical formula or its SMILES structure independent of any specifications. After the query is classified into the specific type, information of the corresponding type in the MRDB is used for matching (i.e. names-to-names, formulae-to-formulae, and masses-to-masses) (Figure 2). All matches obtained within the parameters specified for searches are provided in the output files for viewing and analysis.

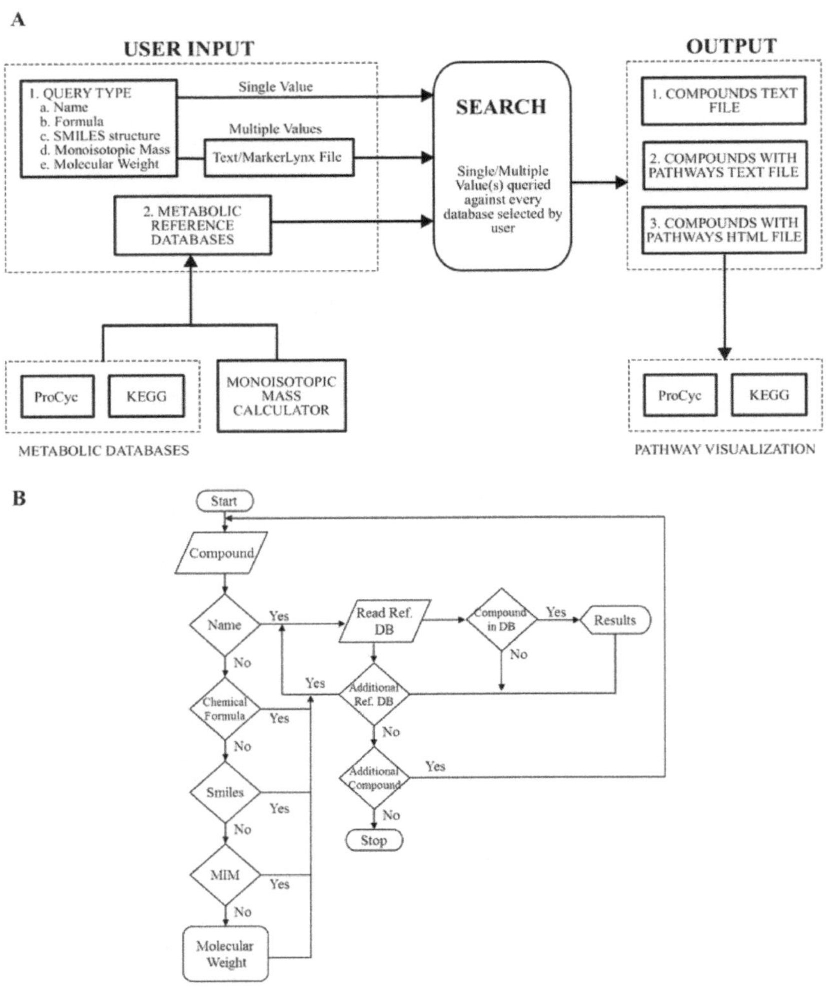

Figure 2: Diagram of the work flow and search operations that underly metabolome searcher to return compounds and pathways. (A) **Metabolome** searcher **workflow.** (B) flowchart of the search operations to find matching compounds.

Output generation

After entering a single query or uploading a query file and specifying the MRDBs along with other MS analysis parameters, the user submits the query. The queries are matched against the MRDBs and the output files are created. Query parameters are printed at the top of all the output files to ensure that the parameters submitted by the user were used for searching the database (Figure 3A).

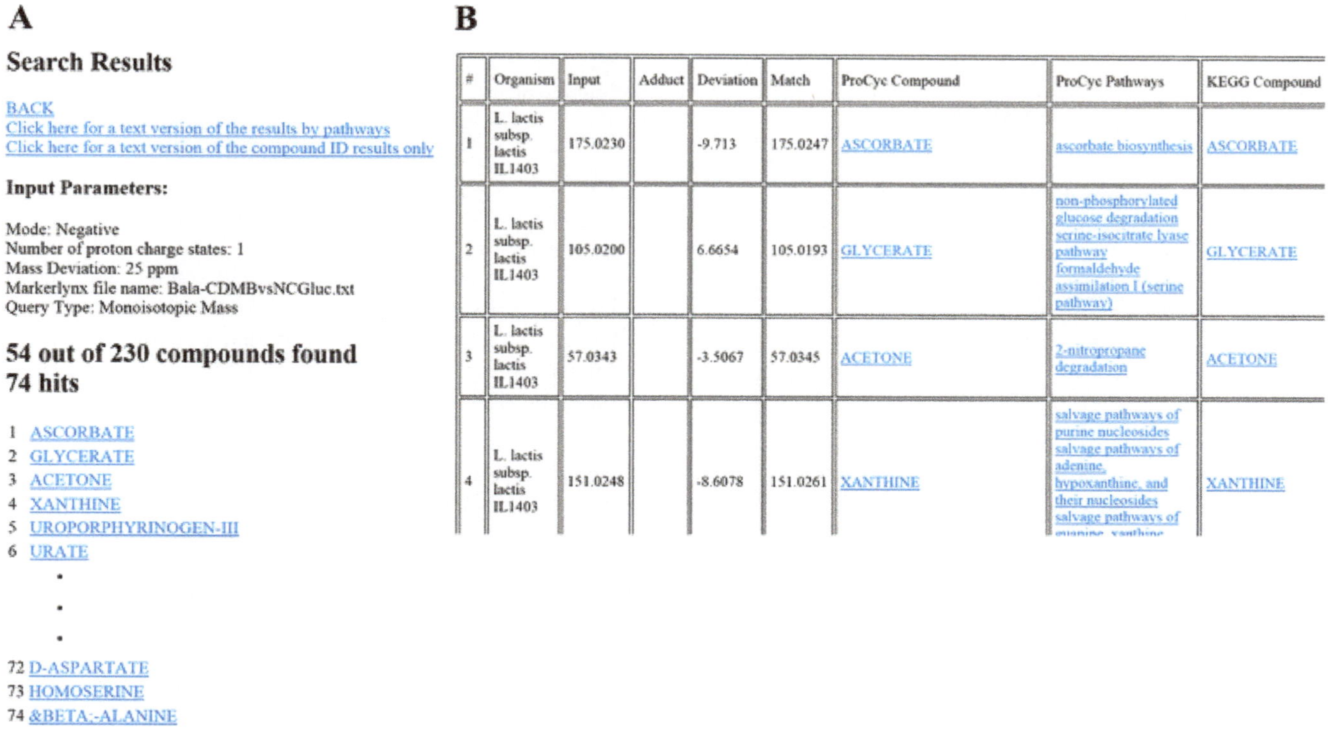

Figure 3: Screenshot of metabolome searcher's output. (A) The top portion of the HTML results page and **(B)** the body of the HTML file demonstrate that sections containing queries, matches, compound and pathway links, and other data and information are provided with in the output.

Three different output files are provided as the result of the analysis, one HTML and two text files. The two text files are embedded as links at the top of the HTML page (Figure 3A) that the user can download. One text file ("compounds file") lists only the matched compounds without any metabolic pathway information, while the other ("pathways file") repeats each compound's data by all the pathways that it belongs to as a metabolite.

All scripts were written in Perl (v5.8.6; www.perl.org). The scripts and the metabolic reference databases for Metabolome Searcher are hosted in an Apple XGrid computational cluster (Panther OS 10.3.9) at the Western Dairy Center at Utah State University as well as University of California, Davis. Web pages for data input and output were created using Perl CGI.

MS data validation

Chemical standards preparation

All compounds used were purchased from Sigma-Aldrich (St. Louis, MO). A chemically defined medium described previously by Ganesan et al. [18] was prepared as a complex mixture for testing Metabolome Searcher's performance. The major components of this medium are 20 amino acids, sodium chloride, citrate, phosphate, 3-(N-morpholino) propane sulfonic acid (MOPS), vitamin solution (containing 15 different compounds), and glucose. Individual standard solutions of selected amino acids, glucose, citrate, and MOPS were also used for molecule identification.

Mass spectrometry

Separation and analysis of standard compound mixtures were done at the mass spectrometry facility in the CIB. The samples were separated by liquid chromatography (2795 LC system; Waters) prior to introduction by electrospray into the mass spectrometer (QTof Premier; Waters) as described by Mortishire-Smith et al. [40]. Briefly, the separation was done for 10 min using a linear gradient of water:acetonitrile from 0-95% using a Symmetry C18 column (Waters). After introduction into the MS by electrospray, the molecules were detected using both positive and negative electrospray conditions, with calibrated settings recommended by the manufacturer. The QTof instrument was operated in W mode throughout MS analysis. For both positive and negative electrospray analysis, the conditions were: desolvation temperature of 250°C, source temperature of 120°C, cone voltage of 40 V, and collision energy of 4 eV. Data acquisition was performed for a mass range of 50–1,000 Da. After acquisition the data were centroided [40] using 1 ng/µl leucine-enkephalin infused at 10 µl/min as a reference, with an m/z of 556.2771 in positive mode and m/z of 554.25 in negative mode. In order to subtract background from the LC column and sample matrix, HPLC-grade water (Thermo Fisher Scientific Inc., Waltham, MA) was injected into the MS as a negative control. All samples were analyzed in technical duplicates.

Peak detection, intensity extraction, and normalization were performed using MarkerLynx software (Waters) to obtain monoisotopic masses and molecule retention times. In this study, only the monoisotopic masses of the markers were used for database searches. The Metabolome Searcher does not support any data analysis of the concentrations or relative measures of compound levels obtained from MarkerLynx.

RESULTS AND DISCUSSION

Metabolomic assessment provides a list of compounds that facilitates the estimation of metabolic flux through both single pathways and networks [41,42]. Metabolome analysis enables determination of abiotic conditions and genetic regulation of metabolic networks. To achieve these purposes a tool that rapidly determines the compound identity, pathways, and metabolic networks was needed [43,44]. The tool accepts queries from common data types and facilitates data integration from independent sources into a unified compound identification and pathway-mapping scheme. To our knowledge such a tool is not available. The Metabolome Searcher addresses these purposes by receiving input from the user, querying the user-selected metabolic reference database(s), and displaying the generated output for further biological interpretation (Figure 3).

Of the Metabolome Searcher's outputs, the compounds file is useful when the user plans to conduct compound classification, data clustering, principal component analysis, analysis of variance, or graphical visualization. The pathways file allows the users to sort the data by pathways and facilitates interpretation of metabolic flux and pathway connections to determine if a compound is an intermediate or an end product. The main feature of the HTML output is that it lists and links compounds to all metabolic pathways in which the metabolite is involved (Figure 3). These links help the user understand the role of that particular metabolite in the organism's metabolic network. The user can click on any one of these links that will navigate them to the PGDBs curated and hosted at ProCyc. The user need not repeat queries on the Metabolome Searcher as the HTML file contains the links to the pathways associated with the returned putative compound IDs. To facilitate obtaining the standard chemicals for verification of retention times, CAS IDs of compounds (where available) are also included in a separate column in the output file.

Verification

For names, formulae, and SMILES structures, any partial matches will also be detected and listed. For example, a query of the word string "glucose" against the MetaCyc database will identify D-glucose and an additional 52 hits (data not shown) that also include alpha-methyl-glucose, NDP-Glucoses, and all other molecules that contain the substring "glucose" in the name. String matching offers the user the ability to obtain partial matches and allows additional control over the query specificity and flexibility for unknown pathways. In most cases, if the specific MetaCyc compound names are used, the results will be restricted to one hit. Searching of word strings was implemented in order that even if other data sources such as GC-MS and LC-MS/MS were provided after identification using other software suites, or even data from standard GC or HPLC analyses based on extractions and retention times under certain conditions was provided, the data can be mapped to metabolites and pathways.

Compound identification from LC-MS or NMR spectrometry data has proven to be a challenge to biologists because the compound databases are limited, especially with respect to the compounds that a specific organism can produce. Based on the user selection of MRDB(s) in Metabolome Searcher, the number of hits is refined and is metabolically relevant to the organism under study, which provides a basis for biological conclusions to be drawn. As an example of the convenience provided by Metabolome Searcher by implementing genome restriction, we initially queried the MetaCyc MRDB with the monoisotopic mass of isocitrate as the search query and used the results for further narrowing the hits by querying organism-specific MRDBs. These genome-restricted results were compared to those hits obtained by querying the monoisotopic mass of isocitrate using Chemspider (Figure 4). The ChemSpider query returned 118 possible compound identifications that included non-biological compounds and required extensive analysis outside the query system to derive possible identifications whereas querying the MetaCyc MRDB provided hits that included 10 compounds with similar monoisotopic masses to that of isocitrate. Each genome (i.e. organism) further reduced the hits to 2–5 compounds that reflected the genetic differences in metabolism, all of which were related to citrate. Combining genome restriction with the MS compound list refined the possible identification list to a low number of compounds that was reasonable for empirical confirmation.

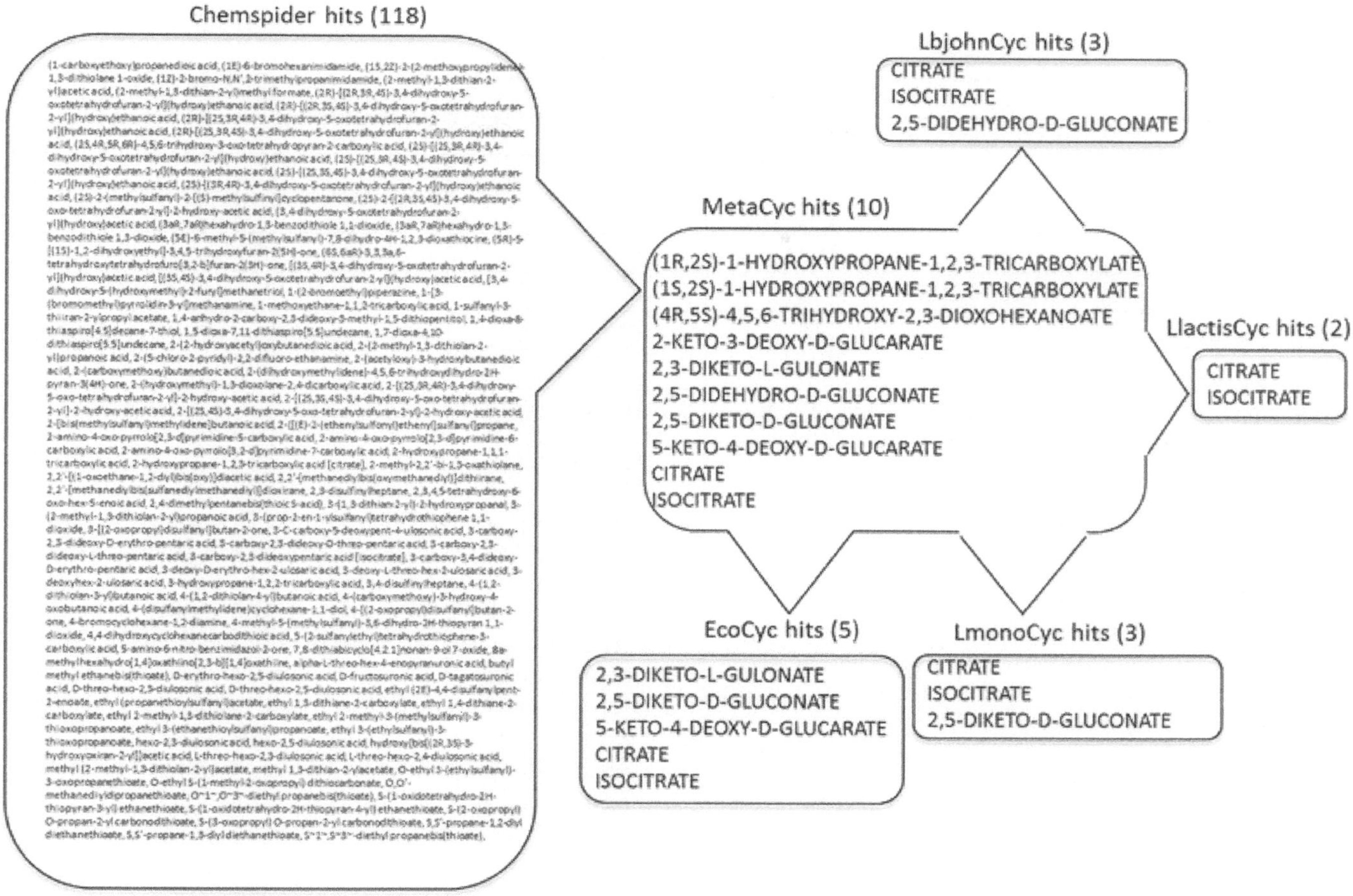

Figure 4: Comparison of results from chemical vs metabolic databases with the monoisotopic mass of isocitrate (192.027±0.001 Da) as the query. Hits to an encyclopedia of genes (MetaCyc), *E. coli* (EcoCyc), *Listeria monocytogenes*(LmonoCyc), *Lactococcus lactis* IL1403 (LlactisCyc), and *Lactobacillus johnsonii* (LbjohnCyc) databases were used to demonstrate multiple genome-restrictions using Metabolome Searcher.

The interface and search function were verified by accessing the database search function using known exact masses and a data set generated from a known mixture of compounds (i.e. a chemically defined bacterial growth medium) from LC-MS output. The resulting markers exported into a MarkerLynx format text file was used to query the compound identification using Metabolome Searcher. All the main ingredients of the growth medium represented in the MetaCyc MRDB were detected during the search (Table 3). MOPS, a buffering salt, was used as a negative control for the chemical challenge, which was done by excluding it from the MetaCyc MRDB. Interestingly, after excluding MOPS, some of the query masses also matched multiple metabolites, many of which were isomeric forms of the metabolites being tested. This allowed further restriction of identification to narrower ranges of mass deviation to obtain better accuracy. However, in nearly 90% of compounds identified the number of hits was limited to <5 metabolites, thus aiding the directed development of protocols for further compound identification. This approach enabled detection of common starting substrates for metabolism and verified that if the compound was in the database, Metabolome Searcher found it.

Table 3: Summary of hits to selected compounds from a chemically defined growth medium determined from a query of monoisotopic masses using the Metabolome Searcher

Compound	Detected by mass match	Number of additional hits*	Number of non-isomeric additional hits*
Tyrosine	Yes	29	25
Pyridoxamine	Yes	0	0
Lysine	Yes	4	1
Thymidine	Yes	0	0
Pyridoxal	Yes	8	5
Glycine	Yes	1	0
Dihydrouracil	Yes	4	0

Arginine	Yes	2	1
Threonine	Yes	7	0
Alanine	Yes	5	0
Serine	Yes	1	0
Asparagine	Yes	5	0
Histidine	Yes	3	1
Tryptophan	Yes	2	1
Aspartate	Yes	2	0
Glutamate	Yes	6	0
Guanosine	Yes	0	0

*Additional hits include all individual compounds that match the query using the provided query settings; non-isomeric hits only include compounds that do not share the same empirical formula.

Uses of metabolome searcher

An example demonstration of the Metabolome Searcher for microbial metabolomics was by collecting metabolomics profiles for both sterile chemically defined media and spent media collected after inoculation with the bacterium *Lactococcus lactis* IL1403 for 16 h. Metabolomics profiles were collected by LC-MS analysis in both positive and negative electrospray modes for the same samples and the masses obtained from MarkerLynx were queried against the *L. lactis* IL1403 MRDB (Table 2). After overlaying the compound identifications we quickly inferred changes in compound classes, such as amino acids (Figure 5), by sorting the compounds file, or pathways file that changed during growth of *L. lactis* IL1403 (Figure 5). This example demonstrated that Metabolome Searcher performed the intended search and enabled the biological meaning to rapidly assign the identified compounds using constructed databases from metabolic reconstruction maps.

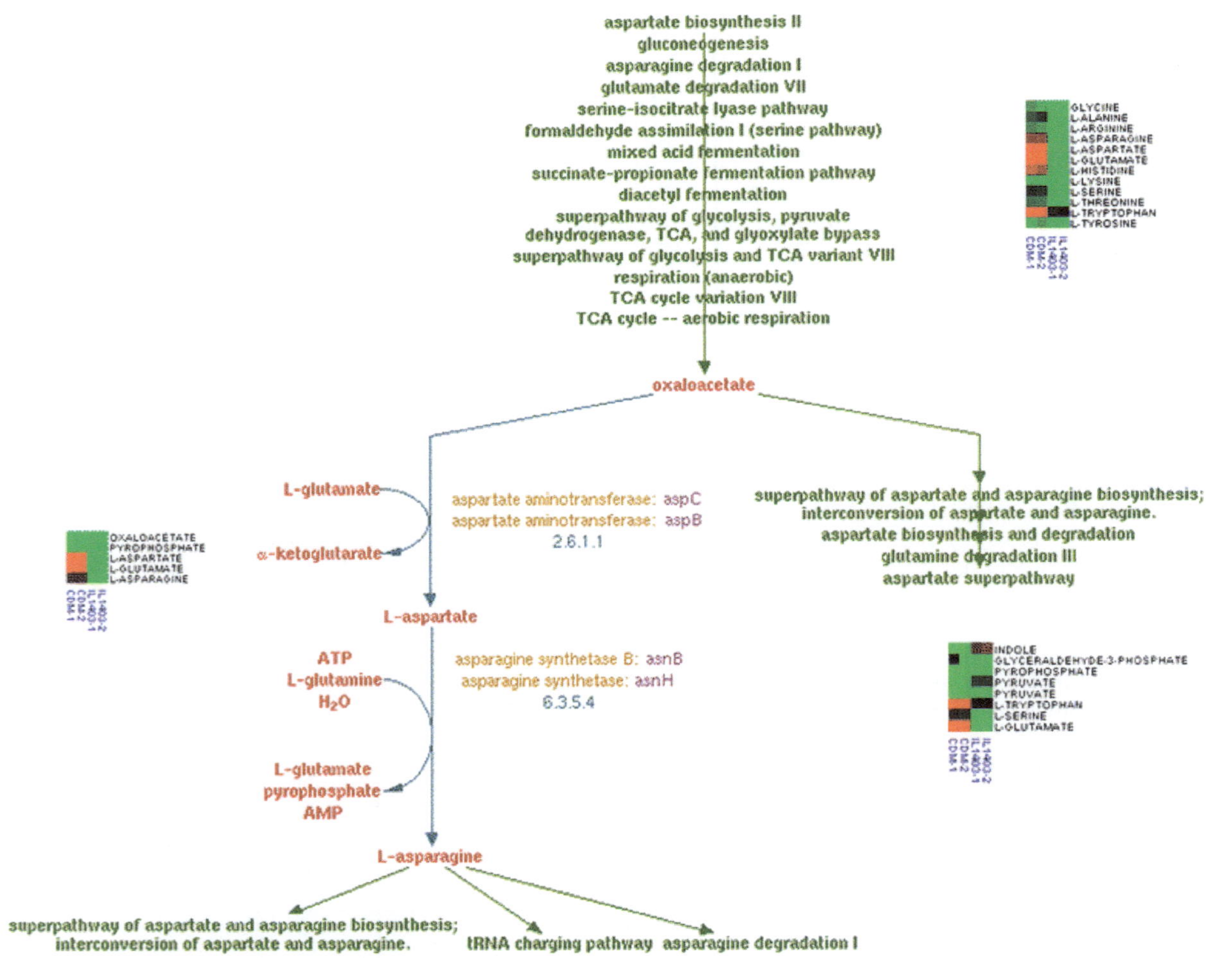

Figure 5: Pathway assembly of metabolome searcher output with heat maps of LC/MS data from the compounds file and the pathways file (tryptophan biosynthesis and asparagine biosynthesis I). This output is obtained by clicking on the pathway link for "asparagine biosynthesis I" inside the HTML file that brings up the pathway page at ProCyc. The heat maps are color-coded with green being 0, black the median value, and red the highest value within that data set. The heatmap number of 1 was at the initial time and number 2 was 24 hours later, which demonstrates the change in metabolites, from the MS list and visualized with the pathway and concentration.

CONCLUSIONS

The Metabolome Searcher provides an automated tool to identify metabolites from MS analyses from metabolic reconstruction of specific genomes. This approach couples long lists of masses to specific genomic-based metabolites for identification and subsequent visualization via metabolic pathways. The tool is flexible so that query types can use many types of data that include names, molecular formulae, or SMILES structures, and monoisotopic masses that are entered singly or in bulk as a text file. The matches to queries are then presented as results along with other input parameters that the user included in the query and the pathways in which the matched metabolites are involved. The versatility of accepted query types and the provision of pathways mapped to queries are unique to the Metabolome Searcher. The Metabolome Searcher's utility and flexibility facilitates rapid advances from metabolomics to biological comprehension.

ACKNOWLEDGEMENTS

Funding for this project was provided by a grant from USDA CSREES 2006-34526-17001 to BCW. Contribution Number 8121of the Utah Agricultural Experimental Station.

COMPETING INTERESTS

The authors declare that they have no competing interests. JLP's work on software development was completed prior to his tenure at his current position, and hence, no commercial interest was involved in the software development.

REFERENCES

1. *Kilara A, Shahani KM. Lactic fermentation of dairy foods and their biological significance. J Dairy Sci. 1978;61:1793–800.*
2. *Sandine WE, Elliker PR. Microbially induced flavours and fermented food flavour in fermented dairy products. J Agr Food Chem. 1970;18:557–62.*
3. *Dickinson JR, Harrison SJ, Hewlins MJE. An investigation of the metabolism of valine to isobutyl alcohol in Saccharomyces cerevisiae. J Biol Chem. 1998;273(40):25751–6.*
4. *Dickinson JR, Lanterman MM, Danner DJ, Pearson BM, Sanz P, Harrison SJ, et al. A 13C nuclear magnetic resonance investigation of the metabolism of leucine to isoamyl alcohol in Saccharomyces cerevisiae. J Biol Chem. 1997;272(43):26871–8.*
5. *Li Y, Horsman M, Wu N, Lan CQ, Dubois-Calero N. Biofuels from Microalgae. Biotechnol Prog. 2008;24(4):815–20. doi:10.1021/bp070371k.*
6. *Singh OV, Harvey SP. Integrating biological processes to facilitate the generation of 'Biofuel'. J Ind Microbiol Biotechnol. 2008;35(5):291–2.*
7. *Vertes AA, Inui M, Yukawa H. Technological options for biological fuel ethanol. J Mol Microbiol Biotechnol. 2008;15(1):16–30.*
8. *Coates PM. Dietary supplements and health: the research agenda. Novartis Found Symp. 2007;282:202–7. discussion 207-218.*
9. *Crowder MW, Spencer J, Vila AJ. Metallo-beta-lactamases: novel weaponry for antibiotic resistance in bacteria. Acc Chem Res. 2006;39(10):721–8.*
10. *Roghmann MC, McGrail L. Novel ways of preventing antibiotic-resistant infections: what might the future hold? Am J Infect Control. 2006;34(8):469–75.*
11. *Yoneyama H, Katsumata R. Antibiotic resistance in bacteria and its future for novel antibiotic development. Biosci Biotechnol Biochem. 2006;70(5):1060–75.*
12. *Deutscher J, Herro R, Bourand A, Mijakovic I, Poncet S. P-Ser-HPr–a link between carbon metabolism and the virulence of some pathogenic bacteria. Biochim Biophys Acta. 2005;1754(1–2):118–25.*
13. *Lemercier G, Espiau B, Ruiz FA, Vieira M, Luo S, Baltz T, et al. A pyrophosphatase regulating polyphosphate metabolism in acidocalcisomes is essential for Trypanosoma brucei virulence in mice. J Biol Chem. 2004;279(5):3420–5.*
14. *Moxley JF, Jewett MC, Antoniewicz MR, Villas-Boas SG, Alper H, Wheeler RT, et al. Linking high-resolution metabolic flux phenotypes and transcriptional regulation in yeast modulated by the global regulator Gcn4p. Proc Natl Acad Sci U S A. 2009;106(16):6477–82.*
15. *Taylor BL, Zhulin IB. In search of higher energy: metabolism-dependent behaviour in bacteria. Mol Microbiol. 1998;28(4):683–90.*
16. *Nedderman AN. Metabolites in safety testing: metabolite identification strategies in discovery and development. Biopharm Drug Dispos. 2009;30(4):153–62.*
17. *Krieger CJ, Zhang P, Mueller LA, Wang A, Paley S, Arnaud M, et al. MetaCyc: a multiorganism database of metabolic pathways and enzymes. Nucleic Acids Res. 2004;32(Database issue):D438–42.*
18. *Ganesan B, Dobrowolski P, Weimer B. C.: Identification of the catabolic pathway of leucine-to-2-methylbutyric acid byLactococcus lactis. Appl Environ Microbiol. 2006;72(6):4264–73.*
19. *Novak L, Loubierre P. The metabolic network of Lactococcus lactis: distribution of 14 C-labelled substrates between catabolic and anabolic pathways. J Bacteriol. 2000;182(4):1136–43.*
20. *Brown M, Dunn WB, Dobson P, Patel Y, Winder CL, Francis-McIntyre S, et al. Mass spectrometry tools and metabolite-specific databases for molecular identification in metabolomics. Analyst. 2009;134(7):1322–32.*
21. *Dettmer K, Aronov PA, Hammock BD. Mass spectrometry-based metabolomics. Mass Spectrom Rev. 2007;26(1):51–78.*
22. *Parr AJ, Mellon FA, Colquhoun IJ, Davies HV. Dihydrocaffeoyl polyamines (kukoamine and allies) in potato (Solanum tuberosum) tubers detected during metabolite profiling. J Agric Food Chem. 2005;53(13):5461–6.*
23. *Whitfield PD, German AJ, Noble PJ. Metabolomics: an emerging post-genomic tool for nutrition. Br J Nutr. 2004;92(4):549–55.*
24. *Feist AM, Herrgard MJ, Thiele I, Reed JL, Palsson BO. Reconstruction of biochemical networks in*

microorganisms. Nat Rev. 2009;7(2):129–43.

25. *Wei X, Sun W, Shi X, Koo I, Wang B, Zhang J, et al. MetSign: a computational platform for high-resolution mass spectrometry-based metabolomics. Anal Chem. 2011;83(20):7668–75.*

26. *Pluskal T, Castillo S, Villar-Briones A, Oresic M. MZmine 2: modular framework for processing, visualizing, and analyzing mass spectrometry-based molecular profile data. BMC Bioinformatics. 2010;11(1):395.*

27. *Clasquin MF, Melamud E, Rabinowitz JD. LC-MS Data Processing with MAVEN: A Metabolomic Analysis and Visualization Engine. Curr Protoc Bioinformatics. 2012;37:14.11.11– 23.*

28. *Benton HP, Wong DM, Trauger SA, Siuzdak G. XCMS2: processing tandem mass spectrometry data for metabolite identification and structural characterization. Anal Chem. 2008;80(16):6382–9.*

29. *Creek DJ, Jankevics A, Burgess KEV, Breitling R, Barrett MP. IDEOM: an excel interface for analysis of LC–MS-based metabolomics data. Bioinformatics. 2012;28(7):1048–9.*

30. *Alonso A, Julià A, Beltran A, Vinaixa M, Díaz M, Ibañez L, et al. AStream: an R package for annotating LC/MS metabolomic data. Bioinformatics. 2011;27(9):1339–40.*

31. *Fernández-Albert F, Llorach R, Andrés-Lacueva C, Perera A. An R package to analyse LC/MS metabolomic data: MAIT (Metabolite Automatic Identification Toolkit). Bioinformatics. 2014;30(13):1937–9.*

32. *Smith CA, Maille GO, Want EJ, Qin C, Trauger SA, Brandon TR, et al. METLIN: a metabolite mass spectral database. Ther Drug Monit. 2005;27(6):747–51.*

33. *Tautenhahn R, Patti GJ, Rinehart D, Siuzdak G. XCMS Online: a web-based platform to process untargeted metabolomic data. Anal Chem. 2012;84(11):5035–9.*

34. *Draper J, Enot DP, Parker D, Beckmann M, Snowdon S, Lin W, et al. Metabolite signal identification in accurate mass metabolomics data with MZedDB, an interactive m/z annotation tool utilising predicted ionisation behaviour 'rules'. BMC Bioinformatics. 2009;10:227.*

35. *Kanehisa M, Goto S, Kawashima S, Okuno Y, Hattori M. The KEGG resource for deciphering the genome. Nucleic Acids Res. 2004;32(Database issue):D277–80.*

36. *Sumner LW, Urbanczyk-Wochniak E, Broeckling CD. Metabolomics data analysis, visualization, and integration. Methods Mol Biol. 2007;406:409–36.*

37. *Calbiani F, Careri M, Elviri L, Mangia A, Zagnoni I. Matrix effects on accurate mass measurements of low-molecular weight compounds using liquid chromatography-electrospray-quadrupole time-of-flight mass spectrometry. J Mass Spectrom. 2006;41(3):289–94.*

38. *Kevin Schug HMM. Adduct formation in electrospray ionization. Part 1: common acidic pharmaceuticals. J Sep Sci. 2002;25(12):759–66.*

39. *Kanehisa M. The KEGG database. Novartis Found Symp. 2002;247:91–101. discussion 101–103, 119–128, 244–152.*

40. *Mortishire-Smith RJ, O'Connor D, Castro-Perez JM, Kirby J. Accelerated throughput metabolic route screening in early drug discovery using high-resolution liquid chromatography/quadrupole time-of-flight mass spectrometry and automated data analysis. Rapid Commun Mass Spectrom. 2005;19(18):2659–70.*

41. *Ando S, Tanaka Y. Mass spectrometric studies on brain metabolism, using stable isotopes. Mass Spectrom Rev. 2005;24(6):865–86.*

42. *Baverel G, Conjard A, Chauvin MF, Vercoutere B, Vittorelli A, Dubourg L, et al. Carbon 13 NMR spectroscopy: a powerful tool for studying renal metabolism. Biochimie. 2003;85(9):863–71.*

43. *Harada K, Fukusaki E, Bamba T, Sato F, Kobayashi A. In vivo 15N-enrichment of metabolites in suspension cultured cells and its application to metabolomics. Biotechnol Prog. 2006;22(4):1003–11.*

44. *Mesnard F, Ratcliffe RG. NMR analysis of plant nitrogen metabolism. Photosynth Res. 2005;83(2):163–80.*

INDEX

N

O

P

Q

R

S

T

U

V